W. L. Steffen, O. T. Denmead (Eds.)

Flow and Transport in the Natural Environment: Advances and Applications

Springer-Verlag Berlin Heidelberg New York
London Paris Tokyo

Dr. W. L. Steffen
Dr. O. T. Denmead
CSIRO Division of Environmental Mechanics
GPO Box 821
Canberra, ACT 2601
Australia

ISBN-13:978-3-642-73847-0 e-ISBN-13:978-3-642-73845-6
DOI: 10.1007/978-3-642-73845-6

Library of Congress Cataloging-in-Publication Data
Flow and transport in the natural environment
(edited by) William L. Steffen, Owen T. Denmead -- (1st ed.)
Based on the International Symposium on Flow and Transport in the Natural Environment, held in Canberra, Australia, in September 1987
ISBN-13:978-3-642-73847-0 (U.S.)
1. Atmospheric physics-- Congresses
2. Transport theory-- Congresses
3. Soil mechanics-- Congresses
I. Steffen, William L.
II. Denmead, Owen T.
III. International Symposium on Flow and Transport in the Natural Environment
(1987, Canberra, A.C.T.)
QC 851. F56 1988
551.5--dc 19 88-24916

Softcover reprint of the hardcover 1st edition 1988

2132/3020 543210 – Printed on acid-free paper

FOREWORD

This volume arises from an International Symposium on Flow and Transport in the Natural Environment held in Canberra, Australia, in September 1987. The meeting was hosted by the CSIRO Division of Environmental Mechanics (now the Centre for Environmental Mechanics) to mark the opening of the second stage of its headquarters, the F.C. Pye Field Environment Laboratory, twenty-one years after the opening of the first stage.

Those twenty-one years have seen much progress in our understanding of the physics of the natural environment and the occasion provided an ideal opportunity to review advances in our knowledge of flow and transport phenomena, particularly with regard to flow and transport in soils, plants and the atmosphere.

The contents of this volume are based very closely on the Symposium's program. Undoubtedly, our choices of topics were idiosyncratic, but we believe that those we have selected exhibit progress, innovation, and much scope for practical application. Rather than being encyclopaedic, we have sought to deal with thirteen selected topics in depth.

As set out in the contents, the general format of the volume is a review of the chosen topic and a commentary by an animateur whose job it was at the Symposium to complement the reviewer's contribution in the light of poster papers presented at the meeting, to summarise strengths and weaknesses in our knowledge of the topic under review, and to point to future developments and research needs. This was a rather difficult task, and our animateurs have done a splendid job of converting their contributions at the Symposium into the written word.

Two points should be made regarding the animateurs' reports. First, when commenting on Symposium poster papers, they have cited literature sources on which the poster paper was based wherever possible. If that could not be done, the poster paper was cited in a footnote. Second, substantive contributions by individuals to the discussion sessions were cited as personal communications.

Finally, we believe that by publishing this volume within twelve months after the meeting itself, the volume retains much of the freshness and liveliness of a most successful Symposium, and presents quality reviews which are still timely.

W.L. Steffen

O.T. Denmead

Contents

SOLUTE TRANSPORT AND DISPERSION

William A. Jury

Department of Soil and Environmental Sciences, University of California, Riverside, CA, 92521, U.S.A.

Summary. Hydrodynamic dispersion represents solute transport by convective velocity fluctuations about the mean solute velocity, which is represented directly in the transport model. Early notions in soil physics that this process is represented best by a convection-dispersion model in which the transport of solute with respect to the mean fluid velocity is a diffusionlike motion are now being reexamined in the light of accumulating field experimental data.

Modern approaches to solute transport, now focused exclusively on the groundwater regime, reformulate the problem as a stochastic process in which the local convection-dispersion model equation CDE is viewed as having a random solute velocity and concentration. A field-scale transport model is then derived by a perturbation expansion of the local equation, which subsequently is volume averaged under the assumption that the velocity fluctuation is a second order stationary random variable.

The resulting groundwater equation possesses time-dependent dispersion coefficients that are functions of the spatial autocorrelation structure of the velocity fluctuations, which are assumed to be measurable from the autocorrelation function of the saturated hydraulic conductivity. Agreement between the predictions of the model and observations of solute dispersion in the Borden aquifer experiments was good.

No corresponding approach has been developed for unsaturated soil. Simpler models based on limiting cases of possible dispersion behaviour are discussed and applied to experimental data from a field study of solute movement over 30 m.

INTRODUCTION

Solute dispersion in porous media is a process of considerable subtlety which has been given a variety of definitions in the literature. The difficulty in arriving at a single unambiguous definition of hydrodynamic dispersion arises at least in part from the fact that it is not a fundamental transport mechanism, but rather is a consequence of averaging the water velocity over a volume scale within which variations in velocity are neglected.

Strictly (ignoring electrical or thermal effects), solute may be transported within a porous medium by two mechanisms only: convection by moving fluid and molecular diffusion within the fluid phase. The solute flux $\underset{\sim}{J}_s$, expressed within the fluid phase as the mass of solute flowing per unit time per unit fluid cross-sectional area, may thus be written as

$$\underset{\sim}{J}_s = -D_d \underset{\sim}{\nabla} C + \underset{\sim}{V} C \ , \qquad (1)$$

where D_d is the coefficient of molecular diffusion, $\underset{\sim}{V}$ is the fluid velocity, and C is the local mass of solute per fluid volume. Equation (1) is intended to represent a volume scale at which the fluid is treated as a continuum.

Within a unit of fluid volume, the continuity equation for a conservative, nonvolatile, nonadsorbing solute is expressed as

$$\frac{\partial C}{\partial t} + \underset{\sim}{\nabla} \cdot \underset{\sim}{J}_s = 0 \ . \tag{2}$$

Combination of (1) and (2) produces, within the fluid phase, the convection-diffusion equation (3)

$$\frac{\partial C}{\partial t} = D_d \underset{\sim}{\nabla}^2 C - \underset{\sim}{V} \cdot \underset{\sim}{\nabla} C \ . \tag{3}$$

It is assumed that the fluid occupies the entire volume over which (3) is written (i.e., within a water-filled pore).

At the Darcy scale, (3) could still be used to describe solute transport through a porous medium, provided that one specified accurately the velocity field and wrote appropriate boundary conditions for all the air-water and solid-water boundaries of the purely liquid volume within which (1) applies. Since this is clearly not tractable for all but the simplest solid-water systems, (1) must be volume averaged to produce a new equation on a scale at which the porous medium is regarded as a continuum and local velocity fluctuations are no longer included in the description of the water flux. As a consequence, the convective portion of the volume-averaged solute flux does not describe completely the transport of solute by convection at the local scale, and the residual convective movement of solute relative to the average fluid motion appears as a chaotic diffusion-like process known as hydrodynamic dispersion. Under restricted conditions, the volume-averaged solute flux may be written as

$$\underset{\sim}{J}_s = -\underset{\approx}{D}_h \cdot \underset{\sim}{\nabla} C - \xi D_d \underset{\sim}{\nabla} C + \underset{\sim}{J}_w C \ , \tag{4}$$

where $\underset{\sim}{J}_w$ is the volume averaged fluid velocity, expressed at the Darcy scale as the volume of water flowing per unit time per unit cross-sectional area of space, $\underset{\sim}{J}_s$ is the mass of solute flowing per unit time per unit cross-sectional area of space, ξ is a tortuosity factor accounting for the increased path length and decreased cross section caused by the solid obstructions, and $\underset{\approx}{D}_h$ is the hydrodynamic dispersion tensor. The concentration C has now been averaged over many pores.

If (4) is combined with the volume averaged solute mass balance equation

$$\frac{\partial(\theta C)}{\partial t} + \underset{\sim}{\nabla} \cdot \underset{\sim}{J}_s = 0 \ , \tag{5}$$

where θ is volumetric water content, one obtains

$$\frac{\partial(\theta C)}{\partial t} = \underset{\sim}{\nabla} \cdot \underset{\approx}{D}_e \cdot \underset{\sim}{\nabla} C - \underset{\sim}{\nabla} \cdot \underset{\sim}{J}_w C \ , \tag{6}$$

where $\underset{\approx}{D}_e = \underset{\approx}{D}_h + \xi D$ is the effective diffusion-dispersion tensor. Equation (6), called the convection-dispersion model, has long been regarded as the appropriate solute transport equation for porous media, even though it has never been proven to be fundamentally correct (Sposito et al. 1983).

Much of the experimental support for the convection-dispersion model has come from miscible displacement experiments in cylindrical columns packed with soil or other porous media. For uniform water flow in homogeneous columns, the transport equation (6), after averaging over the column cross-sectional area, reduces to

$$\frac{\partial c}{\partial t} = D\,\frac{\partial^2 c}{\partial z^2} - V\,\frac{\partial c}{\partial z}\,, \tag{7}$$

where $V = J_w/\theta$ is the average pore water velocity and D is the macroscopic longitudinal diffusion-dispersion coefficient. Equation (7), which will be called the CDE, was very successful in describing outflow concentrations when its parameters D and V were fitted to the observed data (Biggar and Nielsen 1967).

The model equation (7), when applied to solute concentrations observed in groundwater, showed two deviations from the laboratory behaviour. First, the coefficient D required to match the model to observed data from groundwater sampling wells at a given location was significantly higher than values measured in the laboratory columns (Fried 1975). Furthermore, when the model was fitted to outflow observations repeatedly at successive distances from the source of solute entry, D appeared to increase linearly with distance (see review by Gelhar et al. 1985). This has become known as the dispersion scale effect.

SOLUTE TRANSPORT THROUGH A CAPILLARY TUBE

Obviously, some serious limitations exist to the applicability of (5) or (7) to porous media. Early theoretical work established some criteria for evaluating these limitations, which unfortunately were not taken fully into account later in seeking to apply theory to real soils. A classic study which offers insight into the nature of dispersion is the analysis by Taylor (1953) of solute transport through a capillary tube.

Assuming laminar flow of water, the steady state fluid velocity V(r) at any point z within the tube is given by

$$V(r) = V_M\,(1 - r^2/R^2)\,, \tag{8}$$

where V_M is the maximum velocity, r is the radial distance from the centre of the tube and R is the capillary radius. For this situation, the convection-diffusion equation (1) becomes

$$\frac{\partial c}{\partial t} = \frac{D}{r}\frac{\partial}{\partial r}\left(r\,\frac{\partial c}{\partial r}\right) + D\,\frac{\partial^2 c}{\partial z^2} - V(r)\,\frac{\partial c}{\partial z}\,. \tag{9}$$

Taylor analyzed the average concentration emerging from the outflow end of the capillary tube as a function of the relation between two characteristic time scales of the problem: the convection or residence time t_C of the solute in the tube and the radial or transverse mixing time t_R required for solute concentrations to smooth out by radial diffusion. His conclusions were that, when $t_C \gg t_R$, Eq. (9) may be replaced by the area-averaged convection-dispersion equation

$$\frac{\partial c}{\partial t} = D'\,\frac{\partial^2 c}{\partial z^2} - V'\,\frac{\partial c}{\partial z}\,, \tag{10}$$

where $V' = V_M/2$ is the average fluid velocity and $D' = R^2V_M^2/192D$ is the longitudinal hydrodynamic dispersion coefficient. In contrast, when $t_C \ll t_R$, the convection-dispersion model (10) is not valid for the capillary. In fact, in the limiting case of pure convection, the area-averaged solute concentration for a step change in concentration at the inlet end at t = 0 is given by

$$C(z,t)/C_0 = 1 - z^2/V_M{}^2t^2 \qquad t > z/V_M$$
$$= 0 \qquad t < z/V_M \ . \tag{11}$$

Equation (11) is not a solution to (10) for constant D',V'. In fact, if the solution to (10) is fitted to (11), the parameter D will be proportional to z.

The simple example of solute flow through a capillary tube produced an important result. If the lateral variations in concentration within the tube caused by variations in velocity can be smoothed out by transverse mixing during the time of residence between the point of entry and the point of observation, then the convection-dispersion model will be valid for the area-averaged concentration. This principle will prove to be valid for more complex media as well.

REEXAMINATION OF THE LABORATORY EVIDENCE

The one-dimensional convection-dispersion model (7) derived its credibility principally from describing the shapes of breakthrough curves in soil columns. However, any experiment which maintains a fixed distance between the inlet and outlet ends of a solute transport column study and does not monitor the solute concentration except at the outlet end cannot detect a scale effect in D. In respect to the ability of (7) to produce plausible descriptions of solute breakthrough curves, it should be pointed out that other two parameter functions have equal versatility. Figure 1 shows plots of the Fickian distribution f_F

$$f_F(t;z) = z/[2(\pi Dt^3)^{\frac{1}{2}}]\exp[-(z - Vt)^2/4Dt] \ , \tag{12}$$

which is a solution to (7) for a unit δ-function input of solute (Jury and Sposito 1985), together with two other distributions: the gamma f_G

$$f_G(t) = \beta^{1+\alpha}\ t^{\alpha}\ \exp[-\beta t]/\Gamma(1 + \alpha) \ , \tag{13}$$

where Γ is the gamma function, and the lognormal f_L, given by

$$f_L(t) = 1/[(2\pi)^{\frac{1}{2}}\ \sigma t]\exp[-(\ell n\ t - \mu)^2/2\sigma^2] \ . \tag{14}$$

Clearly, each of these distributions could provide nearly equivalent representations of solute outflow data from a single experiment. Thus, a soil column experiment monitoring column effluent concentrations cannot validate a transport model, except insofar as the model represents a transport event between the inlet and outlet ends. For this reason Jury (1982; Jury et al. 1986) proposed a transfer function formulation of solute transport as a fundamental first step in studying solute movement through porous media. In this model, the solute outflow C(L,t) at a distance z = L from the source of solute entry is written as a convolution of the solute input concentration $C_{IN}(t)$ and the travel time probability density function pdf f(t;L) describing the distribution of travel times from 0 to L:

$$C(L,t) = \int_0^{\infty} C_{IN}(t-t')\ f(t';L)dt' \ . \tag{15}$$

In this formulation, (12)-(14) all become parametric representations of f(t:L)

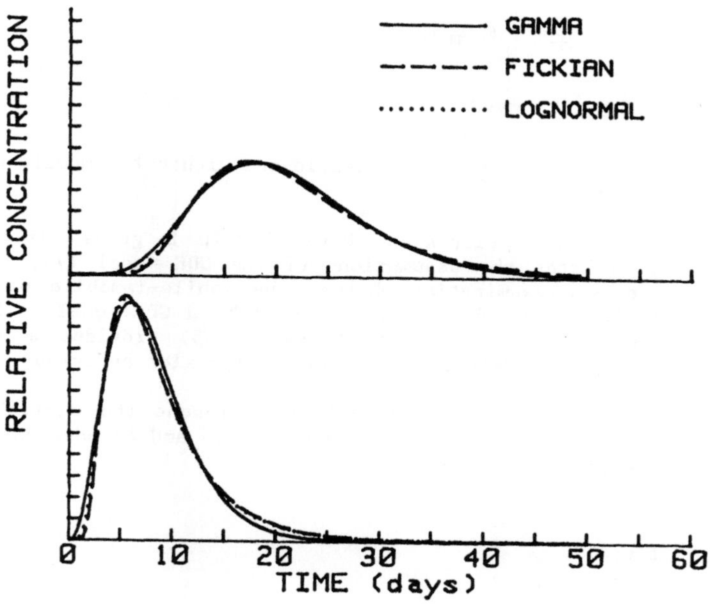

Fig. 1. Plots of different gamma, Fickian and lognormal travel time pdfs whose parameters have been optimized by sum of squares fitting.

and thus are equally valid. Equation (15) is valid for any linear solute transport process obeying the principle of solute mass balance, assuming no reactions or lateral losses or gains between the entry and outflow ends. The function f(t;L), which becomes the characteristic transport property of the soil volume under the conditions studied, may be measured by a tracer experiment (Jury et al. 1986). Equation (15) was also recast with net applied water I instead of time as the independent variable, and was used successfully to predict solute transport within the top 1.8 m after calibration at 0.3 m during a winter of highly variable rainfall (Jury et al. 1982).

Not all column outflow experiments produce breakthrough curves which may be described simply by the two parameter distributions given above. In particular, experiments conducted in undisturbed columns containing aggregates or structural voids tended to produce highly skewed breakthrough curves characterized by early arrival of solute and a slow approach to the final concentration. The most popular model used to describe such curves has been the two-region mobile-immobile water model MIM of Van Genuchten and coworkers (Van Genuchten and Wierenga 1976; Van Genuchten et al. 1977). In this model, solute is convected and dispersed through a portion of the wetted pore space characterized by a mobile water content θ_m, and, in addition, solute can be transferred to and from a stagnant portion of the wetted pore space characterized by an immobile water content θ_i ($\theta_m + \theta_i = \theta$) by a rate-limited exchange process described by the set of equations (16) (assuming steady water flow):

$$\theta_m \frac{\partial C_m}{\partial t} + \theta_i \frac{\partial C_i}{\partial t} + J_w \frac{\partial C_m}{\partial z} = \theta_m D_m \frac{\partial^2 C_m}{\partial z^2} \tag{16a}$$

$$\theta_i \frac{\partial C_i}{\partial t} = \alpha(C_m - C_i) \ , \tag{16b}$$

where D_m is the longitudinal dispersion coefficient within the mobile phase and α is a rate coefficient.

The dispersion scale effect discussed above wherein D grows linearly with distance obviously violates the assumptions of the CDE model (7), but it is not clear without closer examination whether the mobile-immobile water model (16) manifests a scale effect when it is analyzed by a CDE model. The method of moments (Himmelblau 1970; Jury and Sposito 1985) provides a method of analyzing solute transport models for their macrodispersion behaviour.

The travel time moments T_N of the pdfs which represent the distribution of travel times f(t;z) between the source of solute entry and an observation depth z are given by (Jury and Sposito 1985)

$$T_N(z) = \int_0^\infty t^N f(t;z)dt \ , \tag{17}$$

where $T_0 \equiv 1$.

The first two moments of the Fickian pdf (12) are

$$T_1(z) = z/V \tag{18a}$$

$$T_2(z) = z^2/V^2 + 2Dz/V^3 \ , \tag{18b}$$

while the travel time variance S^2 is

$$S^2 = \mathrm{Var}(z) = T_2 - T_1{}^2 = 2Dz/V^3 \ . \tag{19}$$

Thus, for a process which obeys the one-dimensional convection-dispersion model (i.e., has constant D,V), the travel time variance will increase proportionally to z. Conversely, if V is constant but D is proportional to z (a linear scale effect), then S^2 will be proportional to z^2.

An example of a pdf with a linear dispersion scale effect is the convective transfer function model of Jury (1982), in which the travel time pdf f(t;z) at an arbitrary depth z is related to the pdf f(t;L) at a reference or calibration depth z = L by

$$f(t;z) = \frac{L}{z} f(tL/z;L) \ . \tag{20}$$

It is easy to show that for this model the travel time moments at depth z are given by

$$T_N(z) = (z/L)^N T_N(L) \tag{21}$$

and hence that

$$\mathrm{Var}(z) = (z^2/L^2)\mathrm{Var}(L) \ . \tag{22}$$

Thus, if the Fickian pdf (12) were fitted to a process obeying (20) at different depths, the parameters V,D would have the following relation to the moments

$$V \equiv \text{constant} = L/T_1(L) \tag{23a}$$

$$D = (V^3/2z)(z^2/L^2)Var(L) = (zV^3/2L^2)Var(L) = (z/L)D(L) \ . \tag{23b}$$

Jury (1982) described as an example of a process obeying (20), one in which the solute stream tubes remained continuous with depth and isolated from their lateral neighbours. He illustrated the model with the convective lognormal transfer function (CLT)

$$f(t;z) = 1/[(2\pi)^{\frac{1}{2}}\sigma t]\exp[-(\ell n(tL/z)-\mu)^2/2\sigma^2] \ , \tag{24}$$

where μ and σ^2 are the mean and variance of $\ell n\ t$ at $z = L$, defined as the reference depth where the model is calibrated.

It is interesting to compare the CLT with the Fickian pdf (12) from the CDE, since they will virtually superimpose at a given z when their parameters are optimized. Figure 2 shows predicted concentrations at various z for the two models, where their parameters have been optimized at $z = 0.3$ m.

The question of the appropriateness of other models may also be addressed through a moment analysis. Table 1, taken from Jury and Sposito (1985) and Sposito et al. (1986), shows the mean and variance of the travel time pdfs for the CLT, CDE and MIM. It is seen by comparing the variance of the CDE and MIM models that: (i) macrodispersion in the MIM is independent of z and hence does not show a scale effect, and (ii) that part of the macrodispersion in the MIM arises from the rate-limited transfer of solute and not from dispersion within the mobile phase. In fact, if the CDE is fitted to data from a process obeying the MIM, the apparent longitudinal macrodispersion coefficient would be (from Table 1)

$$D = \varphi_m D_m + (1 - \varphi_m)J_w^2/\theta\alpha \ , \tag{25}$$

where $\varphi_m = \theta_m/\theta$.

Table 1. Mean and variance of the travel time distribution for three parametric models of the transport process*

Model	Mean	Variance
CDE	z/V	$2Dz/V^3$
CLT	$(z/L)\exp(\mu+\sigma^2/2)$	$(z/L)^2\exp(2\mu+\sigma^2)[\exp(\sigma^2)-1]$
MIM	$z\theta/J_w$	$\frac{2zD_m\theta^3}{J_w^3}\varphi_m + \frac{2z\theta^2}{J_w\alpha}(1 - \varphi_m)$

* $\varphi_m = \theta_m/\theta$

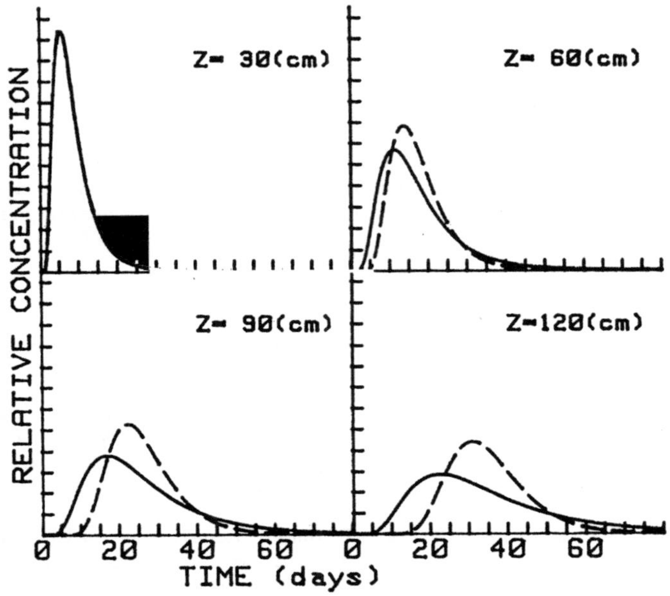

Fig. 2. Predictions of CDE (dashed line) and CLT (solid line) models after calibration at 0.3 m.

GROUNDWATER MODELS OF DISPERSION

Solute moving through groundwater formations has been studied more extensively, both theoretically and experimentally, than that moving through unsaturated soil. A major goal of solute transport theory has been to describe the dispersion characteristics in terms of the measurable hydraulic properties of the aquifer as embodied in the spatial distribution of saturated hydraulic conductivity K_s. Since the exact spatial distribution of K_s is never known, the problem has been addressed with a stochastic formulation using restrictive assumptions about the spatial statistical properties of all variables in the aquifer.

Significant early work in this area was performed by Gelhar and coworkers (Gelhar et al. 1979; Gelhar and Axness 1983), who derived an asymptotic solution to the solute transport equation

$$n \frac{\partial C}{\partial t} = \sum_i \frac{\partial}{\partial x_i} \sum_j D_{ij} \frac{\partial C}{\partial x_j} - q_i C \ , \qquad (26)$$

where n is porosity and q_i is the component of the steady groundwater flux. Gelhar's approach was to treat C and q_i as the random variables expressed as the sum of a mean and zero mean fluctuation

$$C = \langle C \rangle + C' \qquad (27a)$$

$$q_i = \langle q_i \rangle + q_i' \ . \tag{27b}$$

The expressions in (27) are inserted into (26) and solved to first order in the fluctuations under the assumption that the q_i' are zero mean, second-order stationary random variables whose covariance is a function only of separation and not absolute position.

Under these assumptions Gelhar and Axness (1983) showed that the macroscopic convection-dispersion equation will apply asymptotically except under restricted conditions. They derived an infinite-time expression for the macrodispersion tensor $\underset{\sim}{D}^M$ given by

$$D_{ij}^M = \int_{-\infty}^{\infty} \frac{S_{q_i q_j}(\underset{\sim}{k})d\underset{\sim}{k}}{[ik_1 + \alpha_L k_1{}^2 + \alpha_T(k_2{}^2 + k_3{}^2)]q^2} \ , \tag{28}$$

where $S_{q_i q_j}$ is the Fourier transform of the water flux covariance function, α_L and α_T are the longitudinal and transverse dispersivities, q is the mean flux and $\underset{\sim}{k}$ is the wave number vector conjugate to the position vector in the Fourier transform.

Gelhar and Axness (1983) derived asymptotic dispersion coefficients for various models of the covariance structure of the water flow field. They also developed a relation between $S_{q_i q_j}$ and S_{ff}, the Fourier transform of the covariance function of the logarithm of the saturated hydraulic conductivity $f = \ell n K_s$. This allowed asymptotic values of $\underset{\approx}{D}^M$ to be calculated from the statistical structure of the K_s field, inferred from variograms or autocorrelation functions.

Although the Gelhar model provided an estimate of the macrodispersion coefficients for the aysmptotic regime under the assumption of stationarity, it did not make predictions about the preasymptotic state.

Güven et al. (1984) used the Aris (1956) method of moments to derive expressions for the preasymptotic dispersion coefficients of an infinite, perfectly-stratified aquifer for various known distributions of hydraulic conductivity in the vertical direction, finding that in a sinusoidally-varying hydraulic conductivity field large travel distances may be required to reach asymptotic Fickian behaviour when the period of the spatial wave is large. Matheron and de Marsily (1980) used a random motion model to demonstrate that Fickian behaviour would in general never be reached in an infinite, perfectly-stratified aquifer with no convective velocity component perpendicular to the direction of stratification.

The most successful general model of the dispersion process has been derived by Dagan (1982, 1984, 1987), who represented solute transport with a random particle motion model of the form

$$\frac{d\underset{\sim}{X}_t}{dt} = \underset{\sim}{U} + \underset{\sim}{u}(X_t,t) = \frac{d\underset{\sim}{X}_d}{dt} \ , \tag{29}$$

where $\underset{\sim}{X}_t$ is the total particle displacement, $\underset{\sim}{U}$ is the mean fluid velocity, $\underset{\sim}{u}$ is the random velocity fluctuation and $\underset{\sim}{X}_d$ is a Brownian motion. Under the assumption of second order stationarity, Dagan (1984, 1987) derived an expression for the position displacement covariance tensor from which time dependent macrodispersion tensor components could be derived. He also showed that, when the displacement vector follows a multivariate Gaussian distribution, the average solute concentration <C> obeys

$$\frac{\partial \langle C \rangle}{\partial t} + \sum_j \langle U_j \rangle \frac{\partial \langle C \rangle}{\partial x_j} = \sum_j \sum_k D^M_{jk}(t) \frac{\partial^2 \langle C \rangle}{\partial x_j \partial x_k} , \quad (30)$$

where $\underset{\approx}{D}^M(t)$ is the time dependent macrodispersion tensor. Equation (30) applies at all times in the transport process and converges under certain conditions to a macroscopic three-dimensional convection-dispersion model when $\underset{\approx}{D}^M$ is no longer changing with time.

A recent rederivation of the Dagan model by Sposito and Barry (1987) using a cumulant expansion of (5) produced the following elegant expression for the time dependent macrodispersion tensor oriented along the principal axes

$$D^M_{ii}(t) = \int_0^t \int_{-\infty}^{\infty} S_{ii}(k,0)\exp[-(\underset{\sim}{k}\cdot\underset{\approx}{D}\cdot\underset{\sim}{k})t']\cos(\underset{\sim}{V}\cdot\underset{\sim}{k}t')d^3k dt' , \quad (31)$$

where $\underset{\approx}{D}$ is the local macrodispersion tensor and S_{ii} is the Fourier transform of the covariance function of the groundwater velocity $\underset{\sim}{V}$. Equation (31) has been evaluated in the limit that the velocity autocorrelation time is large compared to the local dispersion time $(\underset{\sim}{k}\cdot\underset{\approx}{D}\cdot\underset{\sim}{k})^{-1}$ (Sposito and Barry 1987). Using appropriate special cases of (31), Sposito and Barry (1987) were able to derive the results of the Dagan (1984) model, the Matheron and de Marsily (1980) model, the Güven et al. (1984) model, and the asymptotic macrodispersion coefficients of Gelhar and Axness (1983).

EXPERIMENTAL EVIDENCE

The progress of theoretical solute transport modeling has been hampered until recently by the absence of comprehensive field scale studies of solute movement. In the section to follow, two large scale solute transport studies, one in groundwater and one in the unsaturated zone, will be reviewed in the context of the theories discussed above.

Borden Aquifer

The major groundwater transport study at the Borden aquifer in Waterloo was reported by Roberts and Mackay (1986) and Mackay et al. (1986). In this study a rectangular prism source of solute was added to a shallow, homogeneous groundwater aquifer and monitored for several years by a dense network of sampling wells, from which both solute breakthrough curves at specific sites and spatial concentration distributions at specific points in time were constructed. The Dagan model (1984, 1987) was used successfully to model the growth of the longitudinal and transverse macrodispersion coefficients, using a velocity covariance function inferred from vertical and horizontal transect measurements of K_s (Sudicky 1986). Dagan (1984) concluded that the longitudinal component of $\underset{\approx}{D}^M$ in (30) had reached its asymptotic value during the 600 day experiment, but that the transverse components had not. This was also concluded by Sposito and Barry (1987), who calculated that some 60 years would be required to reach the aysmptotic value of the transverse dispersion coefficients in (31) whereas only one year was required for the longitudinal component to reach a constant value.

Etiwanda Unsaturated Zone Field Study

A major unsaturated zone field study at Etiwanda, California, reported in detail in Butters (1987) and Butters and Jury (1987), consisted of the application of a spatially-uniform solute pulse of KBr to a 0.64 ha field receiving bidaily sprinkler irrigation. The pulse was monitored by solution samplers at 16 sites and six depths (0.3, 0.6, 0.9, 1.2, 1.8, 3.0 m), at six sites at 4.5 m, and by soil cores taken at six sites to 25 m at the conclusion of the study.

Concentrations measured in the replicated solute samplers at each depth were averaged to produce an area-averaged solute concentration (Fig. 3) for the field. These curves were fitted to the two parameter CDE (12) and CLT (24) models discussed above.

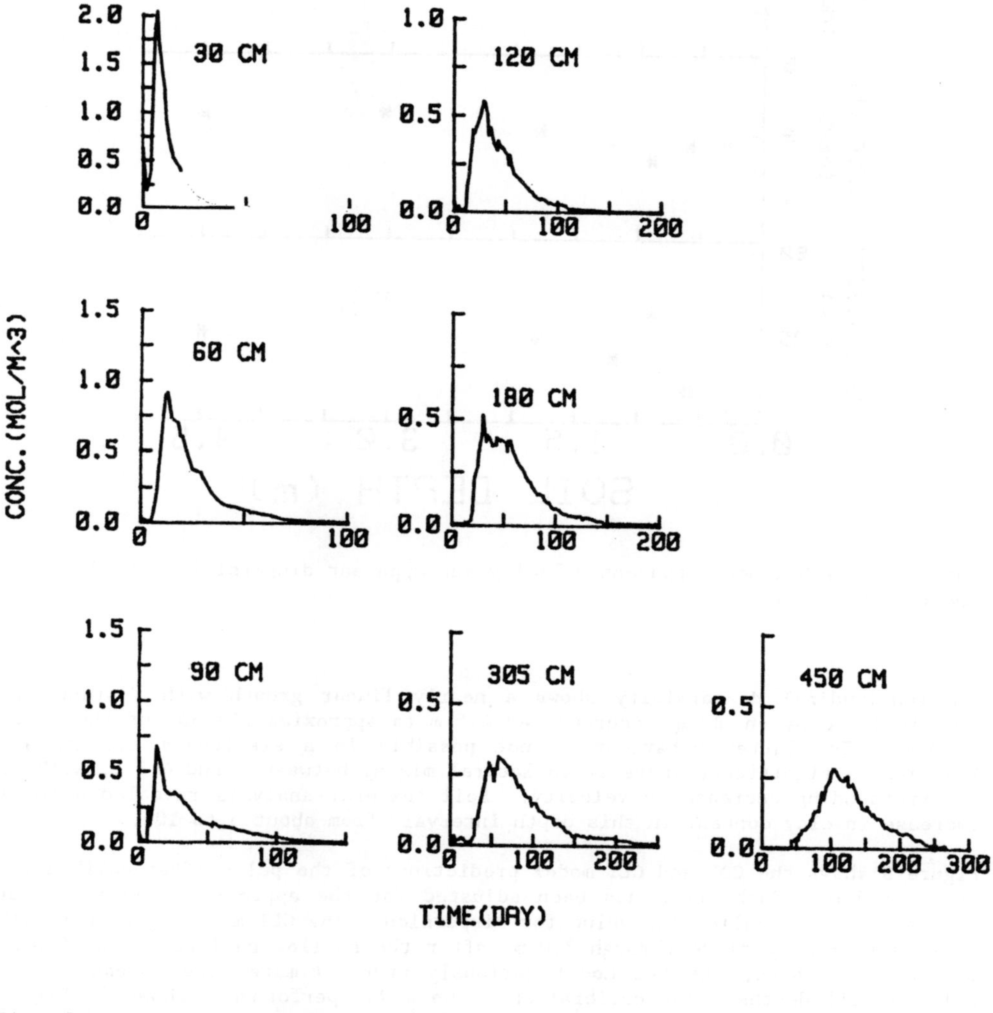

Fig. 3. Area-averaged solute pulses from field study of Butters and Jury (1987).

Figures 4a, b and c show the mass balance, apparent velocity and dispersivity for the seven depths, obtained by fitting the CDE solution (12) to the data averaged over the field area. The mass balance is good at each depth, suggesting that an adequate number of samplers were used. The apparent velocity shows an increase with depth under reasonably steady water application, which is consistent with the decrease in water content with depth measured in the field (Butters and Jury 1987).

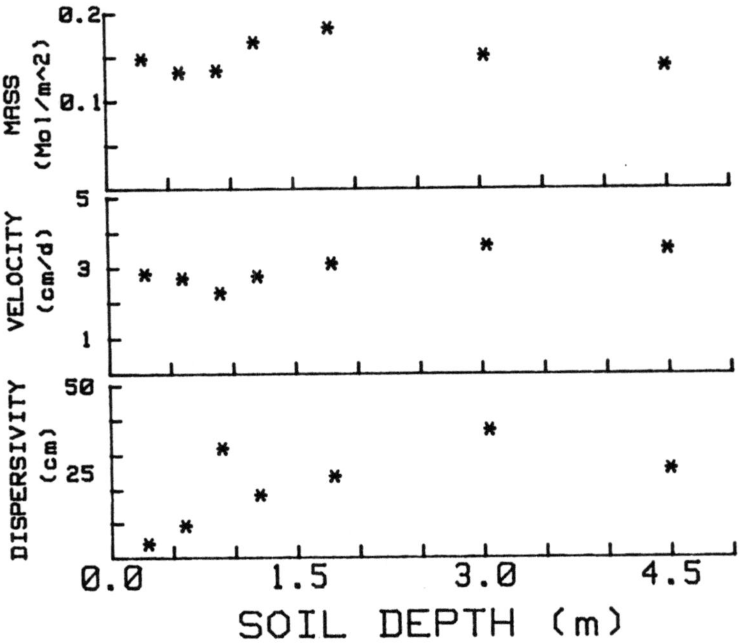

Fig. 4. Mass balance, apparent velocity and apparent dispersivity for data set given in Figure 3.

The longitudinal dispersivity shows a nearly linear growth with distance to 3 m, followed by an abrupt decrease at 4.5 m to approximately 60% of the value at 3 m. The latter behaviour is not possible in a stationary system, and indicates a significant increase in lateral mixing between 3 and 4.5 m, without a corresponding decrease in velocity. Soil textural analysis revealed a small increase in clay content in this depth interval, from about 5 to 10%.

Figure 5 shows the CDE and CLT model predictions of the pulse after calibration at Z = 0.3 m. Each depth has been adjusted for the appropriate velocity but uses the surface calibrated value for dispersion. The CLT model agreement with data is extremely close through 3.0 m, after the shallow calibration at 0.3 m. On the other hand, the CDE model seriously underestimates the spread of the pulse at all depths below calibration. The model performance shown in Fig. 5 offers convincing evidence that the solute transport process on this field was convectively driven through 3.0 m, and began to alter its dispersion behaviour below this depth.

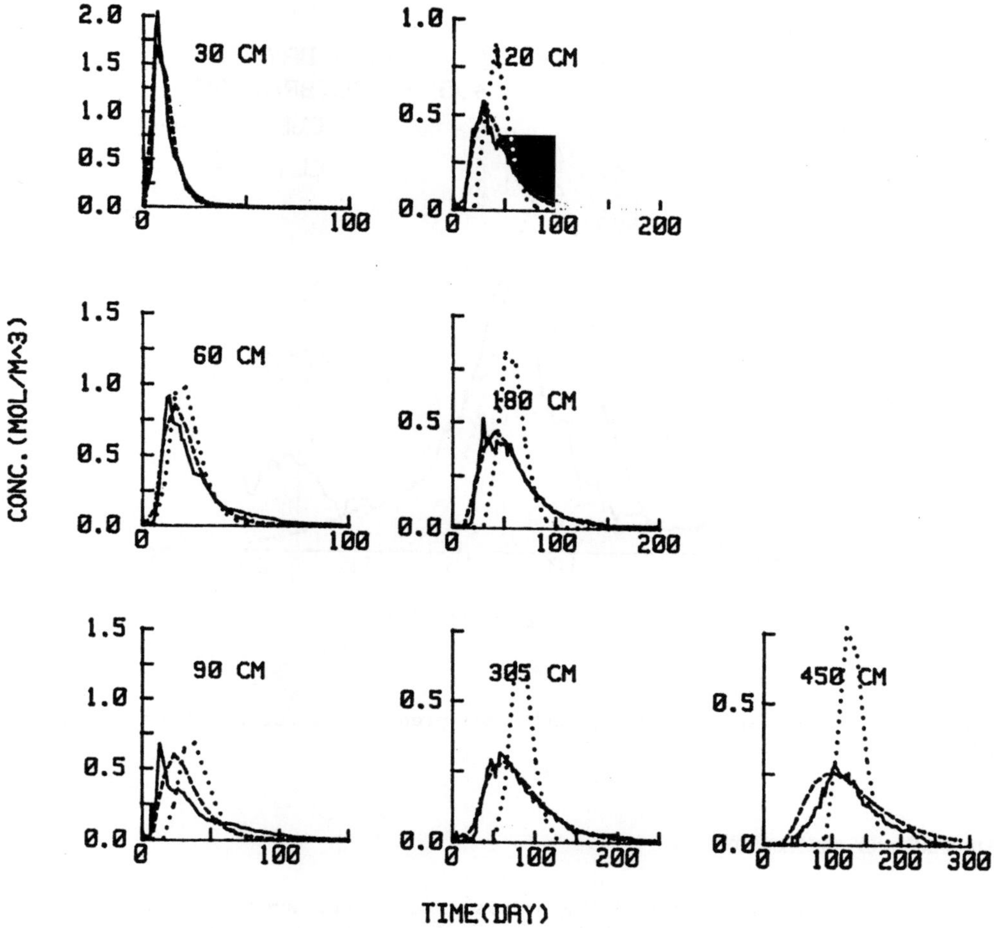

Fig. 5. Predictions of CDE (dotted curve) and CLT (dashed curve) models on data (solid curve) from Fig. 3 after calibration at 0.3 m.

Figure 6 shows the deep soil core average concentrations, together with the model predictions using the CDE and CLT calibrated at 4.5 m. Although the main part of the pulse is represented well by the CDE model, there is still deep movement of part of the pulse which is better described by the purely convective CLT model than by the CDE. Furthermore, the solute concentrations were nonzero at the deepest sampled depths in five of the six cores, implying that some of the deep movement may not have been recovered. Thus, it seems certain that the longitudinal macrodispersion coefficient again grew between 4.5 and the coring depths, and that the CDE model should not be used to describe downward movement on this field at any of the monitored depth intervals.

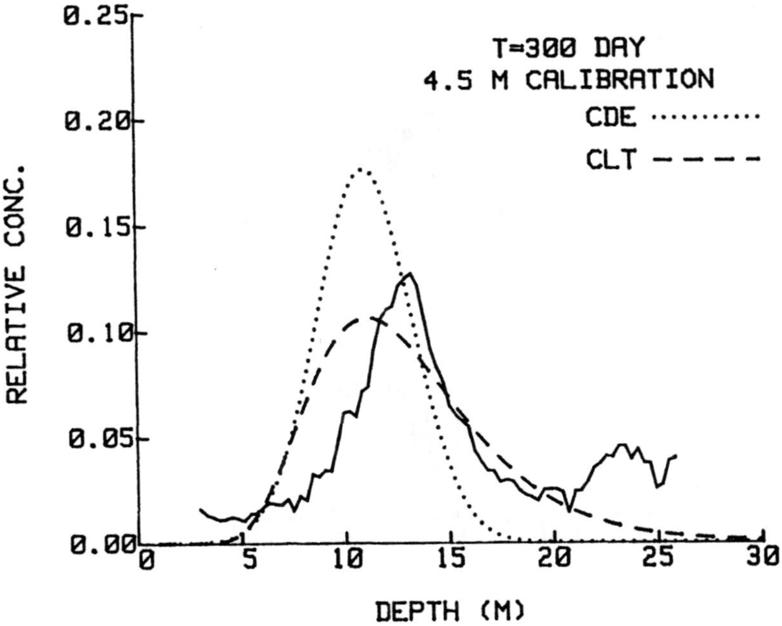

Fig. 6. Field average soil cores and model predictions for field experiment of Butters and Jury (1987).

SUMMARY AND CONCLUSIONS

Solute dispersion in porous media arises from volume averaging over a scale within which local fluctuations in water velocity are too complex to describe directly. As a consequence, convective transport of solute by these fluctuations must be described with a mixing model.

The convection-dispersion model, which assumes that this mixing will be equivalent to a diffusion process when described with respect to the mean fluid motion, represents a special case which will occur only under restricted circumstances. Current groundwater transport theory includes a general formulation for stationary media in which the coefficients describing the dispersion process are time-dependent and are functions of the variance structure of the water velocity field. This model provided a satisfactory description of the single large-scale solute transport study which has been conducted in groundwater.

For nonstationary media there is no theory available. It seems unlikely that a predictive model will ever be capable of describing transport through soil regimes where no information has been obtained in preliminary experiments, if this region embodies new transport features which disrupt the dispersion process. An example of this is the unsaturated zone solute transport study discussed above where a single layer with moderately high clay content caused a 40% decrease in the field scale longitudinal dispersivity, which was followed by a return to a process with convective characteristics.

More studies such as the above are needed to derive measurable characteristics of large-scale field behaviour which will assist in developing improved models of the dispersion process.

ACKNOWLEDGEMENT

The author would like to thank the Electric Power Research Institute, the Southern California Edison Company and the University of California Toxic Substances Research and Training Program for partial financial support of UC Riverside research presented in this paper.

REFERENCES

Aris R (1956) On the dispersion of a solute in a fluid flowing in a tube. Proc R Soc London Ser A 235: 67-77

Biggar JW, Nielsen DR (1967) Miscible displacement and leaching phenomena. Agronomy 11: 938-944

Butters G (1987) Field scale transport of bromide through unsaturated soil. Ph.D. Dissertation, Univ. of Calif., Riverside, in progress

Butters G, Jury WA (1987) Field scale transport of bromide through unsaturated soil. EPRI Topical Report, in review

Dagan G (1982) Stochastic modeling of groundwater flow by unconditional and conditional probabilities. 2. The solute transport. Water Resour Res 15: 835-848

Dagan G (1984) Solute transport in heterogeneous porous formations. J Fluid Mech 145: 151-177

Dagan G (1987) Theory of solute transport by groundwater. Ann Rev Fluid Mech 19: 183-215

Fried JJ (1975) Groundwater pollution. Elsevier, New York

Gelhar LW, Gutjahr AL, Naff RL (1979) Stochastic analysis of macrodispersion in a stratified aquifer. Water Resour Res 15: 1387-1397

Gelhar LW, Axness CL (1983) Three-dimensional stochastic analysis of macrodispersion in aquifers. Water Resour Res 19: 161-180

Gelhar LW, Matutoglou A, Welty C, Rehfeldt KR (1985) A review of field-scale physical solute transport processes in saturated and unsaturated porous media. EPRI Topical Report EA-4190, Electric Power Research Institute

Güven O, Molz FJ, Melville JG (1984) An analysis of dispersion in a stratified aquifer. Water Resour Res 20: 1337-1354

Himmelblau D (1970) Process analysis by statistical methods. Sterling Swift, Manchaca, TX

Jury WA (1982) Simulation of solute transport using a transfer function model. Water Resour Res 18: 363-368

Jury WA, Stolzy LH, Shouse P (1982) A field test of the transfer function model for predicting solute transport. Water Resour Res 18: 369-375

Jury WA, Sposito G (1985) Field calibration and validation of solute transport models for the unsaturated zone. Soil Sci Soc Am J 49: 1331-1341

Jury WA, Sposito G, White RE (1986) A transfer function model of solute transport through soil. I. Fundamental concepts. Water Resour Res 22: 243-247

Mackay DM, Freyberg DL, Roberts PV , Cherry JA (1986) A natural gradient experiment on solute transport in a sand aquifer: 1. Approach and overview

of plume movement. Water Resour Res 22: 2017-2029
Matheron G, de Marsily G (1980) Is transport in porous media always diffusive? A counterexample. Water Resour Res 16: 901-917
Roberts PV, Mackay DM (eds) (1986) A natural gradient experiment on solute transport in a sand aquifer. Tech Report No 292, Department of Civil Engineering, Stanford Univ
Sposito G, Gupta VK, Bhattacharya RN (1983) Foundational theories of solute transport in porous media: A critical review. In: Pinder GF (ed) Flow through porous media, CML Publ Ltd Southampton, UK, pp 76-85
Sposito G, White RE, Darrah PR, Jury WA (1986) A transfer function model of solute transport through soil. 3. The convection-dispersion equation. Water Resour Res 22: 255-262
Sposito G, Barry DA (1987) On the Dagan model of solute transport through groundwater: Foundational aspects. Water Resour Res 23: 1867-1875
Sudicky EA (1986) A natural-gradient experiment on solute transport in a sand aquifer: Spatial variability of hydraulic conductivity and its role in the dispersion process. Water Resour Res 22: 2069-2082
Taylor GI (1953) The dispersion of matter in solvent flowing slowly through a tube. Proc Roy Soc London Ser A 219: 189-203
Van Genuchten MTh, Wierenga PJ, (1976) Mass transfer studies in sorbing porous media. I. Analytical solutions. Soil Sci Soc Am J 40: 473-479
Van Genuchten MTh, Wierenga PJ, O'Connor GA (1977) Mass transfer studies in sorbing porous media. III. Experimental evaluation with 2,4,5-T. Soil Sci Soc Am J 41: 278-285

SOLUTE TRANSPORT AND DISPERSION: COMMENTARY

J.H. Knight

CSIRO Division of Environmental Mechanics, GPO Box 821, Canberra, ACT, 2601, Australia

Summary. The review by Bill Jury in this Symposium (Jury 1988) gave us a comprehensive account of the modern approach to solute transport at the field scale, in which the measured autocorrelation function of the saturated hydraulic conductivity is used to derive a volume-averaged stochastic solute transport model. This model predicts time dependent macrodispersion coefficients which eventually reach asymptotic values, and there is good agreement between the predictions and observations of solute dispersion in the Borden aquifer experiments.

I discuss some investigations of the early stages of Taylor dispersion of solute in simple shear flow in a capillary tube, and survey various postulated holdup mechanisms to explain solute retardation during convective-dispersive transport. Recent experiments designed to measure field dispersion coefficients are discussed, and some experimental and methodological difficulties are explored.

INTRODUCTION

Bill Jury has given us a very comprehensive review of advances during the last two decades in the modelling of solute transport and dispersion in porous media (Jury 1988). In this commentary I make some additional historical remarks, draw parallels with work in other fields, and venture some personal opinions. A common theme of "dispersion of scalar quantities during transport by complicated velocity fields" appeared in many of the presentations at this symposium. Historically, aspects of this generic dispersion problem have been studied by micrometeorologists, chemical engineers and soil physicists working in parallel and largely in isolation from each other, so it is appropriate that all three groups are represented here, taking part in the diffusion (and convection) of scientific knowledge.

THE DIFFUSION APPROXIMATION AND ITS LIMITATIONS

The effect of dispersion in a nonuniform velocity field has usually been modelled by a "diffusion approximation" in which the centre of a cloud of dispersing tracer particles is assumed to move with the "average" velocity, and variations in the velocity field cause the cloud to spread with an "enhanced diffusion" for which the average mass flux is proportional to the average local concentration gradient, and which is much greater than that due to molecular diffusion alone. A common subtheme in each account of an approximate description of dispersion has been the general recognition over the last two decades that the diffusion approximation is only asymptotically true, being valid only if the time and length scales of the transport process are large enough for a typical tracer particle to have experienced the full range of variations in the velocity field. Jury (1988) gave us the example of Taylor's

(1953) beautiful theory of dispersion of a solvent in Poiseuille flow in a tube, in which Taylor quite explicitly showed the limitations of the asymptotic dispersion theory in this situation. Also, Taylor's (1959) review of the theory of turbulent diffusion discusses the use of a mixing length proportional to the average length of path which a tracer particle traverses before its velocity becomes uncorrelated with that which it had at the beginning of the path, and the way in which this length governs the turbulent diffusion of momentum, heat and water vapour in the atmosphere. Raupach (1988) reviewed corresponding advances in the understanding of atmospheric transport processes in plant canopies.

Thurtell (1988) told us of simulations of the effect of a rough wall on the diffusion of gases near it, as a further example of the breakdown of the diffusion approximation. In the "Knudsen region" within a distance of about one mean free path length of the wall, gas molecules are affected more by colliding with the rough wall and rebounding at random angles than by colliding with other molecules (Knudsen 1950; Daniels et al. 1956, p 301). However, for gas flow through fine tubes or porous media, the Knudsen effect has been taken into account within a diffusion model, as I will now indicate. Historically, the nonsteady flow of a gas through a porous medium has been described by the "porous medium equation" (Muskat 1937, p 133)

$$\frac{\partial\rho}{\partial t} = \nabla\cdot(D_0\rho^{\alpha}\,\nabla\rho) \quad , \tag{1}$$

which has the form of a nonlinear diffusion equation for the gas density ρ with diffusion coefficient $D(\rho)$ zero at zero pressure, as it is proportional to the pressure p,

$$D(\rho) = D_0\rho^{\alpha} = D_0 p/p_0 \quad , \tag{2}$$

where D_0, p_0 and α are constants. When the gas pressure is low enough the mean free path length is comparable to the pore size, effects of gas-wall interactions are dominant, and the flow is almost all in the Knudsen region. It has been shown experimentally (Knudsen 1950) and predicted from kinetic theory (Deriagin and Bakanov 1957; Mason et al. 1967) that capillary tube or porous medium gas flow in the Knudsen region still obeys a diffusion equation, provided that we modify the form of the diffusion coefficient to have a minimum (the "Knudsen minimum") at a nonzero value of the pressure and to remain finite as the pressure tends to zero.

If I may digress slightly, I will point out a moral for mathematicians. The singular behaviour of the idealised "porous medium equation" (1) at zero gas pressure effectively changes it there from a parabolic to a locally hyperbolic equation, and changes the character of its solutions from spreading diffusive fronts with theoretically infinite speeds to wavelike fronts with finite speeds, and gives rise to interesting and much studied mathematical problems. (See, for example, Aronson 1986). However, we have seen that for a real gas the true "porous medium equation" is a strictly parabolic diffusion equation which remains nonsingular at zero gas pressure, as the true diffusion coefficient is strictly positive, although strongly varying with pressure above the Knudsen minimum.

SOLUTE TRANSPORT THROUGH A CAPILLARY TUBE

Brenner and Gaydos (1977) demonstrated a formal mathematical analogy between Taylor dispersion in capillary tubes and three-dimensional dispersion phenomena in spatially periodic porous media, so we can learn important lessons from this much-studied problem with simple shear flow. Most studies of dispersion in Poiseuille flow have been from two standpoints: the approach to the asymptotic state of a cloud of solute initially released as an "impulse", or the change of concentration down the tube of solute whose release concentration varies harmonically with time. I will discuss these two strands and their recent combination into a unified theory by Stokes and Barton (1985, 1988), who used integral transforms. Even for this simple shear flow system, working out the approach to the limiting state is a task of considerable complexity and subtlety, which has concerned some very able applied mathematicians.

Taylor (1953) showed that the average solute concentration C across the tube obeyed the equation

$$\frac{\partial C}{\partial t} = D' \frac{\partial^2 C}{\partial z^2} - V' \frac{\partial C}{\partial z} , \tag{3}$$

where a is the tube radius, V' is the mean fluid velocity, D' is the effective dispersion coefficient, t is time, and z is the distance along the tube, and showed that D' had the form $V_m^2 a^2/192D$, with V_m (= 2V') the maximum fluid velocity and D the molecular diffusion constant. Aris (1956) included the effect of longitudinal molecular diffusion, and used the method of moments to derive the effective long time dispersion coefficient

$$D' = D + V_m^2 a^2/192D . \tag{4}$$

It is evident from the form of (4) that the molecular diffusivity D acting across the shear flow is an essential driving mechanism for the process. As emphasized by Brenner (1978), we cannot simply take the limit as D approaches zero and obtain the residual hydromechanical dispersion, as assumed for porous medium dispersion by Fried and Combarnous (1971) and Bear (1972). The early stages of the solute movement were studied by Lighthill (1966), who found the form of the "nose" of the solute cloud at early times. Gill and Sankarasubramanian (1970) showed that the average concentration across the tube approximately satisfied (3) with a time dependent diffusion coefficient D'(t), and gave an explicit exponential series for D'(t). At small times $t < 0.2a^2/D$, their D'(t) can be shown to have a rapid linear increase with time,

$$D'(t) \sim D + V_m^2 t/12 - \ldots , \tag{5}$$

which means that with constant mean velocity, at this stage D' − D will grow approximately linearly with distance. At longer times D'(t) approaches the Taylor-Aris limit,

$$D'(t) = D + (V_m^2 a^2/192D)[\, 1 - 0.971\exp(-14.7Dt/a^2) - \ldots \,] . \tag{6}$$

If the contributions of the first term D to the sums in (5) and (6) are relatively small, then both expressions are proportional to the square of the mean velocity V'. Chatwin (1970) studied the approach of the solute profile to normality in great detail, and Smith (1987) showed how to choose a moving coordinate at a different angle in z − t space from the usual z' = z − V't, to give a more accurate normal approximation.

The important problem of the decay downstream of the amplitude variations of a solute concentration varying harmonically with time was investigated by Philip (1963a,b), who found approximate solutions for the complicated eigenvalue problem which determines the concentration across the tube and the rate of decay of the wave amplitude downstream. Chatwin (1973) gave high- and low-frequency approximations which differed somewhat from those of Philip, and concluded that Philip's solutions were incorrect at high frequencies. Barton (1983) studied the eigenvalue problem in great detail, and found a rich structure in the way the positions of the eigenvalues changed with frequency in the complex plane, including what he regarded as anomalous behaviour of the eigenvalue branches.

Recently Stokes and Barton (1985, 1988) used Laplace transforms in time and Fourier transforms in the longitudinal coordinate z to examine the transient and harmonic problems together, and were able to calculate and invert their transforms semi-analytically, using "numerical complex analysis" where necessary. This gave them a precise appreciation of the complexity of the eigenvalue structure with varying frequency, and furnished accurate transient approximations for the nose and tail of the solute cloud. Their results demonstrated the validity of the Philip (1963a,b) approximations and the incompleteness of the Chatwin (1973) assumptions, sorted out the apparent anomalies of Barton (1983), and improved the Lighthill (1966) result.

LABORATORY SCALE DISPERSION EXPERIMENTS

The Model Identification Problem

Most laboratory experiments on dispersion have consisted in following the breakthrough of a fluid of higher solute concentration as it displaces one of lower concentration from a column of saturated porous material following an initial step change in concentration at the inlet $z = 0$. The measured "breakthrough curve" is a graph of effluent concentration C at the exit $z = L$, plotted against time. As long as the equations governing solute transport are linear in concentration C, we have a principle of superposition, and can define a transfer function g(t) of the system, which is the output for a hypothetical impulse or "delta function" input. Then for a more general input I(t), the output O(t) is given by the convolution integral of the transfer function with the input

$$O(t) = \int_0^t g(\tau)\, I(t-\tau)\, d\tau \quad . \tag{7}$$

Here we follow the convention used by mathematicians, but engineers (e.g. Duffy and Gelhar 1985) call g(t) the "impulse response function" of the system, and use the term "transfer function" only in the transform domain. The "forward problem" of finding the output given the transfer function and the input is mathematically well posed and numerically stable, but the "identification problem" of finding the transfer function g(t) given the input and output is a deconvolution problem, which is well known to be a badly posed "inverse problem" and to be numerically unstable. (See, for example, Philip 1983; Craig and Brown 1986). Conceptually, this means that many different transfer functions applied to the same input will give outputs almost indistinguishable from each other in the presence of measurement errors. Most experimenters realize, at least implicitly, the difficulty of getting enough information to infer reliably the form of a general transfer function g(t), and use simple models with transfer functions depending on a few parameters whose values are estimated from the data. Traditionally, these models had transfer functions derived from the solutions of partial differential equations believed to govern the transport, such as the equation of Lapidus and Amundson (1952) or its generalizations, but more recently Jury (1982) has used probabilistic transfer functions. The proliferation of models for breakthrough curves is a symptom of the ill-posedness of the identification problem, and we should heed the words of Acton (1970, p 255) who is speaking of "a favorite form of lunacy among aeronautical engineers" who attempt to decide what transfer function (corresponding to a differential equation) governs the motion of some physical object, such as a helicopter rotor:

> "In the archtypical experiment, one measures both the input to the blade and its output – that is, motion. Arguing that the system is a "black box" to which they have the input and the output, they now wish to compute the

transfer function, another name for the differential equation. The difficulty here is that many quite different differential equations can give rise to nearly the same outputs from identical inputs. Thus discriminating the "true" equation from its brethren is next to impossible. Computationally speaking, the problem is ill conditioned. I must quickly point out, however, that if one's purpose is to discover an empirical differential equation adequate for limited amounts of prediction, without overmuch worry about its ultimate physical reality, then this fitting approach, properly done with orthogonal functions, can be worthwhile – though still ill conditioned. But arguments about which differential equation represents the Truth, together with their fitting calculations, are wasted time."

Convection-dispersion Equation Models

The equation given by Lapidus and Amundson (1952) included three mechanisms for solute transfer: one dimensional convection, an "effective" longitudinal diffusion, and local adsorption of the solute onto the solid particles

$$\gamma \frac{\partial n}{\partial t} + \frac{\partial C}{\partial t} = D' \frac{\partial^2 C}{\partial z^2} - V' \frac{\partial C}{\partial z} , \tag{8}$$

where n is concentration in the adsorbed phase and γ is a constant, which is equivalent to the capillary tube dispersion equation (3) with an extra term to account for adsorption. They examined adsorption as either an instantaneous process or a first-order linear reaction,

$$n = k_1 C \quad \text{or} \quad \frac{\partial n}{\partial t} = k_2 (C - n) , \tag{9}$$

where k_1 and k_2 are rate constants. Greenkorn and Kessler (1969) describe various authors' derivations of (3) as an equation for dispersion in porous media, using either stochastic models or microscopic "mixing cell" models. On the basis of a mixing cell model, Kramers and Alberda (1953) studied the frequency response of continuous flow systems and concluded that solute transport in packed beds was governed by (3). Hashimoto et al. (1964) showed that instantaneous adsorption in (8) could be considered as mathematically equivalent to a linear change of time scale in (3), and defined the necessary "retardation factor". The solution of (3) subject to appropriate initial and boundary conditions was given by Brenner (1962) in terms of the Pèclet number V'L/D'; van Genuchten and Alves (1982) tabulated analytical solutions for a great number of boundary and initial conditions. Rose and Passioura (1971) discuss graphical methods for estimating this parameter and checking for violations of the model assumptions, while van Genuchten (1980) used a least squares fitting procedure. It is usually much easier to compute the Laplace or Fourier transform directly than to evaluate the time domain solution for the theoretical breakthrough curve, and Aris (1958) showed how to find the time moments from the Laplace transform for comparison with measured values. To reduce the errors in estimating higher moments, Anderssen and White (1971) used a weighted moments method for such parameter estimation. Further advantages of working in the frequency domain were listed by Turner (1972), who measured the frequency response directly by imposing a sine wave input and measuring the attenuation and phase shift of the output sine wave as a function of frequency. Raats (1973) derived the theoretical frequency response of a two-region dispersion model. Duffy and Al-Hassan (1988) also worked in the frequency domain, but used nonideal pulse inputs and Fourier analysed measured inputs and outputs using the fast Fourier transform.

As breakthrough curves for some media were observed not to fit the simple model (3), more complicated nonequilibrium models were introduced and solved. Rose (1977) discusses the interpretation of breakthrough curves in terms of the various models, as do Kool et al. (1987). Valocchi (1985) determined general quantitive criteria to assess the validity of the local equilibrium assumption for some common models. Often, mathematically equivalent models were generated independently in the study of chromatography, heat transfer, petroleum engineering, chemical engineering, and soil physics and chemistry. To account for highly skewed breakthrough curves with long tails, nonequilibrium or delay models with some kind of local reservoirs in which "holdup" occurred were introduced. The delay process could be governed by a local ordinary or partial differential equation, or could be specified by a probabilistic distribution of delay times. Goodknight et al. (1960) and Coats and Smith (1964) postulated the existence of "dead-end pores" containing immobile liquid, with diffusional exchange of solute at the interface between the mobile and immobile phases depending linearly on the difference between their concentrations, the model used by van Genuchten and Wierenga (1976). As pointed out by van Genuchten (1981) and Nkedi-Kizza et al. (1984), this formulation is mathematically equivalent to adsorption with a linear first order exchange process, the model used by Lapidus and Amundson (1952) and many others, which causes difficulty in ascribing proper physical significance to fitted model parameters. Buffham et al. (1970) presented a probabilistic time delay model for local reservoir holdup during flow in packed beds, with convection but no diffusion in the longitudinal direction, and claimed that it was more realistic than a deterministic diffusion model. This was generalised to a unified time delay model for dispersion by Buffham and Gibilaro (1970), and shown to include many previous chemical engineering time delay models as special cases.

An assumption used originally for heat transfer in a heat exchange system is that there is diffusion of heat spatially distributed in the transverse direction within the stationary medium, which interacts with the longitudinal principally convective heat flow in the moving fluid. Turner (1958, 1972) used this model for solute transport and showed that the concentration of solute in the flowing phase and the flux of solute into the reservoir phase were linearly related in the Laplace transform and frequency domains by an admittance, which could be calculated for several realistic reservoir shapes; this work was extended by Aris (1959) and others. Dispersion in aggregated porous media was studied theoretically by Passioura (1971) and experimentally by Passioura and Rose (1971), who explored the effects of fluid velocity and aggregate size. Drummond and McNabb (1972) used Turner-type assumptions in their ground-breaking study of cold water intrusions into a geothermal field, and found Laplace transforms of the solutions for heat flow along and into a fissure and a borehole. (This paper must be unique in the modern era in having no references section and not citing the work of any other authors; unfortunately other authors have returned the compliment). The equivalent problem of solute flow along a cylindrical macropore with transverse diffusion into the soil was investigated by Scotter (1978) and Kanchanasut et al. (1978). Skopp and Warrick (1974) introduced the Turner model to soil-water solute transport, and Raats (1981b) traced the prehistory of the Turner model in heat transfer and its use for transport in structured porous media. He demonstrated that many simpler models could be considered as low-order approximations to the Turner spatially distributed diffusion exchange model. Neretnieks (1980) used essentially the same assumptions and solution procedure as Drummond and McNabb (1972) to describe solute transport in a single fracture. Tang et al. (1981) generalised their model "under consideration of all dispersive and diffusive processes" including dispersion along the fracture, and found an analytical solution as an integral requiring numerical evaluation. Barker (1982) and Sudicky and Frind (1982) (but see Davis and Johnston 1984) applied the model to solute transport in systems of parallel fissures, obtaining Laplace transform solutions which are difficult to invert analytically.

To explain double-peaked breakthrough curves resulting from single-peaked input curves, Skopp et al. (1981) proposed a full two region model with separate longitudinal convection and dispersion in each region, and only a small coupling between regions. Walker (1987) showed how to calculate solutions for this model from solutions of the much simpler Coats and Smith type exchange model, and Walker and Smettem[1] describe an experiment to test this model in a two region column with a hollow cylinder of fine sand around a cylindrical region of glass beads larger than the sand grains.

Most of the above work has been concerned with solute dispersion in saturated steady flow systems with only a limited amount of spatial nonuniformity. However, during unsaturated soil water flow changing or nonuniform application rates can cause large variations in flow and water content, even for homogeneous media, making transient coupled water and solute movement a difficult problem to solve analytically or numerically. Bresler and Hanks (1969) and Bresler (1973) used finite difference numerical methods to estimate simultaneous water and salt transport in unsaturated soils, and Warrick et al. (1971) gave analytical solutions for solute dispersion in some simple unsaturated flow systems. The method of characteristics was used by Wilson and Gelhar (1974, 1981) to model convective solute transport in unsaturated water flow, with a dispersion approximation superimposed on the convective solution. Some examples of dispersion during steady and transient one-dimensional flows were given to illustrate the method. Exchange between mobile and stagnant unsaturated water was proposed by Gaudet et al. (1977) in their theoretical and experimental study. Smiles et al. (1978) and Smiles and Philip (1978) chose boundary and initial conditions for both water and solute flows in order that water and solute profiles would preserve similarity in terms of the variable $\lambda = z/t^{1/2}$, provided that certain conditions were met. Their experiments demonstrated that their dispersion coefficient was independent of flow velocity, and that there was no detectable "immobile water" in their soil, at least under the experimental conditions. Bond and Wierenga[2] carried out dispersion experiments in saturated and unsaturated water flows using the same material for both, and concluded that there was a significant difference in the amounts of "immobile water" needed in each case to explain the shapes of the breakthrough curve and the in situ solute distribution, respectively. Clothier et al.[3] presented experimental results on the simultaneous redistribution of water and solute following three-dimensional absorption from a dripper, and found that advective solute transport was dominant during the absorption phase, but that molecular diffusion also contributed significantly to solute movement during redistribution.

FIELD SCALE DISPERSION EXPERIMENTS

During the last one or two decades there has been widespread recognition by theoreticians of the importance of flow variations caused by macropores, fissures or other large-scale heterogeneities, for solute transport in the field or the aquifer. (See the review by White 1985). I suspect that field practitioners have long been aware of this factor; it would be difficult to improve on the description of preferential solute flow down cracks in clay soil given ninety years ago in a chapter on sewage disposal by Moore (1898, p 429):

> "Another system is that of irrigation, which consists in passing the sewage over land, in order to use its fertilizing properties, and at the same time to purify it before running the liquid into a river or other watercourse.

[1] G Walker and K Smettem (1987) A description of double-peaked breakthrough curves using a two-region model. Symposium poster paper

[2] WJ Bond and PJ Wierenga (1987) Immobile water in unsaturated homogeneous sandy soils. Symposium poster paper

[3] BE Clothier, DR Scotter and TJ Sauer (1987) Redistribution of water and solute following 3-D absorption. Symposium poster paper

Loamy porous soil is the best for sewage irrigation from a sanitary point of view. Unless the subsoil is sand or gravel, it is usual with the denser top soils to provide some subsoil drainage, increasing it in amount and depth with the density of the soil to be utilized, so as to give a free exit for the water, and prevent the ground from getting water-logged.

"Where the land is of a stiff clayey nature, there are considerable difficulties in adapting it for irrigation. In undrained clay land, under ordinary circumstances, cracks one and two inches [2.5 and 5 cm] wide and five feet [1.5 m] deep are sometimes met with, and it has been found that these are intensified in drained land, with the result that direct passage of sewage and surface water into them has occurred on sewage farms of this nature, so that the effluent is not purified as intended."

As Santayana (1905, chapter 12) said: "those who cannot remember the past are condemned to repeat it."

Two approaches to modelling solute transport have been used: the traditional deterministic convection-dispersion equation (CDE) approach in which it is assumed that the velocity field can be measured with sufficient accuracy to calculate advective transport, and that dispersion can be accounted for as a perturbation on this, or alternatively a stochastic approach in which it is assumed that only average velocities and measures of velocity variation are available, giving a probabilistic distribution of travel times to be used in a transfer function model (TFM). For complicated multidimensional flow systems, a boundary value problem for the CDE must be solved on a large computer, using a finite difference or finite element approximation. Pantelis[4] used a simplified asymptotic water flow model involving only the two horizontal coordinates to provide an approximation to the three-dimensional water flow field in the saturated and unsaturated zones, for use in a CDE contaminant transport model. Under field conditions, usually neither time nor resources are available to adequately characterize the water flow, but adopting a stochastic approach does not mean that measurements are not needed. Rather, an ideal stochastic theory should indicate how much data is needed to describe the flow field well enough for the particular application. In principle, detailed local measurements of soil hydraulic conductivity can be made, but since earliest times tracer solutes have been used to infer water flow velocities, as well as parameters of the solute retardation process (Slichter 1905; Rainwater et al. 1987). Installing wells and monitoring solute concentrations in them is expensive and time-consuming; Davis et al.[5] used surface geophysical methods to map the subsurface distribution of a contaminated water plume at a waste site, and compared their results with those from conventional borehole monitoring.

It is fair to say that the application of the deterministic CDE to laboratory scale dispersion was worked out in chemical engineering before soil science or groundwater hydrology, but that the reverse is the case for stochastic descriptions of macrodispersion. Groundwater hydrologists have faced up to their large-scale heterogeneities for some time, but only recently have chemical engineers (Koch and Brady 1987a,b) used a stochastic characterization of spatial nonuniformity to devise "a non-local description of advection-diffusion with application to dispersion in porous media", in which no account is taken of the previous work by groundwater hydrologists.

The problem of identifying the correct solute transport approach on the field

---

[4]G Pantelis (1987) A three-dimensional asymptotic model for the water and contaminant transport in the mine overburden heaps at Rum Jungle. Symposium poster paper

[5]GB Davis, G Buselli and A Zerilli (1987) Preliminary modelling of point source contamination of groundwater utilizing surface geophysics as an aid to model parameter estimation. Symposium poster paper

scale is at least as badly posed as the corresponding model identification problem on the laboratory scale, so we can't expect a universal or "correct" answer. The boundary between CDE and TFM methods is diffuse rather than sharp – the CDE, of course, yields a transfer function which in practice is often treated as an empirical function with parameters to be fitted to the data, and some originally empirical transfer functions and travel time distributions can be shown to be equivalent to physically-based advective or mixing models (Raats 1981a). Raats (1978, 1981a, 1985) discussed transfer functions and travel time distributions for a number of solute transport systems, and Duffy and Gelhar (1985, 1986) worked in the frequency domain to derive frequency responses for three transport models, including a CDE model.

A comparison between the CDE and TFM approaches was given by Jury (1988), and Dyson and White (1987) found that a TFM and a CDE model (with fitted parameters) could be made to match their experimental solute breakthrough curves equally well. Jury (1988) reviewed the stochastic dispersion theories of Gelhar and Axness (1983) and Dagan (1984, 1987) and their successful use by Sudicky (1986) to model measured longitudinal and transverse macrodispersion coefficients in the Borden experiment. (See also comments by Kemblowski 1987; White 1988). White (1988) raised the question of the method of determining "point" values of the saturated hydraulic conductivity K from measurements on repacked core samples, which can produce marked changes in the mean of K and the variance of its logarithm. Recently, Taylor and Howard (1987) performed a careful natural gradient tracer test in a physically homogeneous unconsolidated aquifer composed of highly uniform sands, and found no dependence of the magnitude of dispersion on the scale of measurement along the flow path, despite some observed multiple tracer breakthroughs in the monitoring wells. I believe that it is valid to say that a definitive and comprehensive experimental comparison of the two approaches is still in the future, and that the field is still in a state of flux.

CONCLUDING REMARKS

At the end of the Concluding Summary section of their "state-of-the-art review of dispersion in real porous media" Greenkorn and Kessler (1969) posed a question to be answered for any dispersion problem in a nonideal (heterogeneous, nonuniform, anisotropic) system:

> "What is the limiting step in our analysis of the problem: the mathematical model, the computing techniques, or the experimental data?"

I believe that after two decades of progress we can give the answer – it is now usually the experimental data that are limiting. I quote from the Concluding Remarks section of the review by Dagan (1986):

> "First, field data, existing and newly acquired by carefully designed experiments, have to be collected and analyzed with the aid of the theoretical tools already developed or under development."

REFERENCES

Acton FS (1970) Numerical methods that work. Harper and Row, New York

Anderssen AS, White ET (1971) Parameter estimation by the weighted moments method. Chem Eng Sci 26: 1203-1221

Aris R (1956) On the dispersion of a solute in a fluid flowing through a tube. Proc R Soc London Ser A 235: 67-77

Aris R (1958) On the dispersion of linear kinematic waves. Proc R Soc London Ser A 245: 268-277

Aris R (1959) Diffusion and reaction in flow systems of Turner's structures. Chem Eng Sci 7: 80-87
Aronson DG (1986) The porous medium equation. In: Fasano A, Primicerio M (eds) Nonlinear diffusion problems, Montecatini Terme 1985, Lecture notes in mathematics, vol 1224, Springer-Verlag, Berlin, pp 1-46
Barker JA (1982) Laplace transform solutions for solute transport in fissured aquifers. Adv Water Resour 5: 98-104
Barton NG (1983) The dispersion of solute from time-dependent releases in parallel flow. J Fluid Mech 136: 243-267
Bear J (1972) Dynamics of fluids in porous media. American Elsevier, New York
Brenner H (1962) The diffusion model of longitudinal mixing in beds of finite length: numerical values. Chem Eng Sci 17: 229-243
Brenner H (1978) Dispersion resulting from spatially periodic porous media. In: Fiszdon W, Herczynski R, Bratos M (eds) Fluid dynamics transactions, vol 9, Proc of the XIIIth Symp on Advanced Problems and Methods in Fluid Mechanics, Olsztyn-Kortowo, September 5-10, 1977. Polish Scientific Publishers, Warszawa, pp 7-120
Brenner H, Gaydos LJ (1977) The constrained Brownian movement of spherical particles in cylindrical pores of comparable radius: models of the diffusive and convective transport of solute molecules in membranes and porous media. J Colloid Interface Sci 58: 312-356
Bresler E (1973) Simultaneous transport of solutes and water under transient unsaturated flow conditions. Water Resour Res 9: 975-986
Bresler E, Hanks RJ (1969) Numerical method for estimating simultaneous flow of water and salt in unsaturated soils. Soil Sci Soc Am Proc 33: 827-832
Buffham BA, Gibilaro LG (1970) A unified time delay model for dispersion in flowing media. Chem Eng J Loughborough Engl 1: 31-36
Buffham BA, Gibilaro LG, Rathor MN (1970) A probabilistic time delay description of flow in packed beds. AIChE J 16: 218-223
Chatwin PC (1970) The approach to normality of the concentration distribution of a solute in a solvent flowing along a straight pipe. J Fluid Mech 43: 321-352
Chatwin PC (1973) On the longitudinal dispersion of dye whose concentration varies harmonically with time. J Fluid Mech 58: 657-667
Coats KH, Smith BD (1964) Dead-end pore volume and dispersion in porous media. Soc Petrol Engrs J 4: 73-84
Craig IJD, Brown JC (1986) Inverse problems in astronomy: a guide to inversion strategies for remotely sensed data. Adam Hilger, Bristol
Dagan G (1984) Solute transport in heterogeneous porous formations. J Fluid Mech 145: 151-177
Dagan G (1986) Statistical theory of groundwater flow and transport: pore to laboratory, laboratory to formation, and formation to regional scale. Water Resour Res 22: 120S-134S
Dagan G (1987) Theory of solute transport by groundwater. Ann Rev Fluid Mech 19: 183-215
Daniels F, Mathews JH, Williams JW, Bender P, Alberty RA (1956) Experimental physical chemistry. McGraw-Hill, New York
Davis GB, Johnston CD (1984) Comment on "Contaminant transport in fractured porous media: analytical solutions for a system of parallel fractures" by E.A. Sudicky and E.O. Frind. Water Resour Res 20: 1321-1322
Deriagin BV, Bakanov SP (1957) Theory of the flow of a gas in a porous material in the near-Knudsen region. Sov Phys Tech Phys 2: 1904-1918
Drummond JE, McNabb A (1972) The heating of cold-water intrusions in a geothermal field. N Z J Sci 15: 665-672
Duffy CJ, Al-Hassan S (1988) The time and frequency response of tracer experiments. J Hydrol 97: 59-73
Duffy CJ, Gelhar LW (1985) A frequency domain approach to water quality modeling in groundwater: theory. Water Resour Res 21: 1175-1184
Duffy CJ, Gelhar LW (1986) A frequency domain analysis of groundwater quality fluctuations: interpretation of field data. Water Resour Res 22: 1115-1128

Dyson JS, White RE (1987) A comparison of the convection-dispersion equation and transfer function model for predicting chloride leaching through an undisturbed, structured clay soil. J Soil Sci 38: 157-172
Fried JJ, Combarnous MA (1971) Dispersion in porous media. Adv Hydrosci 7: 169-282
Gaudet JP, Jègat H, Vachaud G, Wierenga P (1977) Solute transfer, with exchange between mobile and stagnant water, through unsaturated sand. Soil Sci Soc Am J 41: 665-673
Gelhar LW, Axness CL (1983) Three-dimensional stochastic analysis of macrodispersion in aquifers. Water Resour Res 19: 161-180
Gill WN, Sankarasubramanian R (1970) Exact analysis of unsteady convective diffusion. Proc R Soc Lond Ser A 316: 341-350
Goodknight RC, Klikoff WA, Fatt I (1960) Non-steady-state flow and diffusion in porous media containing dead-end pore volume. J Phys Chem 64: 1162-1168
Greenkorn RA, Kessler DP (1969) Dispersion in heterogeneous nonuniform anisotropic porous media. Ind Eng Chem 61: 14-32
Hashimoto I, Deshpande KB, Thomas HC (1964) Pèclet numbers and retardation factors for ion exchange columns. Ind Eng Chem Fundam 3: 213-218
Jury WA (1982) Simulation of solute transport using a transfer function. Water Resour Res 18: 363-368
Jury WA (1988) Solute transport and dispersion. This volume, pp
Kanchanasut P, Scotter DR, Tillman RW (1978) Preferential solute movement through larger soil voids. II. Experiments with saturated soil. Aust J Soil Res 16: 269-276
Kemblowski MW (1987) Comment on "A natural gradient experiment on solute transport in a sand aquifer: spatial variability of hydraulic conductivity and its role in the dispersion process" by E.A. Sudicky. Water Resour Res 24: 315-317
Knudsen M (1950) The kinetic theory of gases: some modern aspects. 3rd edition, Methuen, London
Koch DL, Brady JF (1987a) The symmetry properties of the effective diffusivity tensor in anisotropic porous media. Phys Fluids 30: 642-650
Koch DL, Brady JF (1987b) A non-local description of advection-diffusion with application to dispersion in porous media. J Fluid Mech 180: 387-403
Kool JB, Parker JC, van Genuchten MTh (1987) Parameter estimation for unsaturated flow and transport models – a review. J Hydrol 91: 255-293
Kramers H, Alberda G (1953) Frequency response analysis of continuous flow systems. Chem Eng Sci 2: 173-181
Lapidus L, Amundson NR (1952) Mathematics of adsorption in beds. VI. The effect of longitudinal diffusion in ion exchange and chromatographic columns. J Phys Chem 56: 984-988
Lighthill MJ (1966) Initial development of diffusion in Poiseuille flow. J Inst Math Its Applic 2: 97-108
Mason EA, Malinauskas AP, Evans RB (1967) Flow and diffusion of gases in porous media. J Chem Phys 46: 3199-3216
Moore ECS (1898) Sanitary engineering: a practical treatise on the collection, removal and final disposal of sewage and the design and construction of works of drainage and sewerage. BT Batsford, London
Muskat M (1937) The flow of homogeneous fluids through porous media. McGraw-Hill, New York
Neretnieks I (1980) Diffusion in the rock matrix: an important factor in radionuclide retardation? J Geophys Res 85: 4379-4397
Nkedi-Kizza P, Biggar JW, Selim HM, van Genuchten MTh, Wierenga PJ, Davidson JM, Nielsen DR (1984) On the equivalence of two conceptual models for describing ion exchange during transport through an aggregated oxisol. Water Resour Res 20: 1123-1130
Passioura JB (1971) Hydrodynamic dispersion in aggregated media. 1. Theory. Soil Sci 111: 339-344
Passioura JB, Rose DA (1971) Hydrodynamic dispersion in aggregated media. 2. Effects of velocity and aggregate size. Soil Sci 111: 345-351

Philip J (1983) Deconvolution of Gaussian and other kernels. In: Deuflhard P, Hairer E (eds) Numerical treatment of inverse problems in differential and integral equations: Proc of Int Workshop, Heidelberg, Fed Rep of Germany, August 30 – September 3, 1982. Birkhauser, Boston Basel Stuttgart, pp 335-344

Philip JR (1963a) The theory of dispersal during laminar flow in tubes. I. Aust J Phys 16: 287-299

Philip JR (1963b) The theory of dispersal during laminar flow in tubes. II. Aust J Phys 16: 300-310

Raats PAC (1973) Propagation of sinusoidal solute density oscillations in the mobile and stagnant phases of a soil. Soil Sci Soc Am Proc 37: 676-680

Raats PAC (1978) Convective transport of solutes. II. Specific flow problems. Agric Water Manage 1: 219-232

Raats PAC (1981a) Residence times of water and solutes within and below the root zone. Agric Water Manage 4: 63-82

Raats PAC (1981b) Transport in structured porous media. In: Verruyt A, Barends FBJ (eds) Flow and transport in porous media. Proc Euromech 143, Delft 2-4 September 1981. Balkema, Rotterdam, pp 221-226

Raats PAC (1985) Responses of hydrological systems to changes in water quality and quantity. In: Relation of groundwater quantity and quality. Proc Hamburg Symp, August 1983. Internat Assoc of Hydrological Sciences press, Inst of Hydrology, Wallingford, UK, publ no 146, pp 3-12

Rainwater KA, Wise WR, Charbeneau RJ (1987) Parameter estimation through groundwater tracer tests. Water Resour Res 23: 1901-1910

Raupach MR (1988) Canopy transport processes. This volume, pp

Rose DA (1977) Hydrodynamic dispersion in porous materials. Soil Sci 123: 277-283

Rose DA, Passioura JB (1971) The analysis of experiments on hydrodynamic dispersion. Soil Sci 111: 252-257

Santayana G (1905) The life of reason, vol 1, Introduction and reason in common sense. Charles Scribner, New York

Scotter DR (1978) Preferential solute movement through larger soil voids. I. Some computations using simple theory. Aust J Soil Res 16: 257-267

Skopp J, Gardner WR, Tyler EJ (1981) Solute movement in structured soils: two region model with small interaction. Soil Sci Soc Am J 45: 837-842

Skopp J, Warrick AW (1974) A two-phase model for the miscible displacement of reactive solutes in soils. Soil Sci Soc Am Proc 38: 545-550

Slichter CS (1905) Field measurements of the rate of movement of underground waters. U S Geol Surv Water-Supply Pap 140: 1

Smiles DE, Philip JR (1978) Solute transport during absorption of water by soil: laboratory studies and their practical implications. Soil Sci Soc Am J 42: 537-544

Smiles DE, Philip JR, Knight JH, Elrick DE (1978) Hydrodynamic dispersion during absorption of water by soil. Soil Sci Soc Am J 42: 229-234

Smith R (1987) Shear dispersion looked at from a new angle. J Fluid Mech 182: 447-466

Stokes AN, Barton NG (1985) Solution of dispersion in Poiseuille flow by integral transforms. In: Third Australasian Conf on Heat and Mass Transfer, May 1985, University of Melbourne, Melbourne

Stokes AN, Barton NG (1988) The concentration distribution produced by shear dispersion of solute in Poiseuille flow. J Fluid Mech, submitted

Sudicky EA (1986) A natural gradient experiment on solute transport in a sand aquifer: spatial variability of hydraulic conductivity and its role in the dispersion process. Water Resour Res 22: 2069-2082

Sudicky EA, Frind EO (1982) Contaminant transport in fractured porous media: analytical solutions for a system of parallel fractures. Water Resour Res 18: 1634-1642

Tang DH, Frind EO, Sudicky EA (1981) Contaminant transport in fractured porous media: Analytical solution for a single fracture. Water Resour Res 17: 555-564

Taylor GI (1953) Dispersion of soluble matter in solvent flowing slowly through a tube. Proc R Soc London Ser A 219: 186-203
Taylor GI (1959) The present position in the theory of turbulent diffusion. Adv Geophys 6: 101-112
Taylor SR, Howard KWF (1987) A field study of scale-dependent dispersion in a sandy aquifer. J Hydrol 90: 11-17
Thurtell GW (1988) Canopy transport processes: commentary. This volume, pp

Turner GA (1958) The flow-structure in packed beds. Chem Eng Sci 7: 156-165
Turner GA (1972) Heat and concentration waves: analysis and applications. Academic Press, New York
Valocchi AJ (1985) Validity of the local equilibrium assumptions for modeling sorbing solute transport through homogeneous soils. Water Resour Res 21: 808-820
van Genuchten MTh (1980) Determining transport parameters from solute displacement experiments. Research Report No 118, U S Salinity Lab, USDA-SEA-AR, Riverside, CA, 37 pp
van Genuchten MTh (1981) Non-equilibrium transport parameters from miscible displacement experiments. Research Report No 119, U S Salinity Lab, USDA-SEA-AR, Riverside, CA, 88 pp
van Genuchten MTh, Alves WJ (1982) Analytical solutions of the one-dimensional convective-dispersive solute transport equation. U S Dep Agric, Tech Bull No 1661, Washington D C, 151 pp
van Genuchten MTh, Wierenga PJ (1976) Mass transfer studies in sorbing porous media. I. Analytical solutions. Soil Sci Soc Am J 40: 473-479
Walker GR (1987) Solution to a class of coupled linear partial differential equations. IMA J Appl Math 38: 35-48
Warrick AW, Biggar JW, Nielsen DR (1971) Simultaneous solute and water transfer for an unsaturated soil. Water Resour Res 7: 1216-1225
White I (1988) Comment on "A natural gradient experiment on solute transport in a sand aquifer: spatial variability of hydraulic conductivity and its role in the dispersion process" by E.A. Sudicky. Water Resour Res, in press
White RE (1985) The influence of macropores on the transport of dissolved and suspended matter through soil. Adv Soil Sci 3: 95-120
Wilson JL, Gelhar LW (1974) Dispersive mixing in a partially saturated porous medium. Tech Rep 191, Ralph M. Parsons Lab for Water Resour and Hydrodyn, Mass Inst of Technol, Cambridge, MA, 353 pp
Wilson JL, Gelhar LW (1981) Analysis of longitudinal dispersion in unsaturated flow. 1. The analytical method. Water Resour Res 17: 122-130

QUASIANALYTIC AND ANALYTIC APPROACHES TO UNSATURATED FLOW

J.R. Philip

CSIRO Division of Environmental Mechanics, GPO Box 821, Canberra, ACT, 2601, Australia

Summary. We review twenty years' progress in quasianalytic and analytic methods applied to the strongly nonlinear Fokker-Planck equation of flow in unsaturated soils. Specific topics include: the joining problem in one-dimensional infiltration; the utility of the Burgers or Knight soil in one-dimensional flow; the importance of the Broadbridge-White soil in constant-flux one-dimensional infiltration; Broadbridge's result on the nonexistence of linearizing transformations of Burgers' equation in two dimensions (which probably carries over to three); the power of the flux-concentration method; the wide applicability of the quasilinear analysis of multidimensional steady flows, much enhanced by the scattering analogue; and quasilinear results on the watertightness or otherwise to downward unsaturated seepage of macropores, holes, tunnels, and underground cavities.

INTRODUCTION

In a chapter marvellously entitled "The computer: ruin of science and threat to mankind", Clifford Truesdell (1984), in his inimitably pungent style, explores the ultimate vacuity that can follow when man abandons using his mind in favour of the computer. Truesdell, to some degree intoxicated by his own fluency, sometimes overstates and sometimes wanders from his central theme to give backhanders to miscreants incurring his wrath on other counts. For all that, the chapter is a splendid read, and I commend it to all of you as an antidote against the propaganda about computers with which we are bombarded daily.

Truesdell tells us that "for every mathematician who relies on his mind alone, there are now thousands of navvies on computer gangs". This might be close to the truth for natural science as a whole; but, oddly enough, it is not true for soil-water physics and hydrology. I believe this has been due primarily to the accident of persons in this field; but a cynic might suggest that it is at least in part because soil-water physics and hydrology are starved for resources. The huge (often concealed) costs of large-scale computing usually cannot be met, so people, often with some reluctance, have been obliged to try to use their minds. A consequence is that soil-water hydrology has, if anything, been becoming more scientific - at a time when many other fields are losing their way in the computer jungle.

In the soil-water context, the intellectual issue here has been clearly recognized for at least 30 years. In 1969 I wrote (Philip 1969a):

> "The various nonlinear flow equations we have discussed, subject to the conditions we have considered, may, of course, be solved by the brute-force use of finite-difference methods on high-speed computers. There is no special obstacle to using such methods, even for much more elaborate flow

problems with one or more of the complications of spatial heterogeneity, more involved initial and boundary conditions, and the operation of capillary hysteresis.

"A computer attack on problems of this class will always be feasible provided (i) the problem is well-posed in the mathematical sense; (ii) the (well-established) criteria for computational stability are satisfied by the finite-difference scheme adopted. It should be added that the efficiency of the technique demands that the calculator avail himself of analytical methods to establish the behavior of his solutions in the neighborhood of any singularities.

"The ready availability of computer solutions is not, however, so fruitful a potential source of understanding of the theory of infiltration as superficial consideration might suggest. The difficulty is that, in general, each of these nonlinear problems requires for each soil (with its particular hydrological characteristics such as $K(\theta)$ and $\Psi(\theta)$) its own *ad hoc* solution, so that the approach is necessarily a piecemeal one. The possibilities of developing general statements about the phenomena on the basis of calculations of this type are not very good, because they tell us relatively little about the *fundamental structure of the solutions*; and it is upon this which the prospects of useful generalization depend. (It is not my intention to denigrate the use of computers for soil-water calculations; but their most effective rôle in this context seems to me to be as the immensely helpful tools of *ad hoc* investigations)."

QUASIANALYTIC AND ANALYTIC SOLUTIONS

My task is to review quasianalytic and analytic methods applied to problems of water flow in unsaturated soils, with special reference to progress over the last twenty years or so.

We describe as *quasianalytic solutions* those in which the solution is taken a significant part of the way by the methods of mathematical analysis. Typically, the task of solution is reduced from solving a strongly nonlinear partial differential equation to that of solving one or more ordinary equations, of which only one is nonlinear (e.g. Philip 1957a,b), or to performing a convergent iterative sequence of simple integrations (e.g. Philip and Knight 1974).

In some cases, the solutions may be found completely by mathematical analysis. We call these *analytic solutions*. Such solutions necessarily hold only for special forms of the functions that characterize the soil hydrologically. These special forms vary in their ability to represent real soils accurately, but even in the least favourable cases matching procedures can often assure acceptable estimates of integral properties of the solution (all that may be needed in practice).

WATER FLOW IN UNSATURATED SOIL

Just what is so important about water flow in unsaturated soil? And why worry about solving the equation that describes it?

The answer to these questions centres on the fact that in its natural state the soil is normally *unsaturated*: that is, it contains both water and air. Most of the water involved in the hydrologic cycle is located in unsaturated soil between the time of its arrival as rain at the soil surface and that of its return to the atmosphere. A small fraction of precipitation does not enter the soil, but moves overland directly into streams or lakes; and a second small fraction percolates downward through the unsaturated "zone of aeration" and joins the groundwater, i.e. the water in the deep, habitually-saturated strata of soil, alluvia, or rocks. The processes of water movement in unsaturated soil thus play a central part in the scientific study of the terrestrial sector of the hydrologic cycle and in the related problems of irrigated and dry-land agriculture, of plant ecology, and the biology of soil flora and fauna. In addition, they govern transport through soil of materials in solution, such as natural salts, fertilizers, and urban and industrial wastes and pollutants. Specific phenomena of great interest and importance include: infiltration (the entry into a soil mass of water made available at its surface); drainage and retention of water in the soil strata; extraction of soil water by plant roots; and evaporation of water from the soil. The importance of unsaturated flow is greatest in a dry country, such as Australia, where about 93% of precipitation enters the soil and all but 1% of this is returned to the atmosphere by evaporation (Nimmo 1949, Philip 1954).

Many concepts and intuitions of soil and groundwater hydrology evolved in the well-watered regions of Europe and the eastern United States. This humid hydrology, worked out for usually moist environments with a water table near (and even at) the soil surface, has not always taken much account of unsaturated flow. It can, in fact, be quite misleading in arid and semiarid regions such as Australia, Israel, and much of the western United States. In these places the water table may be nonexistent or else at very great depths beneath the soil surface, which is usually rather dry. We need to sharpen our tools for dealing with this arid hydrology; and the study of unsaturated flows is central to this task.

THE GENERAL FLOW EQUATION

The water in unsaturated soils is effectively under suction, because of capillary, adsorptive, and other local forces. Buckingham (1907), a keen disciple of Willard Gibbs, pioneered the physical concepts on which the modern mathematical-physical analysis of the movement of water under these conditions rests. Richards (1931) presented the resulting general nonlinear equation:

$$\frac{\partial \theta}{\partial t} = \nabla.(K\nabla\Psi) - \frac{\partial K}{\partial z} . \tag{1}$$

Here θ is volumetric moisture content, K hydraulic conductivity, t time, and z the vertical space coordinate, positive downward. Ψ is the moisture potential (potential of the forces arising from local interactions between solid matrix and water). The convenient datum for Ψ is free water at atmospheric pressure. Then $\Psi < 0$ in unsaturated soils. Potentials are expressed per unit weight, so Ψ has dimensions [length] and K has dimensions [length] $[\text{time}]^{-1}$. Equation (1) applies to both homogeneous and heterogeneous media. For a homogeneous medium, we may express (1) in more tractable forms. For nonhysteretic processes (e.g. Philip 1969a) K and Ψ are single-valued functions of θ, and we introduce the diffusivity D, such that

$$D = K d\Psi/d\theta; \tag{2}$$

and we rewrite (1) as

$$\frac{\partial \theta}{\partial t} = \nabla.(D\nabla\theta) - \frac{dK}{d\theta}\frac{\partial \theta}{\partial z} . \tag{3}$$

In general the coefficients D and $dK/d\theta$ are strongly varying functions of θ, and (3) is a strongly nonlinear Fokker-Planck equation. Alternatively, with K and D single-valued functions of Ψ, we may take Ψ as the dependent variable; and it is often advantageous to use the Kirchhoff potential (Kirchhoff 1894) Θ, defined by

$$\Theta = \int_{\Psi_0}^{\Psi} K d\Psi \tag{4}$$

as the dependent variable. The flow equation then becomes

$$\frac{1}{D}\frac{\partial\Theta}{\partial t} = \nabla^2\Theta - \frac{1}{K}\frac{dK}{d\Psi}\frac{\partial\Theta}{\partial z} \,. \tag{5}$$

Note that the constant Ψ_0 is often conveniently taken as the "potential at infinity". K decreases so rapidly with Ψ that the integral (4) exists as $\Psi_0 \to -\infty$.

Solutions of (1), or of specialized forms such as (3) and (5), subject to appropriate initial and boundary conditions, are the basis of the physical theory of infiltration and other flow processes in unsaturated soil. The functions characterizing a soil hydrologically, such as $K(\theta)$, $D(\theta)$, $\Psi(\theta)$, are experimentally determined. In some cases they may be approximated analytically, but in general they are available to us in the first instance as obtuse sets of measured values.

The problem of securing relevant solutions seems thus, at first glance, one of bewildering complication. Do we need to crank out a vast array of quite difficult brute-force computer solutions of a complicated nonlinear partial differential equation with a different set for each of the world's million soils?

The challenges which face us are these: Can we bring order out of this chaos? Can we start to understand the shape of the solutions? Can we, perhaps, come up with some simple methods of analysis adequate to guide us in dealing with soils in practice?

In what follows I attempt, within the severe constraints of available space, to answer these questions. I take as our starting point where we were at twenty years ago, and go on to review subsequent progress. I shall deal mostly with infiltration, though I say something about seepage problems of another important class. It is natural and convenient to treat one-dimensional flow problems first; and then to go on to more complicated problems in two and three dimensions.

ONE-DIMENSIONAL INFILTRATION WITH POTENTIAL BOUNDARY CONDITIONS

Series and Travelling Wave Solutions

The earliest quasianalytic infiltration solution was for one-dimensional flow of water supplied to the soil surface at given potential Ψ_1 into a deep homogeneous soil at uniform initial moisture content (Philip 1954, 1957a,b). This was stimulated by the solution of the corresponding horizontal absorption problem by Klute (1952).

The solution was expressible in two forms, a series expansion in $t^{\frac{1}{2}}$ (Philip 1957a), appropriate for small and intermediate values of t, and a travelling wave solution (Philip 1957b) useful for large t, of the form

$$\lim_{t\to\infty} z(\theta,t) = ut + \zeta(\theta) \ . \tag{6}$$

Here u, the velocity of the moisture profile, is given by

$$u = (K_1 - K_0)/(\theta_1 - \theta_0) \ , \tag{7}$$

where θ_1 is the θ-value corresponding to $\Psi = \Psi_1$, θ_0 is the initial moisture content, $K_1 = K(\theta_1)$, and $K_0 = K(\theta_0)$. The function $\zeta(\theta)$ follows from a simple quadrature, but with a constant of integration undetermined. Philip (1957b) estimated this by matching the two solutions at relatively large values of t close to the limits of useful convergence of the series solution, of order 10^6 s in the illustrative numerical example.

The series solution yielded the truncated two-parameter two-term expressions (Philip 1954, 1957d), applicable to small and intermediate times,

$$i(t) = St^{\frac{1}{2}} + At \ , \ v(t) = \tfrac{1}{2}St^{-\frac{1}{2}} + A \ . \tag{8}$$

Here v(t) is the infiltration rate function, i(t) the cumulative infiltration function, and S the sorptivity.

On the other hand integration of (7) gave

$$\lim_{t\to\infty} i(t) = K_1 t + Z \ , \ \lim_{t\to\infty} v(t) = K_1 , \tag{9}$$

with

$$Z = \int_{\theta_0}^{\theta_1} \zeta(\theta)d\theta \ . \tag{10}$$

Evaluation of Z depended on the constant of integration fixing $\zeta(\theta)$ and the assumption that Z was bounded.

Analytic Solutions

The first analytic solutions were for the Green-Ampt (Green and Ampt 1911) and the linear (Philip 1957c, 1966a) soils. The Green-Ampt model is equivalent to taking

$$D(\theta) = \tfrac{1}{2}S^2(\theta_1 - \theta_0)^{-1}\delta(\theta_1 - \theta) \ , \tag{11}$$

where δ is the Dirac delta function (Philip 1957d, 1973b). [Equation (11) means, essentially, that all the diffusivity is concentrated at $\theta = \theta_1$.] For the linear soil, on the other hand,

$$D(\theta) = \text{constant} = \frac{\pi S^2}{4(\theta_1 - \theta_0)^2}, \ \frac{dK}{d\theta} = \text{constant} = \frac{K_1 - K_0}{\theta_1 - \theta_0} \ . \tag{12}$$

We shall discuss these two, plus a third analytic soil, for relatively dry initial states (say $\theta_1 - \theta_0 \geqslant 0.15$), so we shall neglect K_0 as small compared with K_1.

The third soil is the Knight soil. This had been called the Burgers soil, but a referee of Philip (1987b) suggested it be called the Knight soil, since J.H. Knight calculated the relevant solution (Philip 1974); and I agree. Philip (1973a, 1974) pointed out that using certain $D(\theta)$ and $K(\theta)$ functions reduced the one-dimensional form of (3) to the equation of Burgers (1948). The

Hopf (1950) - Cole (1951) transformation linearizes the Burgers equation, so that relevant analytic solutions are readily available. The specialized forms for initially relatively dry soils are

$$D(\theta) = \text{constant} = \frac{\pi S^2}{4(\theta_1 - \theta_0)^2}, \quad \frac{dK}{d\theta} = 2(K_1 - K_0)\left[\frac{\theta - \theta_0}{\theta_1 - \theta_0}\right] . \tag{13}$$

Table 1. The ratio A/K_1 and K_1Z/S^2 for initially relatively dry soils

Author	Soil	$\theta_1 - \theta_0$	A 10^{-8}m s^{-1}	K_1 10^{-8}m s^{-1}	S 10^{-4}m s$^{-\frac{1}{2}}$	Z m	$\frac{A}{K_1}$	$\frac{K_1Z}{S^2}$
Philip (1957a)	Yolo light clay	0.2574	4.654	12.30	1.254	0.067	0.378	0.524
Talsma (1969)	Jondaryan loam	0.24	3.31	8.67			0.382	
	Banna loamy sand	0.35	1.05	3.00			0.350	
	Camarooka clay	0.20	0.305	0.917			0.333	
	Mean of four soils	0.26					0.361	
Philip (1957d)	Green-Ampt soil		$\frac{1}{2}K_1$	K_1	S	∞	0.500	∞
Philip (1966a)	Linear soil		$\frac{2}{3}K_1$	K_1	S	$\frac{\pi S^2}{4K_1}$	0.667	0.785
Philip (1974)	Knight soil		$(1-\frac{2}{\pi})K_1$	K_1	S	$\frac{\pi \ln 2 S^2}{4K_1}$	0.363	0.544

Table 1 compares the three analytic infiltration solutions with exact nonlinear solutions for four real soils. It is of interest to compare their performance for small and moderate t, as measured by the value of A/K_1. It will be recalled that A is the second coefficient of (8). The Knight soil does excellently, giving 0.363, compared with 0.361 as the mean of the four real soils. Green-Ampt gives 0.500 and the linear result is 0.667.

Evolution of the Travelling Wave at Large t

For the Green-Ampt soil Z, the supposed constant in (9), varies in fact like $\ln t$ as $t \to \infty$. For the linear soil, on the other hand, Z is bounded. These results are of little help. Both the Green-Ampt and linear soils are singular (in different ways), and it is unsurprising that they lead to different results on the boundedness of Z and on the approach of v to K_1 as $t \to \infty$.

The severe limitations of both these models at large t are revealed by reference to the differential equation for the travelling wave (Philip 1957b),

$$\frac{d\zeta}{d\theta} = \frac{D}{(K - K_0) - (K_1 - K_0)(\theta - \theta_0)/(\theta_1 - \theta_0)} . \tag{14}$$

The conditions for the *evolution* of the moisture profile into the travelling wave to occur as t increases are

$D \geqslant 0$ in $\theta_0 \leqslant \theta \leqslant \theta_1$, with D positive and finite in at least some finite part of this θ-interval; (15a)

and

$G = 0$ at θ_0 and θ_1, $G < 0$ in $\theta_0 < \theta < \theta_1$. (15b)

Here G is the denominator of the right side of (14), i.e.

$$G = (K - K_0) - (K_1 - K_0)(\theta - \theta_0)/(\theta_1 - \theta_0) .$$

Conditions (15a) and (15b) are both satisfied by the general nonlinear soil, with $D > 0$ throughout $\theta_0 \leqslant \theta \leqslant \theta_1$.

On the other hand, the Green-Ampt model, with a step-function moisture profile for all $t > 0$, cannot embody the *gradual evolution* of the travelling wave shape, which determines the mode of approach of v to K_1 as $t \to \infty$. For the Green-Ampt model (14) is singular: condition (15a) is violated with $D = 0$ everywhere except at $\theta = \theta_1$, where it is infinite.

Secondly, as has been known since its first evaluation (Philip 1957c), the linear soil fails to yield a travelling wave: it gives instead moisture profiles with gradients (at fixed θ) like $t^{-\frac{1}{2}}$ at large t. In this case condition (15b) is violated, with $G \equiv 0$ throughout $\theta_0 \leqslant \theta \leqslant \theta_1$.

Evidently, we can expect reliable guidance to nonlinear behaviour at large t from an exact solution only if it is for a soil satisfying both (15a) and (15b), preferably with $D > 0$ throughout $\theta_0 \leqslant \theta \leqslant \theta_1$. The Knight soil satisfies all these conditions. The upward concavity of $K(\theta)$ ensures that relevant Burgers infiltration solutions evolve into a travelling wave at large t (Philip 1973a).

The Knight soil gives Z bounded and equal to $0.544S^2/K_1$, and the large t expansion for v is

$$v(t) = K_1 + \frac{1}{4}St^{-\frac{1}{2}} \exp\left[-\frac{4K_1^2t}{\pi S^2}\right] + --- . \qquad (16)$$

These are the best estimates of the mode of approach to travelling wave behaviour available from analytic solutions. They are, of course, not exact for the general nonlinear case, but we can be confident that Z is bounded and should not differ much from $0.544S^2/K_1$; and that the second term of the large t expansion for v is exponentially small in t. Note the comparison of values of K_1Z/S^2 for various soils in Table 1. The Knight soil value of 0.544 agrees well with the nonlinear Yolo light clay calculation of 0.524.

Some people worry that the Knight soil has constant D. But the quintessential element in large t behaviour is the travelling wave; and, as Philip (1973a) emphasized, this is the consequence of dynamic equilibrium between a diffusion-like process [generated by a positive D, whether constant or θ-dependent] and a *nonlinear* convection-like process [generated by a concave-upward $K(\theta)$]. Provided D matches the real nonlinear diffusion process adequately in an integral sense [ensured here by relating it to S], we may expect a good *integral* representation of large t behaviour. This is what has been sought here.

The Joining Problem

The problem of how best to join small and intermediate t results such as (8) with large t results such as (9) has interested many people. In practical field terms, however, observational errors, and errors produced by minor heterogeneities and nonuniformities of initial moisture content, will be far larger than the errors of even the crudest joining technique. The practical problem thus warrants only the minimum number of parameters and maximum simplicity.

Philip (1987b) has shown that, for soils which are initially relatively dry, two parameters suffice, and these can be taken as the physically meaningful and measurable sorptivity S and hydraulic conductivity K. Three two-parameter joining techniques were compared. A polynomial fraction method, related to Padé approximants (Baker 1975), failed because it cannot efficiently represent the exponentially rapid decrease of v(t) as $t \to \infty$. A piecewise representation using both (8) and (9) is simple and accurate enough for many purposes. Joining based on the Knight solution gives a continuous representation and greatest accuracy. See Fig. 1.

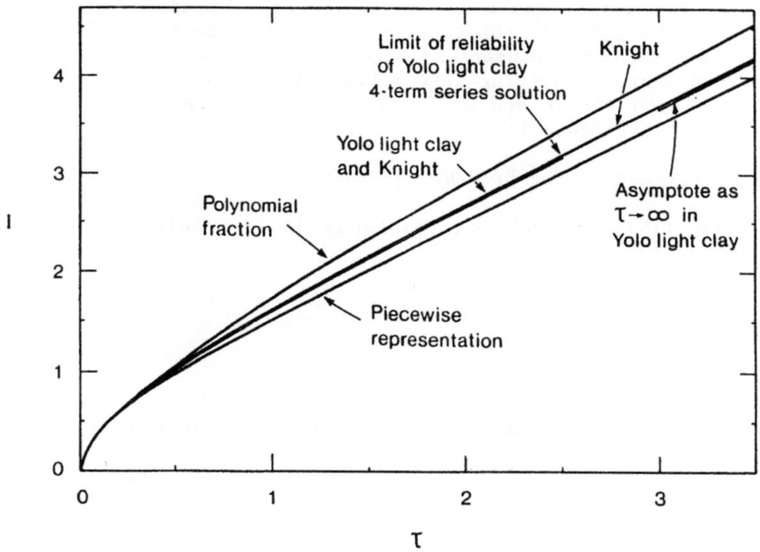

Fig. 1. Comparison of two-parameter joining techniques for infiltration into the Yolo light clay example of Philip (1957a,b). τ is dimensionless time, I is dimensionless cumulative infiltration. The Knight solution and the 4-term series solution for Yolo light clay are indistinguishable for $0 \leqslant \tau \leqslant 2.5$. The truncated Yolo clay series solution is unreliable for $\tau > 2.5$, and so is terminated at $\tau = 2.5$. Note the asymptote as $\tau \to \infty$ for Yolo light clay. Philip (1987b) gives more detail.

ONE-DIMENSIONAL INFILTRATION WITH FLUX BOUNDARY CONDITIONS

The preceding section concerned infiltration of water supplied at a given potential. We now turn to the equally important problems of infiltration of water supplied to the soil surface at a given flux density. We have in mind primarily the case where the soil is homogeneous with initially uniform moisture content.

Since we no longer have the constant θ-interval θ_0 to θ_1 for all $t > 0$, we cannot here take θ as the independent variable, and the quasianalytic series method of Philip (1957a,c) does not apply. Philip (1969a), however, showed that for $D(\theta) \propto (\theta - \theta_0)^n$ $(n > -1)$ a simple similarity solution of the flow equation held for horizontal absorption and also described the early stages of vertical infiltration; and that, provided the surface flux density did not exceed K_1, the constant flux infiltration solution had an asymptotic large t travelling wave solution wholly analogous to (6).

Analytic Solutions

The analytic soil models considered in the preceding section are not well suited to constant flux infiltration problems. For these problems the integral properties of the system of greatest interest are the time course of the θ-value at the soil surface, and of the characteristic depth of wet front penetration. At least for small and moderate t the first of these properties is strongly dependent on the shape of the actual $D(\theta)$ and $K(\theta)$. There is no analogue here to the robust process of exactly representing small t integral behaviour by matching sorptivity S. The Green-Ampt soil is, for this case, totally lacking in physical realism; and here both the linear (Braester 1973) and the Burgers or Knight soil (Clothier et al. 1981) are tenuously based. The Knight soil offers better representation of $K(\theta)$ and, at least, gives when it should a moisture profile evolving into a travelling wave. The linear soil, of course, cannot. Both linear and Knight soils can give analytic solutions for surface flux density variable in time.

Recent Analytic Progress: The Broadbridge-White Soil

Knight and Philip (1974) showed that combining the Kirchhoff transformation (4) and that of Storm (1951),

$$X(x,t) = \int_0^x D(x',t)^{-\frac{1}{2}}\, dx \,, \tag{17}$$

linearized the flow equation for one-dimensional horizontal absorption if, and only if,

$$D(\theta) = a(b - \theta)^{-2} \,, \tag{18}$$

with a, b constants.

Recently Broadbridge and White (1987) have shown that these transformations reduce the one-dimensional *vertical* equation to the Burgers equation when $D(\theta)$ is given by (18) and $K(\theta)$ is of the form

$$K(\theta) = \beta + \gamma(b - \theta) + \lambda(b - \theta)^{-1} \,. \tag{19}$$

Then, as usual, the Hopf-Cole transformation linearizes the Burgers equation. This is a most important and useful result, since the representations (18) and (19) can closely mimic the $D(\theta)$ and $K(\theta)$ functions of real soils. Unfortunately the technique applies only for constant flux boundary conditions. Solutions for special cases of (19) have been given by Fokas and Yortsos (1982) and Rosen (1982).

Broadbridge et al. (1987) have shown that the technique works for finite as well as semi-infinite regions. The calculation for finite regions becomes quite elaborate, involving the Laplace transform boost device of King (1985).

THE FLUX-CONCENTRATION METHOD

Somewhat apart from the approaches considered to this point is the flux-concentration method, an especially powerful, physically based, integral method.

When similarity methods apply, they have the great advantage of reducing the number of independent variables by one. Integral methods achieve this same simplification for many of the problems which stubbornly refuse to fit the Procrustean bed of similarity. In essence, some guessed "shape" of the solution is put into an integral of the equation with respect to the variable to be eliminated. Among the possibilities are that the shape may be a once-and-for-all guess, or it may be subject to iterative improvement. The Green-Ampt model is the primitive integral method which follows when the advancing moisture profile is taken to be a step-function.

The shape used in the flux-concentration method is the dependence of flux density on θ, appropriately normalized. Unlike the moisture profile shape, the flux-concentration function for many problems is very conservative: it does not vary much with soil properties, nor with t during a particular unsteady flow process (Philip, 1973a). Tight bounds can be established for it, and it can be guessed to good accuracy. Simple iterative procedures have been established for various processes (Philip and Knight 1974), but the noniterated first approximations tend to be so good that the iterative procedures have not been much used to date. The method applies to processes involving a single space coordinate, such as one-dimensional absorption and infiltration, and axisymmetric absorption in two and three dimensions. Boundary conditions may be of either potential or flux type, and may involve (monotonic) time-dependence. Development of this method was stimulated by that of Parlange (1971), for which iterations diverge because they do not maintain integral continuity (Knight and Philip 1973), and in which the first approximation, earlier proposed by Macey (1959), fails to use the flux-concentration relation and so is suboptimal.

The full iterative flux-concentration method is powerful, accurate, and versatile. It involves machine computation, but, for a wide class of problems, it is several orders of magnitude more efficient than the customary brute-force methods. It warrants serious attention by even the most devoted soil-water computer jockeys.

MULTIDIMENSIONAL FLOWS

Series Solutions for Unsteady Flows

Quasianalytic series solutions like that for one-dimensional infiltration are available for two- and three-dimensional axisymmetrical absorption, with the three-dimensional solution supplemented by a steady exact solution asymptotically valid for large t (Philip 1966b). Analogous solutions have been found for the early stages of unsteady infiltration from cylindrical and spherical cavities (Philip 1969b).

Analytical Solutions for Unsteady Flows

Solutions are available for unsteady multidimensional infiltration in the linear soil (Philip 1966a, 1966b, 1986c; Warrick 1974; Lomen and Warrick 1974; Warrick and Lomen 1976; Ben-Asher et al. 1978); and these solutions can be matched at large t with the quasilinear steady solutions discussed below (Philip 1986c).

It is clearly highly desirable to establish more realistic analytic solutions, such as for the Burgers or Knight soil. Broadbridge (1986), however, has established the very important result that no transformation analogous to Hopf-Cole exists to linearize Burgers' equation in two dimensions; and it is unlikely that one exists in three.

Quasilinear Analysis of Steady Flows

We turn now to the quasilinear analysis, which is producing a whole budget of new results with relatively little effort. The steady form of (5) is

$$\nabla^2\Theta = \frac{1}{K}\frac{dK}{d\Psi}\frac{\partial\Theta}{\partial z} \, . \qquad (20)$$

We see (Gardner 1958; Philip 1966b, 1968) that, if

$$\frac{1}{K}\frac{dK}{d\Psi} = \text{constant} = \alpha > 0 \, , \qquad (21)$$

(20) reduces to the *linear*

$$\nabla^2\Theta = \alpha\partial\Theta/\partial z \, . \qquad (22)$$

The exponential representation

$$K(\Psi) = K(\Psi_1)e^{\alpha(\Psi - \Psi_1)} \qquad (23)$$

satisfies (21) and so gives (22). We call the flow analysis based on (22) *quasilinear* because, although (22) is linear, it embodies with reasonable accuracy the known strongly nonlinear decrease of K as Ψ decreases through negative values; and it accommodates also the strongly nonlinear dependence of Ψ and θ.

The quantity $2\alpha^{-1}$ is the *sorptive length* and α is the *sorptive number*. Note (Philip 1987a) that α is directly proportional to a characteristic capillary pore radius R. In fact

$$R = 7.4 \times 10^{-6}\alpha$$

when R is in m and α in m^{-1}. The value of α ranges from 0.2 m^{-1} or less in fine-textured soils to 5 m^{-1} or more in coarse ones.

The quasilinear analysis needs only two soil parameters, K_1 and α. In many practical problems $\Psi_1 = 0$ so K_1 is the saturated conductivity, and α tends to be virtually independent of Ψ_0, so long as it is large and negative, as in dry landscapes. Direct field determination of K_1 and α is under active study (Philip 1985b, 1986a; White et al. 1987).

Our further discussion is facilitated by normalizing space coordinates with respect to a, a characteristic length of the water supply surface, thus reducing (22) to

$$\nabla^2\Theta = 2s\partial\Theta/\partial z \qquad (24)$$

with

$$s = \tfrac{1}{2}\alpha a \, . \qquad (25)$$

The dimensionless parameter s is a measure of the relative importance of capillarity and gravity in producing the flow. As $s \to 0$, capillarity dominates the flow process. As $s \to \infty$, gravity dominates.

Numerous solutions of (24) have been established (Philip 1968, 1969a, 1987a; Wooding 1968), some using Green's functions (Philip 1971; Raats 1971) and others based on separation of variables (Raats 1970; Batu 1978; Philip 1983, 1984a,b). They have illuminated many aspects of multidimensional infiltration, but these conventional methods of solution left an important gap.

Separation of variables yields solutions to many important problems in series form. Often these series are readily summable for $0 \leqslant s \leqslant 2$ but the practical limit to summation tends to be about $s = 10$. The range $0 \leqslant s \leqslant 10$ covers many applications, but considerable interest attaches to the solutions as gravity dominates the flow process in the limit as $s \to \infty$. The series solutions thus leave unanswered the question, important in many engineering contexts, of the role of capillary effects in infiltration and seepage problems with large length scales. Conventional methods thus led to something of an *impasse*.

The Scattering Analogue

This obstacle to progress was removed by recognition of the exact analogue between steady quasilinear flows and the scattering of plane waves (Waechter and Philip 1985). Beyond its aesthetic appeal, the analogue leads to saving of scientific effort and to new hydrological insights. A large body of established results on wave scattering (Bowman et al. 1969), and the associated potent mathematical methods, become immediately available for solution and understanding of unsaturated flow problems (Philip 1985b). Perhaps of greatest significance are the asymptotic methods and results for large s.

The first step in establishing the analogue is to notice that the substitution

$$H = \frac{\Theta}{\Theta_1} e^{-sz} \tag{26}$$

transforms (24) to

$$\nabla^2 H = s^2 H . \tag{27}$$

Θ_1 is a characteristic value of Θ at the water supply surface. Equation (27) is exactly the reduced wave (Helmholz) equation for imaginary wave number is. The boundary conditions on (27) for steady infiltration from a cavity supplying water at constant potential Ψ_1 take the same form as those for scattering of a plane wave (wave number is) by an acoustically soft obstacle of the same shape. Hence the analogue.

We are led immediately to many new results in multidimensional unsaturated soil-water flow. An analogue of the optical theorem in scattering enables deduction of cavity discharge, and the far moisture field, from known scattering functions (Philip 1985b).

A notable consequence of the analogue is that we can identify various regions of the soil-water distribution about a cavity corresponding to the optical regions about an illuminated opaque obstacle. Figure 2 is a schematic diagram of flow from circular cylindrical and spherical cavities in which we use the terminology of the optical analogy. Full details are given by Waechter and Philip (1985). The illuminated region is the analogue of that which remains essentially dry; the umbra (or deep shadow) corresponds to the essentially saturated region of the soil; and the penumbra corresponds to the region where infiltration from the cavity significantly wets the soil. In detailed illustrative calculations I defined the umbra as the region with

$0.99 < \theta/\theta_0 \leqslant 1$ and the penumbra as that with $0.01 \leqslant \theta/\theta_0 \leqslant 0.99$. Although Fig. 2 deals specifically with circular cylinders and spheres, very similar results apply to cavities of other shapes.

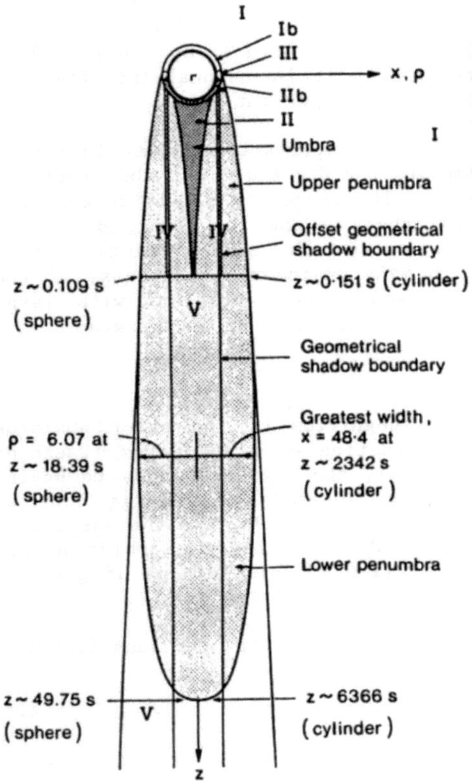

Fig. 2. Steady infiltration from cylindrical and spherical cavities. Schematic diagram illustrating the regions of the flow field, using the terminology of the optical analog. Waechter and Philip (1985) give more detail.

Capillary effects are often ignored in seepage problems with large s. In terms of the optical analogy ignoring capillarity is equivalent to using simple geometrical optics with plane waves casting perfect shadows to infinity. The "gravity only" infiltration solution would consist simply of a saturated column extending downward indefinitely, with no wetting outside the column. Such solutions are singular in the sense that they give a wholly misleading picture of the character and extent of the wetted region for large s. See Fig. 3.

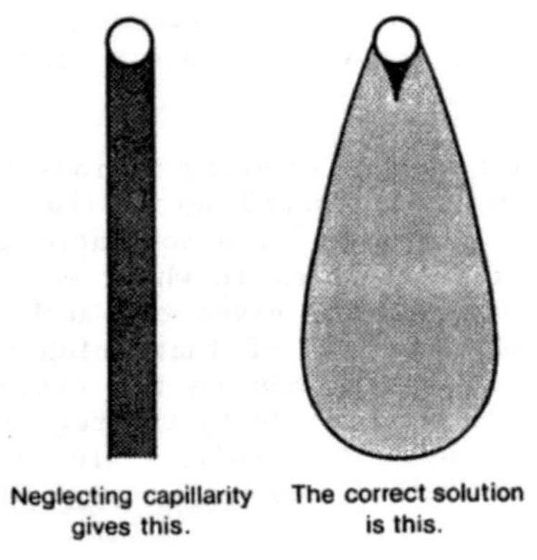

Fig. 3. Steady multidimensional infiltration from finite sources. Cartoon illustrating the singular character of the limit $s \to \infty$. (Waechter and Philip 1985).

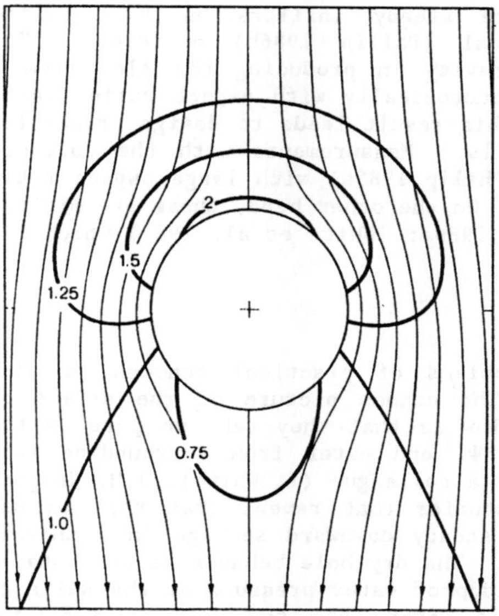

Fig. 4. Steady unsaturated seepage about cylindrical cavities. Map of Kirchhoff potential Θ and stream function for dimensionless radius s = 0.5. Bold curves are equipotentials, with the numerals values of Θ/Θ_0 with Θ_0 the potential at infinity. Fine curves with arrows are streamlines. (J.H. Knight, personal communication; Philip 1987a.)

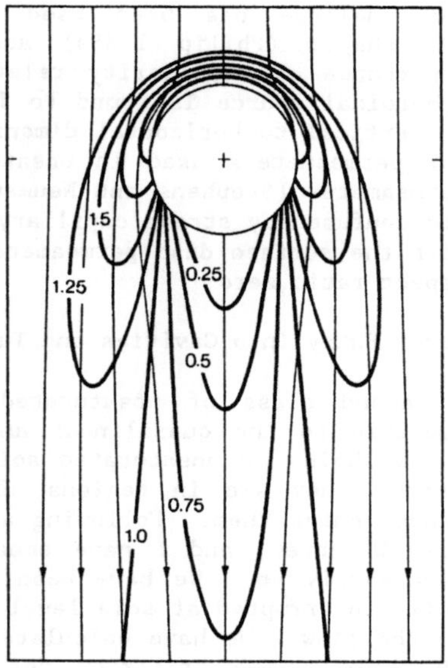

Fig. 5. As Fig. 4 for s = 4. Unnumbered equipotentials are for Θ/Θ_0 = 2,3,5. (J.H. Knight, personal communication; Philip, 1987a.)

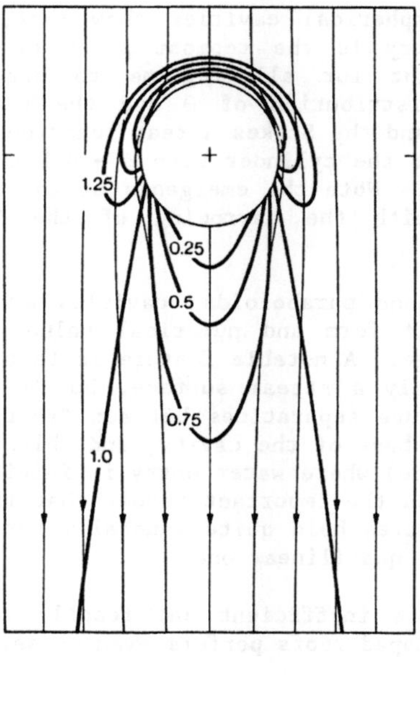

Fig. 6. Steady unsaturated seepage about spherical cavities. Map of Θ and Stokes stream function for s = 4. Bold curves are equipotentials and fine curves with arrows streamlines. Unnumbered equipotentials are for Θ/Θ_0 = 1.5, 2, 3. (J.H. Knight, personal communication; Philip 1987a.)

The analogue has been used to analyze steady infiltration from buried disc-shaped (Philip 1986a) and spheroidal (Philip 1986b) cavities. The importance of capillarity relative to gravity in producing the flow from a spheroidal source is found to increase monotonically with aspect ratio (ratio of vertical to horizontal dimensions). This result leads to design principles for permeameters used in unsaturated soils. Measurements with the borehole permeameter (Stephens and Neuman 1982; Philip 1985a) with large aspect ratio are confused by strong capillary effects. On the other hand, these are minimal with the surface disc permeameter (Philip 1986a; White et al. 1987) which has aspect ratio zero.

Water Entry into Cavities and Tunnels

A second class of unsaturated flow problems of practical concern is also amenable to the quasilinear analysis. The common picture of the effect of buried holes on unsaturated soil water flow is that they take no part in the flow: they are in regions of negative Ψ, and water from surrounding soil cannot enter them. Following a remark by a colleague (I. White), J.H. Knight, R.T. Waechter, and I have embarked on studies that reveal that this is not necessarily so. We have been analyzing steady downward seepage in a uniform soil, interrupted at some level by a hole. The dry hole behaves as an obstacle to the flow. We have calculated the buildup of water pressure at the walls of the hole. The criterion for seepage of water into the hole is that the pressure at some point on the walls reaches that of the local soil atmosphere. The larger the hole, the more prone it is to have water seeping through it.

The applications are numerous. First, the solutions illuminate the role of macropores in unsaturated flow. Second, they apply to caves, tunnels, and underground storage cavities. It is especially interesting that the results lead to design criteria for most efficient prevention of seepage into tunnels and cavities in unsaturated zones. The engineering contexts will be obvious. They include optimal design of underground repositories for nuclear wastes in deep unsaturated seepage zones in arid areas. Essentially, the same mathematical problem arises when we replace the hole by a buried impermeable obstacle, such as a stone, a rock, or an underground structure (Batu 1979).

Solutions for the circular-cylindrical and spherical cavities show that, in these cases, the crucial point of water entry is the topmost point of the cavity roof. I am grateful to J.H. Knight for allowing me to present Figs. 4-6. He has computed maps of the distribution of Θ and the stream function for the circular cylinder, and of Θ and the Stokes stream function for the sphere. Figures 4 and 5 are the maps for the cylinder with $s = 0.5$ and 4 and Fig. 6 is that for the sphere with $s = 4$. Note the emergence of the roof boundary layer and "roof-drip", together with the narrowing of the "dry shadow", as s increases.

The solutions for the parabolic-cylindrical and paraboloidal cavities are of especial interest. They are in simple exact form and numerical values are readily calculable for the full parameter range. A notable feature is that the cavity wall is, for these geometries, not only a stream surface, but also a surface of constant Θ. These shapes constitute separatices between "sharper" shapes where water entry, if any, is near the base of the cavity, and "blunter" ones (like the circular cylinder and the sphere) where water entry is initiated at the roof apex. I have recently established the important result that these properties of parabolic and paraboloidal cavities hold quite generally for the full nonlinear flow equation, not just for the quasilinear one.

Finally we note that flat-topped cavities are inefficient and readily admit water from unsaturated seepage; and that v-shaped roofs perform even worse.

REFERENCES

Batu DK (1979) Steady state moisture flows in unsaturated soil containing a plane, isolated, impermeable barrier. Adv Water Resour 1: 35-46

Batu V (1978) Steady infiltration from single and periodic strip sources. Soil Sci Soc Am J 42: 544-549

Baker GA (1975) Essentials of Padé approximants. Academic Press, New York

Ben Asher J, Lomen DO, Warrick AW (1978) Linear and nonlinear models of infiltration from a point source. Soil Sci Soc Am J 42: 3-6

Bowman JJ, Senior TBA, Uslenghi PLE (1969) Electromagnetic and acoustic scattering by simple shapes. North-Holland, Amsterdam

Braester C (1973) Moisture variation at the soil surface and advance of wetting front during infiltration at constant flux. Water Resour Res 9: 687-694

Broadbridge P (1986) Nonintegrability of nonlinear diffusion-convection equations in two spatial dimensions. J Phys A: Math Gen 19: 1245-1257

Broadbridge P, Knight JH, Rogers C (1987) Constant rate rainfall infiltration in a bounded profile: exact solutions of a nonlinear model. Soil Sci Soc Am J, submitted

Broadbridge P, White I (1987) Constant rate rainfall infiltration: a versatile nonlinear model I. Analytic solution. Water Resour Res 24: 145-154

Buckingham E (1907) Studies on the movement of soil moisture. US Dept Agr Soils Bull 38

Burgers JM (1948) A mathematical model illustrating the theory of turbulence. Adv Appl Mechs 1: 171-199

Clothier BE, Knight JH, White I (1981) Burgers' equation: application to field constant-flux infiltration. Soil Sci 132: 252-261

Cole JD (1951) On a quasi-linear parabolic equation occurring in aerodynamics. Quart Appl Mech 9: 225-236

Fokas AS, Yortas YC (1982) On the exactly solvable equation occurring in two-phase flow in porous media. SIAM J Appl Mech 42: 318-332

Gardner WR (1958) Some steady state solutions of the unsaturated moisture flow equation with application to evaporation from a water table. Soil Sci 85: 228-232

Green WH, Ampt GA (1911) Studies in soil physics, I. The flow of air and water through soils. J Agric Sci 4: 1-24

Hopf E (1950) The partial differential equation $u_t + uu_x = \mu u_{xx}$, Comm Pure Appl Mech 3: 201-230

King MJ (1985) Immiscible two-phase flow in a porous medium : utilization of a Laplace transform boost. J Math Phys 26: 870-877

Kirchhoff G (1894) Vorlesungen über die Theorie der Wärme. Barth, Leipzig

Klute A (1952) A numerical method for solving the flow equation for water in unsaturated materials. Soil Sci 73: 105-116

Knight JH, Philip JR (1973) On solving the unsaturated flow equation: 2. Critique of Parlange's method. Soil Sci 116: 407-416

Knight JH, Philip JR (1974) Exact solutions in nonlinear diffusion. J Eng Math 8: 219-227

Lomen DO, Warrick AW (1974) Time-dependent linearized infiltration: II. Line sources. Soil Sci Soc Am Proc 38: 568-572

Macey RI (1959) A quasi-steady approximation method for diffusion problems: I. Concentration dependent diffusion coefficients. Bull Math Biophys 21: 19-32

Nimmo WHR (1949) The world's water supply and Australia's portion of it. J Inst Eng Aust 21: 29-34

Parlange J-Y (1971) Theory of water movement in soils. 1. One-dimensional absorption. Soil Sci 111: 134-137

Philip JR (1954) Some recent advances in hydrological physics. J Inst Eng Aust 26: 255-259

Philip JR (1957a) The theory of infiltration: 1. The infiltration equation and its solution. Soil Sci 83: 345-357

Philip JR (1957b) The theory of infiltration: 2. The profile at infinity. Soil Sci 83: 435-448
Philip JR (1957c) Numerical solution of equations of the diffusion type with diffusivity concentration-dependent. II. Aust J Phys 10: 29-42
Philip JR (1957d) The theory of infiltration: 4. Sorptivity and algebraic infiltration equations. Soil Sci 84: 329-339
Philip JR (1966a) A linearization technique for the study of infiltration. In: Rijtema RE, Wassink H (eds) Water in the unsaturated zone, vol 1, UNESCO, Paris, pp 471-478
Philip JR (1966b) Absorption and infiltration in two- and three-dimensional systems. In: Rijtema RE, Wassink H (eds) Water in the unsaturated zone, vol 1, UNESCO, Paris, pp 503-525
Philip JR (1968) Steady infiltration from buried point sources and spherical cavities. Water Resour Res 4: 1039-1047
Philip JR (1969a) Theory of infiltration. Adv Hydrosci 5: 215-296
Philip JR (1969b) Early stages of infiltration in two- and three-dimensional systems. Aust J Soil Res 7: 213-221
Philip JR (1971) General theorem on steady infiltration from surface sources, with application to point and line sources. Soil Sci Soc Am Proc 35: 867-871
Philip JR (1973a) Flow in porous media. In: Becker E, Mikhailov GK (eds) Theoretical and applied mechanics, Springer, Berlin, pp 279-294
Philip JR (1973b) On solving the unsaturated flow equation: 1. The flux-concentration relation. Soil Sci 116: 328-335
Philip JR (1974) Recent progress in the solution of nonlinear diffusion equations. Soil Sci 117: 257-264
Philip JR (1983) Infiltration in one, two, and three dimensions. In: Advances in infiltration, Am Soc Agric Eng, St Joseph, Michigan, pp 1-13
Philip JR (1984a) Steady infiltration from circular cylindrical cavities. Soil Sci Soc Am J 48: 270-278
Philip JR (1984b) Steady infiltration from spherical cavities. Soil Sci Soc Am J 48: 724-72
Philip JR (1985a) Approximate analysis of the borehole permeameter in unsaturated soil. Water Resour Res 21: 1025-1034
Philip JR (1985b) Scattering functions and infiltration. Water Resour Res 21: 1889-1894
Philip JR (1986a) Steady infiltration from buried discs and other sources. Water Resour Res 22: 1058-1066
Philip JR (1986b) Steady infiltration from spheroidal cavities in isotropic and anisotropic soils. Water Resour Res 22: 1874-1880
Philip JR (1986c) Linearized unsteady multidimensional infiltration. Water Resour Res 22: 1717-1727
Philip JR (1987a) The quasilinear analysis, the scattering analog, and other aspects of infiltration and seepage. In: Fok Y-S (ed) Infiltration development and application, Water Resour Res Center, Honolulu, Hawaii, pp 1-27
Philip JR (1987b) The infiltration joining problem. Water Resour Res 23: 2239-2245
Philip JR, Knight JH (1974) On solving the unsaturated flow equation: 3. New quasianalytical technique. Soil Sci 117: 1-13
Raats PAC (1970) Steady infiltration from line sources and furrows. Soil Sci Soc Am Proc 34: 709-714
Raats PAC (1971) Steady infiltration from point sources, cavities, and basins. Soil Sci Soc Am Proc 35: 689-694
Richards LA (1931) Capillary conduction of liquids through porous mediums. Physics 1: 318-333
Rosen G (1982) Method of the exact solution of a nonlinear diffusion-convection equation. Phys Rev Letts 49: 1844-1846
Stephens DB, Neuman SP (1982) Vadose zone permeability tests: Summary. J Hydraul Div Am Soc Civ Eng 198: 623-639
Storm ML (1951) Heat conduction in simple metals. J Appl Phys 22: 940-951

Talsma T (1969) *In situ* measurement of sorptivity. Aust J Soil Res 7: 269-276
Truesdell C (1984) An idiot's fugitive essays on science. Springer, Berlin
Waechter RT, Philip JR (1985) Steady two- and three-dimensional flows in unsaturated soil: the scattering analog. Water Resour Res 21: 1875-1887
Warrick AW (1974) Time-dependent linearized infiltration: I. Point sources. Soil Sci Soc Am Proc 38: 383-386
Warrick AW, Lomen DO (1976) Time-dependent linearized infiltration: III. Strip and disc sources. Soil Sci Soc Am J 40: 639-643
White I, Perroux KM, Sully MJ (1987) The disc permeameter: a rapid nondestructive method for measuring soil hydraulic properties. Water Resour Res, in preparation
Wooding RA (1968) Steady infiltration from a shallow circular pond. Water Resour Res 4: 1259-1273

QUASIANALYTIC AND ANALYTIC APPROACHES TO UNSATURATED FLOW: COMMENTARY

P.A.C. Raats

Institute for Soil Fertility, PO Box 30003, 9750 RA Haren, The Netherlands.

Summary. Analysis of movement of water in unsaturated soils is a basis for water management and for understanding the environment in which soil mechanical, chemical, and biological processes occur. This commentary presents further examples of quasianalytic and analytic approaches to unsaturated flow. Three approaches to one-dimensional problems are discussed in some detail: (1) approximate methods subject to integral constraints, (2) methods based on transformations of nonlinear equations to linear equations, and (3) solutions for vertical flows dominated by gravity. Also, some further examples of quasilinear analysis of steady multidimensional flows are given, with emphasis on flows from shallow circular ponds.

ON THE ROLE OF WATER IN UNSATURATED SOILS

Spatial and temporal distribution of soil water is the central theme of soil physics, not only in an arid country like Australia, but also in a humid country like The Netherlands. Even in humid climates temporary shortages of water occur. The planning and execution of supplemental irrigation requires detailed analysis of movement of water in the unsaturated zone, especially concerning capillary rise, as an alternative source of water supply to crops. A certain degree of unsaturation is necessary for a soil to be sufficiently aerated and trafficable. Poor aeration may lead to impaired root growth and functioning and to rapid denitrification. The trends of mechanization in agriculture have stimulated a worldwide interest in the timing of soil tillage crop care and harvest operations.

Soil water plays a key role with regard to the dynamics of dissolved plant nutrients and contaminants. For transport over distances of 0.1 m or more, convective transport with the water dominates, resulting in significant average displacement and dispersion. Diffusive transport of solutes can be the prime mode of transport over distances of 0.01 m or less, as is the case for solutes being absorbed or exuded by plant roots. Availability of nutrients for plants is strongly influenced by the water status of an unsaturated soil, primarily as a consequence of the low values of the solute diffusivity at low water contents.

The list of publications of the former Agricultural Physics Section of the CSIRO Division of Plant Industry and the current CSIRO Division of Environmental Mechanics is an amazingly rich source of information on movement of water in unsaturated soil (Anonymous 1987). A large fraction of the existing quasianalytic and analytic approaches were originated, or stimulated, or critically evaluated by J.R. Philip and his numerous past and present co-workers. I acknowledge the recurrent inspiration derived from the work of this group.

RICHARDS' EQUATION

Richards (1931) put into rigorous mathematical-physical form the concepts of water flow in unsaturated soil due primarily to Buckingham (1907), but also partially formulated by others seemingly unaware of Buckingham's work. By now, the general flow equation is firmly established by reasoning and experimental evidence. Of course, Richards' theory had to be extended to cover complications arising from vapour transport, lack of continuity of soil air, limited mobility of soil air, deformation of the solid phase, structure of the solid phase, and deviations from Darcy's law associated with thermal and osmotic pressure gradients. These extensions in many directions have helped to put Richards' theory in a proper perspective. The modern theory of mixtures provides clear guidance for rational extensions of Richards' theory (Raats 1984b).

In the context of Richards' theory, specific soils are characterized by the dependencies of the pressure head Ψ and hydraulic conductivity K upon the volumetric water content θ. Most quasianalytic and analytic approaches are restricted to classes of specific soils. Examples are (i) the quasilinear analysis of steady, multidimensional flows in soils with an exponential relationship between the hydraulic conductivity and the pressure head, and (ii) unsteady flows in the broad Broadbridge-White class bridging a whole range from the Knight class to the Green-Ampt class.

All quasianalytic and analytic methods for solving Richards' equation involve simplifications. Problems can first of all be classified in geometric categories of vertical versus nonvertical flows and in temporal categories of steady versus nonsteady flows. Further useful simplifications may arise from considering the relative importance of capillary and gravitational components of the flux of the water. Thus far analytic solutions in the presence of a nonzero distributed sink λ have been obtained only for certain types of steady flows with specific spatial distributions of λ (Raats and Warrick 1981).

Many processes of flow of water in unsaturated soils so far have been, and perhaps will always be, amenable only to numerical methods. This is especially true for processes resulting from boundary conditions operating outdoors. Generally these processes are affected by hysteresis of the soil water retentivity curve. Nevertheless, anyone equipped with insights derived from quasianalytic and analytic approaches will be in a good position to describe and explain the main features in results of detailed calculations.

ONE-DIMENSIONAL PROBLEMS

More than one-third of Philip's (1988) review is devoted to one-dimensional problems. In the following, I firstly comment on what Philip calls the flux-concentration method, secondly, expand on Philip's remarks about solutions based on transformations of nonlinear equations to linear equations, and, thirdly, present some results for vertical flow dominated by gravity. The last class of flows has been an important source of inspiration for field studies over the past 15 years (e.g., references given in Raats 1983).

Approximate Methods Subject to Integral Contraints

Several time-honoured solutions of the Richards' equation are of the form

$$z = z(\theta,t) \ , \quad z = z(\Psi,t) \ , \tag{1}$$

where z is the depth of the soil, positive downward, and t is time. Examples

$$Z = \gamma z \quad , \quad T = \gamma^2 t \ . \tag{7}$$

Transformations of Nonlinear Equations to Linear Equations

The approximate methods subject to integral constraints apply to any soil. Philip (1988) also reviewed some results for particular classes of soils, namely Green-Ampt, linear, Knight, and Broadbridge-White soils. For these particular classes of soils, solutions covering the whole range from very small to very large times can be derived for certain boundary conditions. The Green-Ampt and linear soils do not allow proper large-time travelling waves, but this is not too disturbing since these classes are not realistic models of real soils anyway.

For the Knight class, solutions for infiltration with a wide range of boundary conditions are known. Philip showed that the special solution for infiltration with constant potential boundary condition is suitable for filling the gap between the series expansion in $t^{1/2}$ for small/intermediate time and the travelling, time-invariant wave for large time.

For the Broadbridge-White class the dependence of soil-water diffusivity D on relative water content θ is given by (Broadbridge and White 1988):

$$D = b(C)C(C - 1)\{S[\Delta\theta(C - \Theta)]\}^2 \ , \tag{8}$$

and the dependence of hydraulic conductivity K on θ is described by:

$$(K - K_n)\Delta K = \Theta^2(C - 1)/(C - \Theta) \ . \tag{9}$$

Here $\Theta = (\theta - \theta_n)/\Delta\theta$, with θ_n the initial soil-water content, $\Delta\theta = \theta_s - \theta_n$, with θ_s the value at field saturation, $S = S(\theta_s, \theta_n)$ is the sorptivity, $K_n = K(\theta_n)$, $\Delta K = K(\theta_s) - K_n$, C is a single free parameter of the model, and b(C) is a constant which depends on C. The parameter C is constrained to the range $1 < C < \infty$ with corresponding range $1/2 < b(C) < \pi/4$ for the constant b(C). As $C \to 1$, the hydraulic properties (8) and (9) approach those of a highly nonlinear Green-Ampt-like model. At the other end of the parameter range, as $C \to \infty$, (8) and (9) approach the properties of the weakly nonlinear Knight soil. For most soils C is confined to the range $1.01 < C < 2$.

Figure 1 shows results of calculations based on an analytical solution for constant rate rainfall infiltration into Broadbridge-White soils. The solution is parameterized in terms of the dimensionless rainfall rate $R_* = (R - K_n)/\Delta K$, dimensionless time $t_* = t/[bS/(\Delta K)^2]$, θ, and C. Figure 1 shows the evolution of soil saturation profiles for both nonponding, $R_* < 1$, and ponding, $R_* > 1$, rain and for two values of C. Figure 2 shows results for the time to incipient ponding.

Vertical Flows Dominated by Gravity

Solutions of the form (1) also arise for purely gravitational flows satisfying

$$\frac{\partial\theta}{\partial t} = - \frac{dK}{d\theta}\frac{\partial\theta}{\partial z} \tag{10}$$

(Raats 1983). Equation (10) is a kinematic wave equation with wave speed $dK/d\theta = \kappa(\theta)$. According to (10), surfaces of constant θ will propagate downward at speeds $dK/d\theta = \kappa(\theta)$. The ratio R of the wave speed $dK/d\theta$ to the speed K/θ of a parcel of water due to the gravitational force is given by

$$R = \frac{\theta}{K}\frac{dK}{d\theta} = \frac{d\ell nK}{d\ell n\theta} \ . \tag{11}$$

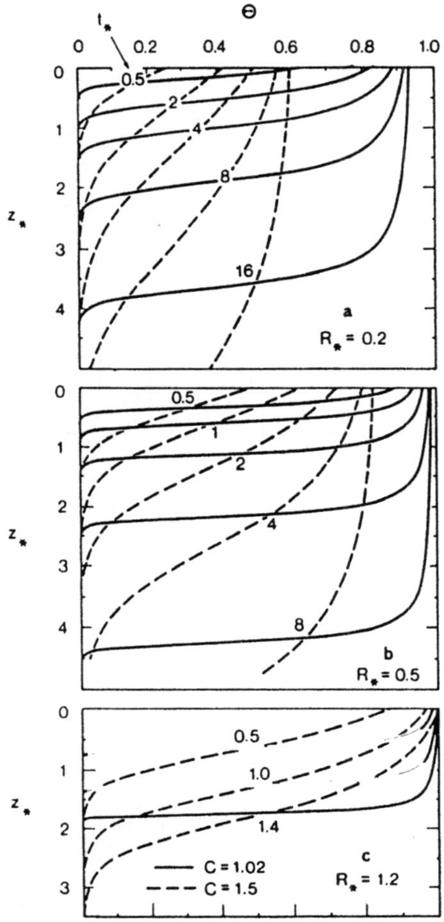

Fig. 1. Dimensionless saturation profiles at various dimensionless times t_* during constant-rate rainfall for C = 1.02 and 1.5 with (a) $R_* = 0.2$, (b) $R_* = 0.5$, and (c) $R_* = 1.2$. (Broadbridge and White 1988).

Thus, R is the ratio of the relative rates of change of K and of θ (see Raats 1984a).

Consider the drainage of a uniform profile with a deep water table given the initial condition

$$\theta(z,0) = \theta_n \,, \quad K(z,0) = K_n \,, \quad \text{for } z > 0 \,, \tag{12}$$

where θ_n and K_n are the uniform initial water content and hydraulic conductivity in the soil profile. Sisson et al. (1980) observed that if one supplements (12) with a fictitious condition

$$\theta(z,0) = \theta_m \,, \quad K(z,0) = K_m = 0 \,, \quad \text{for } z < 0 \,, \tag{13}$$

where θ_m is a fictitious, uniform water content in the region $z < 0$, chosen such that the hydraulic conductivity K_m corresponding to θ_m is zero, then conditions (12) and (13) describe an initial shock, and solving (10) subject to these conditions describes the decay of this shock. The solution of this type of problem is due to Lax (1972). Values of the water content in the range

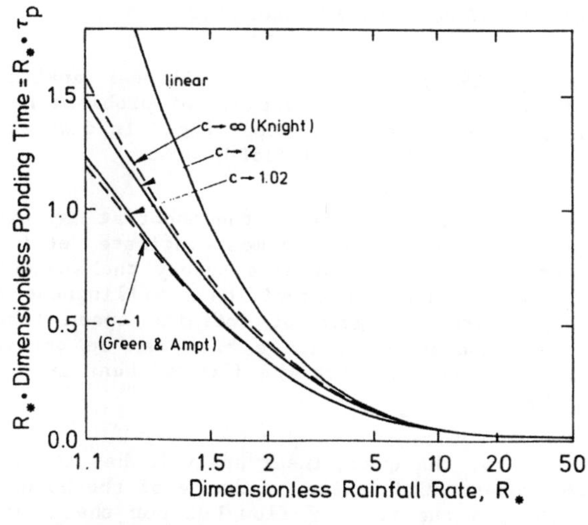

Fig. 2. Comparison of the dependence of the dimensionless ponding time upon the dimensionless rainfall rate for soils with the parameter C ranging from 1 (Green-Ampt soil) to ∞ (Knight soil), and for linear soils. (Broadbridge and White 1987).

$\theta_m < \theta < \theta_n$ will propagate downward from the soil surface at speeds $\kappa(\theta)$. This means that the depth z at which the water content will be θ at time t is given by

$$z(\theta,t) = \kappa(\theta)t \ . \tag{14}$$

For $z > \kappa(\theta)t$, the water content will be θ_n. Note that (14) is yet another solution of the form (1).

It can be shown (Raats 1983) that the average water content *above* depth z is related to the water content *at* depth z by

$$\theta = [1 - R^{-1}(\theta)]\theta \ . \tag{15}$$

It is remarkable that this relationship does not involve the depth z and time t explicitly.

Explicit expressions for $\theta(z,t)$ can be easily derived for various classes of soils, for example, for the class of mildly nonlinear soils with a linear retention curve and an exponential dependence of the hydraulic conductivity upon the water content. For this particular class of soils, it is possible also to evaluate the influence of capillarity upon the drainage process (Raats 1983).

Wooding (1975) analyzed flows induced by a variable rate of infiltration by the method of matched asymptotic expansions. In relatively uniform regions the distribution of the water content is represented asymptotically by an outer expansion, the first term of which takes into account the effect of gravity only and represents a kinematic wave. Higher order terms of the outer expansion include the small effect of capillarity. However, in regions where the gradient of water content is large, the effects of gravity and capillarity are of the same order of magnitude. Wooding shows that in such regions an appropriate inner expansion can be found.

QUASILINEAR ANALYSIS OF STEADY MULTIDIMENSIONAL FLOWS

The last third of Philip (1988) deals with quasilinear analysis of steady flows. Within this class of flows a wide variety of problems has been solved in the last two decades. This is not too surprising. In discussing Fourier's linear theory of heat conduction, Truesdell (1980) wrote

> By paring the physical model to the bone, the theorist may extract from it incredibly precise predictions about the most intimate detail of bodies of the most complicated shapes. In Fourier's theory the student must pay as the price for such detail and such precision a willingness to admit that differences of temperature propagate at infinite speed through material bodies and that the temperature within a body can be determined without taking into account the changes which the flow of heat into it may effect upon its size and shape.

Like Fourier's linear theory, the quasilinear analysis has its limitation: for a given geometry of the unsaturated region, a change of the boundary conditions by a constant factor changes the rate of flow but not the flow pattern. If $\alpha = K^{-1}dK/d\Psi$ is not constant, then the flow pattern is not subject to such invariance.

If a change of the boundary conditions causes a shift of the boundary between the saturated and unsaturated regions, then the flow pattern changes even if α in the unsaturated region is constant. Interest in flow in such adjoining saturated and unsaturated regions has been stimulated by the borehole permeameter (Philip 1985; Reynolds et al. 1985; Pullan 1988). For flow from cavities, the saturated-unsaturated system which results from allowing for the variation of hydrostatic pressure over the cavity surface is more physically realistic than the purely unsaturated system (Pullan 1988). The presence of the saturated region about a surface cavity removes the singularity in the flow velocity that occurs at the edges of the cavity when the head is ignored. Another effect of including the head is to increase the steady state flux from the cavity. This is illustrated in Fig. 3 for the case of a circular cylindrical cavity buried at an infinite depth, where the flux from the coupled system F_c divided by the flux from the purely unsaturated system F_u is plotted as a function of the reduced cavity radius s. For $s > 1$, it is difficult to evaluate F_c accurately due to the increasing size of the saturated bulb (for $s = 1$ the bulb already extends to a depth of 7.9 times the radius of the cavity). However, these results are still sufficient to indicate the large errors that can arise when the head of water is neglected.

Another problem of great practical and current interest is infiltration from a shallow circular pond. The surface of the pond is an equipotential surface, while the remainder of the soil surface is a stream surface. Wooding (1968) derived a formal solution of the linearized flow equation subject to these mixed boundary conditions using a Hankel transform. This solution involves an unknown function f(p) to be determined from the system of dual integral equations arising from the mixed boundary conditions. Wooding uses Tranter's method involving a series expansion of the unknown function f(p). A natural choice for this expansion follows from the fact that at large distance from the disc source the flow should be indistinguishable from the flow due to a point source. The resulting system of linear algebraic equations has been approximately inverted by an analytical/numerical method for $s < 6$ (Wooding 1968) and by a wholly analytic method for $s < 0.15$ (Weir 1987).

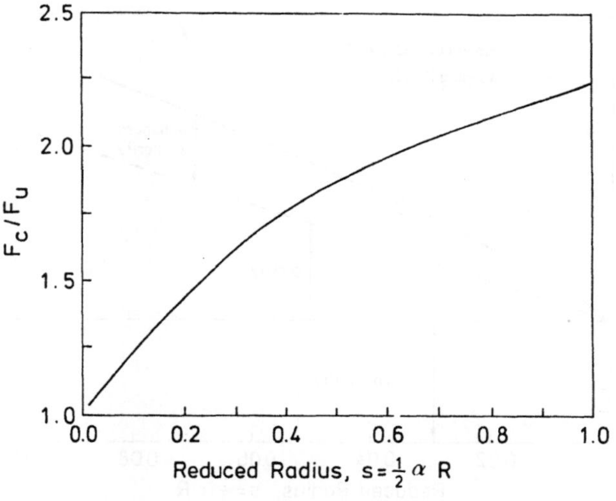

Fig. 3. Increase in flux from a coupled saturated-unsaturated system F_c over that of a purely unsaturated system F_u for the case of a circular cylindrical cavity buried at an infinite depth. (Pullan 1988).

Figure 4 shows the total flux from a disc source as a function of the dimensionless size of the disc. It is interesting that the total flux is larger than the sum of a purely capillary flux and a purely gravitational flux, although such a simple summation is a quite satisfactory approximation for large s. These results suggest that the results of Scotter et al. (1982) based on this approximation should be re-evaluated.

In an invited lecture at the 19th New Zealand Mathematics Colloquium, Philip (1984) described Wooding's analysis of 1968 as "an heroic piece of mathematics". Perhaps this assessment induced several New Zealand mathematicians to extend Wooding's analysis (Weir 1986, 1987; Pullan 1987, 1988; Pullan and Collins 1987).

Pullan (1987, 1988) used the boundary element method to calculate the transient quasilinearized infiltration with constant diffusivity from a shallow circular pond. Figure 5 shows the approach of the flux F to its steady value F_∞ for several values of s. The dimensionless time in the horizontal axis involves only the properties α and D of the soil. Therefore, the family of curves compares the approach to steady state for ponds of different sizes on the same soil. A comparison of ponds of the same size on soils with different α but the same D is obtained by shifting the curves in Fig. 5 a distance $\log\alpha^2 D^2$ along the time axis. If the soils being compared are similar in the sense of Miller and Miller (1956), then the diffusivity D should change by the same factor as α (Raats 1983). The required shift is then $\log\alpha^4$.

Finally, we note that the quasilinear analysis of steady flows can also be carried through for flow to sinks (Raats 1977; Warrick and Amoozegar-Fard 1977). Such flows are of interest in connection with flow to plant roots, and the operation of porous cup water samplers and suction lysimeters.

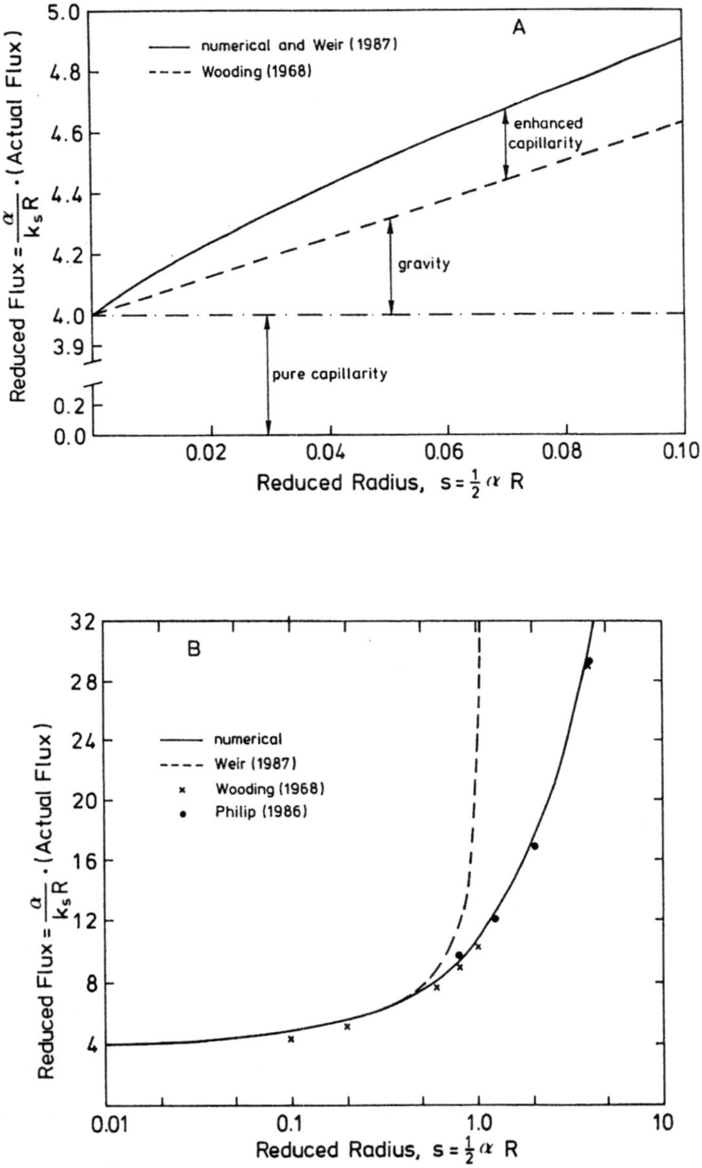

Fig. 4. Reduced flux from a shallow circular pond as a function of the reduced radius of the pond with (a) $0 < s < 0.1$, and (b) $0.01 < s < 10$.

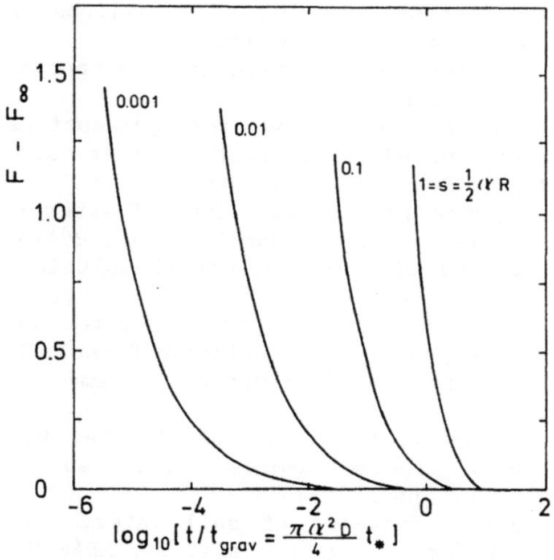

Fig. 5. Approach of the flux F from an axisymmetric flat pond to its steady value F_∞ expressed as a function of t/t_{grav}, where t/t_{grav} is the dimensionless time and $t_{grav} = (\pi s^2)^{-2}$. (Pullan 1987, 1988).

REFERENCES

Ahuja LR, Swartzendruber D (1973) Horizontal water intake through a thin zone of reduced permeability. J Hydrol 19: 71-89

Anonymous (1987) Publications List. Division of Environmental Mechanics, CSIRO, Canberra, Australia, 27 pp

Broadbridge P, White I (1987) Time-to-ponding: comparison of analytic, quasi-analytic and approximate solutions. Water Resour Res 23: 2302-2310

Broadbridge P, White I (1988) Constant rate rainfall infiltration: a versatile nonlinear model I. Analytic solution. Water Resour Res 24: 145-154

Buckingham E (1907) Studies on the movement of soil moisture. US Dept Agric Soils Bull 38

Hornung U, Parlange J-Y, Hogarth WL, Connell LD, Peters P (1987) Water movement in a finite layer: absorption for constant water content at the surfaces. Soil Sci Soc Am J 51: 557-562

Lax PD (1972) The formation and decay of shock waves. Am Math Monthly 79: 227-241

Miller EE, Miller RD (1956) Physical theory for capillary flow phenomena. J Appl Phys 27: 324-332

Parlange J-Y (1980) Water transport in soils. Ann Rev Fluid Mech 12: 77-102

Perroux KM, Smiles DE, White I (1981) Water movement in uniform soils during constant-flux infiltration. Soil Sci Soc Am J 45: 237-240

Philip JR (1969) The theory of infiltration. Adv Hydrosci 5: 215-296

Philip JR (1984) Mathematics, soil, and water. New Zealand Mathematical Society Newsletter, No 31, pp 28-35

Philip JR (1985) Approximate analysis of the borehole permeameter in unsaturated soil. Water Resour Res 21: 1025-1034

Philip JR (1986) Steady infiltration from buried discs and other sources. Water Resour Res 22: 1058-1066

Philip JR (1988) Quasianalytic and analytic approaches to unsaturated flow. This volume, pp

Pullan AJ (1987) Boundary element solutions of quasilinearized time dependent infiltration. Appl Math Modelling, in press

Pullan AJ (1988) Quasilinearised infiltration and the boundary element method. PhD thesis, University of Auckland, NZ

Pullan AJ, Collins IF (1987) Two- and three-dimensional steady quasi-linear infiltration from buried and surface cavities using boundary element techniques. Water Resour Res 23: 1633-1644

Raats PAC (1977) Laterally confined, steady flows of water from sources and to sinks in unsaturated soils. Soil Sci Soc Am J 41: 294-303

Raats PAC (1983) Implications of some analytical solutions for drainage of soil water. Agric Water Manage 6: 161-175

Raats PAC (1984a) Accumulation and transport of water and solutes in the saturated and unsaturated zones. In: Eriksson E (ed) Proc. Uppsala Symp. on Hydrochemical Balances of Freshwater Systems. IAHS-AISH Publ. no. 150, pp 343-357

Raats PAC (1984b) Applications of the theory of mixtures in soil physics. In: C Truesdell, Rational thermodynamics, 2nd ed., Appendix 5D, Springer-Verlag, New York, pp 327-343

Raats PAC, Warrick AW (1981) Dynamics of soil water. In: ME Jensen (ed) Design and operation of farm irrigation systems, ASAE Monograph No 3, ASAE St. Joseph, Michigan, USA, Chapter 4, Part 2, pp 125-142

Reynolds WD, Elrick DE, Clothier BE (1985) The constant head well permeameter: effect of unsaturated flow. Soil Sci 139: 172-180

Richards LA (1931) Capillary conduction of liquids through porous mediums. Physics 1: 318-333

Scotter DE, Clothier BE, Harper ER (1982) Measuring saturated hydraulic conductivity and sorptivity using twin rings. Aust J Soil Res 20: 295-304

Sisson JB, Ferguson AH, van Genuchten MTh (1980) Simple method for predicting drainage of field plots. Soil Sci Soc Am J 44: 1147-1152

Smiles DE, Knight JH, Perroux KM (1982) Absorption of water by soil: the effect of a surface crust. Soil Sci Soc Am J 46: 476-481

Truesdell C (1980) The tragicomical history of thermodynamics 1822-1854. Springer-Verlag, New York

Warrick AW, Amoozegar-Fard A (1977) Soil water regimes near porous cup samplers. Water Resour Res 13: 203-207

Weir GJ (1986) Steady infiltration from large shallow ponds. Water Resour Res 22: 1462-1468

Weir GJ (1987) Steady infiltration from small shallow circular ponds. Water Resour Res 23: 733-736

White I, Smiles DE, Perroux KM (1979) Absorption of water by soil: the constant flux boundary condition. Soil Sci Soc Am J 43: 659-664

Wooding RA (1968) Steady infiltration from a shallow circular pond. Water Resour Res 4: 1259-1273

Wooding RA (1975) Unsaturated seepage flow from a horizontal boundary. Q Appl Math 33: 143-159

Youngs EG (1968) An estimation of sorptivity for infiltration studies from moisture moment considerations. Soil Sci 106: 157-163

Zaslavsky D, Ravina I (1965) Measurement and evaluation of hydraulic conductivity through the moisture moment method. Soil Sci 100: 104-107

MEASUREMENT OF SOIL PHYSICAL PROPERTIES IN THE FIELD

I. White

CSIRO Division of Environmental Mechanics, GPO Box 821, Canberra, ACT, 2601, Australia

Summary. Increasing efforts have been spent in the past two decades on field measurement of soil physical properties and on the development of transport models. Here a summary is given of the characteristics of field soils gleaned from such measurements. The principle feature of field soils is their variability over many length and time scales. Issues involved in measurements in a heterogeneous environment are considered. These include the interaction of measurement scale with perceived variability, soil-imposed weighting of measurements, the Heisenberg uncertainty principle, the use of correlations, inverse measurement techniques, and soil macrostructure and biological activity. Some criteria for field measurements are identified. Predictive frameworks for interpreting and using field data are discussed. These range from the fully deterministic to stochastic theories. The successes and limitations of these are considered. Finally, some remaining problem areas are identified.

INTRODUCTION

The last decade has seen a rapidly increasing demand for quantitative information on the physical properties of porous materials, and particularly those which govern mass and energy transport in the environment. The pressing needs to manage effectively water, soil, plant and mineral resources, waste disposal, sustainable agricultural systems, construction industries and a variety of other activities all contribute to this demand.

The problems faced in making the required measurements in field soils are similar to those arising in quantitative descriptions of the atmosphere or of water bodies. All involve temporal and spatial variability, a plethora of time and length scales and nonsteady, multidimensional velocity fields. There is also a suspicion that the theoretical frameworks underpinning the measurements may be less than adequate. Soils present an additional challenge in that they are composed of at least two, and in most cases, three phases, one of which is the solid phase. This makes *in situ* measurement difficult. As well, transport equations in even simple situations are highly nonlinear.

Perhaps because of the difficulty in making field measurements, there has been a tendency in the past to determine the physical properties of isolated cores or remixed soil samples in the laboratory. While these measurements tell us a great deal about the properties of isolated or remixed samples, there is no guarantee that they provide any information on the properties of soils as they occur in the landscape (Gardner 1974). Twenty-one years ago Nielsen and Biggar (1967) concluded that "... physically characterizing the soil *in situ* offers a challenge that few have ever accepted." Their belief was that the necessary theory was already in place; what was lacking was the courage to apply it. With the 20/20 vision of hindsight it is clear that theory extant then was not adequate to describe *in situ* soil behaviour.

Many in the intervening years have accepted the field measurement challenge enthusiastically. In some cases the questions being addressed seem obscured by the act of measurement or the sheer weight of data or the elegance of its manipulation. John Philip is said to have once peered down a very large piezometer being hand-dug into the heavy clay soil of the Riverine Plain of N.S.W. (Australia) by an energetic hydrologist in order to determine its hydraulic conductivity. He apparently asked the luckless individual ... "If God came down and told you what the hydraulic conductivity of this soil was, what would you do with it?" What properties we need to know and how we measure them are related directly to the questions we are attempting to answer.

The general soil physical properties which are, in my view, important are those which govern the transport and storage of energy (heat, electromagnetic radiation, subatomic particles), momentum (soil rheology, shear, tension and compression) and mass (liquids, gases, solutes, particulates). Space and my competence exclude a general discussion of all these aspects. Instead, I shall slant my remarks predominantly towards soil-water properties. This is not to denigrate the importance of other properties. Rather, it serves to confirm Truesdell's (1984) belief that "Science today is much like government or big business: scientists are specialists not only in a single science but even in a single problem". It must be added, however, that in many cases, soil-water properties govern gas, solute, and heat transport in the soil. Aspects relevant to solute movement have been covered by Jury (1988) and Knight (1988) in this Symposium. De Vries (1987) has summarised elsewhere progress in heat and in coupled heat and mass tranport in porous materials.

In discussing soil-water flow properties it is helpful to introduce the properties listed in Table 1.

Table 1. Some soil-water properties.

	Fundamental Properties	Dimensions
θ:	volumetric water content	$[L^3L^{-3}]$
$V(\theta,x,y,z,t)$:	soil-water flux density	$[LT^{-1}]$
$\Psi(\theta)$:	soil-water potential, potential/unit weight water	$[L]$
$K(\Psi)$, $K(\theta)$:	hydraulic conductivity	$[LT^{-1}]$
	Derived Properties	**Dimensions**
$D(\theta) = Kd\theta/d\Psi$:	soil-water diffusivity	$[L^2T^{-1}]$
$S(\Psi_0,\Psi_n)$:	sorptivity	$[LT^{-1/2}]$
$\lambda_c(\Psi_0,\Psi_n) = [K(\Psi_0) - K(\Psi_n)]^{-1}\int_{\Psi_n}^{\Psi_0} K(\Psi)d\Psi$:	macroscopic capillary length	$[L]$

Here x and y are horizontal space coordinates, z the vertical coordinate, positive downwards, and t time.

The dynamics of soil-water are most commonly treated through the Darcy-Buckingham equation which connects the soil-water flux with gradients in the driving forces, capillarity and gravity. For nonswelling media,

$$V = -K\nabla\Psi + z \quad , \tag{1}$$

with z the vertical space coordinate positive downwards. In the absence of sources or sinks in the soil, (1) combined with the statement of continuity leads to Richards' equation

$$\frac{\partial \theta}{\partial t} = \nabla.(K\nabla\Psi) - \frac{\partial K}{\partial z} \quad . \qquad (2)$$

The assumptions embodied in (1) and (2) and their limitations have been discussed by Philip (1969, 1970) and more recently by Sposito (1986). Because K is strongly dependent on Ψ, (2) is highly non linear in general. Equations (1) and (2) not only form the basis of deterministic treatments of soil-water flow (Philip 1969), but are employed also in stochastic approaches (e.g. Mantoglou and Gelhar 1987a).

I will not present here a conspectus of recent technical advances. Rather, I examine and illustrate some of the issues involved. My belief is that progress in field measurement is intimately coupled with advances in theory, and vice versa. We commence by summarising some of the relevant features of field soils.

CHARACTERISTICS OF FIELD SOILS

Heterogeneity

Soils, to even the most casual observer, are layered, intersected with cracks and holes, replete with gravel and stones, bounded by thin impermeable crusts, subject to shrinking and swelling and the vicissitudes of man, biota and climate. Their genesis and evolution guarantee an inherent spatial and temporal heterogeneity. Philip (1986b) identified three types of heterogeneity: stochastic (Gelhar 1974; Freeze 1975), in which soil properties vary in an imperfectly known fashion in space and time; deterministic, where the variations are everywhere known; and flawed homogeneity, in which a material with known properties is flawed by the presence of cracks or holes. These distinctions are, of course, artificial and contain implicit assumptions about the scale of observation.

Length and Time Scales

The identification of length and time scales is of fundamental importance to measurements in field soils. Dagan (1986) identified three fundamental length scales associated with porous materials, to which we may add one more, the pore scale. These are listed in Table 2.

Table 2. Fundamental length scales of field soils

Scale	Dimension of Averaging Volume (m)
Pore	10^{-6} - 10^{-3}
Laboratory	10^{-2} - 10^{-1}
Local	10^{0} - 10^{1}
Regional	10^{2} - 10^{4}

For aquifers, Gelhar (1986) adds another, larger, length scale. Time scales associated with these lengths via characteristic transport velocities range from approximately 10^{-3} to 10^{10} s. In addition to these time scales, the genesis, removal, biological activity, and management of soils, and even the transport processes themselves, can impose their own time scales.

The separate hierachy of length scales in Table 2 is in essence more an article of faith then fact. There is a possibility that there may exist a succession of nonseparable length scales (Burrough 1983; Jury 1985). If so, soils may be fractal media (Mandelbrot 1983) where the only discernible length scales are those of the lower cut off limit, the measurement volume scale, and the upper cut off, the largest scale of interest (Philip 1986b).

Variability

The measured range of variability of soil properties is well-documented (Nielsen et al. 1973; Warrick and Nielsen 1980). A summary of frequently found coefficients of variation of soil physical properties relevant to water flow is given in Table 3.

Table 3. Approximate ranges of coefficients of variation and possible probability distribution functions of selected soil physical properties

Soil Property	Coefficient of Variation (%)	Approximate Probability Distribution Function
Bulk density	5 - 10	normal
Water Content (at field saturation)	5 - 20	normal
Hydraulic conductivity (at field saturation)	10 - 400	log-normal
Sorptivity	10 - 70	normal or log-normal
Macroscopic Capillary Length	30 - 100	log-normal

Lists such as these serve to illustrate the difficulties involved in using field measured data in purely deterministic models. In addition, they suggest that a large number of measurements will be required. In order to find appropriate mean values of soil physical properties, their probability distribution functions need to be known. The approximate probability distribution functions of some soil properties are also given in Table 3. Statistical tests of the appropriateness of distribution functions are carried out only rarely since they require numerous measurements (Rao et al. 1979). Instead, distribution functions are chosen more or less as a matter of convenience. The stochastic treatment of water flow, for example, has $K(\Psi)$ log-normally distributed (Gelhar 1986); so there is little incentive to search for other distribution functions. Typical frequency histograms found for field measured sorptivity (a property which is a measure of the capillarity of soils and governs the early stages of flow (Philip 1957)) is given in Fig. 1 from Clothier and White (1981). We see here a typical feature of field hydraulic properties measured at or close to "field" saturation; the occurrence of "outliers" with large values. Note, however, that changes in the boundary condition, or sample size, can dramatically alter the frequency distribution and coefficient of variation (see, e.g., Sisson and Wierenga 1981).

Spatial Correlation

Measured soil properties are not always random in space, but frequently exhibit discernible spatial correlation. Figure 2 shows a 200 m transect of volumetric water content of the top 0.15 m of soil across "patterned" ground (Zegelin and White, unpublished data, 1986).

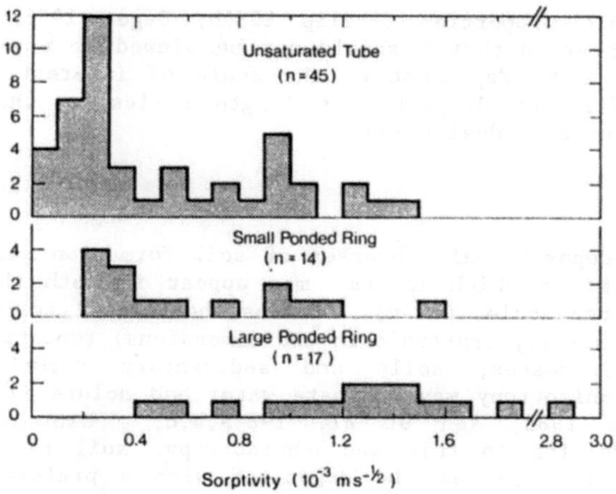

Fig. 1. Frequency histograms for field measured sorptivity. (Clothier and White 1981). Large rings 300 mm diameter, small rings 90 mm. Unsaturated $\Psi_0 = -40$ mm; ponded $\Psi = +15$ mm.

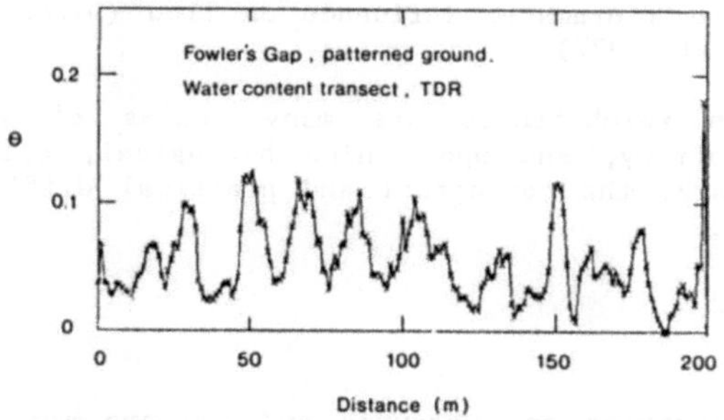

Fig. 2. Transect of volumetric soil-water content across "patterned" ground.

Troughs of low water content correspond to bare stony areas, peaks to vegetated swelling areas. The periodic spatial correlation is apparent. For such data we can define a spatial autocorrelation length L,

$$L = \sigma^{-2} \int_0^\infty C(r)dr \ , \tag{3}$$

where σ is the variance of the measured values, C(r) the covariance and r the spatial separation. The actual upper and lower limits of r in practice are the maximum range of interest and the characteristic length of the measurement volume. In essence, L is considered a fundamental heterogeneity length scale of porous material properties. Values between 10^{-1} and an astounding 10^4 m have been reported (Hoeksema and Kitanidis 1985). The θ data in Fig. 2 have a "local" L of 2 m together with a wavelength of about 20 m. Field measurements are taken to be point values when the characteristic sample dimension is much less than L.

Different soil properties measured on the same soil site usually have different correlation lengths. The value of L then cannot be considered characteristic of a site or soil type, only of the property. In addition, the work of Burrough (1983) and Jury (1985) has shown a distinct relationship between estimated values of L for soil properties and sample spacing. Hoeksema and Kitanidis (1985) have exposed similar problems in aquifers. This points to a multiplicity of length scales and raises the question of the statistical stationarity of field properties (Philip 1986b; Dagan 1986; Gelhar 1986). Gelhar (1986) has observed that L should not be viewed in any absolute sense but rather be seen to be dependent on the scale of interest. This assumes pragmatically a sufficient disparity of length scales so that correlations appear superimposed on a gradual trend.

Anisotropy

The weathering and depositional processes of soil formation generally lead to layered soil profiles in which layers may appear in both the small scale, $O(1\text{ mm})$, and the large scale, $O(1\text{ m})$. Because horizontal correlation lengths (themselves not necessarily isotropic in two dimensions) tend to be larger than vertical layer thicknesses, soils and sedimentary rocks are generally anisotropic. This anisotropy may dominate water and solute flow in the field (Gelhar and Axness 1983; Yeh et al. 1985a,b,c; Mantoglou and Gelhar 1987a,b,c). In addition to this macro-anisotropy, soil particles within a relatively homogeneous layer may be deposited with a preferred orientation, giving rise to anisotropy in flow. It has also been pointed out that anisotropy may depend on Ψ for unsaturated flows (Yeh et al. 1985b).

Additional processes leading to anisotropy in homogeneous soil masses include shrinkage cracks or the burrows of soil fauna or old root channels. These tend to be vertical and have a dramatic influence on flow (Dixon 1972; Beven and Germann 1982; Bouma et al. 1977).

In an environment in which there are many scales of variation, large variability, and anisotropy, and upon which biological, climatic and human factors leave their mark, the conceptual and practical difficulties of field measurement are evident.

FIELD MEASUREMENT

What soil-physical properties one measures and how one measures them depend both on the questions at hand and the conceptual framework used to tackle the questions. Additionally, it seems good practice that the measurements should, in some sense, mimic the problem at hand. This implies that there may be no unique method to determine a soil physical property. The experimenter may need to select appropriate techniques and conceptual frameworks for particular applications. For water flow this usually involves assuming that (1) or (2) or solutions of (2) hold over the measurement volume. It is here that the analytic solutions described by Philip (1988) and Raats (1988) in these proceedings have great utility. Not only do they provide a framework for data analysis but they may be used to suggest new measurement techniques. The question of the required measurement volume is obviously important.

Measurement Volume and Time Scale

The central issue here is whether measurements made at one scale may be used to infer behaviour at larger scales. The related problem of what sample dimension is necessary in order to make representative measurements has long been recognised. When a sample volume is such that any increase in sample size does not alter the measured mean value, the sample volume is said to be the representative elementary volume or rev (see, e.g., Bear 1972). Examples of the effect of scale of measurement on the magnitude and variability of field soil properties are given, for example, in Ritchie et al. (1972); Sisson and Wierenga (1981); Hawley et al. (1982) and Hassan et al. (1983).

Dagan (1986) has presented a statistical view of sample size. In this, when the mean measured quantity and the ensemble average are interchangeable, the sample volume is representative. This approach assumes the ergodic hypothesis and is essentially untestable. Cushman (1987) used a set-theoretic approach to define a rev ensemble which is claimed to be independent of the ergodic assumption. While of mathematical interest, the concept does not seem to bear directly on field scale transport (Sposito et al. 1987).

Cushman (1984) attempted to incorporate the effect of the instrument scale on measurements of a stochastic process. Sample size was treated as a filter on the possible spatial scales and conventional signal analysis theory was used to examine the problem. The approach appears to offer few solutions to existing practical problems.

The imposed boundary conditons may influence both the required sample volume and the perceived variability. Figure 3 shows the dependence of the coefficient of variation, c.v., of sorptivity on water-supply potential, Ψ_0, for a single set of 0.2 m diameter field samples (Perroux and White 1988). This dependence is compared with that for a repacked laboratory material. The rapid drop and stabilization of c.v. is due to the removal of macropores from the flow process by the decrease of Ψ_0. Note that at the more negative Ψ_0 the c.v. of the field data approaches that of the laboratory data. The stabilization of c.v. below $\Psi_0 = -35$ mm may indicate that the 0.2 m size samples are close to a representative elementary volume when pores of diameter greater than 1.5 mm are excluded from flow.

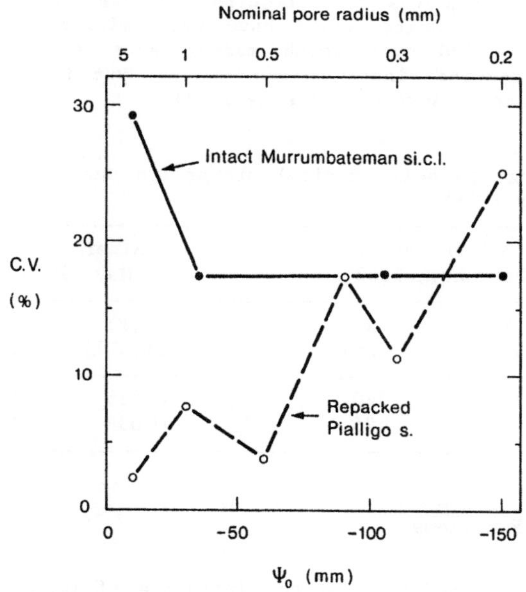

Fig. 3. Effect of supply potential, Ψ_0, on coefficient of variation of sorptivity. Field results compared with repacked samples. (Perroux and White 1988)

At potentials close to zero the random intersection of samples and large macropores (d > 1.5 mm) appears to cause increases in the c.v.. In some observations the measured variance is found to increase with sample size (Burrough 1983). For such cases the concept of representativeness appears meaningless.

Two disparate approaches to sample size are identifiable. In one, measured quantities are taken as point values. Here the assumption is that the measurement scale is much larger than any heterogeneities within the sample but much smaller than those of the scale of interest. Such an approach is required by the stochastic treatment of flow (Freeze 1975; Dagan 1986; Gelhar 1986). *A priori* identification of the necessary measurement scale therefore poses a problem. In the other, one essentially attempts to select a sample size which is a rev. In pragmatic approaches, the sample size may approach the entire scale of interest (Jackson 1988; Jury 1988). Thus infiltration from an entire furrow or bay may, for example, be used to estimate the water intake of a field. Such an approach is of value in determining the integral behaviour of a field for a particular boundary condition but the results may not be readily transferrable to other boundary conditions.

A problem intimately connected with sample dimension is sample number. Where properties are uncorrelated, the relation between sample number and observed variance is straightforward (e.g., Warrick and Nielsen 1980). McBratney and Webster (1983); Burgess et al. (1981) and Vieira et al. (1981) have used geostatistics to address the problem of sample number when the data are spatially correlated.

The question of the appropriate time scale for measurements is less well studied. If measurement times are of the same order as the time scale of change of physical properties (a growing season?), the data may be suspect. To illustrate the difficulties that can occur, we show in Table 4 measurements of hydraulic conductivity of a surface soil made before and after drought-breaking rains. Had the time required to make these measurements spanned that period, the results would have been misleading. A related question is the time period over which measurements should be averaged. Deterministic theories require "instantaneous" rates, yet in some cases daily averaged properties appear to suffice (Ritchie 1972).

Table 4. Temporal change in soil physical properties due to drought-breaking rains (White and Perroux 1988).

	Before rains February, 1983		After rains May, 1983	
$-\Psi$ (mm)	$K(\Psi)$ ($\mu m\ s^{-1}$)	λ_m (μm)	$K(\Psi)$ ($\mu m\ s^{-1}$)	λ_m (μm)
22.5	1.25	240	0.235	99
92.5	0.067	26	0.039	23

Soil-imposed Weighting Functions

We must recognise that in addition to the influence of measurement scale or instrument "size" on the perceived values and variabilities of field measured properties, the properties of the soil themselves impose a weighting function. We know, for example, that the extent of the sphere of influence of the neutron soil-water probe is a function of the water content being measured (see, e.g., Greacen 1981). This is also true when time domain reflectometry (Topp et al. 1980) is used to estimate θ. The wetted zone around any water source used in two- and three-dimensional flow measurement is similarily a function of the soil macroscopic capillary length and hydraulic conducitivty which are being determined by the flow measurement. These soil-imposed weighting functions are largely overlooked.

The Heisenberg Uncertainty Principle

The nature of field soil and the use of techniques which have principally evolved from the laboratory mean that almost all field measurements involve a degree of soil disturbance. The very act of measurement may influence both the observed magnitude and variability of the property under study. This is a macroscopic, hydrologic analogy of the Heisenberg Uncertainty Principle in quantum mechanics. The sensitivity of soil-water flow measurement to disturbance is readily understood. Hydraulic conductivity, for example, is proportional to the square of a characteristic pore size. Any pathways caused by the insertion or removal of a core sampler, for example, could totally swamp the effect of the naturally occurring pore sequences.

To illustrate this, Table 5 shows the coefficient of variation of hydraulic conductivity of Bungendore fine sand measured with a three-dimensional technique which involved no soil disturbance during flow measurement (White, Sully and Perroux, unpublished measurements, 1985). These are compared with the c.v. of K for one-dimensional excised cores of identical diameter taken from precisely the same locations. Values are shown for two water supply potentials.

Table 5. Effect of disturbance on the coefficient of variation of hydraulic conductivity. *In situ* Bungendore fine sand.

Method	c.v. (%)	
	$\Psi_0 = -40$ mm	$\Psi_0 = -10$ mm
Three-dimensional disc permeameter (no soil disturbance)	18	21
One-dimensional excised cores (removed from soil profile)	69	75

Soil disturbance due to measurement can clearly influence our perception of soil variability. The measurement techniques we select ought therefore to be those in which soil disturbance is minimised.

Sidestepping the Problem

One method of sidestepping the problem of measurement disturbance is to avoid the measurement. Instead, one property is inferred from another simpler-to-measure parameter or property using either an exact relationship or an empirical correlation.

The most studied relation is that between soil particle size distribution or texture and hydraulic conductivity. The best known of these is the Kozeny-Carman equation, which semi-empirically relates hydraulic conductivity to mean particle size. While this approach works well for very simple materials such as graded sands, it is generally inappropriate for field soils. This failure is hardly surprising since flow is governed by the geometry of the pore space, which in structured field soils may have little dependence on soil texture.

Despite this, Duffy et al. (1981) (see also Byers and Stevens 1983; Bouma and Denning 1975) have reported reasonable agreement between measured infiltration rates and those estimated from soil survey texture data and an empirical correlation for a site comprised of eight soil series. Most of these were fairly coarse textured soils and the agreement at low infiltration rates was poor. The use of soil texture and mathematically simple parametric forms for $K(\Psi)$ or $K(\theta)$ and $\Psi(\theta)$ to produce generalized, "universal" relations has some following (Brakensiek et al. 1980; Nielsen et al. 1983). Such universal relations may provide, at best, order-of-magnitude estimates in freshly cultivated soils. Their utility in areas where soil biota are active or where soil structural collapse occurs appears doubtful (Talsma 1988; White and Sully 1987).

There are two examples in which the use of porosity and the analysis of soil composition to infer soil properties do appear to be successful. They are in estimating the thermal conductivity of soil (de Vries 1987) and in calculating the calibration curve of neutron moisture meters (Wilson and Ritchie 1986). In both cases expensive, time consuming soil compositional analysis is required. In general, there are no "free lunches". Sidestepping the measurement will merely result in estimates in which one can have little confidence.

Inverse Techniques

The pioneering efforts in field soil measurements tended to transfer laboratory techniques almost directly to the field. The amount of time and soil disturbance necessary for these has led to the use of less direct, inverse techniques for determining soil physical properties. Such techniques are common in other areas in geoscience. The principal problem is that most inverse measurement schemes are mathematically ill-posed. In order to deconvolute the problem, additional information is required, such as the functional dependence of the property under study.

For soil hydraulic properties simple functional forms abound (e.g., Brooks and Corey 1966; Mualem 1976; van Genuchten 1980; Broadbridge and White 1988). These involve a measured value of hydraulic conductivity at field saturation, K_s, and several adjustable parameters. The model of Broadbridge and White (1988) has only one free parameter and uses measured values of both K_s and sorptivity. In addition, it has analytic solutions for (2) for certain boundary conditions.

One inverse problem which can be solved exactly is that of estimating soil hydraulic properties from sorptivity (Smiles and Harvey 1973; Parlange 1975; Dirksen 1975; White and Perroux 1987). Because sorptivity governs the early stages of flow, it may be determined rapidly and "integrates" over shallow soil depths. This means it is useful for examining the properties of layered field soils.

The functional dependence of sorptivity on soil hydraulic properties is best seen in Philip and Knight's (1974) integral representation (see also Smiles et al. 1981; White and Perroux 1987, 1988).

$$S(\Psi_0,\Psi_n) = \left[2 \int_{\Psi_n}^{\Psi_0} [(\theta - \theta_n)K(\Psi)/F(\Psi,\Psi_0,\Psi_n)]d\Psi\right]^{\frac{1}{2}} \quad . \tag{4}$$

Here Ψ_n and Ψ_0 are the initial and supply soil-water potentials, $\theta_n = \theta(\Psi_n)$ and $F(\Psi,\Psi_0,\Psi_n)$ is the dimensionless soil-water flux (Philip 1973).

The introduction of field devices in which Ψ_0 can be easily controlled (Clothier and White 1981; Perroux and White 1988) means that we have a method of deconvoluting (4) to find $K(\Psi)$. Differentiation of (4) with respect to Ψ_0 and a suitable "guess" for F leads to an approximation, $\hat{K}(\Psi)$, for $K(\Psi)$

$$\hat{K}(\Psi_0) = (S_0/\Delta\theta_0)\partial S_0/\partial\Psi_0 - (S_0/[2\Delta\theta_0])^2 d\theta_0/d\Psi_0 \tag{5}$$

with $\Delta\theta_0 = \theta(\Psi_0) - \theta_n$. For field applications the finite difference form of (5) is more useful. With measurements at two supply potentials Ψ_{01}, Ψ_{02} we have

$$\hat{K}(\overline{\Psi}_0) = (\overline{S}_0/\overline{\Delta\theta}_0)\Delta S_0/\Delta\Psi_0 - (\overline{S}_0/[2\overline{\Delta\theta}_0])^2\Delta\theta_0/\Delta\Psi_0 \qquad (6)$$

where $\overline{S}_0 = (S_{01} + S_{02})/2$, $\overline{\Psi}_0 = (\Psi_0 + \Psi_{02})/2$,

$$\Delta S_0 = S_{01} - S_{02}, \quad \Delta\Psi_0 = \Psi_{01} - \Psi_{02} \quad .$$

Equations (4), (5) and (6) are valid for both positive (ponded) and negative (unsaturated) supply heads.

Table 6 shows a comparison of the saturated hydraulic conductivity of Yolo light clay with those calculated from sorptivity and (6) at positive Ψ_0 (White and Perroux 1987). We see the maximum deviation is less than one percent. Comparison of the approximation (5) with the exact solution for the model of Broadbridge and White (1988) shows that we may expect a maximum deviation of less than 5% over the entire Ψ-range for most soils. For field measurement this suffices. Iterative procedures based on (5) may be used to refine estimates to any desired accuracy. The use of inverse techniques coupled with simple parametric forms for transport properties and analytic or quasianalytic solutions (Philip 1988) has considerable potential for field applications.

Table 6. Estimation of K_s from pairs of sorptivity measurements; test of approximation (6) for Yolo light clay*. (White and Perroux 1987)

$\overline{\Psi}_0$ (mm)	$\Delta\Psi_0$ (mm)	$\overline{S}$ (mm $h^{-1/2}$)	K_s (mm h^{-1}) Eq. (6)	K_s (mm h^{-1}) Exact	% Error
50	100	8.216	0.4428	0.4428	0
175	150	9.783	0.4426		0.045
375	250	11.848	0.4388		0.9
750	500	14.954	0.4448		0.45
1500	1000	19.755	0.4431		0.068

* S data from Philip (1958)

Soil Macrostructure and Biological Activity

In many field measurement techniques the soil is treated essentially as an inert, structureless material. It has long been recognised, however, that soil biota and macrostructure profoundly influence transport in soils (see, e.g., Moore 1898). Only a few techniques have been developed to cope specifically with the preferential flow pathways which predominate in many undisturbed soils (see, e.g., Dixon 1972; Ritchie et al. 1972; Scotter 1978; Kanchanasut et al. 1978; Bouma et al. 1977; Bouma and Wosten 1979; Clothier and White 1981, 1982; Perroux and White 1988). These preferential pathways do not merely determine the magnitude and variance of flow properties but also influence their functional dependence (Clothier and White 1981, 1982; Clothier et al. 1981a). As an example, Fig. 4 shows the contrast between soil-water diffusivities for intact and repacked Bungendore fine sand. The differences are attributed to continuous biopores (Clothier and White 1982).

Biological activity, whether floral, faunal or microbiological, is usually confined to the top 3 m of the soil profile. Its influence on transport and temporal change of properties needs to be accounted for in field measurements in this zone.

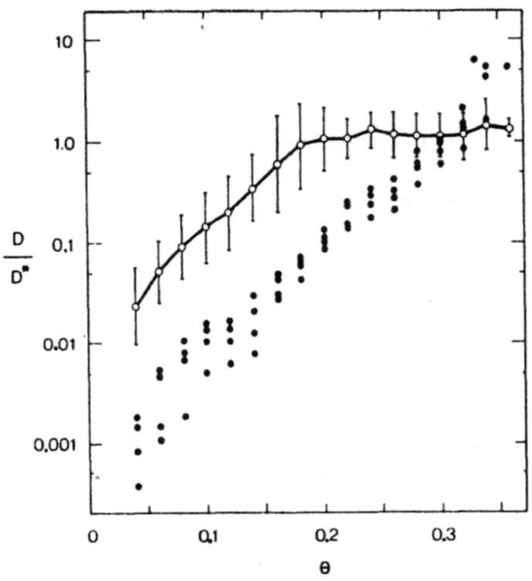

Fig. 4. Comparison of normalized soil-water diffusivities for intact (open points) and repacked (solid points) soil cores. (Clothier and White 1982).

Criteria for Measurement Techniques

We may use the foregoing considerations to develop criteria for field measurement techniques. Such criteria are necessarily idiosyncratic. I belive field measurements should:

(i) be done *in situ* where possible;

(ii) involve minimum disturbance - so that the measurements reflect actual soil properties rather than our enthusiasm with spade and auger;

(iii) possess the 4 R's; rapid, reliable, routine and robust - the variability of field soil requires many measurements;

(iv) be three-dimensional, but weighted in the direction of interest - flows in the field are three-dimensional;

(v) be able to be scaled - the effect of instrument size and soil-imposed weighting function needs to be accounted for;

(vi) be capable of measuring over shallow layers - layered soils are endemic;

(vii) have physical similarities with the process being investigated - correlations with other properties are second best;

(viii) allow for preferential transport where necessary;

(ix) be cheap - obvious in these straightened times.

An example of a field technique designed with these criteria in mind follows.

The Disc Permeameter and the Measurement of Soil Hydraulic Properties

In 1979 during field work on a uranium mine overburden dump (White, Clothier and Perroux, unpublished measurements, 1979) it became obvious that there were serious difficulties with traditional one-dimensional techniques for measuring hydraulic properties, which involved driving metal cylinders into the surface. This problem occurs not only in mining dumps but in stony soils, in soils with fragile crusts and in hard setting soils. A degree of disturbance is also seen when rings are driven into "well-behaved" soils. During our measurements we showed that even when the ring was only sealed to the surface of the mining dump, sorptivity could be measured during the early stages of three-dimensional flow. We have used this fact and the work of Wooding (1968) on steady state flow from surface ponds to develop a technique for determining the sorptivity and hydraulic conductivity $K(\Psi_0)$ of surface soils containing preferential flow paths (White, Sully and Perroux, unpublished manuscript, 1987).

The disc permeameter shown in Fig. 5a (Perroux and White 1988) is designed so that the pressure at the supply surface may be accurately controlled. This permits macropores of various sizes to be included in or excluded from the flow process. Both positive and negative Ψ_0 versions are available. The disc is merely placed against the soil surface (with suitable contact material where necessary) and three-dimensional flow commences as in Fig. 5b. Sorptivity may be determined from the early stages of flow. At large times the steady state flow rate, q_∞, is determined by both gravity and capillarity, as seen in Wooding's (1968) approximation for discs of large radius, r;

$$q_\infty / [\pi r^2(K_0 - K_n)] = 1 + 4\lambda_c/(\pi r) \qquad (7)$$

with $K_0 = K(\Psi_0)$ the conductivity at the supply pressure, and $K_n = K(\Psi_n)$ that at the antecedent pressure (usually $K_0 >> K_n$), and λ_c, the macroscopic capillary length (Table 1). The first term on the right-hand side of (7) corresponds to the influence of gravity on flow and the second that of capillarity (see also Weir 1986, 1987).

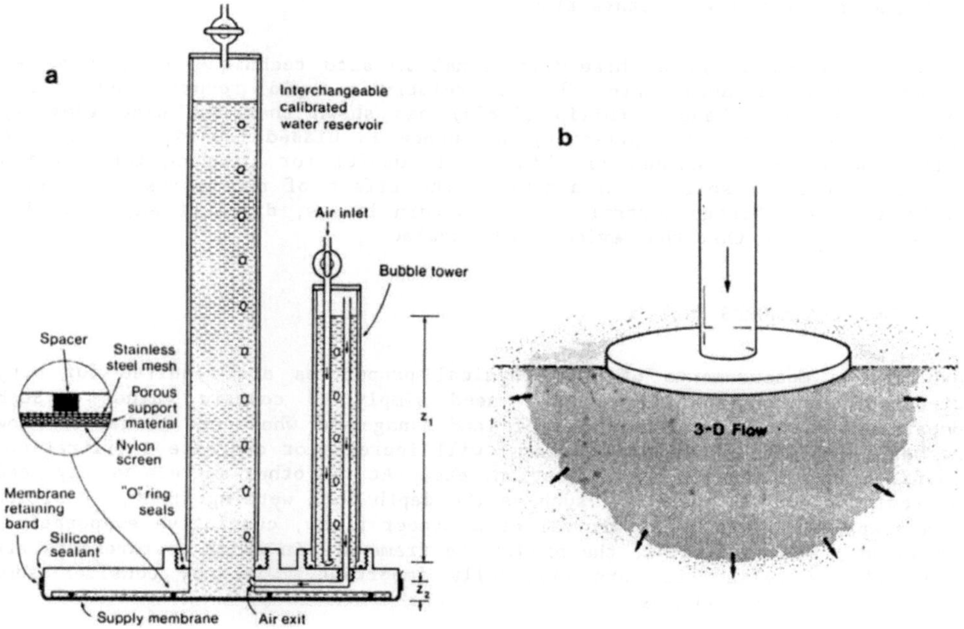

Fig. 5. (a) Disc permeameter for unsaturated flow measurement in field soils. Supply potential controlled by bubble tower (Perroux and White 1988). (b) Three-dimensional surface flow from the disc.

Using the relationship between λ_c and both sorptivity, S_0, and hydraulic conductivity (White and Sully 1987), we may rewrite (7) as

$$q_\infty/(\pi r^2) = K_0 - K_n + 4bS_0{}^2/[\pi(\theta_0 - \theta_n)r] \quad , \qquad (8)$$

where b is a constant confined to $\frac{1}{2} < b < \pi/4$ and $\theta_0 - \theta_n = \theta(\Psi_0) - \theta(\Psi_n)$. For field soils $b \approx 0.55$ (White and Sully 1987). Having found S_0 from the early stages of flow, (8) together with measurements of $\theta_0 - \theta_n$, r and q_∞ permits us to estimate K_0. Typically measurements take about one hour. The approach to steady state flow for a cracking clay soil is shown in Fig. 6.

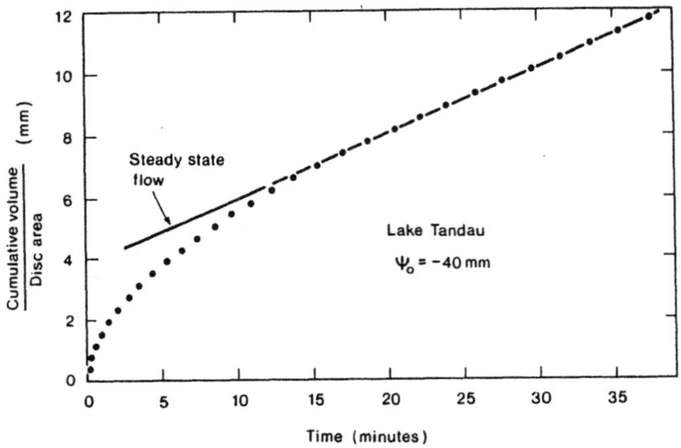

Fig. 6. Cumulative flow from a disc permeameter on a cracking clay soil showing approach to steady state flow.

The disc permeameter is a three-dimensional *in situ* technique which involves minimum soil disturbance during flow, is relatively rapid, robust, easy to use and is relatively cheap. Philip (1986a) has shown that the disc geometry minimises the effect of capillarity and hence is biased towards flow in the vertical direction. Because of this, it is useful for studying infiltration processes in field soils. In addition, the effect of macroporosity on flow may be assessed through control of Ψ_0. Within limits, discs of any dimension may be produced so that the device can be scaled.

PREDICTIVE FRAMEWORKS

Quantitative measurements of soil physical properties are required for many purposes. At one extreme we may need simply to compare values. Such comparisons are common in the area of land management where one wishes to know whether a particular management scheme will increase or decrease infiltration, runoff, erosion, leaching, salinization, etc. At the other extreme we may need to predict material transport, such as the depth to a wetting front, position of a seepage face, time of arrival of a tracer plume, cumulative evaporation, and so on. For the latter, the predictive framework in which measurements are used and its reliability are critically important. We now consider some existing predictive schemes.

Homogeneous Deterministic

In this approach the soil is assumed homogeneous, and deterministic descriptions of transport such as (1) and (2) are assumed to apply. Here the "effective" transport parameters are assumed to be appropriate means of measured values. The measurement variance and any spatial correlations are ignored. Instead, a spatially averaged view is produced. Such assumptions are used in almost all flow measurements for determining "point" values where small scale variability is "swept under the rug".

The severe limitations of replacing a heterogeneous soil mass by a homogeneous one with effective mean properties has long been recognised. Freeze (1975) showed that for one-dimensional saturated flows, such a model is of little use. For one-dimensional capillarity-dominated flows perpendicular to soil layers, Philip (1980) concluded that no conceivable averaging process could give a useful estimate of the flow behaviour of the total system. Bresler and Dagan (1983) found that the concept of an equivalent uniform soil only held in the particular case of steady saturated flow.

It must be admitted that the constraint to one-dimensional flow in both systems is artificial and that the problem is less drastic in a three-dimensional world (Dagan 1976; Freeze 1976; Philip 1980). Additionally, when heterogeneities occur in parallel to flow (Warrick et al. 1977), effective mean properties do permit reasonable prediction of spatially averaged quantities. Such formations are, however, rare in soils.

It is pertinent to ask under what circumstances, if any, will this simplistic model of soil water flow be valid? Since soils are, in general, layered and because most top soils or cultivated soils are relatively uniform to depths of order 0.2 m, we might expect that flows confined to the top soil could be described reasonably by this model. For rainfall infiltration in semi-arid or arid zones, this may be particularly so. There, rainfall quantities and durations are such that fronts seldom penetrate beyond 0.2 m during rainfall. Moreover, for this situation, up to the time of surface ponding, preferential flow appears not to be a major factor.

Clothier et al. (1981a,b) and White and Broadbridge (1988) have successfully used the homogeneous assumption to predict the evolution of water content profiles and the time to incipient ponding for constant rate rainfall infiltration in field soils using independently measured soil physical properties in quasi-analytic or analytic solutions to (2). Figure 7 shows the level of agreement found.

It does seem that there is a limited class of problems in which the homogeneous deterministic model is applicable.

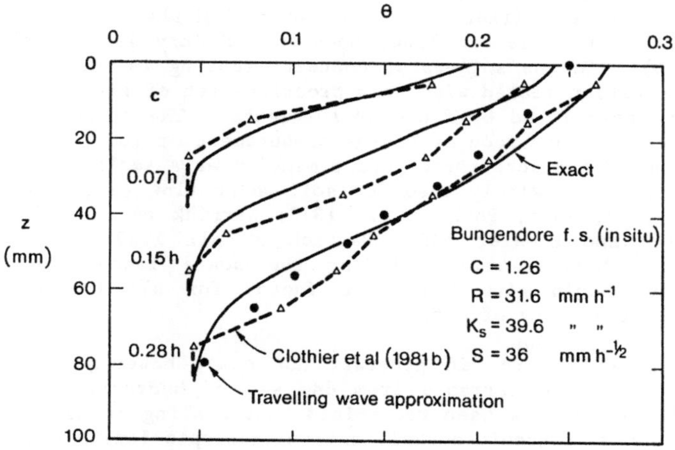

Fig. 7. Evolution of soil-water profiles during constant rate rainfall infiltration in a field soil: comparison of analytic solution with measurement. (White and Broadbridge 1988).

Heterogeneous Deterministic

Theoretically, if soil physical properties were everywhere known, then the application of deterministic forms such as (1) and (2) together with a catalogue of spatial information in a large computer model would suffice to solve soil transport problems. Practically, the collection of such detailed information is impossible except for some limited cases. Where soil is composed of relatively uniform layers, or when a uniform top soil is capped with a crust or intersected by cracks and macropores, such an approach is possible. With the last, however, detailed information on crack and macropore geometry is required. The amount of information necessary for this model severely limits its usefulness.

Deterministic Stochastic

In this approach the stochastic features of measured data are retained but their spatial distributions are ignored (Warrick et al. 1977; Dagan and Bresler 1979, 1983; Bresler and Dagan 1979, 1981, 1983). The soil is treated as essentially a collection of independent, noninteracting columns whose properties have the probability distribution functions, means and variances of the measured properties. For water movement, flow is one-dimensional in the columns. This approach seeks to predict averaged soil-water distribution together with its variance.

The assumption here is that the spatial correlation length of soil properties is much greater in the vertical than in the horizontal direction. Usually the reverse is found. While no exacting tests have been performed, we must expect that the physically improbable model of columns of noninteracting one-dimensional flow paths should considerably overestimate depths to wetting fronts.

Scaling

A technique which relies on the deterministic stochastic approach is that of the scaling of soil properties, a procedure which has attracted renewed interest in soil science (Miller 1980; Warrick and Nielsen 1980; Youngs and Price 1981; Tillotson and Nielsen 1984; Sposito and Jury 1985, 1986; Cushman 1986; Sposito 1986). In the soil-water context, scaling is used to produce a generalized flow equation scaled with an appropriate set of site-specific scale factors which have been found by field measurements. The characterisation of any particular location then reduces to the measurement of local scale factors. The notion of using a characteristic pore dimension as a scale factor (Miller and Miller 1956) has been widely used for soil-water flow (e.g., Philip 1967; Reichardt et al. 1972, 1975; Peck et al. 1977; Warrick et al. 1977; Simmons et al. 1979; Russo and Bresler 1980; Warrick et al. 1985). Provided the materials are scale-heterogeneous, finding the scale factor for one soil hydraulic property should give the scale factor for all the rest. This promises considerable economy of effort.

Unfortunately field soils are, in general, not scale heterogeneous, so that scale factors are not transferrable from one set of hydraulic properties to another. In addition, it has been recognised that scaling of macroscopic flow equations requires a macroscopic as well as a microscopic length (Miller 1980). An inherent soil-based scaling length which seems to fulfill both macroscopic and microscopic requirements is the macroscopic capillary length, λ_c, in Table 1 (Myers and van Bavel 1963; Bouwer 1964, 1966; Raats and Gardner 1971; Philip 1985). This length can be related directly to the field measurable quantities sorptivity and hydraulic conductivity (White and Sully 1987).

$$\lambda_c = b\ S_0^2/([\theta_0 - \theta_n][K_0 - K_n]) \quad , \tag{9}$$

with b as in (8). Philip (1987) suggested that simple capillary theory could be used to relate λ_c to a characteristic microscopic pore dimension, λ_m,

$$\lambda_m = \sigma/(\rho g \lambda_c) \quad . \tag{10}$$

For architecturally simple materials we have been able to show that (10) gives remarkably good estimates of mean particle size or characteristic pore dimensions <d> (obtained using the Kozeny-Carman equation) as shown in Fig. 8 (White and Sully 1987).

The utility of using λ_c to scale the flow equation has yet to be fully explored. It has been suggested (White and Sully 1987) that λ_m may be an *in situ* measure of soil structure. As an example, values of λ_m before and after drought breaking rains are given in Table 4. There drought breaking rains effected the larger λ_m dramatically but not the smaller λ_m.

Not only, then, does λ_c provide a length scale for both Ψ and distance, it may also reveal the flow-weighted, mean, microscopic pore dimension. Finally, as a bonus it also appears as a critical parameter in the quasi-linear approach to soil-water flow (Gardner 1958; Philip 1988).

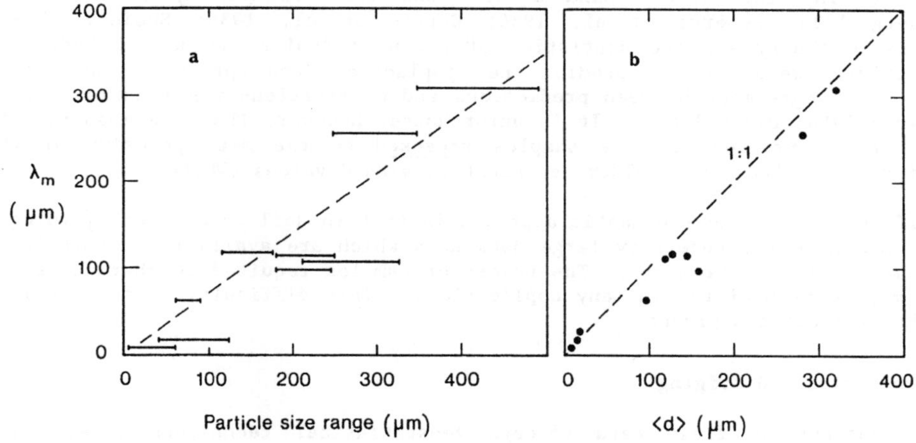

Fig. 8. Comparison of estimates of characteristic pore dimension λ_m with (a) particle size range and (b) mean pore size <d> from data of Youngs and Price (1981). (White and Sully 1987).

Stochastic

It is normally possible only to make measurements on a small sample of any naturally occurring porous material. Because of this it seems reasonable to consider the measured physical properties as random processes or spatial random fields. The stochastic approach to groundwater flow and dispersion has been summarised recently by Gelhar (1986) and Dagan (1986, 1987).

Information required by stochastic models are the probability distribution function, the mean and variance of the suitably transformed variable, and its correlation length(s). The ergodic hypothesis is of central importance in this treatment. In the present context this is equivalent to assuming that a spatially averaged soil property may be represented probabilistically by the ensemble average over a large number of realizations of soils having the same statistical properties. Clearly at any time, we have but one physical realisation of any soil site (Philip 1986b). Pragmatically, the proponents of this model treat situations in which heterogeneities are of smaller scale than the scale of interest as satisfying the ergodic hypothesis. Larger scale variations are treated as trends to be removed. Some evidence exists, however, (Burrough 1983; Jury 1985) to indicate that the correlation length may be a function of sample spacing, which would point to the absence of any preferred length scale. While the ergodic hypothesis may hold for dispersion processes, its validity for unsaturated flow is arguable despite recent progress (Yeh et al. 1985a,b,c; Mantoglou and Gelhar 1987a,b,c).

The techniques employed in the stochastic model involve decomposition of properties into mean and fluctuating parts, finding the expected value of the deterministic flow equation and linearising about the mean (provided fluctuations are small). These are essentially borrowed from turbulence theory or time series analysis. It is tacitly assumed that the mean, variance and correlation lengths are error free. This is never so with the small data sets commonly available (Russo and Jury 1987a,b).

The most extensive and critical test of the stochastic theory of transport was the Borden natural gradient experiment, in which the fate of a plume of chemicals injected into a sand aquifer was followed (Mackay et al. 1986; Freyberg 1986; Roberts et al. 1986; Curtis et al. 1986; Sudicky 1986). Stochastic theory and the statistics of measured "point" values of hydraulic conductivity were used to predict the displacement and spread of the tracer plume. The agreement between predictions and observations was remarkably good (Sudicky 1986; Dagan 1986). It is unfortunate, however, that the measurements of K were performed on core samples repacked to the mean porosity of the aquifer. Repacked cores seldom represent *in situ* K values (White 1988).

A difficulty with the stochastic approach is that in soil science or hydrology, we seldom have the necessary large data sets which are available in turbulence or time series measurements. The number of samples required to characterise a site may be prohibitive in many applications. This difficulty is not confined to the stochastic approach.

Geostatistics and Kriging

The treatment of field data through geostatistical techniques (see, e.g., Matheron 1971; Gutjahr 1985; Journel 1986) has become extremely fashionable in soil science. In essence, geostatistics is the spatial equivalent of time-series analysis but with change in terminology so that, for example, structure functions become semivariograms. Much has been claimed for geostatistics and particularly for the procedure of kriging. In kriging measured values and estimates of the spatial structure obtained through the semivariogram are used to interpolate estimates of properties and the variance at intermediate but unmeasured sites. It has been claimed that this procedure is unbiased and is, in a sense, an "optimal" interpolator. Moreover, kriging has been said to be able to identify both the number and the location of additional measurements needed to increase accuracy. Philip and Watson (1986) have pointed out that the construction of the experimental semivariogram averages out local fluctuations so that "... variation is everywhere averaged and nowhere known". Journel (1986) has acknowledged that the procedure does not give measures of local accuracy.

In addition to this, the problems associated with the somewhat arbitrary procedures used to fit smooth functions through generally noisy variograms (part of the requirement for kriging) compounded with larger scale trends have been highlighted recently (Russo and Jury 1987a,b). Despite earlier claims, kriging is not "optimal" but merely one of many methods of interpolating spatially (or serially) correlated data.

Transfer Functions

Jury (1988) has reviewed the concepts and use of the transfer function model. The principal attraction of this model for solute transport is that it is essentially agnostic about the details of flow. Instead, it pragmatically uses the distribution of solute at a selected depth, measured at some time (or infiltrated volume of water) to predict the distribution at a later stage in the process. Rather than use measurements at a scale of at least one order less than the scale of interest, this approach involves measurements at only a slightly smaller scale for predictions. As Jury (1988) points out, this approach is valid for linear transport processes with no sources or sinks of material within the region of interest.

PROBLEM AREAS

Any list of problem areas runs the risk of being simply a reflection of the predilections of the writer. There are, however, some problems in the field measurement of soil-water properties which need to be addressed (Warrick 1983; Nielsen et al. 1986).

Length and Time Scales

In any transport process the identification of appropriate length and time scales is of fundamental importance. These scales are not merely properties of the material but of the process or questions being addressed. Appropriate, reliable and efficient techniques for identifying these scales and incorporating them in predictive frameworks need to be found. As an example, in the prediction of plant production it appears that soil-water balance and climate data, collected on a daily basis, may be sufficient (Ritchie 1983).

Soil-Water Flux Meter

Unlike other areas in environmental science, and despite some earlier attempts (Cary 1973; Dirksen 1972), no satisfactory three-dimensional soil-water flow meter is currently available.

Soil-Water Potential, Soil-Water Content

Soil-water potential is recognised as being of fundamental importance, yet its measurement in the field poses two problems. The first is that the characteristic size of the measurement volume of current devices is small relative to most scales of interest or other measurements, and the second is that existing devices do not cover the complete range of water potential.

For soil water content, the use of neutron scattering to monitor water content is widely accepted. Reliable calibration of neutron moisture meters is still a problem although a method of calculating calibration curves from the scattering cross-section of the soil constituents is now available (Wilson and Ritchie 1986). This procedure, however, is expensive and time consuming. Time domain reflectometry (Topp et al. 1980) appears, to a reasonable approximation, to have a universal calibration curve. The introduction of new field probes should increase its accuracy (Zegelin et al. 1988). It should be noted, however, that the response function of TDR probes seems heavily weighted towards the region close to the probe surfaces[1].

Disturbance and Measurement Error

The problem of the effect of soil disturbance due to the act of measurement on the perceived value of a field soil property and its variability is of importance. So too is the assessment of actual measurement errors on perceived spatial variability (Sully et al. 1987).

Unstable Flows or Viscous Fingering

Viscous fingering (Chuoke et al. 1959), which occurs when one fluid displaces a more viscous fluid in a porous material such as in petroleum reservoirs, may also occur in field soil-water flows (Hill and Parlange 1972; Raats 1973; Philip 1975a,b; White et al. 1976, 1977). Three of the conditions leading to unstable infiltration flows (crusted soils, layered soils, entrapped air) occur in the field. The importance of viscous fingering in field situations is unknown and current theories are physically unsatisfactory. For unsaturated flows in layered or crusted materials the possiblity exists that the flow will switch from stable to unstable or vice versa depending on the magnitudes of Ψ at the layer interface and the relative magnitude of $K(\Psi)$ in each layer. In these situations the textural descriptions "coarse" and "fine" are misleading. The interaction between heterogeneity and the dimensionality of the flow raises other interesting possibilities for unstable flows.

Swelling Soils

Smiles and Kirby (1988) have summarised the successes of the macroscopic theory of one-dimensional water movement in two-phase swelling materials. Despite its triumphs, the extension to three-phase swelling soils is by no means straightforward. Also, the inclusion of electrolyte effects and catastrophic structural collapse presents difficulties.

Air Entrapment

The movement and trapping of air (or other nonmiscible fluids) during water flow in soil is neglected in many descriptions of soil transport. For some situations, such as large area ponded infiltration, the inability of air to escape from the soil profile can dramatically throttle flow. Approximate theories exist (Morel-Seytoux 1973) but the importance and extent of air-regulated flow in the field are largely unexplored.

Microscopic and Macroscopic Hysteresis

It has long been known that repacked soil columns exhibit hysteresis in $\Psi(\theta)$ as a consequence of pore geometry. Models for hysteresis exist (e.g., Mualem 1974). The influence and importance of this microscopic hysteresis on wetting and draining of field soils is difficult to assess. In addition, Philip (1986b) and Mantoglou and Gelhar (1987b,c) have pointed out that soil heterogeneity can give rise to a macroscopic hysteresis. The importance of this remains to be demonstrated.

---

[1]JH Knight (1987) The response of time domain reflectometry measurements to lateral variations in moisture content. Symposium poster paper

The Interaction of Biology and Physics

Soils, unlike aquifers, support a teeming biotic population. This means that the physical properties of soil cannot be considered static in time. Describing both theoretically and experimentally the interaction of soil-water, solutes and heat with plant and micro- and macro-organisms presents a considerable challenge. In this area, for example, the possibility exists of using plants as appropriate spatial and temporal "integrators" of soil physical properties. The recent finding that plants appear to respond to soil-water content rather than water potential (Gollan et al. 1986) presents a serious challenge to the Darcy-Buckingham view of water movement in the soil/plant environment (see Clothier 1988).

CONCLUDING REMARKS

Over two millenia ago, the author of the book of Job (Ch 5), wrote

> "As for the earth, out of it comes bread, but underneath
> it is churned up as if by fire ... and where is the
> place of understanding?"

In the last twenty years we have gone some way to quantifying how "churned up" is the soil. Mathematically sophisticated theories of transport in porous material have been proposed which have given some glimmer of understanding. What progress has been made stems from the interplay of theory and experiment. The difficulties of making reliable field measurements at an appropriate scale and using them in physically realistic predictive models are undiminished. Innovative and determined experimenters and perceptive theorists are required more than ever.

ACKNOWLEDGEMENTS

I wish to thank my colleagues Mike Sully, Keith Perroux, John Knight and John Philip for their contributions, suggestions and discussions.

REFERENCES

Bear J (1972) Dynamics of fluids in porous media. Elsevier, New York

Beven K, Germann PF (1982) Macropores and water flow in soils. Water Resour Res 18: 1311-1325

Bouma JA, Denning JG (1975) A comparison of hydraulic conductivities calculated with morphometric and physical methods. Soil Sci Soc Am Proc 38: 124-127

Bouma JA, Jongerius A, Boersma O, Jager A, Schoonderbeek D (1977) The function of different types of macropores during saturated flow through four swelling soil horizons. Soil Sci Soc Am J 41: 945-950

Bouma JA, Wosten JHL (1979) Flow patterns during extended saturated flow in two undisturbed swelling clay soils with different macro sturctures. Soil Sci Soc Am J 43: 16-22

Bouwer H (1964) Unsaturated flow in ground-water hydraulics. J Hydraul Div Am Soc Civ Eng 90(HY5): 121-144

Bouwer H (1966) Rapid field measurement of air entry value and hydraulic conductivity of soil as significant parameters in flow system analysis. Water Resour Res 2: 727-738

Brakensiek DL, Engleman RL, Rawls WJ (1980) Variation within texture classes of soil water parameters. Paper No 2006 Am Soc Agr Eng Summer Meeting, San Antonio, Texas

Bresler E, Dagan G (1979) Solute transport in unsaturated heterogeneous soil at field scale: 2. Applications. Soil Sci Soc Am J 43: 467-472

Bresler E, Dagan G (1981) Convective and pore scale dispersive solute transport in unsaturated heterogeneous fields. Water Resour Res 17: 1683-1693

Bresler E, Dagan G (1983) Unsaturated flow in spatially variable fields: 2. Application of water flow models to various fields. Water Resour Res 19: 421-428

Broadbridge P, White I (1988) Constant rate rainfall infiltration: A versatile nonlinear model: 1. Analytic solution. Water Resour Res 24: 145-154

Brooks RH, Corey AT (1966) Properties of porous media affecting fluid flow. J Irrig Drain Div Am Soc Civ Eng 92(IR2): 61-88

Burgess TM, Webster R, McBratney, AB (1981) Optimal interpolation and isarithmic mapping of soil properties: IV. Sampling strategy. J Soil Sci 32: 643-659

Burrough PA (1983) Problems of superimposed effects in statistical study of the spatial variation of soil. Agric Water Manage 6: 123-143

Byers E, Stephens DB (1983) Statistical and stochastic analysis of hydraulic conductivity and particle size in a fluvial sand. Soil Sci Soc Am J 47: 1072-1081

Cary JW (1973) Soil water flow meters with thermocouple outputs. Soil Sci Soc Am Proc 37: 176-181

Chuoke RL, van Meurs P, van der Poel C (1959) The instability of slow, immiscible, viscous liquid-liquid displacements in permeable media. Trans Am Inst Min Metall Pet Eng 216: 188-194

Clothier BE (1988) Measurement of soil physical properties in the field: Commentary. This volume, pp

Clothier BE, Knight JH, White I (1981a) Burgers' equation: Application to field constant-flux infiltration. Soil Sci 132: 255-261

Clothier, BE, White I (1981) Measurement of sorptivity and soil water diffusivity in the field. Soil Sci Soc Am J 45: 241-245

Clothier BE, White I (1982) Water diffusivity of a field soil. Soil Sci Soc Am J 46: 636-640

Clothier BE, White I, Hamilton GJ (1981b) Constant-rate rainfall infiltration: Field experiments. Soil Sci Soc Am J 45: 245-249

Curtis GP, Roberts PV, Reinhard M (1986) A natural gradient experiment on solute transport in a sand aquifer: 4. Sorption of organic solutes and its influence on mobility. Water Resour Res 22: 2059-2067

Cushman JH (1984) On unifying the concepts of scale, instrumentation, and stochastics in development of multiphase transport theory. Water Resour Res 20: 1668-1678

Cushman JH (1986) On measurement, scale and scaling. Water Resour Res 22: 129-134

Cushman JH (1987) More on stochastic models. Water Resour Res 23: 750-752

Dagan G (1976) Comment on 'A stochastic-conceptual analysis of one dimensional ground water flow in nonuniform homogeneous media' by RA Freeze. Water Resour Res 12: 567

Dagan G (1986) Statistical theory of groundwater flow and transport: Pore to laboratory, laboratory to formation, and formation to regional scale. Water Resour Res 22: 120S-143S

Dagan G (1987) Theory of solute transport by groundwater. Ann Rev Fluid Mech 19: 183-215

Dagan G, Bresler E (1979) Solute dispersion in unsaturated heterogeneous soil at field scale: 1. Theory. Soil Sci Soc Am J 43: 461-467

Dagan G, Bresler E (1983) Unsaturated flow in spatially variable fields: 1. Derivation of models of infiltration and redistribution. Water Resour Res 19: 413-420 de Vries DA (1987) The theory of heat and moisture transfer in porous media revisted. Int J Heat Mass Transfer 30: 1343-1350

Dirksen CA (1972) A versatile soil water flux flow meter. In: Proc 2nd Symp on Fundamentals of Transport Phenomena in Porous Media, vol 2, IAHR, ISSS, University of Guelph, Ontario, pp 425-442

Dirksen CA (1975) Determination of soil water diffusivity by sorptivity measurement. Soil Sci Soc Am Proc 39: 22-27

Dixon RM (1972) Controlling infiltration in biomodal porous solids: Air-earth interface concept. In: Proc 2nd Symp on Fundamentals of Transport Phenomena in Porous Media, vol 2, IAHR, ISSS, University of Guelph, Ontario, pp 107-117

Duffy C, Wierenga PJ, Kselik RA (1981) Variations in infiltration rate based on soil survey information and field measurements. New Mexico Agr Exper Stat Bull 680: 1-40

Freeze RA (1975) A stochastic conceptual analysis of one-dimensional groundwater flow in nonuniform homogeneous media. Water Resour Res 11: 725-741

Freeze RA (1976) Reply. Water Resour Res 12: 568

Freyberg DL (1986) A natural gradient experiment on solute transport in a sand aquifer: 2. Spatial moments and the advection and dispersion of nonreactive tracers. Water Resour Res 22: 2031-2046

Gardner WR (1958) Some steady state solutions of the unsaturated moisture flow equation with application to evaporation from a water table. Soil Sci 85: 228-232

Gardner WR (1974) The permeability problem. Soil Sci 117: 243-249

Gelhar LW (1974) Stochastic analysis of phreatic aquifers. Water Resour Res 10: 539-545

Gelhar LW (1986) Stochastic subsurface hydrology from theory to applications. Water Resour Res 22: 135S-145S

Gelhar LW, Axness CL (1983) Three-dimensional stochastic analysis of macro-dispersion in aquifers. Water Resour Res 19: 161-180

Gollan T, Passioura JB, Munns R (1986) Soil water status affects the stomatal conductance of fully turgid wheat and sunflower leaves. Aust J Plant Physiol 13: 459-464

Greacen W (1981) (ed) Soil water asssessment by the neutron method. CSIRO, Melbourne

Gutjahr A (1985) Spatial variability: geostatistical methods. In: Nielsen DR, Bouma J (eds) Soil spatial variability, Pudoc, Wageningen, pp 9-28

Hassan HM, Warrick AW, Amoozegar-Fard A (1983) Sampling volume effects on determining salt in a soil profile. Soil Sci Soc Am J 47: 1265-1267

Hawley ME, Richard HM, Thomas JJ (1982) Volume-accuracy relationship in soil moisture modeling. J Irrig Drain Proc ASCE 108: 1-11

Hill DE, Parlange J-Y (1972) Wetting front instability in layered soils. Soil Sci Soc Am Proc 36: 697-702

Hoeksema RJ, Kitanidis PK (1985) Analysis of spatial structure of properties of selected aquifers. Water Resour Res 21: 563-572

Jackson RD (1988) Surface temperature and surface energy balance. This volume, pp

Journel AG (1986) Geostatistics: models and tools for earth sciences. Math Geol 18: 119-139

Jury WA (1985) Spatial variability of soil physical parameters in solute migration: A critical literature review. Top Rep EA 4228. Elec Power Res Inst, Palo Alto, California, 80 pp

Jury WA (1988) Solute transport and dispersion. This volume, pp

Kanchanasut P, Scotter DR, Tillman RW (1978) Preferential solute movement through soil voids: 2. Experiments with saturated soil. Aust J Soil Res 16: 269-276

Knight JH (1988) Solute transport and dispersion: Commentary. This volume, pp

Mackay DM, Freyberg DL, Roberts PV, Cherry JA (1986) A natural gradient experiment on solute transport in a sand aquifer: 1. Approach and overview of plume movement. Water Resour Res 22: 2017-2029
Mandelbrot BB (1983) The fractal geometry of nature. WH Freeman, San Francisco
Mantoglou A, Gelhar LW (1987a) Stochastic modeling of large-scale transient unsaturated flow systems. Water Resour Res 23: 37-46
Mantoglou A, Gelhar LW (1987b) Capillary tension head variance, mean soil moisture content, and effective specific soil moisture capacity of transient unsaturated flow in stratified soils. Water Resour Res 23: 47-56.
Mantoglou A, Gelhar LW (1987c) Effective hydraulic conductivities of transient unsaturated flow in stratified soils. Water Resour Res 23: 57-67
Matheron G (1971) The theory of regionalized variables and its applications. Ecole des Mines de Paris, Fontainebleau, France
McBratney AB, Webster R (1983) How many observations are needed for regional estimation of soil properties? Soil Sci 135: 177-183
Miller EE (1980) Similitude and scaling of soil-water phenomena. In: Applications of soil physics, Academic Press, New York, pp 300-318
Miller EE, Miller RD (1956) Physical theory for capillary flow phenomena. J Appl Phys 27: 324-333
Moore ECS (1898) Sanitary engineering: a practical treatise on the collection, removal and final disposal of sewage and the design and construction of works of drainage and sewerage. BT Batsford, London
Morel-Seytoux HJ (1973) Two-phase flows in porous media. Adv Hydrosci 9: 119-202
Mualem Y (1974) A conceptual model of hysteresis. Water Resour Res 10: 514-520
Mualem Y (1976) A new model for predicting the hydraulic conductivity of unsaturated media. Water Resour Res 12: 513-522
Myers LE, van Bavel CH (1963) Measurement and evaluation of water table elevations. Paper presented at 5th Congress, International Committee on Irrigation and Drainage, Tokyo
Nielsen DR, Biggar JW (1967) The physical characterization of field soils. In: Bradley EF, Denmead OT (eds) The collection and processing of field data. Interscience (Wiley), New York
Nielsen DR, Biggar JW, Erh KT (1973) Spatial variability of field-measured soil-water properties. Hilgardia 42: 215-259
Nielsen DR, Reichardt K, Wierenga PJ (1983) Characterization of field-measured soil-water properties. Intern Atomic Energy Agency Symp on Isotopes and Radiation Technology. In: Soil physics and irrigation studies, Aix-en-Provence, France, pp 18-22
Nielsen DR, van Genuchten MTh, Biggar JW (1986) Flow and solute transport processes in the unsaturated zone. Water Resour Res 22: 89S-108S
Parlange J-Y (1975) Determination of soil water diffusivity by sorptivity measurements. Soil Sci Soc Am Proc 39: 1011-1012
Peck AJ, Luxmore RJ, Stolzy JL (1977) Effects of spatial variability of soil hydraulic properties in water budget modelling. Water Resour Res 13: 348-354
Perroux KM, White I (1988) Designs for disc permeameters. Soil Sci Soc Am J, in press
Philip GM, Watson DF (1986) Matheronian geostatistics - quo vadis? Math Geol 18: 93-117
Philip JR (1957) The theory of infiltration: 4. Sorptivity and algebraic infiltration equations. Soil Sci 84: 257-265
Philip JR (1958) The theory of infiltration: 7. Soil Sci 85: 333-357
Philip JR (1967) Sorption and infiltration in heterogeneous media. Aust J Soil Res 5: 1-10
Philip JR (1969) Theory of infiltration. Adv Hydrosci 5: 215-296
Philip JR (1970) Flow in porous media. Ann Rev Fluid Mech 2: 177-204

Philip JR (1973) On solving the unsaturated flow equation: 1. The flux concentration relation. Soil Sci 116: 328-335

Philip JR (1975a) Stability analysis of infiltration. Soil Sci Soc Am Proc 39: 1042-1049

Philip JR (1975b) The growth of disturbances in unstable infiltration flows. Soil Sci Soc Am Proc 39: 1049-1053

Philip JR (1980) Field heterogeneity: Some basic issues. Water Resour Res 16: 443-448

Philip JR (1985) Reply to Comments on "Steady infiltration from spherical cavities". Soil Sci Soc Am J 49: 788-789

Philip JR (1986a) Linearized unsteady multidimensional infiltration. Water Resour Res 22: 1717-1727

Philip JR (1986b) Issues in flow and transport in heterogeneous porous media. Transp Porous Media 1: 319-338

Philip JR (1987) The quasilinear analysis, the scattering analog, and other aspects of infiltration and seepage. In: Fok Y-S (ed) Infiltration development and application, Water Resour Res Centre, Honolulu, Hawaii, pp 1-27

Philip JR (1988) Quasianalytic and analytic approaches to unsaturated flow. This volume, pp

Philip JR, Knight JH (1974) On solving the unsaturated flow equation: 3. New quasi-analytic technique. Soil Sci 117: 1-13

Raats PAC (1973) Unstable wetting fronts in uniform and non-uniform soils. Soil Sci Soc Am Proc 39: 1049-1053

Raats PAC (1988) Quasianalytical and analytical approaches to unsaturated flow: Commentary. This volume, pp

Raats PAC, Gardner WR (1971) Comparison of empirical relationships between pressure head and hydraulic conductivity and some observations on radially symmetric flow. Water Resour Res 7: 921-928

Rao PV, Rao PSC, Davidson JM, Hammond LC (1979) Use of goodness-of-fit tests for characterizing the spatial variability. Soil Sci Soc Am J 43: 274-278

Reichardt K, Libardi PL, Nielsen DR (1975) Unsaturated hydraulic conductivity determination by a scaling technique. Soil Sci 120: 165-168

Reichardt K, Nielsen DR, Biggar JW (1972) Scaling of horizontal infiltration into homogeneous soils. Soil Sci Soc Am Proc 36: 241-245

Ritchie JT (1972) Model for predicting evaporation from a row crop with incomplete cover. Water Resour Res 8: 1204-1213

Ritchie JT (1983) Efficient water use and crop production: Discussion on the generality of relations between biomass production and evapotranspiration. In: Limitations to efficient water use in crop production, Am Soc Agron, Special Publication, pp 29-44

Ritchie JT, Kissel DE, Burnet E (1972) Water movement in undisturbed swelling clay soils. Soil Sci Soc Am Proc 36: 874-879

Roberts PV, Goltz MN, Mackay DM (1986) A natural gradient experiment on solute transport in a sand aquifer: 3. Retardation estimates and mass balances for organic solutes. Water Resour Res 22: 2047-2058

Russo D, Bresler E (1980) Scaling soil hydraulic properties of a heterogeneous field. Soil Sci Soc Am J 44: 681-684

Russo D, Jury WA (1987a) A theoretical study of the estimation of the correlation scale in spatially variable fields: 1. Stationary fields. Water Resour Res 23: 1257-1268

Russo D, Jury WA (1987b) A theoretical study of the estimation of the correlation scale in spatially variable fields: 2. Nonstationary fields. Water Resour Res 23: 1269-1279

Scotter DR (1978) Preferential solute movement through larger soil voids: 1. Some computations using simple theory. Aust J Soil Res 16: 257-267

Simmons CS, Nielsen DR, Biggar JW (1979) Scaling of field-measured soil-water properties: I. Methodology. Hilgardia 47: 77-102

Sisson JB, Wierenga PJ (1981) Spatial variability of steady-state infiltration rates as a stochastic process. Soil Sci Soc Am J 45: 699-704

Smiles DE, Kirby JM (1988) One-dimensional solid-liquid separation. This volume, pp
Smiles DE, Harvey AG (1973) Measurement of moisture diffusivity in wet swelling systems. Soil Sci 116: 391-399
Smiles DE, Perroux KM, Zegelin SJ (1981) Absorption of water by soil: some effects of a saturated zone. Soil Sci Soc Am J 44: 1153-1158
Sposito G (1986) The "physics" of soil-water physics. Water Resour Res 22: 82S-88S
Sposito G, Jury WA (1985) Inspectional analysis in the theory of water flow through unsaturated soil. Soil Sci Soc Am J 49: 791-798
Sposito G, Jury WA (1986) Group invariance and field scale solute transport. Water Resour Res 22: 1743-1748
Sposito G, Jury WA, Gupta VK (1987) Reply. Water Resour Res 23: 753-754
Sudicky EA (1986) A natural gradient experiment on solute transport in a sand aquifer: Spatial variability of hydraulic conductivity and its role in the dispersion process. Water Resour Res 22: 2069-2082
Sully MJ, Flocchini RG, Nielsen DR (1987) Linear distribution of naturally occurring radionuclides in a mollic xerofluvent. Soil Sci Soc Am J 51: 276-281
Talsma T (1988) Evaluation of the well permeameter as a field method of measuring hydraulic conductivity. Aust J Soil Res 25: 361-368
Tillotson PM, Nielsen DR (1984) Scale factors in soil science. Soil Sci Soc Am J 48: 953-959
Topp GC, Davis JL, Annan AP (1980) Electromagnetic determination of soil water content: Measurements in coaxial transmission lines. Water Resour Res 16: 574-582
Truesdell C (1984) An idiot's fugitive essays on science. Springer, Berlin
van Genuchten MTh (1980) A closed-form equation for predicting the hydraulic conductivity of unsaturated soils. Soil Sci Soc Am J 44: 892-898
Vieira SR, Nielsen DR, Biggar JW (1981) Spatial variability of field-measured infiltration rate. Soil Sci Soc Am J 45: 1040-1048
Warrick AW (1983) Parameters in infiltration equations. In: Advances in infiltration, ASAE, New York, pp 69-81
Warrick AW, Lomen DO, Yates SR (1985) A generalized solution to infiltration. Soil Sci Soc Am J 49: 34-38
Warrick AW, Mullen GW, Nielsen DR (1977) Scaling of field-measured soil hydraulic properties using a similar media concept. Water Resour Res 13: 355-362
Warrick AW, Nielsen DR (1980) Spatial variability of soil physical properties in the field. In: Applications of soil physics, Academic Press, New York, pp 319-344
Weir GJ (1986) Steady infiltration from large shallow ponds. Water Resour Res 22: 1462-1468
Weir GJ (1987) Steady infiltration from small shallow ponds. Water Resour Res 23: 733-736
White I (1988) Comment on "A natural gradient experiment on solute transport in a sand aquifer: spatial variability of hydraulic conductivity and its role in the dispersion process" by Sudicky EA. Water Resour Res, in press
White I, Broadbridge P (1988) Constant rate rainfall infiltration: A versatile nonlinear model. 2. Applications of solutions. Water Resour Res 24: 155-162
White I, Colombera PM, Philip JR (1976) Experimental study of wetting front instability induced by sudden change of pressure gradient. Soil Sci Soc Am J 40: 824-829
White I, Colombera PM, Philip JR (1977) Experimental studies of wetting front instability induced by gradual change of pressure gradient and by heterogeneous porous media. Soil Sci Soc Am J 41: 483-489
White I, Perroux KM (1987) Use of sorptivity to determine field soil hydraulic properties. Soc Sci Soc Am J 51: 1093-1101
White I, Perroux KM (1988) Estimation of unsaturated hydraulic conductivity from field sorptivity measurements. Soil Sci Soc Am J, in press

White I, Sully MJ (1987) Macroscopic and microscopic capillary length and time scales from field infiltration. Water Resour Res 23: 1514-1522

Wilson DJ, Ritchie AIM (1986) Neutron moisture meters - the dependence of their response on soil parameters. Aust J Soil Res 24: 11-23

Wooding RA (1968) Steady infiltration from a shallow circular pond. Water Resour Res 4: 1259-1273

Yeh T-C, Gelhar LW, Gutjahr AL (1985a) Stochastic analysis of unsaturated flow in heterogeneous soils. 1. Statistically isotropic media. Water Resour Res 21: 447-456

Yeh T-C, Gelhar LW, Gutjar AL (1985b) Stochastic analysis of unsaturated flow in heterogeneous soils. 2. Statistically anisotropic media with variable α. Water Resour Res 21: 457-464

Yeh T-C, Gelhar LW, Gutjahr AL (1985c) Stochastic analysis of unsaturated flow in heterogeneous soils. 3. Observations and applications. Water Resour Res 21: 465-471

Youngs EG, Price RI (1981) Scaling of infiltration behaviour in dissimilar porous materials. Water Resour Res 17: 1065-1070

Zegelin SJ, White I, Jenkins DR (1988) Improved field probes for soil-water content and electrical conductivity measurement using time domain reflectometry. Water Resour Res, submitted

MEASUREMENT OF SOIL PHYSICAL PROPERTIES IN THE FIELD: COMMENTARY

B.E. Clothier

DSIR Plant Physiology Division, Private Bag, Palmerston North, New Zealand

Summary. The review lecture has demonstrated how, with a deterministic leap of faith, it is possible to measure appropriately soil physical properties in the field (White 1988). This commentary considers, in broader terms, how the articles of the *uniform* theoretical faith have to be modified to accommodate secular concerns such as macropore flow and preferential transport through soil. White (1988) reveals that the prospects appear bright for an increase in our ability to describe and understand the processes of water movement in the field. However, we currently lack an ability to incorporate rationally the plant into our theoretical schemes. Plants, and their associated soil flora and fauna, often critically control water flow and chemical transport in the natural environment of surface soil.

INTRODUCTION

With a sense of place we congregated at the Australian Academy of Science, in a lecture theatre surrounded by the Jaeger Room, to listen to the 'New Testament' of Philip. Forty years ago, J.C. Jaeger analysed linear forms of a Fokker-Planck type equation to predict the conduction of heat in solids (Carslaw and Jaeger 1946). Six years later, J.R. Philip was studying a nonlinear Fokker-Planck type equation to describe the unsaturated flow of water through soil. That analysis culminated in publication of the 'Old Testament' of the "Theory of Infiltration" (Philip 1969). Now at this Symposium we have been brought up to date with some of the recent quasianalytical and analytical descriptions of unsaturated flow (Philip 1988).

With a sense of timing J.R. Philip's theoretical lecture was followed by Ian White's Symposium address on the measurement of soil physical properties in the field (White 1988). Although such a close association of theory and measurement should be expected, it is surprisingly rare. Indeed, Ian White has demonstrated the synergistic benefit of combining theoretical insights with the field measurement of soil physical properties. Whereas Ian White has highlighted many of the specific advances that have accrued from this combined approach, I wish to comment more generally on the respective roles of theory and field measurement on our perception of flow and transport in the natural environment.

In the first instance I will discuss the belated theoretical recognition of the prime role played by surface-vented, preferential water-flow pathways on the rapid transport of water, and other matter, through soil. The recent combination of theoretical description and appropriate measurement, as described by Ian White, gives a more realistic appreciation of saturated/unsaturated water flow in field soil. The prospects for future progress look bright.

I conclude by lamenting our current inability to incorporate, in a rational way, biological or biochemical processes in either our descriptions or measurements of flow and transport in the soil. Here the prospects for rapid progress appear bleaker.

THEORY AND FIELD MEASUREMENT

Soil physicists, like other natural scientists, appear quite distinctly divided into experimentalists and theoreticians. Thus, consideration by soil physicists of the mechanisms of flow and transport is conditioned by their proclivity to either experiment or theory (Fig. 1). Field experimentalists derive their perception of water flow and transport mechanisms from direct field measurement. This link gives them an intimate acquaintance with the

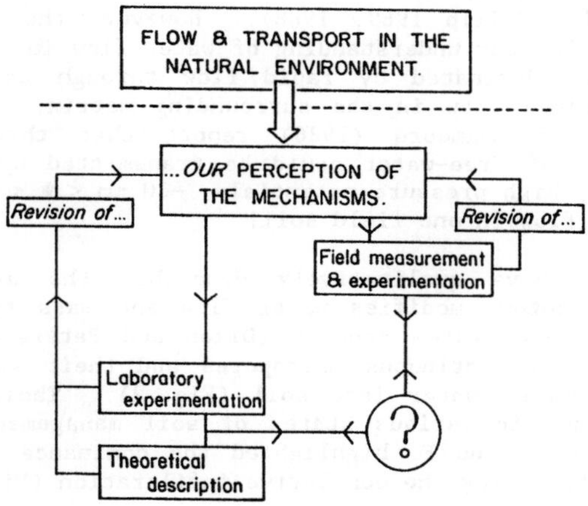

Fig. 1. Observation of the mechanisms of flow and transport in the natural environment of the soil.

field processes that control water flow in natural and anthropogenic soils. However, the view they obtain via such direct measurement may sometimes be only qualitative, and it may be myopic, for often it is specific to the local environment studied. In contrast, the theoretician treats analytically the mechanisms presumed to operate in model laboratory experiments, or 'field' processes simply inferred by dead reckoning. This affords the theoretician a broader perspective and more general view of soil water flow mechanisms. Theoreticians have successfully concentrated on solving the analytically solveable. In doing so they may have developed a degree of presbyopia by their failure to treat appropriately the detail of the processes actually controlling water flow in field soil.

Both camps of soil physicists seek to understand and describe the same mechanisms of flow in the soil; however, there is a philosophical difference between them that often prohibits easy cooperation. For example, all soil physicists, irrespective of ilk, believe that the nonlinear, Fokker-Planck type, Richards' equation,

$$\frac{\partial\theta}{\partial t} = \nabla.(D\,\nabla\theta) - \frac{dk}{d\theta}.\frac{\partial\theta}{\partial z}\,, \tag{1}$$

describes the flow and transport of water in soil. However, this universal belief may well be because "... experimentalists think that it is a mathematical theorem, while mathematicians believe it to be an experimental fact"! (The borrowed quote was made by Gabriel Lippmann to Henri Poincaré, on another matter; Thompson 1917).

The difference between the theoretician and the field experimenter can be highlighted by noting the timing of their respective analyses of water flow in field soil containing a complex of continuous channels. Such macroporous structure is the rule, not the exception, in field soil. The crucial role for surface-vented, connected pores in the transmission of surface free-water through soil has long been recognised in the field (Moore 1898; see Knight 1988; Burger 1922). Meanwhile at the turn of the century theoreticians were just beginning a long struggle with the description of uniform, unsaturated water movement through the matrix of homogeneous soil in response to simple boundary conditions (Buckingham 1907; Richards 1931). Undoubtedly this quest has revealed the underlying structure of the processes of unsaturated flow in uniform porous media (Philip 1969, 1988). However, theory has not been as successful in advancing our understanding of water flow in field soil, which is more often than not dominated by rapid flow through connected macropores, rather than unsaturated flow in the surrounding matrix. For their forested watershed, Watson and Luxmoore (1986) report that three-quarters of the surface-applied flux of free-water could be transmitted by only 0.04% of the soil volume, at very high pressure potentials, -30 mm $< \Psi < 0$. So much for the role of unsaturated flow in one field soil!

Field experimenters have qualitatively described the means by which the macropore-matrix dichotomy modifies water flow and mass transport in natural soils. Dixon's channel system concept (Dixon and Peterson 1971) accentuates the control exerted by continuous macropores and their surface vents on the entry and penetration of water into soil (Fig. 2). Their field experiments with Miami silt loam, in various states of soil management thought to mimic channel conditions A, D and F, highlighted the dominance of macropores, over soil texture, in determining the cumulative infiltration (Fig. 3).

Field workers have struggled to develop infiltrometers capable of discriminating between macropore and matrix flow (Bouma and Denning 1972; Dixon 1975; Dirksen 1975; Clothier and White 1981). Despite novel results, general interpretation of these measurements was made difficult by lack of any appropriate theoretical framework describing water flow in macroporous soil.

Analytical help was soon on the way. Various solutions of Eq. (1), subject to boundary conditions emulating surface-vented macropores, have provided basic insight into the relative transport roles of macropore dimensions and matrix properties (Scotter 1978; Edwards et al. 1979; Beven and Germann 1981; Davidson 1984; Beven and Clarke 1986; Smettem 1986; White 1988). The macropore effects qualitatively displayed in Figs. (2) and (3) can now be quantified. Indeed, there is a graphic similarity between the figures in Davidson' (1984) paper (Figs. 4 and 5 here) and those in Dixon and Peterson (1971) (Figs. 2 and 3 here). Figure 4 shows the Green-Ampt soil, with a single crack of depth a, into which Davidson (1984) predicted water entry. This Green-Ampt soil has matrix conductivity K, wet-front potential Ψ_c, and wet minus dry water content $\theta_s - \theta_n$. Predictions of the relative effects of macropore depth a and spacing 2b and matrix properties K, Ψ_c, $\theta_s - \theta_n$ on cumulative infiltration are given in Fig. 5. Dixon's observations of the impact of macropores in Miami silt loam are quantified.

This, and Ian White's current analyses and observations (White 1988), demonstrate a recent convergence between field and theoretical soil physics. The distinction between theory and practice has become appropriately blurred: in this regard Ian is a very blurry soil physicist! His treatise demonstrates

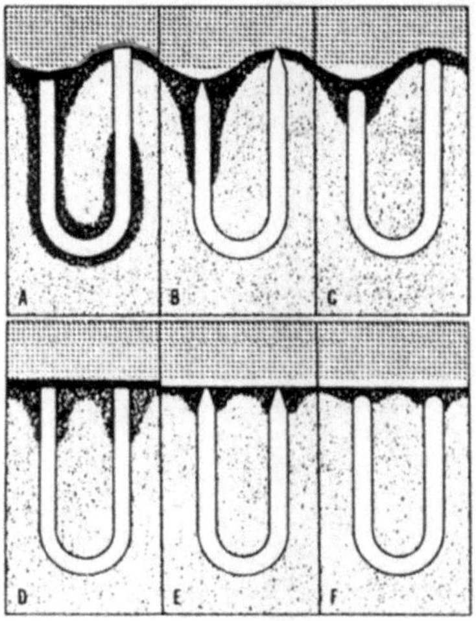

Fig. 2. Dixon's idealized channel system states with six combinations of surface roughness and openness. Rough surfaces with open, constricted and closed orifices (A, B and C) are contrasted with smooth surface equivalents (D, E and F). (Dixon and Peterson 1971).

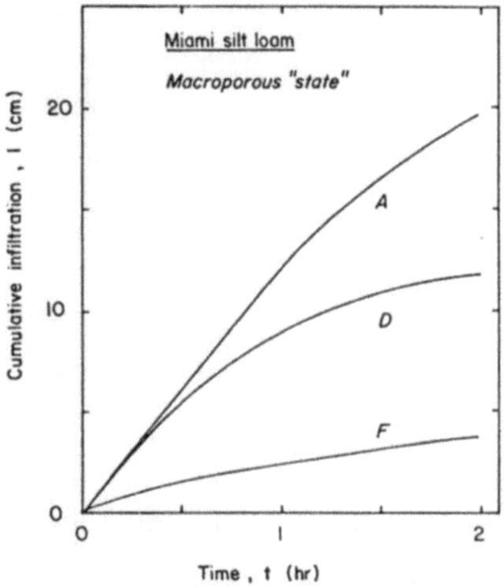

Fig. 3. Cumulative infiltration in Miami silt loam in various states thought to mimic macropore conditions A, D and F in Fig. 2. (Dixon and Peterson 1971).

how analytical and quasianalytical theory (Philip 1986) can in the first instance be used to design better infiltrometers (Perroux and White 1988); and how theory (Smiles and Harvey 1973; Parlange 1975; Philip and Knight 1974) can then be used to interpret in a rational way results obtained in the field (White and Perroux 1987; White and Sully 1987). It is salutary to note that these novel and practical field techniques have only been made possible by the earlier, esoteric theoretical analyses of Eq. (1) applied to flow in homogeneous porous media under simple initial and boundary conditions. The new field results are quite startling. The characteristic capillary length scale, λ_m, a property directly linked to the 'mean' pore dimension, is much larger than expected (circa 100 mm), and not related to soil texture (White and Sully 1987). Rather, λ_m appears dependent upon preferential (macro- and meso-pore) flow paths. As an index, λ_m has been found useful in discriminating between various 'structural' states of surface soils (I White and MJ Sully, personal communication 1987). Additional field measurements of physical properties show natural soils to be sufficiently plagued by biopores that the $\Psi(\theta)$ relation lacks a distinct air-entry value (White and Perroux 1987). This explains the flat diffusivity $D(\theta)$ relation commonly found in field soil (Clothier and White 1981; Hamblin 1982; Kutílek 1984), in stark contrast to the enormous nonlinearity found in repacked laboratory media (Bruce and Klute 1956).

Soil water physics is, I consider, in fine fettle. The prospects look bright. Ian White has shown how the fruitful combination of appropriate theory and field experience can lead to a more realistic perception of the mechanisms of water flow and transport in the 'physical' environment of the soil. Indeed, the current bane of field variability might well be lessened if the effects of macropores are realistically considered and appropriately measured (Bouma 1981). Unfortunately, such optimism does not transfer to the prospects for progress in describing the flow and transport of water and other matter in the 'biological' environment of the soil.

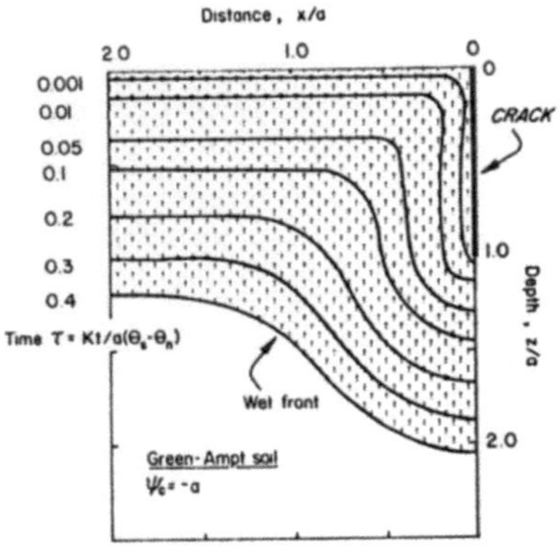

Fig. 4. Water entry into a Green-Ampt soil ($\Psi_c = -a$) with a single crack of depth a, and crack spacing 4a. (Davidson 1984).

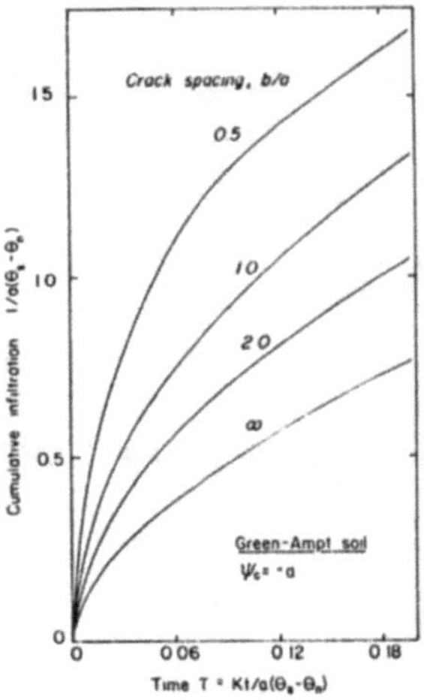

Fig. 5 Infiltration into the soil of Fig. 4 for various crack depths a, and spacings 2b. (Davidson 1984).

BIOLOGICAL 'SOIL PHYSICS'

A primary reason for the scientific study of water flow and chemical transport in soil is because plants of economic and environmental value grow in it. Plants are supported by the soil. They grow by exacting sustenance from it in the form of water and nutrients. Nonetheless, the plant is a persuasive inhabitant of the soil for it indeed creates, exacerbates or maintains many continuous macropores. This macroporous structure is the *sine qua non* of field soil. It is this facet of heterogeneity, more than any other factor, that has required soil physicists to revoke their uniformitarian views when applying porous media theory in the field.

Presently we can only incorporate simplistically the plant actively into our mathematical descriptions of water and chemical movement in the soil.

For the one-dimensional flow of water in the root zone of plants we modify Eq. (1),

$$\frac{\partial \theta}{\partial t} = \frac{\partial}{\partial z}\left(D\frac{\partial \theta}{\partial z}\right) - \frac{dk}{d\theta}\cdot\frac{\partial \theta}{\partial z} - S_w , \qquad (2)$$

and state naively that $S_w(z,t)$ is the sink term representing water extraction by roots. The water content change symbolized by S_w may often be first-order, yet its evaluation is currently far more reliant upon empirical knowledge than theoretical insight.

The equivalent formulation for the one-dimensional flow of noninteracting chemical,

$$\frac{\partial(\theta c)}{\partial t} = \frac{\partial}{\partial z}\left(\theta D_s\frac{\partial c}{\partial z}\right) - \frac{\partial(vc)}{\partial z} - S_c , \qquad (3)$$

includes the sink term $S_c(z,t)$ which itself is simply a sink for our inability to describe more effectively root absorption, precipitation, volatilization, or indeed any other mechanism, often biologically engendered, by which mass of the chemical species in solution is produced or consumed.

A large part of the reason for our struggle with water and chemical flow in the presence of roots is methodological (Böhm 1979). Certainly it is difficult to obtain easily measurements and quantitative descriptions of rooting in the field. Roots grow in a topologically complex manner, often with longitudinal variation in root permeability (Fig. 6). These features are difficult to describe quantitatively at a macroscopic level, yet they markedly affect the spatial and temporal pattern of S_w and S_c. Description of the physical system is further complicated for it appears that the plant 'senses' soil dryness (Gollan et al. 1986), and maybe chemical toxicity, and via hormonal messengers may alter its S_w and S_c strategy (Cowan 1988). Monteith (1986) has recently suggested we turn the whole subject of crop water supply and demand on its head. Rather than atmospherically reference plant water-use to a potential transpiration rate, for annual crops, he feels we should use the velocity of 'root-front' penetration to infer crop water-use. This emphasises the importance of roots in the water and chemical balance of natural soils.

At the microscopic level apparent success has been achieved in analysing water flow to roots. Gardner's (1960) SCI ® *Classic* ® described quantitatively the radial flow of water to a single root, represented by a line sink. Twenty five years on, Wilford Gardner notes that this "... approach is essentially the same all computer models...now follow [and while] this was probably a useful start...it has eventually led us to a dead end" (Gardner 1985).

Fig. 6. The topological complexity typical of (kiwifruit) rooting, showing a decrease in the degree of root suberization towards the younger roots at the right.

Tom Denmead in his *Classic* ® commentary reiterates Gardner's lament, and considers that we do not know "...enough about water movement in crop root zones" (Denmead 1987).

There was comment from this Symposium audience that our understanding of S_w and S_c is in fact quite reasonable. Further, it was hinted that the water uptake problem is tautological; the answer is achieved automatically once we 'tease' out the roots into an equivalent surface area of absorption, A(z). Certainly theoretical solutions to Eqs. (2) and (3) can be obtained for water and solute transport in the root zone for analytically specified forms of S_w and S_c (Raats 1981). My concern is that we do not actually know how to tease out our observations of rooting into appropriate sink functions $S_w(z,t)$ and $S_c(z,t)$ for uniform crops (or $S_w[r,\phi,t]$ and $S_c[r,\phi,t]$ functions for isolated plants). We cannot yet adequately explain many of our observations of root water extraction (Gardner 1983), although some interesting hypotheses are being mooted (Monteith 1986). Our nescience should challenge physicists to learn more biology, and stimulate physiologists to quantify their understanding more rigourously. Progress will only come from within the realm of 'biological' soil physics.

There is some light on the horizon. As with most scientific progress, it will arise via a sharper perception of the mechanisms brought about by an increase in our ability to measure and observe the real world.

Measurement prospects are brightening. In the laboratory, computer-assisted tomographic (CAT) representation of plant water extraction using X-rays and gamma-ray attenuation may provide us with a clearer picture of the small-scale mechanisms operating in the rhizosphere (Hainsworth and Aylmore 1986; LAG Aylmore et al., personal communication 1987). Field measurement of soil water content with greater spatial resolution, especially near the soil surface, is now possible with the new TDR technique (Topp and Davis 1981; JH Knight, personal communication 1987; SJ Zegelin et al., personal communication 1987). This will allow better macroscopic observation of the processes of water movement operating in the root zone. In terms of plant measurement, automatic (soil-core) root-washing devices are becoming available, and the use of periscopes is giving a clearer picture of rooting *in situ*. The provision of better data on the character of rooting, and on the pattern of water extraction, is a necessary condition for an improved analytical treatment of the mechanisms of flow and transport in the root environment. Computer simulation is not sufficient; indeed, it could ruin the science (Truesdell 1984).

CONCLUSION

We have been provided substantial theoretical understanding of the flow of water and transport of chemicals from simple sources through homogeneous porous media (Philip 1988). Laboratory experimentation on repacked soil has complemented and verified these theories. The payoff from this investment in 'uniform' soil physics theory has been an ability to appreciate rigourously the means by which small-scale heterogeneities and continuous macropores can subvert, distort and control the flow of water, and other chemicals, in the field (White 1988). An added bonus has been that perceptive theoretical understanding has led to the development of novel field devices that will now provide more appropriate measurement of those physical properties controlling flow and transport in the field (White 1988).

Nevertheless, there is currently an inability to incorporate adequately biological processes into our theoretical descriptions. Plants, and their associated flora and fauna, are not passive; rather they are persuasive inhabitants of the natural soil environment (Cowan 1988). Many of the all-important, small-scale heterogeneities characteristic of field soil are actually engendered by plants. Progress will follow when soil physicists assess more realistically the critical role the plant plays in controlling flow and transport, and when biologists more rigourously quantify and analyse their observations in the natural environment of the soil.

REFERENCES

Beven KJ, Germann PF (1981) Water flow in soil macropores. II A combined flow model. J Soil Sci 32: 15-29

Beven KJ, Clarke RT (1986) On the variation of infiltration into a homogeneous soil matrix containing a population of macropores. Water Resour Res 22: 383-388

Böhm W (1979) Methods of studying root systems. Springer-Verlag, New York

Bouma J (1981) Soil morphology and preferential flow along macropores. Agric Water Manage 3: 235-250

Bouma J, Denning JL (1972) Field measurement of unsaturated conductivity by infiltration through gypsum crusts. Soil Sci Soc Am Proc 36: 846-847

Bruce RR, Klute A (1956) The measurement of soil moisture diffusivity. Soil Sci Soc Am Proc 20: 458-462

Buckingham E (1907) Studies on the movement of soil moisture. USDA, Bureau of Soils Bull. 38, pp 61

Burger H (1922) Physikalische Eigenschaften von Wald- und Feianboden. Mitteilungen der schweizerischen Centralanstalt für das forstliche Versuchswesen XIII: 1-221

Carslaw HS, Jaeger JC (1946) Conduction of heat in solids. 1st Ed. Oxford Univ. Press (Clarendon), London

Clothier BE, White I (1981) Measurement of sorptivity and soil water diffusivity in the field. Soil Sci Soc Am J 45: 241-245

Cowan IR (1988) Stomatal physiology and gas exchange in the field. This volume, pp

Davidson MR (1984) A Green-Ampt model of infiltration in a cracked soil. Water Resour Res 20: 1685-1690

Denmead OT (1987) Citation classic. Current Contents/Agric, Biol and Environ Sci 18(32): 20 (August 10th)

Dirksen C (1975) Determination of soil water diffusivity by sorptivity measurement. Soil Sci Soc Am Proc 39: 22-27

Dixon RM (1975) Design and use of closed-top infiltrometers. Soil Sci Soc Am Proc 39: 755-763

Dixon RM, Peterson AE (1971) Water infiltration control: A channel system concept. Soil Sci Soc Am Proc 35: 968-973

Edwards WM, van der Ploeg RR, Ehlers W (1979) A numerical study of the effect of non-capillary-sized pores upon infiltration. Soil Sci Soc Am J 43: 851-856

Gardner WR (1960) Dynamic aspects of water availability to plants. Soil Sci 89: 63-73

Gardner WR (1983) Soil properties and efficient water use: An overview. In: Limitations to efficient water use in crop production, Am Soc Agron, Madison, Wisc, pp 45-65

Gardner WR (1985) Citation classic. Current Contents/Agric, Biol and Environ Sci 16(3): 20 (September 2nd)

Gollan T, Passioura JB, Munns R (1986) Soil water status affects the stomatal conductance of fully turgid wheat and sunflower leaves. Aust J Plant Physiol 13: 459-464

Hainsworth JM, Aylmore LAG (1986) Water extraction by single plant roots. Soil Sci Soc Am J 50: 841-848

Hamblin AP (1982) Soil water behaviour in response to changes in soil structure. J Soil Sci 33: 375-386

Knight JH (1988) Solute transport and dispersion: Commentary. This volume, pp

Kutílek M (1984) Some theoretical and practical aspects of infiltration in clays with D = constant. In: Bouma J, Raats PAC (eds) Proc. ISSS Symp. Water and Solute Movement in Heavy Clay Soils. ILRI Pub 37, Wageningen, pp 114-128

Monteith JL (1986) How do crops manipulate water supply and demand? Phil Trans R Soc Lond A316: 245-259

Moore ECS (1898) Sanitary engineering: a practical treatise on the collection, removal and final disposal of sewage and the design and construction of works of drainage and sewerage, BT Batsford, London
Parlange J-Y (1975) Determination of soil water diffusivity by sorptivity measurements. Soil Sci Soc Am Proc 39: 1011-1012
Perroux KM, White I (1988) Designs for disc permeameters. Soil Sci Soc Am J, in press
Philip JR (1969) Theory of infiltration. Adv Hydrosci 5: 215-296
Philip JR (1986) Steady infiltration from buried discs and other sources. Water Resour Res 22: 1058-1066
Philip JR (1988) Quasianalytical and analytical approaches to unsaturated flow. This volume, pp
Philip JR, Knight JH (1974) On solving the unsaturated flow equation: 3. New quasi-analytical technique. Soil Sci 117: 1-13
Raats PAC (1981) Residence times of water and solutes within and below the root zone. Agric Water Manage 3: 63-80
Richards LA (1931) Capillary conduction of liquids through porous mediums. Physics 1: 318-333
Scotter DR (1978) Preferential solute movement through larger soil voids. 1. Some computations using simple theory. Aust J Soil Res 16: 257-267
Smettem KRJ (1986) Analysis of water flow from cylindrical macropores. Soil Sci Soc Am J 50: 1139-1142
Smiles DE, Harvey AG (1973) Measurement of moisture diffusivity in wet swelling systems. Soil Sci 116: 391-399
Thompson D'Arcy (1917) On growth and form. Cambridge Univ Press
Topp GC, Davis J (1981) Detecting infiltration of water through soil cracks by time-domain reflectometry. Geoderma 26: 13-23
Truesdell C (1984) An idiot's fugitive essays on science. Springer, Berlin
Watson KW, Luxmoore RJ (1986) Estimating macroporosity in a watershed by use of a tension infiltrometer. Soil Sci Soc Am J 50: 578-582
White I (1988) Measurement of soil physical properties in the field. This volume, pp
White I, Sully MJ (1987) Macroscopic and microscopic capillary length and time scales from field infiltration. Water Resour Res 23: 1514-1522
White I, Perroux KM (1987) Use of sorptivity to determine field soil hydraulic properties. Soil Sci Soc Am J 51: 1093-1101

CANOPY TRANSPORT PROCESSES

M.R. Raupach

CSIRO Division of Environmental Mechanics, GPO Box 821, Canberra, ACT, 2601, Australia

Summary: Over the last twenty years, paradigms for atmospheric transfer processes in plant canopies have altered in two main ways. Firstly, simple gradient-diffusion theory has ceased to be acceptable as a description of the turbulent transfer of scalars and momentum, and is being replaced by physically sounder theories. Secondly, the turbulence in and above the canopy has been recognised as coherent and organised rather than random and chaotic, and the character of this organised motion is being elucidated. This review outlines the development and implications of these changes.

INTRODUCTION

The transfer of scalars and momentum through the air in a vegetation canopy exerts control over many aspects of the canopy micro-environment. When the canopy is wet, atmospheric transfer provides the dominant control over the evaporation rate of intercepted rainwater; when the canopy is dry and the water vapour flux is dominated by transpiration from leaves, the evaporation rate is influenced by the efficacy of atmospheric transfer, together with the state of the leaf stomata and the supply of radiant energy. Similar factors determine the distributions of leaf and air temperatures within the canopy. Likewise, the deposition and distribution of spores, pollen, contaminants and aerosol droplets from the air to the canopy is controlled by atmospheric transfer and the boundary conditions at the receptor surfaces. These examples concern scalar transfer, but momentum transfer and the turbulent wind field are equally important, not only because the turbulence sustains scalar transfer but also because of direct effects such as wind throw in forests, lodging in cereal crops and erosion on sparsely vegetated surfaces.

This wide range of applications has led to the development of two broad groups of theories, known as "single-layer" and "multi-layer", respectively. The multi-layer theories consider explicitly the structured nature of the microclimate within the canopy, and the fact that different canopy surface elements are exposed to different environmental conditions. An essential part of such a consideration, and the main focus of this review, is an examination of the dynamics of atmospheric transfer within and just above the canopy. Single-layer theories, on the other hand, consider the entire canopy to be a single transferring surface; all transfer dynamics are encapsulated into semi-empirical bulk transfer coefficients, usually expressed as bulk aerodynamic resistances or conductances, which combine the effects of molecular transfer through element boundary layers (together with form or pressure-induced drag, in the case of momentum) at the individual-element level, and of turbulent transfer at the whole-canopy level. The archetypal single-layer theory is the combination equation for evaporation from a whole canopy, or Penman-Monteith equation (Penman 1948; Monteith 1965), but the approach is also widely used for other scalars (e.g. Garratt and Hicks 1973) and for momentum: in either case, it amounts to the characterisation of a rough surface by a single roughness length. We will not consider single-layer theories further, except to remark that they are actually simple parameterizations of a multi-layer description. The connection between the two kinds of theory was first explored systematically in the important papers of Cowan (1968) and Thom (1971, 1972), and has been discussed recently by Finnigan and Raupach (1987).

This review first examines some historical developments over the last three decades. We then summarize the basic observed properties of canopy turbulence to provide the necessary background for a discussion of scalar transfer, primarily from a Lagrangian standpoint: in other words, by considering the motion of scalar-bearing fluid particles under the influence of a given turbulent wind field. Finally, we treat the properties of the turbulent wind field in more detail, leading to a discussion of (Eulerian) turbulence models.

HISTORICAL PERSPECTIVES

Two definitive early works on the plant canopy microclimate were the study of weather in wheat by Penman and Long (1960) and the review of Inoue (1963). These papers addressed a problem which remains of central concern today: for a scalar entity such as heat, water vapour or CO_2, exchanged under steady conditions between the atmosphere and a horizontally uniform canopy in which z is the height, what is the relationship between the mean specific concentration profile $\bar{c}(z)$ and the source (or sink) density profile S(z)? The problem is also applicable for momentum, $\bar{c}$ being the mean streamwise wind velocity component $\bar{u}$, and S the mean drag force per unit volume in the canopy. To solve this problem, one requires a theory or hypothesis about the turbulent transfer process in the canopy. The early work, and most that has followed since, was based on the assumption that scalar transfer in the canopy satisfies a simple gradient-diffusion postulate, or "K-theory":

$$F(z) = -\rho K(z)\frac{d\bar{c}}{dz} \quad , \tag{1}$$

where ρ is the air density, F(z) the vertical scalar flux density and K(z) the vertical eddy diffusivity. Combining (1) with the scalar conservation equation for steady, horizontally uniform conditions,

$$dF/dz = S(z) \quad , \tag{2}$$

we obtain the steady-state diffusion equation

$$\frac{d}{dz}\left[\rho K(z)\frac{d\bar{c}}{dz}\right] + S(z) = 0 \quad , \tag{3}$$

which relates $\bar{c}(z)$ to S(z) if one knows the diffusivity K(z). One can then use measurements of $\bar{c}(z)$ to deduce S(z) and vice versa, or solve for both $\bar{c}(z)$ and S(z) explicitly by specifying a relationship between S and $\bar{c}$ at the individual-leaf level. Alternatively, if both $\bar{c}(z)$ and S(z) are known, K(z) can be deduced.

From the early sixties on, this analysis provided a basis for both experimental and theoretical activities. Many experimental investigations of the canopy microclimate aimed to relate the physical environment to the biological activity of the canopy; to do this, it was (and remains) particularly important to infer the profile of photosynthetic rate, which is given by S(z) for CO_2.

The usual approach was to find K(z) by measuring both $\bar{c}(z)$ and S(z) (or F(z)) for some suitable entity, momentum and energy being the simplest and commonest choices, and then to infer S(z) for some other entity, such as CO_2, from measurements of the appropriate concentration profile $\bar{c}(z)$. The details are summarized in the recent review by Denmead and Bradley (1987). Examples include the work of Penman and Long (1960), in which water vapour was used to infer K(z); Uchijima (1962), Denmead (1964), Brown and Covey (1966), Denmead and McIlroy (1970), Johnson et al. (1976) and Denmead et al. (1976), in which available energy was employed; Uchijima and Wright (1964), Wright and Brown (1967), Lemon and Wright (1969) and Thom (1971), using momentum; and Druilhet et al. (1971), Legg (1975) and Legg and Long (1975), using natural or introduced biologically inert gases.

On the theoretical side, models for the canopy microclimate were developed by writing and solving (1) to (3) for temperature, humidity and possibly CO_2 and other entities as well, with the aid of relationships specifying S(z) for each entity in terms of ambient conditions (examples of expressions for S(z) appear later). The diffusivity K(z) was obtained either by direct assumption or from a solution of (1) to (3) for momentum (with $\bar{c}$ replaced by $\bar{u}$) and an assumed relationship between K(z) and $\bar{u}(z)$. The early microclimate models were analytic (Philip 1964; Cowan 1968; Thom 1972) but the inevitable pressure towards greater complexity led rapidly to the development of numerical models (Waggoner and Reifsnyder 1968; Waggoner et al. 1969; Goudriaan and Waggoner 1972; Furnival et al. 1975; Goudriaan 1977; Halldin and Lindroth 1986).

Amid all this endeavour, there were growing doubts about the validity of K-theory within the canopy. Disquiet was expressed from the outset, as can be seen from the recorded discussion on the review of Inoue (1963), which begins with the following prescient observations (quoted with minor paraphrasing):

> "The chairman [W.C. Swinbank] opened the discussion by emphasising the difficulty of transferring to the plant-air layer in the field such understanding and concepts as have been developed for the atmospheric boundary layer. In the latter the mean air motion is basically horizontal and the various transfers essentially one-dimensional, whereas processes in the plant-air layer are greatly complicated by both the complex geometry of the system and by the presence of distributed sources and sinks. These views were supported by de Vries, Inoue and Philip."

With the luxury of hindsight, it is easy to say that this and similar warnings were insufficiently heeded. However, two factors contributed to the continuing use of K-theory: there was no clear alternative, and there was no direct evidence of its invalidity. Firm evidence that K-theory was in serious error accumulated only gradually from several sources, both indirect and direct.

The first indirect evidence was that different experiments were unable to reach any repeatable conclusion about the form of the K(z) profile, or its relationship with canopy morphology or turbulence structure. Several widely different possible forms for K(z) were reviewed by Thom (1975). Legg and Long (1975) measured a negative K for momentum at the mid-canopy level in a wheat crop, whereas simultaneously measured K values for energy and nitrous oxide (which they introduced from a ground-level source) were positive.

Legg and Long's observations were typical of those leading to a second difficulty: observed mean wind profiles within canopies are often S-shaped, displaying a characteristic bulge in the lower canopy (Shaw 1977). An example appears in the mean wind profile for Uriarra Forest in Fig. 2a. The bulge implies countergradient momentum transfer or negative K values for momentum, provided advection and horizontal pressure gradients can be excluded as momentum sources, and provided the measurements are accurate. Exclusion of extraneous momentum sources can usually be demonstrated fairly convincingly for extensive canopies on good, uniform sites (Shaw 1977; Wilson and Shaw 1977), but measurement accuracy is a problem because most wind profiles within canopies are measured with cup anemometers, which have well-documented overspeeding properties in flows with high turbulence intensity (e.g. Coppin 1982). In a canopy, the turbulence intensity can reach extremely high values of 6 or more, leading to cup anemometer overspeeding by factors of more than 2 (Wilson et al. 1982). The inference of countergradient momentum transfer from bulges in cup-anemometer wind profiles is therefore equivocal, a fact which was used by some to defend K-theory.

A third argument follows from observations of turbulence length scales. Measurements by several techniques show that, within and just above the canopy, the length scale L_w of vertical eddy motions is of the order of the canopy height h, so the vertical mixing cannot be considered to be "fine-grained" with respect to the canopy or to concentration profiles within and just above the canopy. From this experimental result, one may argue along either Eulerian or Lagrangian lines. The Eulerian argument, clearly set out by Corrsin (1974) and applied to transfer in plant canopies by Finnigan (1979b), runs thus: $L_w \sim h$ implies that L_w is of the same order as the vertical length scale for change in the mean concentration gradient $d\bar{c}/dz$, which is less than h within the canopy where gradients change rapidly with height. A K-theory cannot describe vertical transfer in this situation. The Lagrangian argument stems from the observation of Taylor (1959) that an eddy diffusivity is an "illogical conception". Taylor used the phrase because, for vertical dispersion in a steady, horizontal turbulent wind, the effective diffusivity for scalar dispersion (or flux-to-gradient ratio) is a strongly increasing function of the downwind travel distance of the dispersing plume, at least for distances less than or of order L_w/i_w where i_w is the vertical turbulence intensity. Hence, sources released at different times, or at different distances upwind of an observer, cannot be described by a unique value of the diffusivity. Precisely this situation occurs in a plant canopy, where sources are distributed in space and where the distance L_w/i_w is of order h; hence, it is not reasonable to expect turbulent transfer in a canopy to be described by a unique (that is, source-independent) eddy diffusivity profile K(z). This Lagrangian argument proves to be very useful, and is developed at length later.

Direct evidence for the failure of K-theory was obtained in an experiment undertaken by EF Bradley, OT Denmead and GW Thurtell at Uriarra Forest, ACT, Australia, in 1975; further data were gathered by Bradley, Denmead and E Ohtaki in 1981. Preliminary accounts of the findings have been given by Denmead (1984) and Denmead and Bradley (1985, 1987). The original motivation for this experiment came from GW Thurtell, who perceived the relevance of Taylor's Lagrangian dispersion theory for transfer in canopies.

In the Uriarra Forest experiment, the flux densities F(z) for heat, water vapour and CO_2 were measured directly as eddy covariances:

$$F = \rho\overline{w'c'} \quad , \tag{4}$$

where w is the instantaneous vertical velocity component and c the instantaneous specific concentration, with overbars and primes denoting time averages and departures therefrom, respectively; we ignore the negligible molecular contribution to the flux density. Figure 1, from Denmead and Bradley (1987), shows hourly-averaged direct measurements of F at two heights (one within, the other above the canopy), together with simultaneously measured concentration profiles $\bar{c}(z)$, for heat and water vapour. Above the canopy, the temperature and humidity gradients are both negative and the measured heat and water vapour fluxes positive, in agreement with the gradient-diffusion expectation of transfer down a concentration gradient. However, the measured heat flux and the measured temperature gradient within the canopy (at height

11 m) are both positive, so the heat transfer is locally countergradient and the diffusivity K(z) in (1) is negative. The transfer of water vapour is also countergradient where the humidity gradient is positive just below the bump in the humidity profile. This situation was typical of the daytime microclimate in the forest (Denmead and Bradley 1985), and was observed in CO_2 fluxes and gradients as well as those for heat and water vapour (Denmead and Bradley 1987).

The Uriarra experiment showed conclusively that scalar transfer within a canopy does not obey simple gradient-diffusion theory, (1), with any well-behaved or predictable K(z). We now turn to theories for scalar and momentum transfer which can replace the part played by K-theory in the analysis of (1) to (3). The possibilities run in two directions, Lagrangian and Eulerian. The Lagrangian approach proves to be the more informative, although it is restricted in principle to scalar transfer; accordingly, it is treated first.

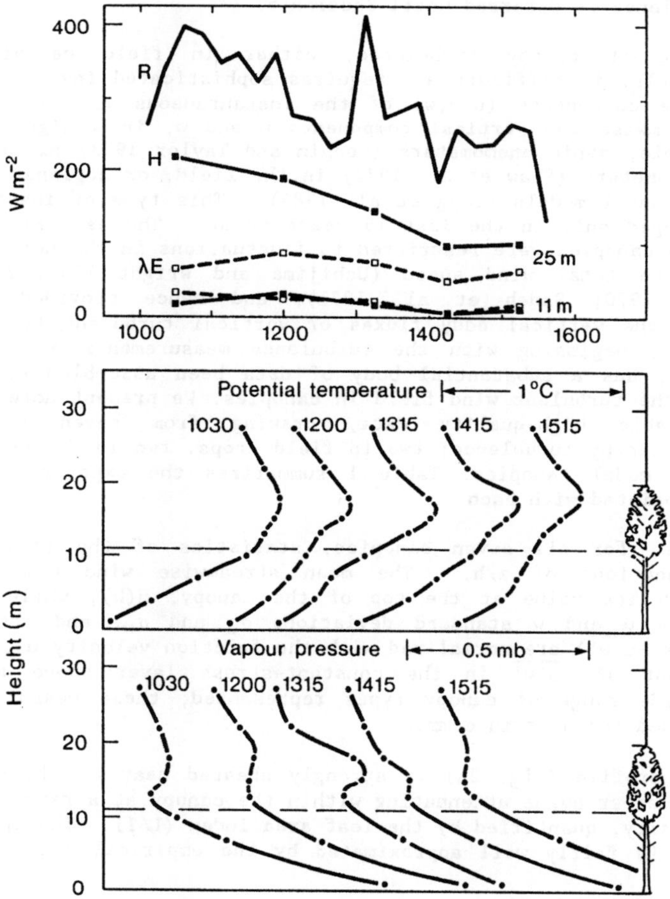

Fig.1. (Top) Time histories of net radiation R above Uriarra Forest and sensible and latent heat fluxes, H and λE, at 25 m and 11 m (canopy height h = 20 m). (Bottom) Corresponding profiles of potential temperature and vapour pressure. (Denmead and Bradley 1987).

Lagrangian analyses predict the transfer of a scalar entity by considering the motion of wandering fluid particles. Of all the particles in the fluid, only some come into contact with the scalar source: these constitute an "ensemble of marked fluid particles" and thereafter carry the scalar. The prediction of scalar transfer amounts to predicting the statistics of the motion of these marked fluid particles, in a turbulent wind field which is supposed to be specified to any necessary degree of detail. Hence, before dealing with the Lagrangian analysis itself, the basic properties of the turbulence within the canopy must be considered.

BASIC PROPERTIES OF CANOPY TURBULENCE

Our aim here is to summarize the available data on the simplest and most fundamental properties of canopy turbulence. Attention will be restricted to steady flow in a horizontally uniform canopy on level ground, so that the turbulence statistics depend only on the height z. A more detailed discussion of canopy turbulence is deferred until Section 5.

The measurement of canopy turbulence, either in field canopies or in wind-tunnel models, is difficult and requires sophisticated instrumentation to sense all three components (u,v,w) of the instantaneous wind vector, or at least the streamwise and vertical components u and w, in a highly turbulent flow; for example, sonic anemometers (Coppin and Taylor 1983) or servo-driven split-film anemometers (Shaw et al. 1973) in the field, or coplanar triple hot wires in wind-tunnel models (Legg et al. 1984). This type of instrumentation has been developed only in the last 15 years or so. The earliest turbulence measurements in canopies were restricted to fluctuations in the horizontal wind component or the total wind speed (Uchijima and Wright 1964; Allen 1968; Meroney 1968, 1970; Sadeh et al. 1971), and hence provided no direct information on the vertical eddy fluxes or vertical turbulent transfer. Only fairly recently, beginning with the turbulence measurements of Shaw et al. (1974) in corn, has a substantial body of data been assembled on all three components of the turbulent wind field in canopies. We present here a synopsis of the available high-quality data, drawing from seven comprehensive experiments on canopy turbulence: two in field crops, two in forests and three in wind-tunnel model canopies. Table 1 summarizes the experiments and some parameters associated with each.

Figure 2 shows, for all seven canopies, statistics of the first two wind moments as functions of z/h. The mean streamwise wind component $\bar{u}$ is normalized with its value at the top of the canopy, $\bar{u}(h)$, while turbulence statistics (the u and w standard deviations σ_u and σ_w, and the kinematic Reynolds stress $-\overline{u'w'}$) are normalized with the friction velocity u_*, defined as the square root of $-\overline{u'w'}$ in the constant-stress layer above the canopy. Despite the wide range of canopy types represented, these measurements have some well-defined features in common.

The mean wind profile (Fig. 2a) is strongly sheared near z = h, with both $\bar{u}$ itself and the shear $d\bar{u}/dz$ attenuating within the canopy at a rate depending on the canopy density, quantified by the leaf area index (LAI). The upper part of the $\bar{u}$ profile is fairly well approximated by the empirical "exponential wind profile":

$$\bar{u}(z)/\bar{u}(h) = \exp(-\gamma(1 - z/h)) \quad , \tag{5}$$

where the coefficient γ tends broadly to increase with LAI, as shown in Table 1. However, the scatter in the relationship between γ and LAI is high.

Table 1. Physical and aerodynamic properties of seven canopies for which high-quality turbulence data exist. WT denotes wind tunnel. Sensors: C = cup anemometer, F = split-film servo-driven anemometer, X = X-configuration hot-wire anemometer, T = coplanar triple hot-wire anemometer, S1 = single-dimension (vertical) sonic anemometer, S3 = three-dimensional sonic anemometer. * denotes papers in preparation.

Canopy	Site	Reference	h	LAI	$\overline{u}(h)/u_*$	γ	Sensors: mean	turbulence
Strips	WT	Raupach et al. (1986)	60 mm	0.23	3.3	0.84	T	T
Wheat	WT	Finnigan, Brunet and Raupach*	47 mm	0.50	4.1	1.5	T	T
Rods	WT	Seginer et al. (1976)	19 cm	1.0	5.0	1.7	X	X
Corn	Elora	Shaw et al. (1974)	260 cm	3.0	3.6	2.4	C	F
Corn	Elora	Wilson et al. (1982)	225 cm	2.9	3.2	4.1	C,F	F
Forest	Moga	Raupach, Finnigan and Brunet*	12 m	1.0	2.9	1.7	C,S3	S3
Forest	Uiarra	Denmead and Bradley (1987)	16,20 m	4.0	2.5	1.7	C	S1

The standard deviations σ_u and σ_w (Figs. 2c,d) both scale with u_*. The ratios σ_u/u_* and σ_w/u_* approach typical surface-layer values (2.5 and 1.25, respectively) only well above z = h; both ratios fall with decreasing z above the canopy, taking values at z = h of about 2.0 and 1.1, respectively. Within the canopy, σ_u/u_* and σ_w/u_* fall more strongly as z decreases. Much of the scatter between canopy types is understandable in terms of the differences in canopy morphology: for example, the extra σ_w between 0.8h and h for the "strips" wind-tunnel model canopy, relative to other canopies, is generated by vortices in the wake at the top of each bluff, strip-like element.

The profile of $-\overline{u'w'}$ (Fig. 2b) exhibits a constant-stress layer above the canopy and rapid attenuation as z decreases within the canopy. The correlation coefficient $r_{uw} = \overline{u'w'}/(\sigma_u\sigma_w)$, which is about -0.3 in the surface layer well above the canopy, increases with decreasing z above the canopy to a maximum magnitude of about -0.45 at z = h. With further decrease in z, r_{uw} attenuates rapidly within the canopy. Hence, the turbulence at the top of the canopy is very efficient at downward momentum transfer, but deep within the canopy, the turbulence loses its ability to transfer momentum as well as its overall strength as measured by σ_u and σ_w; in other words, $-\overline{u'w'}$ decays even faster than the product $\sigma_u\sigma_w$.

Another important property of canopy turbulence emerges from the higher (third and fourth) velocity moments. Figure 3 shows the u and w skewnesses Sk_u and Sk_w. Despite variations imposed by morphological differences, the clear overall trend is for Sk_u to be strongly positive and Sk_w strongly negative within the canopy, indicating that the strongest turbulent events there are downward-moving gusts. The u and w kurtoses within the canopy (not shown here) are both greater than the Gaussian value of 3 which is approached well above the canopy (Maitani 1979; Raupach et al. 1986), indicating that the turbulent events within the canopy are much more intermittent than above.

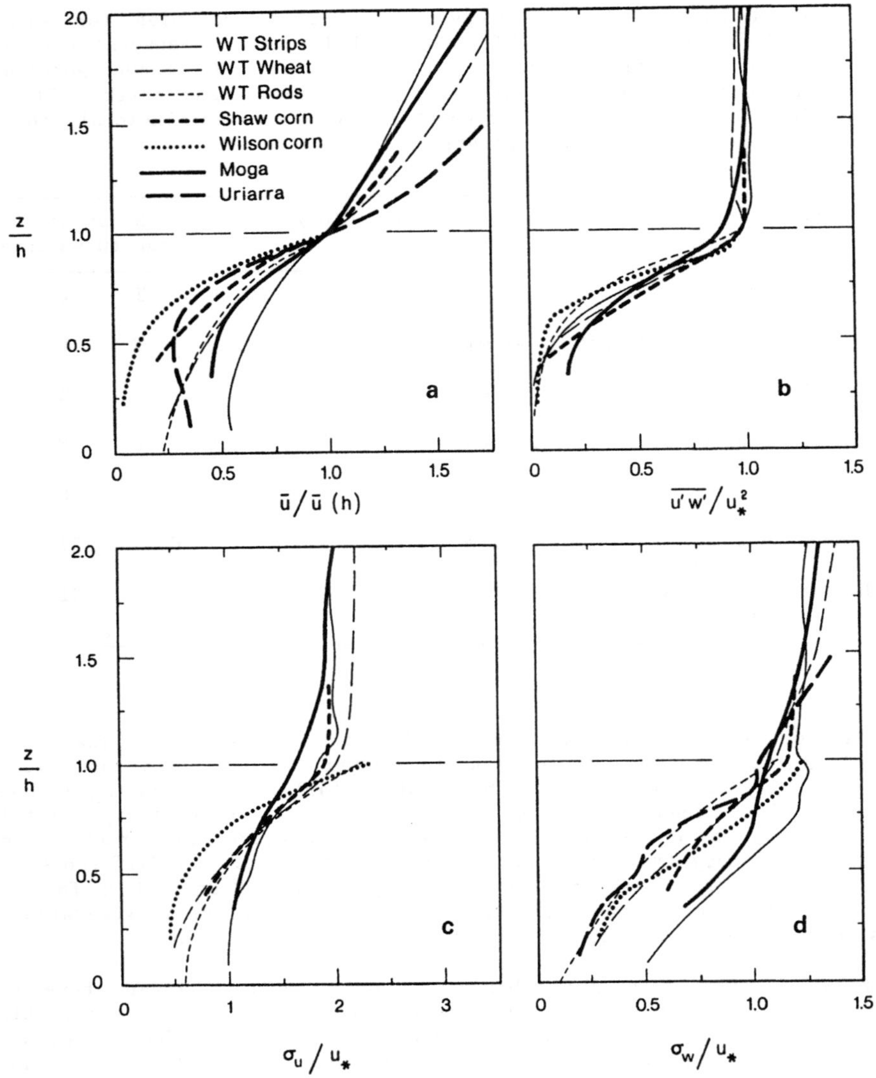

Fig. 2. Profiles with height of (a) $\overline{u}$, (b) $\overline{u'w'}$, (c) σ_u and (d) σ_w for canopies listed in Table 1.

The turbulence length and time scales are just as fundamental in specifying the flow as the velocity moments. Figure 4 shows measurements of the single-point Eulerian length scales L_u and L_w, which are obtained from the integral time scales for single-point time histories u(t) and w(t) by applying Taylor's frozen-turbulence hypothesis. Formally,

$$L_w = (\overline{u}/\sigma_w{}^2) \int_0^\infty \overline{w'(s)w'(s+t)}\, dt \quad , \tag{6}$$

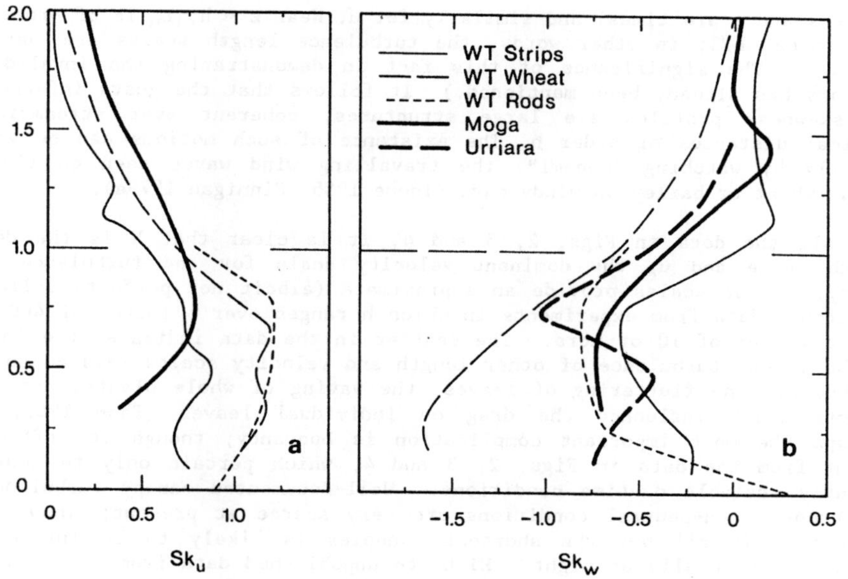

Fig. 3. Profiles with height of (a) Sk_u and (b) Sk_w for canopies listed in Table 1.

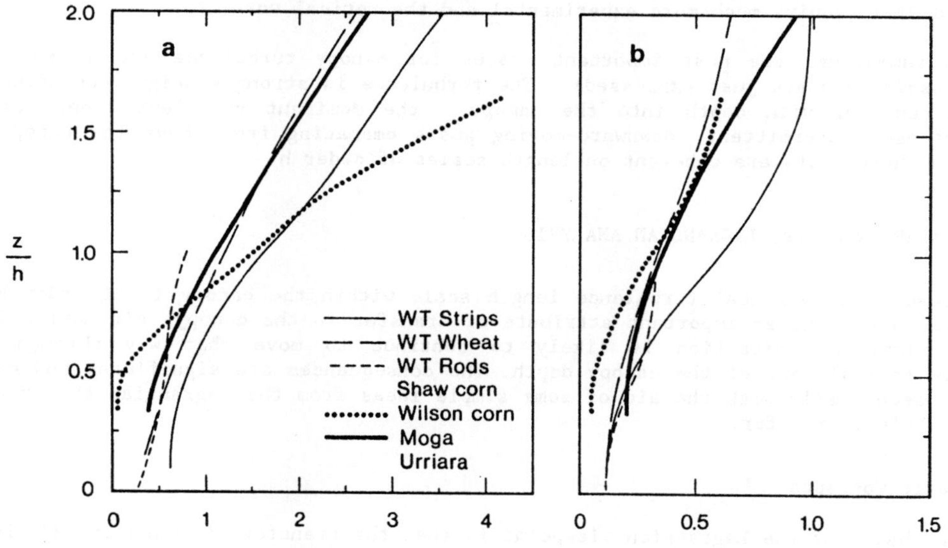

Fig. 4. Profiles with height of (a) L_u and (b) L_w for canopies listed in Table 1.

(where s and t are times) and similarly for u. Near z = h, L_u is of order h and L_w of order h/3; in other words, the turbulence length scales are comparable with h. (The significance of this fact in demonstrating the invalidity of K-theory has already been mentioned.) It follows that the gusts inferred from the skewness profiles are large structures, coherent over streamwise and vertical distances of order h. The existence of such motions can be verified visually by watching "honami", the travelling wind waves seen on fields of grass, wheat or barley on windy days (Inoue 1955, Finnigan 1979a).

From all the data in Figs. 2, 3 and 4, it is clear that h is the dominant length scale and u_* the dominant velocity scale for the turbulence in the canopy. These scales provide an approximate (albeit not perfect) collapse of turbulence data from experiments in which h ranges over a factor of 400 and u_* over a factor of 10 or more. The scatter in the data indicates the influence on the canopy turbulence of other length and velocity scales related to canopy morphology, the fluttering of leaves, the waving of whole plants, and viscous effects which influence the drag on individual leaves (Thom 1968, 1971). Perhaps the most important complication is buoyancy, though its effects are absent from the data in Figs. 2, 3 and 4, which pertain only to neutral or slightly unstable daytime conditions. Well-documented canopy turbulence data in strongly non-neutral conditions are very scarce at present; however, the turbulence in all but the shortest canopies is likely to be influenced by buoyancy, especially at night. Hitherto unpublished data from a recent forest experiment at Camp Borden (Ontario, Canada) suggest that at night, the flow in a forest canopy tends to decouple thermally from the stable flow above, leading to much lower σ_w/u_* values within the canopy than in daytime neutral or lapse conditions (RH Shaw and G den Hartog, personal communication 1987). The problem is complicated because the temperature gradients within and above a canopy tend to be opposite during both day and night (the gradient is positive deep within and negative above the canopy by day, and the reverse at night), and will require much more experimental and theoretical work.

To summarize, the most important scales for canopy turbulence are h and u_* (despite caveats just expressed). The turbulence is strongly height-dependent, attenuating with depth into the canopy; the dominant turbulent events are intense, intermittent, downward-moving gusts emanating from above the canopy; and these gusts are coherent on length scales of order h.

SCALAR TRANSFER: LAGRANGIAN ANALYSIS

Because the vertical turbulence length scale within the canopy is of order h, *persistence* is an important attribute of transfer in the canopy: air moving in a particular direction is likely to continue to move that way through a substantial part of the canopy depth. The consequences are significant and can be seen easily with the aid of some simple ideas from the Lagrangian theory of turbulent transfer.

Basic Concepts

The basis of the Lagrangian viewpoint is that the transfer of a scalar additive in a turbulent flow is statistically equivalent to the dispersion of an ensemble of marked fluid particles. This is a justifiable proposition provided the Peclet number UL/κ (U being a turbulent velocity scale, L a turbulent length scale and κ the molecular diffusivity for the scalar) is high enough for molecular transfer to be neglected relative to turbulent transfer (Monin and Yaglom 1971). In a canopy, such a requirement is easily satisfied, provided only that we regard the thin boundary layers of individual leaves, through which scalar transfer is predominantly molecular, as parts of the scalar source (Raupach 1988).

Restricting attention to scalar transfer or dispersal in the vertical direction, let $Z(t)$ be the height and $W(t) = dZ/dt$ the vertical Lagrangian velocity component of a particular marked fluid particle. For a source density $S(z,t)$ and constant air density ρ, the ensemble-averaged scalar specific concentration is (Monin and Yaglom 1971; Lamb 1980):

$$[c](z,t) = \frac{1}{\rho} \iint P(z,t:z_o,t_o)\, S(z_o,t_o)\, dz_o\, dt_o \quad , \tag{7}$$

where the square brackets denote an ensemble average over many independent realizations of the flow, and $P(z,t;z_o,t_o)$ is the particle transition probability, defined as the conditional probability density that $Z(t) = z$, given $Z(t_o) = z_o$. Since P depends only on the turbulent wind field, (7) specifies the mean scalar concentration in terms of the source density and the turbulence. The problem is to specify P from measureable Eulerian turbulence statistics.

The case of steady, homogeneous turbulence (with zero mean velocity in the z-direction, for convenience) provides a touchstone, since here P can be specified analytically with the aid of some classical arguments. In this flow, moments of the Lagrangian velocity $W(t)$ and the Eulerian velocity $w(z,t)$ are equal and constant (Tennekes and Lumley 1972), so $[W^2] = \sigma_w^2$. If a scalar additive or an ensemble of marked particles is released from an instantaneous point source (at $t = 0$ and $z = 0$, say), then the ensemble-averaged depth at time t of the resulting cloud, $\sigma_Z(t) = [Z^2(t)]^{1/2}$, is related to the velocity field by the kinematic theorem of Taylor (1921):

$$\frac{d\sigma_Z^2}{dt} = 2\sigma_w^2 \int_0^t r_L(s)\, ds \quad , \tag{8}$$

where $r_L(s)$ is the autocorrelation function of W at time lag s, $[W(t)W(t+s)]/\sigma_w^2$. A good approximation for $r_L(s)$ is the exponential form $\exp(-s/T_L)$, where T_L is the integral Lagrangian time scale (Tennekes and Lumley 1972; Csanady 1973; Sawford 1984). Consequently, the square of the cloud depth is:

$$\sigma_Z^2(t) = 2\sigma_w^2 T_L^2\, (t/T_L - 1 + \exp(-t/T_L)) \quad . \tag{9}$$

Since the distributions of both W and Z are both very nearly Gaussian (Batchelor 1949), the transition probability P, or mean scalar concentration from an instantaneous unit point source, is determined as

$$P(z,t;0,0) = \frac{1}{(2\pi)^{1/2}\sigma_Z(t)} \exp\left[\frac{-z^2}{2\sigma_Z^2(t)}\right] \quad . \tag{10}$$

Taylor's theorem, (8), implies that the dispersion obeys different dynamics in the limits of small and large travel time t, relative to T_L. In the large-time or *far-field* limit ($t \gg T_L$), the ensemble of particle paths constitutes a random walk and the cloud depth grows with $t^{1/2}$. More formally, $Z(t)$ can be approximated by a Markov process and the mean scalar concentration in the cloud obeys a diffusion (Fokker-Planck) equation

$$\partial P/\partial t = K_f\, \partial^2 P/\partial z^2 \quad , \tag{11a}$$

where K_f, the far-field eddy diffusivity, is defined by

$$K_f = \sigma_w^2 T_L \tag{11b}$$

and is a property of the turbulence alone (for details, see Durbin 1983). In this sense, dispersion in the far field can be described as *diffusive*. By contrast, the small-time or *near-field* limit ($t \ll T_L$) is the region where the persistence of the turbulent motions causes particle paths to be nearly straight lines, so the cloud depth grows linearly with t. Formally, Z(t) cannot be approximated by a Markov process and the mean scalar concentration does not obey a Fokker-Planck equation. Therefore, the near field can be described as *nondiffusive*. Instead, as shown by Batchelor (1949), the cloud concentration in both the near and far fields is described by an "effective diffusivity" $K_e(t)$ which depends on the travel time t and asymptotically approaches K_f in the far field:

$$K_e(t) = K_f(1 - \exp(-t/T_L)) \quad . \tag{12}$$

The dependence of K_e on t is the cause of the "illogical conception" identified by Taylor (1959): scalar clouds released from point sources at different times require different diffusivities. Moreover, the term "effective diffusivity" carries no implication that scalar dispersion in the near field has a diffusive character in the sense that Z(t) is Markovian (Durbin 1983).

The analytic results in this brief summary, (8) to (12), are restricted to the case of steady, homogeneous turbulence. However, the concepts of the near-field and far-field limits are applicable also in the general case of unsteady, inhomogeneous turbulence. It is true in general that scalar dispersion is nondiffusive in the near field, $t \ll T_L$, and diffusive (meaning that Z(t) is approximately Markovian and P obeys a Fokker-Planck equation) in the far field, $t \gg T_L$ (Durbin 1984). The existence of a diffusive far field also requires that T_L exists, in other words that $r_L(s)$ has a convergent integral, which is a condition violated in systems with no bounding large length scale (Philip 1987); however, such cases are not our concern here.

A Simple Paradigm for Scalar Transfer in a Canopy

Scalars such as heat, water vapour and CO_2 in a canopy can be considered as being emitted from or absorbed by a large number of approximately point sources or sinks, the individual leaves. The scalar plume emitted from each leaf passes through both the near-field and far-field dispersion regimes, so the overall scalar concentration at any one point in the canopy is a superposition of contributions from plumes at all stages of evolution between the near-field and far-field limits. The transition between the two regimes occurs at a travel time $t \sim T_L$, or at a distance downwind of the emitting leaf $\Delta x \sim \bar{u}T_L$, which can be estimated by noting that $T_L \sim L_w/\sigma_w$ (Corrsin 1963), with $L_w \sim h/3$ (Fig. 4) and $\bar{u}/\sigma_w \sim 2$ (Fig. 2). It follows that $\Delta x \sim 2h/3$, so the scalar from an individual leaf must travel through the canopy for a typical horizontal distance of order h before the near-field, nondiffusive effects of the persistence of the canopy turbulence are lost. Since the leaves nearest to any point in the canopy make the strongest contributions to the scalar concentration there, it is likely that the near-field regime has a dominant influence on scalar concentrations in the canopy.

To test this hypothesis, Raupach (1987) calculated the scalar field in a hypothetical canopy in which the turbulence is homogeneous, so that analytic solution for the concentration field is possible with (7) to (12) and their extensions. This drastic simplification of the real turbulence structure within the canopy eliminates the height dependence of the turbulence and the special part played by downward-moving gusts, but retains the correct order of magnitude for the vertical turbulence length scale and also allows arbitrary source density profiles S(z). The homogeneous turbulence was assumed to have a Lagrangian time scale $T_L = 0.4h/\sigma_w$ and a streamwise advection velocity

$\bar{u} = 2.5\sigma_w$, on the basis of measurements from the "strips" wind-tunnel model canopy (see Table 1 and Figs. 2 and 8). Figure 5 shows calculated profiles of scalar mean concentration $\bar{c}(z)$, vertical flux density $F = \rho\overline{w'c'}$ and eddy diffusivity K(z) (defined as $-\overline{w'c'}/(d\bar{c}/dz)$), together with measurements of $\bar{c}$, F and K from an experiment on scalar dispersion from an elevated plane source at a height of 0.8h in the same wind-tunnel model canopy (Coppin et al. 1986). To simulate the experimental source distribution, S(z) for the calculations was assumed to have a single, narrow Gaussian peak centred at 0.8h. Significant features of the experimental data, reproduced by the calculated results, include the substantial peak in $\bar{c}$ and the low values of K near the source height. The governing mechanism in both cases is the persistence of canopy turbulence and the resulting influence of the near-field dispersion: scalar plumes with travel times t less than or of order T_L are narrower and more concentrated than they would be if far-field diffusion operated at all travel times, so they impose locally high magnitudes of $\bar{c}$ and $d\bar{c}/dz$ close to the source height. Furthermore, the low effective diffusivities of the near-field plumes (see (12)) bias the overall diffusivity to low values near the source height.

The homogeneous-turbulence model reveals the mechanism underlying the countergradient flux phenomenon. If the source density profile S(z) is bimodal - the usual daytime situation for heat or water vapour in a forest, where most scalar is emitted in the tree crown region with a small extra contribution from the ground - then the extra ground-level source causes the vertical flux density F(z) to be positive or upward throughout the canopy, from (2). However, the peak in $\bar{c}(z)$ near the upper (crown-level) source maximum causes $d\bar{c}/dz$ to be positive for a significant part of the canopy depth, so that K(z) is negative and the vertical flux is countergradient. This scenario is illustrated in Fig. 6 by calculated $\bar{c}(z)$ and F(z) profiles for four bimodal S(z) profiles, which show a significant region of countergradient flux when the crown-level source is sufficiently strong and localized (Raupach 1987). Comparison of Figs. 1 and 6 shows a qualitative similarity between the homogeneous-turbulence model results and observations from Uriarra Forest, with countergradient fluxes being associated in both cases with peaks or bulges in $\bar{c}(z)$ near source maxima.

This simple paradigm already yields several general conclusions (others will be mentioned later). Firstly, near-field dispersion is important in a canopy, and must be accounted for if observations of scalar transfer are to be reconciled with theory. Secondly, countergradient fluxes in canopies are caused by the large local concentration gradients in near-field scalar plumes. Hence, in the canopy context, it is more appropriate to think of a "local counterflux gradient" than a "countergradient flux". Thirdly, the eddy diffusivity K(z) depends very strongly on S(z) as well as on the structure of the turbulence. Variation of the form for S(z) has a profound effect on K(z), and therefore on the presence and extent of any countergradient region; a substantial countergradient flux is only seen when S(z) has a strong and localized maximum above the ground and an additional small contribution at or near the ground.

Lagrangian Modelling of Scalar Transfer in Realistic Canopy Turbulence

Although the homogeneneous-turbulence model yields some useful qualitative insights, its predictions cannot aspire to be quantitatively correct. As already shown, the vertical velocity field in a canopy is vertically inhomogeneous; further, it is dominated by intense, intermittent downward motions associated with gusts, which cause the w skewness to be negative and the distribution of w to be highly nonGaussian. Therefore, a realistic Lagrangian description of scalar transfer in a canopy, based on (7), requires knowledge of the transition probability P in inhomogeneous, nonGaussian turbulence. The problem of finding P is decidedly nontrivial, not only in the

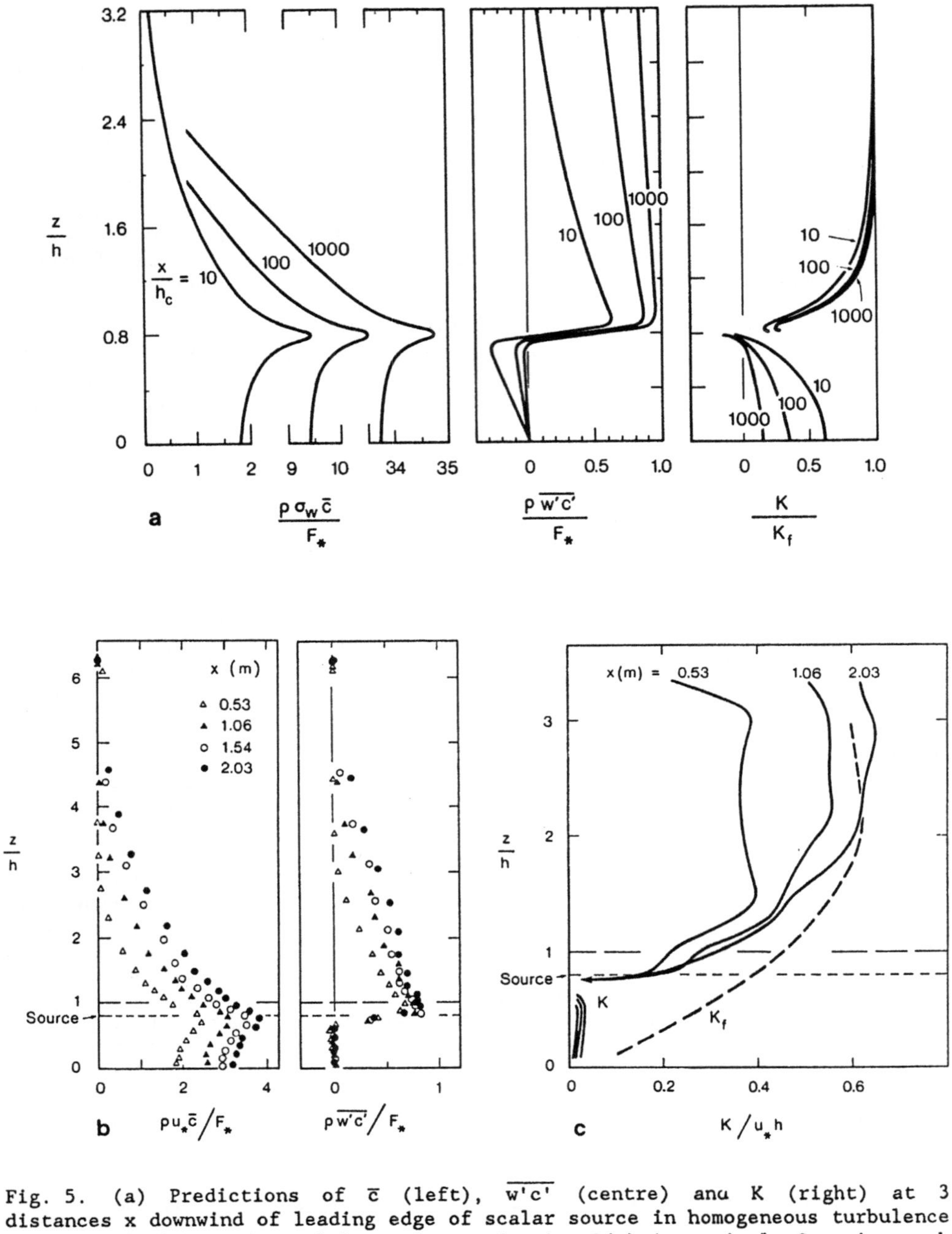

Fig. 5. (a) Predictions of $\bar{c}$ (left), $\overline{w'c'}$ (centre) and K (right) at 3 distances x downwind of leading edge of scalar source in homogeneous turbulence with $T_L = 0.4h/\sigma_w$ and $\bar{u} = 2.5\sigma_w$. Source density S(z) is a single Gaussian peak (mean 0.8h, standard deviation 0.04h); ground z = 0 is a plane of reflection. Normalization is with total source strength F_*, the integral of S(z) from z = 0 to z = h. After Raupach (1987). (b) Measured profiles of $\bar{c}$ (left) and $\overline{w'c'}$ (right) from a wind-tunnel experiment on scalar (trace heat) dispersion from a plane source at height 0.8h in the "strips" wind-tunnel model canopy (Table 1). (c) Measured profiles of K and the far-field diffusivity K_f in the same canopy. (Parts (b) and (c) from Coppin et al. 1986 and Legg et al. 1986).

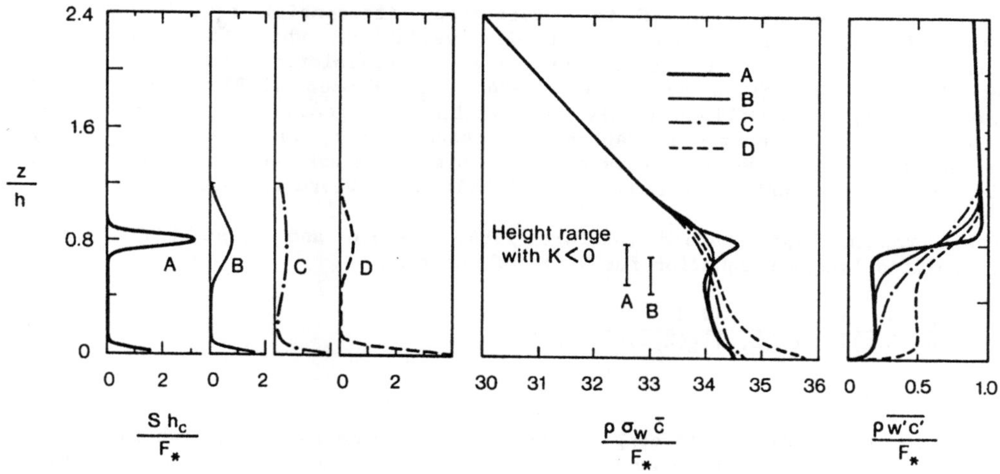

Fig. 6. Predictions of $\overline{c}$ and $\overline{w'c'}$ for four bimodal source distributions in the homogeneous turbulence of Fig. 5a. A countergradient flux is observed in cases A and B. After Raupach (1987).

present context but also for dispersion in inhomogeneous turbulent flows in general, especially the atmospheric convective boundary layer (CBL), as discussed in this volume by Weil (1988).

Here we consider "random-flight" or "stochastic" theories for finding P, which have yielded substantial progress over the last decade and promise more in the future. These theories seek the statistics of the velocities W(t) and trajectories Z(t) of an ensemble of marked fluid particles. A stochastic differential equation - that is, an equation including a statistically specified random term - is constructed for W(t), and solved to find the statistics of Z(t), and P in particular. The calculations are usually performed numerically by computing random single-particle trajectories in a large number of realizations (typically at least 10^4). Random flight theories have a long history (Sawford 1985), but development has surged during the past decade. Although much of the motivation has come from the study of dispersion in the CBL, work in the canopy context has also contributed (Wilson et al. 1981a,b,c; Legg and Raupach 1982; Legg 1983; Wilson et al. 1983). It is likely that random-flight theories will continue to be important aids to understanding and modelling transfer in canopies.

The basic assumption of random-flight theories is that the single-particle Lagrangian velocity sequence $W(n\Delta t)$ (n being the integer sequence) is Markovian. This will be true when $T_K \ll \Delta t \ll T_L$, T_K being the acceleration time scale and T_L the integral time scale of W(t). Since $T_K \sim T_L Re^{-1/2}$, where Re is the Reynolds number of the turbulence (Tennekes and Lumley 1972), it follows that $T_K \ll T_L$ in atmospheric flows where Reynolds numbers are high, so there is no difficulty in choosing an appropriate time interval Δt (Monin and Yaglom 1971; Sawford 1984).

The task of finding the correct stochastic differential equation for W(t) has proved difficult and remains a subject of active research. Both the construction and solution of the stochastic differential equation require application of the calculus of random variables, which has properties substantially different from ordinary deterministic calculus; good introductions are given by Arnold (1974), van Kampen (1981) and Gardiner (1983). Early random-flight models were of limited generality, but a general framework began to emerge in papers by Thomson (1984), van Dop et al. (1985) and Sawford (1986). Models stemming from this framework have been applied to dispersion in the CBL by de Baas et al. (1986) and Sawford and Guest (1987).

Recent random-flight models (Sawford 1986; Sawford and Guest 1987) use a generalized Langevin equation for W(t), of the form

$$d\left[\frac{W}{\sigma_w(Z,t)}\right] = \frac{-W\,dt}{\sigma_w(Z,t)T_L(Z,t)} + d\Omega(Z,t) \quad , \tag{13a}$$

$$dZ = Wdt \quad , \tag{13b}$$

where $d\Omega$ is a random forcing increment. The moments $[d\Omega^n]$ (where square brackets denote an ensemble average, as before), and therefore the distribution of $d\Omega$, depend on z and t via the (specified) Eulerian velocity moments and the time scale T_L. We may write $[d\Omega^n] = M_n(z,t)dt$.

In the simplest and most tractable cases, $d\Omega$ is Gaussian. In this case, M_n is nonzero only for n = 2 (and possibly n = 1), so $d\Omega \sim (dt)^{1/2}$. Herein lies the root of the peculiarities of random-variable calculus: second-order infinitesimals cannot be discarded relative to first-order. In more general cases, though, $d\Omega$ is nonGaussian and M_n is nonzero for all positive integers n, so that all moments of $d\Omega$ are of order dt.

Procedures for finding $[d\Omega^n]$ have been evolved by the work of Thomson (1984), van Dop et al. (1985) and Sawford (1986). From the simple criterion that an initially well-mixed scalar distribution (c = constant) be preserved, one may derive for the first three moments $[d\Omega^n]$ (n = 1, 2 and 3):

$$M_1 = \frac{d\sigma_w}{dz} \; , \; M_2 = \frac{2}{T_L} + \frac{d}{dz}(\sigma_w Sk_w) \; , \; M_3 = \frac{3Sk_w}{T_L} + \frac{d}{dz}(\sigma_w(Kt_w - 3)) \quad , \tag{14}$$

where Sk_w and Kt_w are the w skewness and kurtosis, respectively. These expressions are equivalent to those of Sawford (1986).

It is useful to consider (13) and (14) on three levels of increasing generality. At the simplest level, when the turbulence is steady and homogeneous, M_n is zero for all n except n = 2, so $d\Omega$ is Gaussian with mean 0. Equation (13a) reduces to a simple Langevin equation with Gaussian random forcing which can be solved analytically, yielding a continuous, stationary Markov process W(t) and a position process Z(t) which satisfies (9) and (10) (Arnold 1974; Legg and Raupach 1982; Durbin 1983). Hence, the classical homogeneous-turbulence dispersion results are special cases of (13) and (14). At the next level, when the turbulence is inhomogeneous but with a Gaussian Eulerian velocity field (so $Sk_w = 0$ and $Kt_w = 3$), $d\Omega$ is Gaussian but with a nonzero mean. Equation (13a) becomes a Langevin equation in $W(t)/\sigma_w(Z(t))$, but analytic solution is no longer possible, so numerical, random-number techniques must be used. In the canopy context, this case represents vertically inhomogeneous but Gaussian turbulence within the canopy, so the vertical structure of the canopy can be accounted for but the gustiness of the turbulence cannot. Finally, at the most general level, the turbulence is nonGaussian (either homogeneous or inhomogeneous). This case allows description of both the vertical structure and gustiness of canopy turbulence, and is also relevant to the CBL where w is strongly positively skewed. Equation (14) and its generalization now show that $d\Omega$ is nonGaussian with nonvanishing moments at all orders, which implies that the solution W(t) of

(13) is a discontinuous Markov process which may exhibit jumps (Gardiner 1983 p 51-54). Numerical methods must always be used; the resulting solutions can only ever be approximate, not only because of statistical uncertainity but also because only the first few (usually 3) moments of $d\Omega$ can be specified numerically.

Figure 7 shows the result of an application of (13) and (14) to the calculation of scalar dispersion from an elevated plane source in a canopy, together with the measurements by Coppin et al. (1986) of dispersion from an elevated plane source in the "strips" wind-tunnel model canopy, already cited in Fig. 5. The assumed profiles of σ_w, Sk_w and T_L, also shown in Fig. 7, were based on experimental measurements (see Figs. 2, 3 and 8). To test the effect of Sk_w on the predicted $\bar{c}$ profile, the maximum Sk_w in the canopy, say Sk_{w0}, was varied through the values 0, -0.5 and -1.0, with the value -0.5 being the closest to the measured skewness. The agreement in Fig. 7 between calculated and measured profiles of $\bar{c}$ is fair when Sk_{w0} takes the values 0 and -0.5 (which give insignificantly different $\bar{c}$ profiles) but $Sk_{w0} = -1$ produces a significantly different prediction which does not apparently agree with the data. This trial calculation shows, firstly, that a numerical random-flight model based on (13) and (14) can successfully describe dispersion in typical canopy turbulence, using σ_w, Sk_w and T_L as parameters specifying the turbulence structure; secondly, that the effect of Sk_w is modest (in comparison with a calculation for unskewed turbulence) provided that $-Sk_w$ is less than or of order 0.5; thirdly, that w skewness substantially changes the dispersion pattern when $-Sk_w$ is as high as 1. Since such high Sk_w values are observed in some canopies (Fig. 3), it is clear that w skewness must be accounted for in general. These conclusions are broadly similar to those from the CBL dispersion calculations of Sawford and Guest (1987), who found good agreement between a random-flight model based on (13) and (14) and dispersion data from a laboratory model of the CBL (Deardorff and Willis 1975).

Despite some success of this model in both CBL and canopy dispersion, there are several significant difficulties. The need to truncate the moment sequence has already been mentioned. Much more seriously, the model is mathematically inconsistent because it is not always realizable: for some choices of the turbulence parameters σ_w, Sk_w and T_L, the moments specified by (14) do not all exist (de Baas et al. 1986; Thomson 1987). This is a consequence of the requirement for nonGaussian forcing in (13), which carries with it a panoply of mathematical difficulties associated with the fact that all moments of $d\Omega$ are of order dt. Hence, further development is urgently needed.

A promising avenue has been suggested very recently by Thomson (1987), who writes a general random-flight model of the form

$$dW = a(W,Z,t)dt + b(W,Z,t)d\omega \quad , \tag{15}$$

where $d\omega$ is always a Gaussian increment (mean 0, variance dt) and the functions a and b can be nonlinear, in contrast to (13a) where a is linear in W. Because (15) uses Gaussian rather than nonGaussian forcing, mathematical tractability is recovered and realizability problems disappear.

Specifying the Canopy Turbulence in Lagrangian Models

Having said that the profiles of σ_w, Sk_w and T_L are sufficient for a Lagrangian description of scalar transfer in the canopy, we must discuss how these profiles are determined. The first point is that the turbulent transfer process is not particularly sensitive to the precise forms of $\sigma_w(z)$, $Sk_w(z)$ and $T_L(z)$, provided their overall magnitudes are of the right order. This is shown by the success of the homogeneous-turbulence model described earlier and by the insensitivity of the predictions in Fig. 7 to Sk_w (at least, for $-Sk_w < 0.5$).

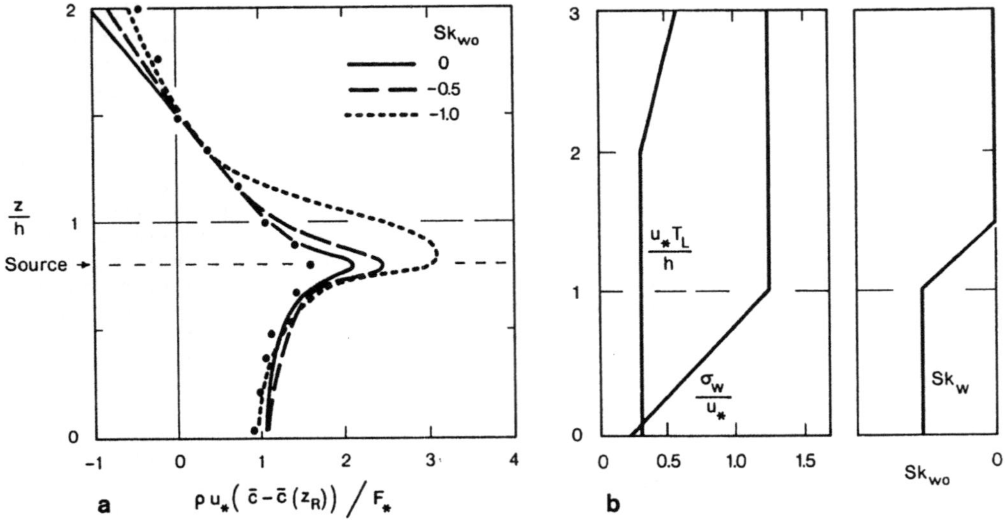

Fig. 7. (a) Assumed profiles of σ_w, T_L and Sk_w for predictions in (b). (b) Random-flight predictions, using (13) and (14), of $\bar{c}$ in a canopy with inhomogeneous turbulence. Source profile S(z) as for Fig. 5. Points are the measurements of Coppin et al. (1986). Reference height $z_R = 1.5h$.

Hence, rough descriptions of the profiles, like those in Fig. 7, may often be adequate. Both σ_w and Sk_w are measureable with modern turbulence sensors, in particular with sonic anemometers. Moreover, the data collected in Figs. 2 and 3 suggest that σ_w, at least, is not particularly sensitive to canopy structure, so that measurements at two heights or even a single height, together with an assumed profile shape based on existing data, should usually suffice. In the absence of measurements, scaling an assumed profile shape with u_* is a reasonable procedure. The main problem with this approach will occur in stable conditions in large canopies such as forests, as already mentioned in the previous section.

The time scale T_L is more difficult as it is impossible to measure directly and must be inferred. One line of inference is to apply (11b) in the form $T_L = K_f/\sigma_w^2$ to measurements of σ_w and a tracer flux density and concentration gradient in the far field, which give K_f. This method was used by Legg et al. (1986) to infer T_L in the "strips" wind-tunnel model canopy (Table 1), from measurements of trace heat released from a lateral line source at height 0.8h. They found that T_L was approximately constant within the canopy, such that $T_L u_*/h = 0.3$.

A second way of inferring T_L is based on the estimate of Corrsin (1963) that $T_L = \beta L_w/\sigma_w$, where β is a dimensionless constant of order 1. Tennekes and Lumley (1972) gave a rough argument, based on assumed Eulerian and Lagrangian spectral shapes, which showed that $\beta = 4/3$, while Snyder and Lumley (1971) found $\beta = 1$ from dispersion measurements in grid turbulence. Corrsin's estimate for T_L leads to the relationship

$$T_L u_*/h = \beta(u_*/\sigma_w)(L_w/h) \tag{16}$$

between (dimensionless) T_L and L_w. Figure 8 shows profiles of T_L from (16) with $\beta = 1$, calculated using available data from the canopies listed in Table 1. Also shown is the dispersion-based T_L profile of Legg et al. (1986). Although there is considerable difference in the height variation of the various measurements, the usual trend is for a bow-shaped T_L profile within the canopy, with an average value for $T_L u_*/h$ of about 0.3; a reasonably harmonious result, given the substantial uncertainties involved in both routes for inferring T_L.

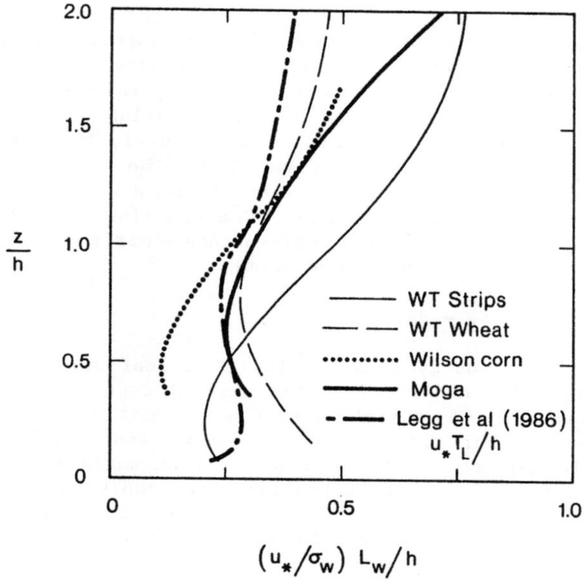

Fig. 8. Profiles of $(u_*/\sigma_w)(L_w/h)$ for canopies listed in Table 1. Symbols as in Fig. 2. Also shown is the dispersion-based profile of $T_L u_*/h$ from Legg et al. (1986).

It is plausible to extend the assumption of constant T_L from within to just above the canopy, allowing the complete T_L profile to consist of two straight lines as in Fig. 7. The upper line is the surface layer requirement $T_L = ku_*(z-d_o)/\sigma_w^2$ (d_o being the zero-plane displacement of the canopy and k the von Karman constant) which causes the far-field diffusivity $K_f = \sigma_w^2 T_L$ to match the well-known eddy diffusivity $ku_*(z-d_o)$ for the neutral atmospheric surface layer well above an underlying rough surface (Hunt and Weber 1979; Raupach 1983). The two lines intersect at about $z = 2h$, producing a layer just above the canopy (between h and 2h) where K_f exceeds its surface-layer value. Such behaviour accords with observations of diffusivities for momentum, heat and water vapour just above rough surfaces, which show values enhanced above the surface layer expectation $ku_*(z-d_o)$ in a "roughness sublayer" extending typically to z/h from 2 to 5 (Thom et al. 1975; Garratt 1978, 1980; Raupach 1979; Raupach et al. 1980; Chen 1987).

Lagrangian Modelling and the Canopy Microclimate

This section has thus far centred on the problem of finding the concentration $\bar{c}(z)$ from a given source density S(z). However, in a real canopy, S(z) is dependent on $\bar{c}(z)$ for most scalars of interest, and also for momentum. The task of a canopy microclimate model is to solve for the coupled $\bar{c}(z)$ and S(z) profiles, given appropriate external conditions, by combining a source-concentration relationship at the individual-leaf level with a description of turbulent transfer through the canopy. We have already cited some of the numerous microclimate models developed during the last two decades that have used the framework of K-theory to describe turbulent transfer. It is now necessary to consider how microclimate models can incorporate more realistic transfer descriptions such as those reviewed above.

It suffices to discuss the coupled problems of water vapour and heat transfer, for which the scalar specific concentrations are the latent and sensible heat contents per unit mass of air, λq and $c_p\theta$ (where q is specific humidity, θ potential temperature, λ the latent heat of vapourization of water and c_p the specific heat of air at constant pressure). In a level, horizontally uniform canopy under steady conditions, (2) shows that the corresponding source densities are $S_q = d\lambda E/dz$ and $S_\theta = dH/dz$, where E and H are the vertical flux densities of water vapour and sensible heat. The coupling between water vapour and heat transfer arises through the leaf-surface energy balance condition, which can be written in the two equivalent forms

$$\lambda E_o + H_o = A_o \quad , \quad S_q + S_\theta = dA/dz = \Lambda A_o \quad , \tag{17}$$

where A_o is the available energy flux density at a leaf surface (incident net irradiance less energy flux density into the leaf storage and photosynthesis), E_o and H_o the water vapour and sensible heat flux densities at a leaf surface, Λ the leaf area density per unit volume of canopy and A the vertical flux density of available energy in the canopy. Leaf-surface flux densities, denoted by the subscript o, are normalized per unit area of leaf and horizontally averaged through the canopy. The source densities S_q and S_θ are related to the ambient conditions within the canopy by the well-known leaf combination equation:

$$S_q/\Lambda = \lambda E_o = \frac{\epsilon A_o + \rho\lambda g_b(q_s(\theta)-q)}{\epsilon + 1 + g_b/g_s} \quad , \tag{18}$$

where $\epsilon = (\lambda/c_p)dq_s/d\theta$ is the ratio of latent to sensible heat increase with temperature for saturated air, $q_s(\theta)$ the saturated specific humidity at temperature θ, g_b the leaf boundary-layer conductance and g_s the leaf stomatal conductance. The history, derivation and properties of this equation are reviewed by Monteith (1981) and Finnigan and Raupach (1987), among many others. From (18) and (17), S_q and S_θ both depend not only upon the leaf-surface available energy, which is externally specified (essentially by the profile of net irradiance through the canopy), but also upon the concentrations q and θ, and therefore upon the turbulent transfer process in the canopy.

This problem can be solved in practice with matrix methods similar to those used in the past in diffusivity-based microclimate models (e.g. Furnival et al. 1975). The canopy and the air immediately above it are divided into N layers; let c_i, S_i, Δz_i and z_i be, respectively, the average concentration, average source density, thickness and height of layer i, and let the top layer coincide with a reference level so that c_N is a given, reference concentration above the canopy. Suppose that some scalar tracer is released uniformly in layer j with a given source density s, so that $S_i = s$ when $i = j$ and 0 otherwise. The resulting concentration profile defines a matrix

$$C_{ij} = \frac{c_i - c_N}{s\Delta z_j} \quad , \tag{19}$$

which is the concentration (relative to the reference level) produced in level i by a unit source in level j. This matrix is a function only of the turbulence structure (specifically, of $\sigma_w(z)$, $Sk_w(z)$ and $T_L(z)$) which can be calculated from a theory of turbulent transfer through the canopy. The dimension of C_{ij} is inverse velocity, suggesting that C_{ij} is a generalized aerodynamic resistance; this can be verified by noting that, for K-theory transfer, C_{Nj} is precisely equal to the integral of $1/K(z)$ between levels j and N, which is the aerodynamic resistance in the layer of constant flux $F = s\Delta z_j$ between z_j and z_N (Thom 1975).

Once C_{ij} has been found from turbulent transfer calculations, the concentration profile from an arbitrary (possibly concentration-dependent) source profile can be written from the principle of linear superposition embodied in (7), which enables the concentration fields from separate source densities (here, separate layers) to be summed. Hence, we have:

$$c_i - c_R = \sum_j S_j(c_j)C_{ij}\Delta z_j \quad , \tag{20}$$

which is a set of N equations relating c_i to S_j. When the source-concentration relationship S(c) is specified at the individual-leaf or single-layer level, (20) can be solved explicitly for c and S. In the (common) case where the S(c) relationship is linear, the equations are linear and can be solved algebraically; however, when the S(c) relationship is nonlinear, recursive solution is necessary. A linear solution is possible for heat and water vapour transfer, since (17) and (18) specify a linear S(c) relationship which leads to 2N coupled linear equations, N for heat and N for water vapour. Alternatively and preferably, one can define two new concentration variables, linear combinations of q and θ, which separate the 2N equations into two sets of N equations each (McNaughton 1976).

The role of this kind of solution method is to combine a theory of turbulent transfer (which fixes C_{ij}) with knowledge of the behaviour of individual leaves (which fixes S(c)). The individual-leaf and turbulent-transfer parts of the microclimate problem appear in separate factors in (20), and so are amenable to separate detailed study and rational amalgamation.

MOMENTUM TRANSFER: EULERIAN ANALYSIS

The momentum field is not predicted by the Lagrangian analyses of the previous section, which require external specification of some simple statistics of the turbulence. This section discusses momentum and turbulence at a deeper level than the basic overview given earlier, and then briefly looks at models for the momentum field, which are naturally Eulerian, at least at the present stage in the development of the subject.

The Structure of Canopy Turbulence

Earlier, it was shown by simple considerations of velocity moments and length scales that canopy turbulence is characterized by coherent, gusting motions with length scales of the order of the canopy height. The nature of these motions has been further elucidated by several kinds of more sophisticated analysis developed for basic investigations of turbulence in the laboratory and introduced into the study of canopy turbulence by Finnigan (1979a,b). Firstly, conditional sampling of u and w with the "quadrant-hole" technique has confirmed not only that the momentum transfer is associated with gusts but also

also that the process is highly intermittent, with most of the momentum transfer occuring in a small fraction (typically a few percent) of the total time (Finnigan 1979b, Raupach 1981, Shaw et al. 1983). The same technique has been applied to heat transfer: evidence to date indicates that gusts dominate heat as well as momentum transfer in near-neutral conditions, but that this nexus is broken with increasing instability (Coppin 1985; Grant et al. 1986).

A second technique is short-term profile analysis, which shows graphically how the transfer process behaves. For daytime conditions in a forest canopy, this technique has revealed coherent gust motions transporting heat and water vapour deficits downward from well above the canopy to deep within it (Denmead 1984, Denmead and Bradley 1985).

Thirdly, two-point correlation measurements (Finnigan 1979b, Crowther and Hutchings 1985) are beginning to reveal the shape of a typical eddy motion in and just above the canopy. A recent wind-tunnel experiment by Y Brunet, JJ Finnigan and the author (on the wind-tunnel model wheat canopy in Table 1) has shown that the dominant eddies above and within the canopy are paired counter-rotating vortices, sloping with the mean wind direction at an angle of about 30 deg above the canopy, with a characteristic spacing between vortices of several canopy heights; in short, structures very like the classic double-roller eddies proposed by Townsend (1976) as the dominant turbulent motions in a boundary layer. The coherent gusting motions in the canopy represent the "footprints" of these larger structures, though the precise relationship is still being investigated.

The Maintenance of Turbulent Energy in the Canopy

Further insight into the nature of turbulence in the canopy, as well as an introduction to some common Eulerian modelling techniques, can be obtained from the budget equation for turbulent kinetic energy per unit mass of air, or TKE:

$$\mathrm{TKE} = \overline{u_i'u_i'}/2 = (\overline{u'^2} + \overline{v'^2} + \overline{w'^2})/2 \quad , \tag{21}$$

where $u_i' = (u',v',w')$ is the fluctuating velocity vector in tensor notation, and the summation convention is employed (repeated indices are summed over values 1, 2 and 3). The TKE budget, an account of the processes creating and destroying TKE within an infinitesimal control volume and also moving TKE into and out of the control volume, is derived by averaging the second moment of the Navier-Stokes equations of motion for a viscous, dynamically incompressible fluid subject to the Boussinesq approximation (Lumley and Panofsky 1964). For the canopy environment, it is necessary to average the turbulence not only in time in the usual way, but also over a horizontal plane to remove the small-scale variability imposed by individual canopy elements (Wilson and Shaw 1977; Raupach and Shaw 1982; Finnigan 1985; Raupach et al. 1986). The horizontal average of a quantity φ will be denoted by $\langle\varphi\rangle$, and the fluctuation by φ'', so that $\varphi = \langle\varphi\rangle + \varphi''$. The TKE budget in the canopy may be written:

$$\partial\langle\mathrm{TKE}\rangle/\partial t = \text{advection} + P_s + P_w + P_b + T_t + T_p - D \quad , \tag{22}$$

where the terms on the right are defined below. Discussion is restricted to steady conditions and an extensive, uniform canopy on level ground: then the left-hand side of (22) is zero and the advection term vanishes.

The shear production term, $P_s = -\langle\overline{u'w'}\rangle\, d\langle\bar{u}\rangle/dz$, is the generation rate of TKE from the kinetic energy of the mean flow (MKE) by the interaction of the turbulent stress with the mean strain; the resulting "shear turbulence" has a vertical length scale of order h within and just above the canopy, and dominates momentum and scalar transfer. The wake production term, P_w, is unique to the canopy environment and does not occur in the free atmosphere: it is the generation rate of TKE in the wakes of canopy elements by interaction of local turbulent stresses and strains, and is approximately equal to the rate of working of the mean flow against canopy drag, $-\langle\bar{u}\rangle\, d\langle\overline{u'w'}\rangle/dz$ (Raupach and Shaw 1982). Like P_s, P_w represents conversion of MKE to TKE, but the length scale of the "wake turbulence" generated by P_w is of order d, the element diameter, and much smaller than the length scale (~ h) of the shear turbulence. The buoyant production term P_b (where g is gravitational acceleration and T_o a reference temperature) represents the conversion rate of buoyant potential energy to TKE. The transport terms, $T_t = -\, d\langle\overline{w'u_i'u_i'}/2\rangle/dz$ and $T_p = -(1/\rho)\, d\langle\overline{w'p'}\rangle/dz$ (where p is the pressure), account for the movement of TKE into and out of the control volume by the fluctuating velocity and pressure fields, respectively. The dissipation term, D, accounts for the conversion of TKE to heat by viscosity.

Figure 9 shows profiles with height of the terms in (22), for the "strips" wind-tunnel model canopy and the Moga forest canopy (see Table 1). Pressure transport T_p could not be measured in either experiment, and is treated as part of the residual. The main features are the large shear production P_s near and just above z = h, the large wake production P_w in the upper part of the canopy, and the major role of turbulent transport T_t in carrying TKE from its level of production near z = h down to deeper levels in the canopy: T_t is a strong sink near z = h and the major source lower down. This occurs because the turbulent TKE flux $\overline{w'u_i'u_i'}/2$ is large and downward near and just above z = h (Maitani 1978, 1979). Hence, most of the TKE deep in the canopy is not generated locally, but imported from above. The importance of turbulent transport in the TKE budget is directly related to the gusty, intermittent nature of the canopy turbulence, since both require that the velocity third moments or skewnesses be significant. In daytime conditions, buoyant production P_b is generally small (as in Fig. 9b), because the shear and the downward momentum flux just above such a rough surface as a canopy are both large enough to make $P_s \gg P_b$. However, P_b will be important when stratification modifies the turbulence structure in the canopy.

Several features of the TKE budget do not emerge in Fig. 9. Firstly, pressure transport T_p was not measured but is probably significant. There is very little data on pressure fluctuations in canopies, because of difficulties in measuring the fluctuating static pressure. On dimensional grounds, the pressure standard deviation σ_p is of order ρu_*^2 if u_* is the dominant fluctuating velocity scale. Sigmon et al. (1983) measured $\sigma_p/\rho u_*^2$ to be of order 2 above a forest canopy (a value inferred from their Figs. 4 and 5) and also found that σ_p attenuates little within the canopy, the main dampening being in the high-frequency fluctuations. Hence, enough pressure fluctuation exists for T_p to be significant if w and p are well correlated as seems probable. Maitani and Seo (1985) estimated $\overline{w'p'}$ from surface pressure measurements, concluding that the pressure-driven TKE flux $\overline{w'p'}/\rho$ is downward in the canopy and about half the magnitude of the turbulent TKE flux.

Next, (22) conceals the fact that wake production removes energy not only from the mean flow but also from the large-scale (shear) turbulence, converting both to small-scale wake TKE which is quickly dissipated by viscosity. The conversion rate of shear TKE to wake TKE by this route is approximately the rate of working of the fluctuating part of the flow against canopy drag (recall that P_w, the conversion rate of MKE to wake TKE, is approximately the rate of working of the mean flow against drag). The result is a sink for large-scale, shear TKE and an equal and opposite source of small-scale, wake TKE, so no terms representing this process appear in (22) which is integrated over all scales of turbulence. However, the effect is to short-circuit the normal eddy cascade process by which large-scale TKE is handed down through a sequence of successively smaller eddies before being dissipated as heat. The dissipation of canopy turbulence is therefore much more rapid than for free turbulence with similar velocity and length scales.

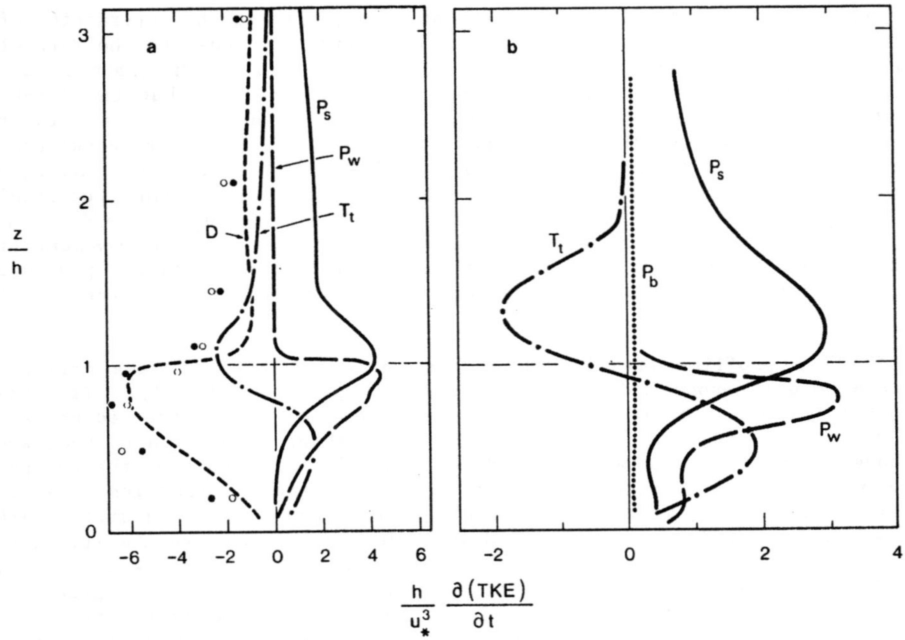

Fig. 9. Terms in the TKE budget, (22), for (a) the "strips" wind-tunnel model canopy (h = 60 mm) and (b) the Moga forest canopy (h = 12 m, $\bar{u} \approx 3.5$ m s^{-1}, slightly unstable daytime conditions). In (a), curve D is the residual $- P_s - P_w - T_t$; points are values of D from inertial-subrange spectral levels (open: u spectrum; solid: w spectrum).

Lastly, the formal derivation of (22) produces two kinds of flux density: conventional eddy fluxes, which appear as covariances of temporal fluctuations, and also "dispersive fluxes", or spatial covariances of time-averaged quantities (Wilson and Shaw 1977; Raupach and Shaw 1982; Finnigan 1985). For example, the eddy and dispersive vertical flux densities of streamwise momentum are $\langle \overline{u'w'} \rangle$ and $\langle \bar{u}''\bar{w}'' \rangle$, respectively. Although dispersive fluxes arise from the spatial averaging operation, experiments have failed to produce evidence that they are significant in comparison with eddy fluxes. Indeed, several wind tunnel experiments have shown that dispersive fluxes are negligible near z = h over model canopies (Mulhearn 1978; Raupach et al. 1980; Raupach et al. 1986). Until the contrary is demonstrated, it seems safe to neglect dispersive fluxes.

Higher-Order Closure Models for Canopy Turbulence

K-theory is sometimes called "first-order closure" because unknown second moments, the eddy fluxes, are parameterized with a gradient-diffusion assumption in terms of gradients of first moments, as evident in the argument leading to (3). The principle of higher-order closure models is to replace the gradient-diffusion assumption with explicit equations for the second moments themselves, derived from the conservation equations by algebraic manipulation and averaging. The TKE equation, (22), offers an example. Because of the loss of information in the averaging process, equations for second moments contain unknown third moments, and so on. Hence, the need for parameterization is never eliminated, but merely shifted away from the first-moment equations; this is a manifestation of the closure problem of turbulence. The process of writing extra equations can be carried on indefinitely, producing equations of rapidly blossoming number and complexity, but it is usual to carry explicit equations only for first and second moments (a second-order model) or third moments as well (a third-order model). A turbulence length or time scale equation may also be included, usually in the form of an equation for the dissipation rate D. Since the skill of the model is largely dependent on the physical insight embodied in its parameterizing assumptions, it is far from automatic that invoking more equations implies higher skill. Over the last two decades, numerous higher-order closure models of many forms have been developed and applied in engineering (e.g. Launder and Spalding 1972) and meteorology (e.g. Donaldson 1973).

Recognising the inapplicability of K-theory in the canopy, Wilson and Shaw (1977) first applied these techniques to develop a second-order closure model for canopy flow. Since then, several different higher-order canopy flow models and applications have appeared (Periera and Shaw 1980; Albini 1981; Shaw and Periera 1982; Yamada 1982; Meyers and Paw 1986), while recently, Meyers and Paw (in preparation) have extended their model to include scalar transfer.

These models differ substantially in detail, but the structure of the equations they employ is common. Table 2 shows the nature of the terms appearing in each equation, together with the requirements for parameterization. The second-order turbulence models (Shaw and Periera 1982; Yamada 1982) carry equations for five wind moments (the mean, three variances and the uw covariance); the inclusion of scalar transfer at second order requires a mean concentration equation and one for the wc covariance, with additional equations for the $w\theta$ and θc covariances and the variances of θ and c if buoyancy is to be incorporated (Finnigan 1985). Third order models (Meyers and Paw 1986) carry additional equations for the third moments. The Albini (1981) model is a simplified second-order model based on the questionable assumption that all terms involving pressure fluctuations can be neglected.

Table 2. Structure of equations for first and second moments in a higher-order closure model for the momentum and scalar fields in a horizontally uniform canopy. Terms: P_s - shear production, P_w - wake production, P_b - buoyant production, T_t - turbulent transport, T_p - pressure transport, φ_p - pressure-gradient interaction, D - molecular dissipation, S - source-sink; $\overline{\theta}$ is potential temperature and c the concentration of another scalar; the $\overline{c}$, $\overline{w'c'}$ and $\overline{c'\theta'}$ equations are carried for each such scalar. Symbols: 0 - term known, P - parameterization required in both second and third order models, E - parameterization required at second order and extra equation carried at third order. No symbol implies that a term is either identically zero or negligible in practice.

Equation	Term P_s	P_w	P_b	T_t	T_p	φ_p	D	S
$\overline{u}$				0				0
$\overline{u'^2}$	0	0		E	P	P	P	
$\overline{v'^2}$		0		E	P	P	P	
$\overline{w'^2}$		0	0	E	P	P	P	
$\overline{u'w'}$	0		0	E	P	P		
$\overline{\theta}$				0				0
$\overline{w'\theta'}$	0		0	E	P	P		
$\overline{c}$				0				0
$\overline{w'c'}$	0		0	E	P	P		
$\overline{c'\theta'}$	0		0	E			P	

Some of the predictions of the higher-order closure models are very good, others less so. On the positive side, the early model of Wilson and Shaw (1977) gave predictions of the mean wind profile and the $\overline{u'w'}$ profile in excellent agreement with data from a corn canopy. The model also predicted profiles of terms in the TKE budget, (22), in qualitative agreement with the measurements in Fig. 9 (Wilson and Shaw did not have TKE budget data for their corn canopy). Success at modelling observed $\bar{u}$ and $\overline{u'w'}$ profiles was also obtained by Meyers and Paw (1986) with a third-order model which they tested against measurements from a number of canopies. On the negative side, though, the models tend to overpredict the measured variances of u, v and w, especially near z = h where a spurious peak in variance is evident in both the second-order (Wilson and Shaw) and third-order (Meyers and Paw) predictions. Surprisingly, the overprediction is slightly worse for the third-order than for the second-order model, perhaps indicating that increasing complexity does not necessarily lead to increasing skill.

Finally, we mention a connection between the Eulerian and Lagrangian approaches. In a higher-order closure model, the transport terms T_t are usually parameterized at the highest order of the model with a gradient-diffusion assumption. For example, the term T_t in (22) would be parameterized by expressing the turbulent flux of TKE in terms of the vertical TKE gradient:

$$\overline{w'u_i'u_i'}/2 = -K\, d(\overline{u_i'u_i'}/2)/dz \quad , \tag{23}$$

where K is an eddy diffusivity which is either directly specified externally, or modelled as UL where the velocity and length scales U and L are derived from other equations. Deardorff (1978) noted that the validity of this type of assumption can be checked directly for the special case of scalar transfer from an instantaneous point source in homogeneous turbulence, since an essentially exact analytic Lagrangian solution of the concentration field is available. For dispersion from an instantaneous source in one dimension in homogeneous turbulence, Deardorff showed that the Lagrangian solution implies that:

$$\overline{w'c'} = -K_e(t)\, d\bar{c}/dz \quad , \tag{24a}$$

$$\overline{w'^2c'} = -K_e(t)\, d\overline{w'c'}/dz \quad , \tag{24b}$$

$$\overline{w'^3c'} = 3\overline{w'^2}\;\overline{w'c'} - K_e(t) d\overline{w'^2c'}/dz \quad , \tag{24c}$$

where $K_e(t)$ is the effective diffusivity defined by (12) and t is the travel time (time since release) of the scalar cloud. The first of these is simply the first-order flux-gradient relationship incorporating a time- dependent effective diffusivity, constant in the far field ($t \gg T_L$) and attenuating toward 0 in the near field as t approaches 0; this is the cause of Taylor's "illogical conception" and the countergradient flux phenomenon in a canopy, as discussed in the previous section. The other two of (24) show that the "illogical conception" is not restricted to first order but appears at second and third order as well, since the diffusivities required in gradient-diffusion models of the higher velocity-concentration moments depend on travel time just as at first order. This implies that, in a canopy where the scalar-bearing fluid particles at any point encompass the full gamut of travel times from near to far fields, the second and third order diffusivities do not depend only on the turbulence but also on the source density profile S(z), just as for the first-order diffusivity. Calculations with the homogeneous-turbulence model show that the extent of the S(z) dependence increases with the order of the diffusivity (Raupach 1987). Although these conclusions are based on an analysis of scalar transfer in homogeneous turbulence, the same physics operates on higher-order diffusivities in inhomogeneous turbulence and for the momentum field as well. Hence, while closure assumptions of the form of (23) may be defensible in the free atmosphere where near-field effects are not dominant, there are grounds for doubting their applicability in a canopy.

An alternative approach for third-order models is closure by means of the "quasi-Gaussian approximation", in which fourth moments are parameterized by neglecting the last (gradient-diffusion) term in (24c). However, the resulting lack of a diffusing mechanism can cause unrealistic oscillatory or wave-like behaviour in model predictions (Deardorff 1978) and appears not to be supported by observations in plant canopies (Shaw and Seginer 1987).

From a homogeneous-turbulence analysis, one may calculate terms in the $\overline{w'c'}$ budget. Figure 10 shows the results, together with measurements of the $\overline{w'c'}$ budget by Coppin et al. (1986) for an elevated plane scalar source in the "strips" wind-tunnel model canopy. Both the calculations and the measurements show the same dominant role for turbulent transport (T_t) as in the TKE budget (Fig. 9), with T_t redistributing $\overline{w'c'}$ downwards into the canopy from its region of strongest production near z = h. In the homogeneous-turbulence analysis, T_t is large because of the influence of near-field dispersion in the canopy, and not because of the gustiness and skewness of canopy turbulence, since the model turbulence is unskewed. In the TKE budget, on the other hand, the importance of T_t is directly related to skewness as noted in the discussion of Fig. 9. It seems reasonable to infer from the similarity between Figs. 9 and 10 that the same mechanism is responsible for the importance of T_t in both the TKE and $\overline{w'c'}$ budgets; namely, near-field influence arising from the fact that the length scales of canopy turbulence are of order h. This implies that the role of T_t and the skewness of canopy turbulence are both kinematic consequences of the near-field influence, and therefore ultimately of the length scale properties of the flow.

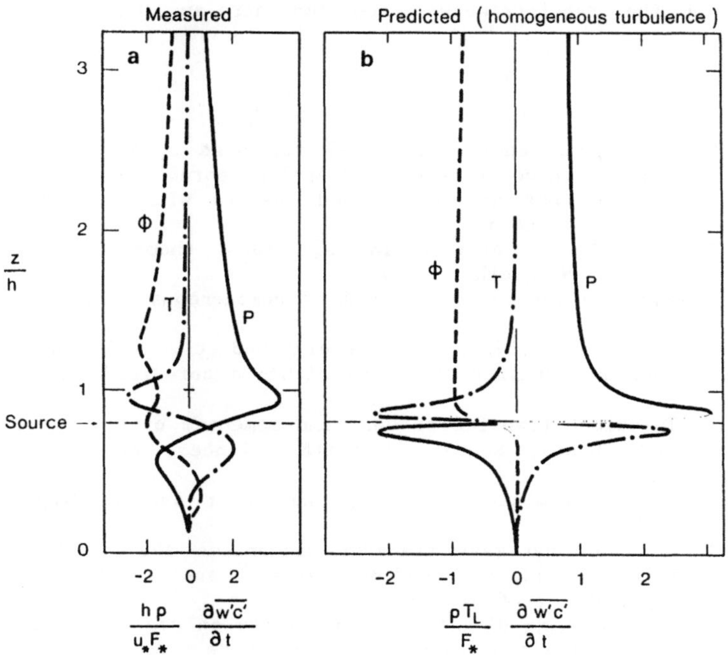

Fig. 10. Terms in the $\overline{w'c'}$ budget for an elevated plane scalar source at height z = 0.8h: (a) measurements by Coppin et al. (1986) and (b) predictions for homogeneous turbulence as in Fig. 5a. The budget is $\partial\overline{w'c'}/\partial t = 0 = P + T + \varphi$ with production $P = -\overline{w'^2}d\bar{c}/dz$, transport $T = -d\overline{w'^2c'}/dz$ and destruction $\varphi = -(1/\rho)\overline{c'\partial p'/\partial z}$.

CONCLUSIONS

We have concentrated on the basic but very important situation of a uniform canopy on level ground, reviewing knowledge of the structure of the turbulent flow in and above the canopy and the way in which the turbulence transfers scalars between the canopy and the atmosphere. Developments over the last two decades have led to an extensive database and a quite detailed understanding which can be expressed both in terms of simple paradigms and also predictive models for the canopy microclimate. Lagrangian and Eulerian models are both being developed rapidly and their continued improvement in the immediate future seems assured.

Looking further ahead and over a wider field, several thorny problems await systematic physical investigation. The problem of buoyancy effects in canopies has been mentioned several times; both comprehensive experimental data and theoretical work are required, the latter drawing from improvements over the last decade in knowledge of the stable atmospheric boundary layer. Another area ripe for development is that of advection, edge effects and changes in surface type. Again, the extension of current canopy modelling techniques to two and three dimensions of flow variation can draw on analytic and numerical techniques established for the atmospheric boundary layer. Finally, the most difficult future problem appears to be that of canopies in hilly terrain. In this volume, Finnigan (1988) discusses the current understanding of flow over hills, much of which is based on the idea that the underlying surface (usually a plant canopy of some kind) behaves on the hill as it does on level ground; in particular, the canopy on the hill is ascribed a "flat-earth" roughness length and drag coefficient. Work on the interaction of canopy and hill flows will help to refine this assumption, thus contributing to knowledge of larger-scale flows as well as the transfers between the atmosphere and the canopy itself.

REFERENCES

Albini FB (1981) A phenomenological model for wind speed and shear stress profiles in vegetation cover layers. J Appl Meteorol 20: 1325-1335

Allen LH (1968) Turbulence and wind speed spectra within a Japanese Larch plantation. J Appl Meteorol 7: 73-78

Arnold L (1974) Stochastic differential equations: theory and applications. Wiley-Interscience, New York, 228 pp

Batchelor GK (1949) Diffusion in a field of homogeneous turbulence. Aust J Sci Res 2: 437-450

Brown KW, Covey W (1966) The energy-budget evaluation of the micrometeorological transfer processes within a cornfield. Agric Meteorol 3: 73-96

Chen F (1987) Turbulent transfer of heat and momentum over a rough vegetated surface. PhD Thesis, School of Earth Sciences, Flinders University, South Australia

Coppin PA (1982) An examination of cup anemometer overspeeding. Meteorol Rdsch 35: 1-11

Coppin PA (1985) Heat and mass transfer mechanisms above and within plant canopies. Third Australasian Conf on Heat and Mass Transfer, University of Melbourne, May 1985, pp 465-472

Coppin PA, Raupach MR, Legg BJ (1986) Experiments on scalar dispersion within a plant canopy, part II: An elevated plane source. Boundary-Layer Meteorol 35: 167-191

Coppin PA, Taylor KJ (1983) A three-component sonic anemometer-thermometer system for general micrometeorological research. Boundary-Layer Meteorol 27: 27-42

Corrsin S (1963) Estimates of the relations between Eulerian and Lagrangian scales in large Reynolds number turbulence. J Atmos Sci 20: 115-119

Corrsin S (1974) Limitations of gradient transport models in random walks and in turbulence. Adv Geophys 18A: 25-60

Cowan IR (1968) Mass, heat and momentum exchange between stands of plants and their atmospheric environment. Q J R Meteorol Soc 95: 523-544

Crowther JM, Hutchings NJ (1985) Correlated vertical wind speeds in a spruce canopy. In: Hutchison BA, Hicks BB (eds) The forest-atmosphere interaction, D Reidel, Dordrecht, pp 543-561

Csanady GT (1973) Turbulent diffusion in the environment. D Reidel, Dordrecht, 248 pp

Deardorff JW (1978) Closure of second- and third-moment rate equations for diffusion in homogeneous turbulence. Phys Fluids 21: 525-530

Deardorff JW and Willis GE (1975) A parameterization of diffusion into the mixed layer. J Appl Meteorol 14: 1451-1457

de Baas AF, van Dop H, Nieuwstadt FTM (1986) An application of the Langevin equation for inhomogeneous conditions to dispersion in a convective boundary layer. Q J R Meteorol Soc 112: 165-180

Denmead OT (1964) Evaporation sources and apparent diffusivities in a forest canopy. J Appl Meteorol 3: 383-389

Denmead OT (1984) Plant physiological methods for studying evapotranspiration: problems of telling the forest from the trees. Agric Water Manage 8: 164-189

Denmead OT, Bradley EF (1985) Flux-gradient relationships in a forest canopy. In: Hutchison BA, Hicks BB (eds) The forest-atmosphere interaction, D Reidel, Dordrecht, pp 421-442

Denmead OT, Bradley EF (1987) On scalar transport in plant canopies. Irrig Sci 8: 131-149

Denmead OT, Freney JR, Simpson JR (1976) A closed ammonia cycle within a plant canopy. Soil Biol Biochem 8: 161-164

Denmead OT, McIlroy IC (1970) Measurements of non-potential evaporation from wheat. Agric Meteorol 7: 282-302

Donaldson CduP (1973) Construction of a dynamic model of the production of atmospheric turbulence and the dispersal of atmospheric pollutants. In: Haugen DA (ed) Workshop on micrometeorology, Am Meteorol Soc, Boston, pp 313-392

Druilhet A, Perrier A, Fontan J, Laurent JL (1971) Analysis of turbulent transfer in vegetation: use of thoron for measuring the diffusivity profiles. Boundary-Layer Meteorol 2: 173-187

Durbin PA (1983) Stochastic differential equations and turbulent dispersion. NASA reference publication 1103

Durbin PA (1984) Comments on papers by Wilson et al. (1981) and Legg and Raupach (1982). Boundary-Layer Meteorol 29: 409-411

Finnigan JJ (1979a) Turbulence in waving wheat. I. Mean statistics and honami. Boundary-Layer Meteorol 16: 181-211

Finnigan JJ (1979b) Turbulence in waving wheat. II. Structure of momentum transfer. Boundary-Layer Meteorol 16: 213-236

Finnigan JJ (1985) Turbulent transport in flexible plant canopies. In: Hutchison BA, Hicks BB (eds) The forest-atmosphere interaction, D Reidel, Dordrecht, pp 443-480

Finnigan JJ (1988) Air flow over complex terrain. This volume, pp

Finnigan JJ, Raupach MR (1987) Transfer processes in plant canopies in relation to stomatal characteristics. In: Zeiger E, Farquhar GD, Cowan IR (eds) Stomatal function, Stanford University Press, Stanford, CA, pp 385-429

Furnival GM, Waggonner PE, Reifsnyder WE (1975) Computing the energy budget of a leaf canopy with matrix algebra and numerical integration. Agric Meteorol 14: 405-416

Gardiner CW (1983) Handbook of stochastic methods for physics, chemistry and the natural sciences. Springer-Verlag, Berlin, 442 pp

Garratt JR (1978) Flux profile relations above tall vegetation. Q J R Meteorol Soc 104: 199-212

Garratt JR (1980) Surface influence upon vertical profiles in the atmospheric near-surface layer. Q J R Meteorol Soc 106: 803-819

Garratt JR, Hicks BB (1973) Momentum, heat and water vapour transfer to and from natural and artificial surfaces. Q J R Meteorol Soc 99: 680-687

Goudriaan J (1977) Crop micrometeorology: a simulation study. Pudoc, Wageningen

Goudriaan J, Waggoner PE (1972) Simulating both aerial microclimate and soil temperature from observations above the foliar canopy. Neth J Agric Sci 20: 104-124

Grant RH, Bertolin GE, Herrington LP (1986) The intermittent vertical heat flux over a spruce forest canopy. Boundary-Layer Meteorol 35: 317-330

Halldin S, Lindroth A (1986) Pine forest microclimate simulation using different diffusivities. Boundary-Layer Meteorol 35: 103-123

Hunt JCR, Weber AH (1979) A Lagrangian statistical analysis of diffusion from a ground-level source in a turbulent boundary layer. Q J R Meteorol Soc 105: 423-443

Inoue E (1955) Studies of the phenomena of waving plants ("Honami") caused by wind. Part I. Mechanism and characteristics of waving plants phenomena. J Agric Meteorol (Japan) 11: 18-22

Inoue E (1963) The environment of plant surfaces. In: Environment control of plant growth, Academic Press, New York, pp 23-32

Johnson CE, Biscoe PV, Clark JA, Littleton EJ (1976) Turbulent transfer in a barley canopy. Agric Meteorol 16: 17-35

Lamb RG (1980) Mathematical principles of turbulent diffusion modelling. In: Longhetto A (ed) Atmospheric boundary-layer physics, Elsevier, pp 173-210

Launder BE, Spalding DB (1972) Lectures in mathematical models of turbulence. Academic Press, London, 169 pp

Legg BJ (1975) Turbulent diffusion within a wheat canopy: I. Measurement using nitrous oxide. Q J R Meteorol Soc 101: 597-610

Legg BJ (1983) Turbulent diffusion from an elevated line source: Markov chain simulations of concentration and flux profiles. Q J R Meteorol Soc 109: 645-660

Legg BJ, Coppin PA, Raupach MR (1984) A three-hot-wire anemometer for measuring two velocity components in high-intensity turbulent boundary layers. J Phys E 17: 970-976

Legg BJ, Long IF (1975) Turbulent diffusion within a wheat canopy: II. Results and interpretation. Q J R Meteorol Soc 101: 611-628

Legg BJ, Raupach MR (1982) Markov-chain simulation of particle dispersion in inhomogeneous flows: the mean drift velocity induced by a gradient in Eulerian velocity variance. Boundary-Layer Meteorol 24: 3-13

Legg BJ, Raupach MR, Coppin PA (1986) Experiments on scalar dispersion within a plant canopy, part III: an elevated line source. Boundary-Layer Meteorol 35: 277-302

Lemon ER, Wright JL (1969) Photosynthesis under field conditions. XA. Assessing sources and sinks of carbon dioxide in a corn (Zea Mays L) crop using a momentum balance approach. Agron J 61: 405-411

Lumley JL, Panofsky HA (1964) The structure of atmospheric turbulence. Interscience, New York, 239 pp

McNaughton KG (1976) Evaporation and advection I: evaporation from extensive homogeneous surfaces. Q J R Meteorol Soc 102: 181-191

Maitani T (1978) On the downward transport of turbulent kinetic energy in the surface layer over plant canopies. Boundary-Layer Meteorol 14: 571-584

Maitani T (1979) A comparison of turbulence statistics in the surface layer over plant canopies with those over several other surfaces. Boundary-Layer Meteorol 17: 213-222

Maitani T, Seo T (1985) Estimates of velocity-pressure and velocity-pressure gradient interactions in the surface layer over plant canopies. Boundary-Layer Meteorol 33: 51-60

Meroney RN (1968) Characteristics of wind and turbulence in and above model forests. J Appl Meteorol 7: 780-788

Meroney RN (1970) Wind tunnel studies of the air flow and gaseous plume diffusion in the leading edge and downstream regions of a model forest. Atmos Environ 4: 597-614

Meyers T, Paw UKT (1986) Testing of a higher-order closure model for modelling air flow within and above plant canopies. Boundary-Layer Meteorol 37: 297-311

Monin AS, Yaglom AM (1971) Statistical fluid mechanics: mechanics of turbulence, Vol 1. Lumley JL (ed), MIT Press, Cambridge, Mass, 769 pp

Monteith JL (1965) Evaporation and environment. Symp Soc Exp Biol 19: 205-234

Monteith JL (1981) Evaporation and surface temperature. Q J R Meteorol Soc 107: 1-27

Mulhearn PJ (1978) Turbulent flow over a periodic rough surface. Phys Fluids 21: 1113-1115

Penman HL (1948) Natural evaporation from open water, bare soil and grass. Proc R Soc London, Ser A 193: 120-146

Penman HL, Long IF (1960) Weather in wheat: an essay in micrometeorology. Q J R Meteorol Soc 86: 16-50

Pereira AR, Shaw RH (1980) A numerical experiment on the mean wind structure inside canopies of vegetation. Agric Meteorol 22: 303-318

Philip JR (1964) Sources and transfer processes in the air layers occupied by vegetation. J Appl Meteorol 3: 390-395

Philip JR (1987) Issues in flow and transport in heterogeneous porous media. Transp Porous Media 1: 319-338

Raupach MR (1979) Anomalies in flux-gradient relationships over forest. Boundary-Layer Meteorol 16: 467-486

Raupach MR (1981) Conditional statistics of Reynolds stress in rough-wall and smooth-wall turbulent boundary layers. J Fluid Mech 108: 363-382

Raupach MR (1983) Near-field dispersion from instantaneous sources in the atmospheric surface layer. Boundary-Layer Meteorol 27: 105-113

Raupach MR (1987) A Lagrangian analysis of scalar transfer in vegetation canopies. Q J R Meteorol Soc 113: 107-120

Raupach MR (1988) Turbulent transfer in plant canopies. In: Russell G, Marshall B, Jarvis PG (eds) Plant canopies - their growth, form and function, Cambridge University Press, Cambridge, in press

Raupach MR, Coppin PA, Legg BJ (1986) Experiments on scalar dispersion within a plant canopy, part I: The turbulence structure. Boundary-Layer Meteorol 35: 21-52

Raupach MR, Shaw RH (1982) Averaging procedures for flow within vegetation canopies. Boundary-Layer Meteorol 22: 79-90

Raupach MR, Thom AS, Edwards I (1980) A wind tunnel study of turbulent flow close to regularly arrayed rough surfaces. Boundary-Layer Meteorol 18: 373-397

Sadeh WZ, Cermak JE, Kawatani T (1971) Flow over high roughness elements. Boundary-Layer Meteorol 1: 321-344

Sawford BL (1984) The basis for, and some limitations of, the Langevin equation in atmospheric relative dispersion modelling. Atmos Environ 18: 2405-2411

Sawford BL (1985) Lagrangian statistical simulation of concentration mean and fluctuation fields. J Climate Appl Meteorol 24: 1152-1166

Sawford BL (1986) Generalized random forcing in random-walk turbulent dispersion models. Phys Fluids 29: 3582-3585

Sawford BL, Guest FM (1987) Lagrangian stochastic analysis of flux-gradient relationships in the convective boundary layer. J Atmos Sci 44: 1152-1165

Seginer I, Mulhearn PJ, Bradley EF, Finnigan JJ (1976) Turbulent flow in a model plant canopy. Boundary-Layer Meteorol 10: 423-453

Shaw, RH (1977) Secondary wind speed maxima inside plant canopies. J Appl Meteorol 16: 514-521

Shaw RH, Kidd G, Thurtell GW (1973) A miniature three-dimensional anemometer for use within and above plant canopies. Boundary-Layer Meteorol 3: 359-380

Shaw RH, Pereira AR (1982) Aerodynamic roughness of a plant canopy: a numerical experiment. Agric Meteorol 26: 51-65

Shaw RJ, Seginer I (1987) Calculation of velocity skewness in real and artificial plant canopies. Boundary-Layer Meteorol 39: 315-332

Shaw RH, Silversides RH, Thurtell GW (1974) Some observations of turbulence and turbulent transport within and above plant canopies. Boundary-Layer Meteorol 5: 429-449

Shaw RH, Tavangar J, Ward DP (1983) Structure of the Reynolds stress in a canopy layer. J Climate Appl Meteorol 22: 1922-1931

Sigmon JT, Knoerr KR, Shaughnessy EJ (1983) Microscale pressure fluctuations in a mature deciduous forest. Boundary-Layer Meteorol 27: 345-358

Snyder WH, Lumley JL (1971) Some measurements of particle velocity autocorrelation functions in a turbulent flow. J Fluid Mech 48: 41-71

Taylor GI (1921) Diffusion by continuous movements. Proc London Math Soc A20: 196-211

Taylor GI (1959) The present position in the theory of turbulent diffusion. Adv Geophys 6: 101-112

Tennekes H, Lumley JL (1972) A first course in turbulence. MIT Press, Cambridge, Mass, 300 pp

Thom AS (1968) The exchange of momentum, mass and heat between an artificial leaf and airflow in a wind tunnel. Q J R Meteorol Soc 94: 44-55

Thom AS (1971) Momentum absorption by vegetation. Q J R Meteorol Soc 97: 414-428

Thom AS (1972) Momentum, mass and heat exchange of vegetation. Q J R Meteorol Soc 98: 124-134

Thom AS (1975) Momentum, mass and heat exchange of plant communities. In: Monteith JL (ed) Vegetation and the atmosphere, Vol 1, Academic Press, London, pp 57-109

Thom AS, Stewart JB, Oliver HR, Gash JHC (1975) Comparison of aerodynamic and energy budget estimates of fluxes over a pine forest. Q J R Meteorol Soc 101: 93-105

Thomson DJ (1984) Random walk modelling of diffusion in inhomogeneous turbulence. Q J R Meteorol Soc 110: 1107-1120

Thomson DJ (1987) Criteria for the selection of stochastic models of particle trajectories in turbulent flows. J Fluid Mech 180: 529-556

Townsend AA (1976) The structure of turbulent shear flow. 2nd Ed, Cambridge University Press, 429 pp

Uchijima Z (1962) Studies on the microclimate within the plant communities: 1. On the turbulent transfer coefficient within plant layers. J Agric Meteorol (Tokyo) 18: 1-10

Uchijima Z, Wright JL (1964) An experimental study of air flow in a corn plant-air layer. Bull Nat Inst Agric Sci (Japan) A11: 19-65

van Dop H, Nieuwstadt FTM, Hunt JCR (1985) Random walk models for particle displacements in inhomogeneous unsteady turbulent flows. Phys Fluids 28: 1639-1653

van Kampen NG (1981) Stochastic processes in physics and chemistry. North-Holland, Amsterdam, 419 pp

Waggoner PE, Furnival GM, Reifsnyder WE (1969) Simulation of the microclimate in a forest. For Sci 15: 37-45

Waggoner PE, Reifsnyder WE (1968) Simulation of temperature, humidity and evaporation profiles in a leaf canopy. J Appl Meteorol 7: 400-409

Weil JC (1988) Atmospheric dispersion - observations and models. This volume, pp

Wilson JD, Legg BJ, Thomson DJ (1983) Calculation of particle trajectories in the presence of a gradient in turbulent velocity variance. Boundary-Layer Meteorol 27: 163-169

Wilson JD, Thurtell GW, Kidd GE (1981a) Numerical simulation of particle trajectories in inhomogeneous turbulence, I: systems with constant turbulent velocity scale. Boundary-Layer Meteorol 21: 295-313

Wilson JD, Thurtell GW, Kidd GE (1981b) Numerical simulation of particle trajectories in inhomogeneous turbulence, II: systems with variable turbulent velocity scale. Boundary-Layer Meteorol 21: 423-441

Wilson JD, Thurtell GW, Kidd GE (1981c) Numerical simulation of particle trajectories in inhomogeneous turbulence, III: comparison of predictions with experimental data for the atmospheric surface layer. Boundary-Layer Meteorol 21: 443-463

Wilson JD, Ward DP, Thurtell GW, Kidd GE (1982) Statistics of atmospheric turbulence within and above a corn canopy. Boundary-Layer Meteorol 24: 495-519

Wilson NR, Shaw RH (1977) A higher-order closure model for canopy flow. J Appl Meteorol 16: 1198-1205

Wright JL, Brown KW (1967) Comparison of momentum and energy balance methods of computing vertical transfer within a crop. Agron J 59: 427-432

Yamada T (1982) A numerical model study of turbulent flow in and above a forest canopy. J Meteorol Soc (Japan) 60: 439-454

CANOPY TRANSPORT PROCESSES: COMMENTARY

G.W. Thurtell

Department of Land Resource Science, University of Guelph, Guelph, Ontario, N1G 2W1, Canada

Summary: The theory of canopy transport processes has progressed rapidly during the last few years and the inadequacy of K-theory, which is based on simple linear diffusion concepts, is now clear. Unfortunately the new concepts are more complex and remain difficult to exploit for the description of diffusion in field situations.

The development of readily usable canopy transport models is dependent on the availability of high quality data sets. The current increase in the number of canopy diffusion experiments is appropriate and continued progress is anticipated.

OVERVIEW

The theoretical development of first order, higher order, and Lagrangian canopy transport models is progressing more rapidly than field experimentation. Data sets suitable for using, testing and comparing these models remain rare, and the need exists for more comprehensive data sets as the models become more sophisticated and their inputs more varied. Fortunately, economically priced field data systems are now available which can not only store reliably very large quantities of raw data, but also perform the extensive on-line analyses so important for quality control. Unfortunately, the cost of sensors remains high.

A cooperative field experiment was performed at Borden, Ontario, in 1986/87. The participants included R.H. Shaw, University of California, Davis; G. den Hartog and H.H. Neumann, Atmospheric Environment Service, Canada; and G. Kidd, M.Y. Leclerc, G. Shi, and G.W. Thurtell, University of Guelph. A total of 10 three-dimensional sonic anemometers were mounted within and above the approximately 20 m high forest on two towers, the tallest of which was 43 m. Standard supporting micrometeorological data were also collected. Preliminary samples of the data are presented below.

Figure 1 (Leclerc 1987) shows examples of half-hour mean potential temperature profiles in the forest canopy. Two of these profiles show maxima within the canopy even though the heat fluxes were measured to be positive at all levels. Countergradient transport was a common occurrence within this forest. Figures 2 and 3 (RH Shaw, personal communication 1987) show profiles of turbulence intensities for different stability conditions where the stability classes were determined by the Monin-Obukhov length at the canopy top. While the large effects that stability has on canopy flow are clear, a suitable parameter for defining stability within the canopy remains to be developed. The complex relationships which exist between sources, fluxes and gradients complicate this problem.

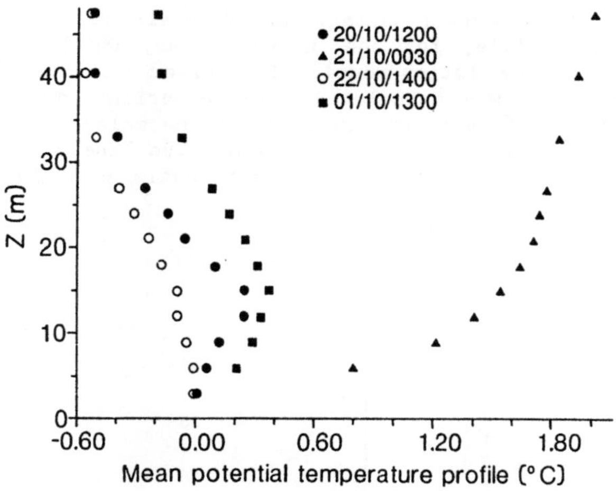

Fig. 1. Mean potential temperature (°C). (Leclerc 1987).

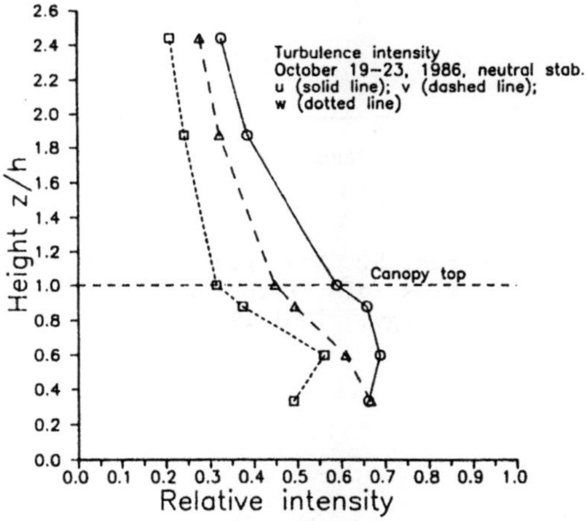

Fig. 2. The intensity of turbulence under neutral conditions (RH Shaw, personal communication 1987).

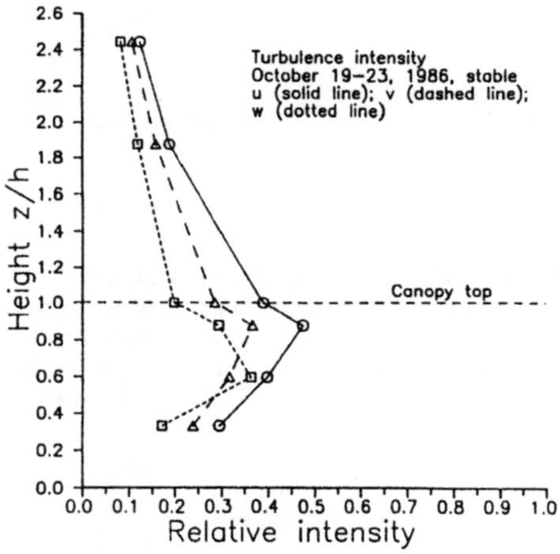

Fig. 3. The intensity of turbulence under stable conditions (RH Shaw, personal communication 1987).

The importance of near- and far-field effects to the diffusion of scalars within plant canopies has been clearly stated. However, I would like to indicate that these concepts (Taylor 1921) apply to a wide range of diffusion problems. Figure 4 shows the results of molecular diffusion occurring within a very small box (100 molecular diameters long) containing two gases which are diffusing in opposite directions (Thurtell and Thurtell 1987). The circles show the concentration profile, for one of the gases, which was calculated using a molecular dynamics simulation model. The curved line, which shows good agreement with the circles, was calculated using equations which include the concepts of Taylor. It illustrates the analogy, at the molecular scale, to the near- and far-field effects in turbulent diffusion. The linear solution to the familiar relationship $F = -K\,dc/dz$ would not satisfactorily model this transport process.

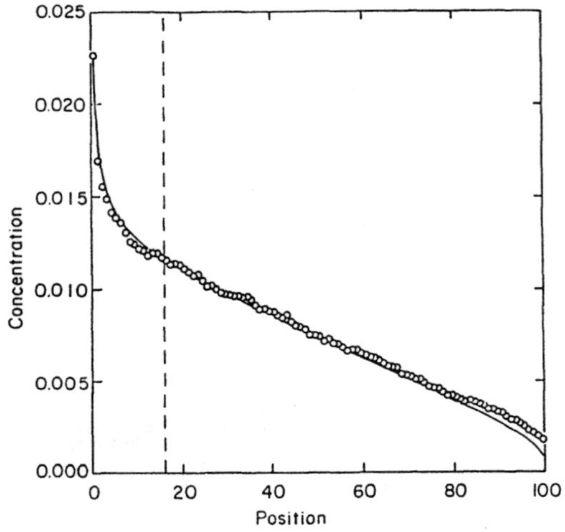

Fig. 4. The predicted (–) and simulated (o) concentration profiles for molecular diffusion. The vertical dashed line indicates a distance of one mean free path from the source. (Thurtell and Thurtell 1987).

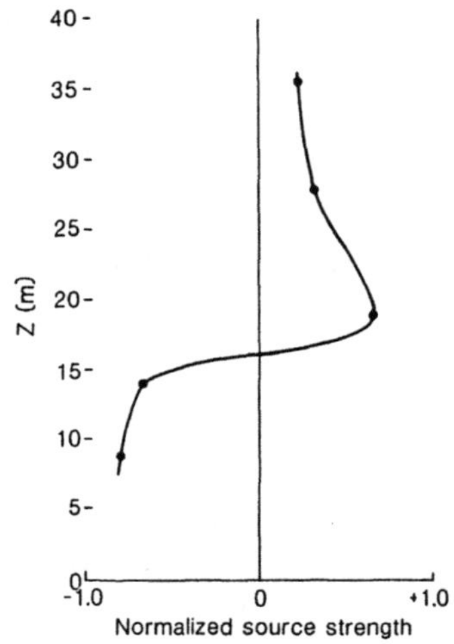

Fig. 5. The gradient of the flux of turbulent kinetic energy. (Shi 1987).

Figure 5 shows an example of the diffusion source strength (Shi 1987) of the turbulent kinetic energy, within and above the forest canopy. We can see that a diffusion source of turbulent kinetic energy existed near the canopy top but that at lower levels a diffusion sink occurred. Turbulent kinetic energy produced near the top of the canopy was transported to both higher and lower levels by the diffusion process. Thus, the near- and far-field concepts should be included when these and similar terms are modelled in higher order models.

ON COUNTERGRADIENT TRANSPORT

Peter Taylor (personal communication 1987) has referred to the use of the term "countergradient" in describing (heat) fluxes within plant canopies. He goes on to say that the atmospheric turbulence and micrometeorological communities know well what is meant and implied by this phrase, but some of our colleagues in soil physics or other disciplines may see this as a violation of the laws of thermodynamics. He asks if we can make it clear that the flux concerned is only counter to the *local* gradient and suggests that "nonlocal gradient transfer" might be a better description for the process.

I agree that the words "countergradient" do not generally describe the transfer process we have been discussing. Encountering theoretical problems when trying to relate turbulent fluxes to concentration gradients is very common, especially in plant canopies, while observing countergradient fluxes is less common. The fluxes only become countergradient in extreme cases, i.e., the same difficulties, although less obvious, exist even though the flux is in the familiar direction, but the solutions are not unique. As Peter Taylor indicates, this process is now widely accepted; however, this was not always the case and many attempts were made to measure counter-gradient canopy fluxes because this would provide proof of the inadequacy of K-theory (e.g., the Uriarra forest experiment, Denmead and Bradley (1985)) and the term has persisted. I am not sure that "nonlocal gradient transfer" is an improvement.

Brian Legg (personal communication 1987) observes that K-theory has been used to describe turbulent transport within crop canopies for 25 years and is still being used for applications that include the dispersion of spray droplets and the uptake of pollutant gases. He suggests that it has been easy and convenient to use, but we must now agree that the concept of diffusion in response to local concentration gradients is not valid and may cause serious errors. The full implementation of second order closure models or Lagrangian statistical models, however, is very complex and unwieldy. He asks if there are any alternative approaches that are convenient to use, but which provide a valid means of deriving concentration profiles from a knowledge of source distribution and vice versa.

In response to this, Mike Raupach notes that the demise of K-theory in canopies leaves two options for practical predictions. The first is to ignore the mixing process within the canopy entirely; this amounts to assuming a uniform scalar concentration within the canopy, as done by Thom (1972), or equivalently (in resistance terminology) to ignore the turbulent transfer resistance within the canopy in comparison with other contributions to the total aerodynamic resistance. Such a drastic simplification is permissible only in circumstances where the aerodynamic resistance is not a dominant influence on overall transfer; for example, when the stomatal resistance dominates. However, it is much simpler than K-theory and probably no worse an approximation.

His second option is a matrix formulation of the canopy transfer process for a scalar, with linearization of the source-to-atmosphere concentration relationship. This was done by Waggoner and others in the 70's with a K-theory transfer model (Waggoner 1975), but is readily adapted to any description of the transfer process, such as Lagrangian or random flight. Models of this type should become widely available in the next few years.

REFERENCES

Denmead OT, Bradley EF (1985) Flux-gradient relationships in a forest canopy. In: Hutchison BA, Hicks BB (eds) The forest-atmosphere interaction, D Reidel, Dordrecht, Netherlands, pp 421-443

Leclerc MY (1987) Turbulence and turbulent diffusion inside and above vegetation. PhD Thesis, Univ of Guelph

Shi G (1987) Studies of atmospheric turbulence within and above a deciduous forest. MSc Thesis, Univ of Guelph

Taylor GI (1921) Diffusion by continuous movements. Proc London Math Soc Ser 2 20: 196-211

Thom AS (1972) Momentum, mass and heat exchange of vegetation. Q J R Meteorol Soc 98: 124-134

Thurtell JH and GW Thurtell (1987) Adsorption and diffusion at rough surfaces: A comparison of statistical mechanics, molecular dynamics, and kinetic theory. J Chem Phys, submitted.

Waggoner PE (1975) Micrometeorological models. In: Monteith JL (ed) Vegetation and the atmosphere, Vol 1, principles, Academic Press, London, p 205

SURFACE TEMPERATURE AND THE SURFACE ENERGY BALANCE

Ray D. Jackson

U.S. Department of Agriculture, Agricultural Research Service, Phoenix, AZ 85040, U.S.A.

Summary. Surface temperatures, determined from measurements of emitted radiation, can be obtained at scales ranging from a few square millimetres to the global hemisphere. The ability to measure temperatures over large areas has led to the development of techniques for evaluating the surface energy balance at regional scales. In addition to surface temperatures, some techniques require inputs of surface-measured meteorological parameters. Others model the surface fluxes and use a comparison of predicted and measured surface temperatures to keep the models on track. Because of the large scale, validation of the models is difficult. On a smaller scale, it is possible to use both remotely-sensed data and ground-based data to evaluate the energy balance, with validation being somewhat easier. In this review, both regional- and local-scale techniques are discussed. An experiment in which remotely-derived results were compared with Bowen ratio data is described in detail. It is shown that the remote technique will produce adequate values of latent heat flux for uniform surfaces, but may yield erroneous values for heterogeneous surfaces such as partial canopies.

INTRODUCTION

The evaporation of water from the earth's surface is of profound importance because of its influence on atmospheric circulation and the water balance of the earth. Because of its ubiquitousness, evaporation processes (and thus the surface energy balance) have been studied by researchers representing a wide variety of scientific disciplines, at scales ranging from the global hemisphere to a few square millimetres of a plant leaf. The tools available to evaluate the components of the energy balance are different for different scales, and the accuracy to which the components can be determined may depend on scale. It is intuitively obvious that a method that measures some meteorological parameters over, say, a plant canopy, may well provide a reasonably accurate estimate of evaporation from that canopy, but would not necessarily be appropriate for evaluating evaporation over a large area of diverse, nonuniformly distributed vegetation.

The temperature of a surface from which water is evaporating tends to an equilibrium value when the loss of latent heat is balanced by the supply of heat through processes such as radiation, convection and conduction. It is conventional to consider the net flux density of radiation R_n as the main component of energy supply, balanced by losses of latent heat λE, sensible heat H, and storage in the substrate G (Monteith 1981). Thus, the heat balance of a surface can be written as

$$R_n = \lambda E + H + G , \tag{1}$$

where the latent heat is the product of the heat of vaporization λ and the rate of evaporation E. All terms in Eq. (1) have units of W m^{-2}.

Of the many methods by which evaporation from surfaces such as soil and vegetation can be estimated, eddy correlation methods (Garratt 1984; Kanemasu et al. 1979), and Bowen ratio techniques (Spittlehouse and Black 1980) are recognized as reliable ways to measure vertical flux density at a point. However, the instrumental and sampling problems involved in extending these techniques to give time- and spatially-averaged fluxes at large scales are formidable (Denmead and Bradley 1987).

In extending a measurement of evaporation beyond a point, a means of area averaging is required. Surface temperatures, inferred from measurements of emitted infrared radiation, can be averaged over a large range of scales, with the scale determined by instrument characteristics and distance from the surface. Thus, there have been numerous attempts to estimate evaporation over a wide range of scales using surface temperatures obtained from infrared measurements.

At present, radiometrically-measured surface temperatures offer a reasonably viable means of obtaining spatially-averaged fluxes over relatively large areas. The technique determines λE as the residual in Eq. (1) by evaluating R_n, G, and H, with H written as

$$H = \rho c_p(T_s - T_a)/r_a \ , \qquad (2)$$

where ρc_p is the volumetric heat capacity (J m^{-2} $^oC^{-1}$), T_s the surface temperature (^oC), T_a the air temperature (^oC), and r_a an aerodynamic resistance (s m^{-1}). The residual method can be applied at different scales, provided that a reliable measurement of area-averaged surface temperatures can be made and an accurate value of r_a be obtained (Garratt 1984). These provisos are often ignored; and attempts are made to estimate evaporation using questionable assumptions, under conditions where validation is difficult, if not impossible. It is easy to criticize these attempts, yet when one considers the importance of practical problems to which they are directed, the necessity of approximation becomes apparent.

This review examines efforts to evaluate components of the surface energy balance using radiometrically-determined surface temperatures as a major measured parameter at areal scales ranging from small research plots of a few square metres to large (400 × 400 km) grids of global circulation models (GCM's).

BACKGROUND

Although the earth's surface is a composite of soils, rocks, vegetation, water bodies, and man-made structures, a large majority of the reports reviewed were concerned with vegetated surfaces. These surfaces are perhaps the most complex from the point of view of evaluating the energy balance, and are very important from the practical point of view of evaluating water losses. It is appropriate to begin with a look at some early work concerning plant temperature measurements, to review some principles of radiometry, and to scrutinize some assumptions and mention some precautions concerning the use of radiometric measurements in evaluating the surface energy balance.

Surface Temperature of Vegetation

Nearly a century and a half ago, Rameaux (1843), as cited by Ehlers (1915), placed a number of leaves on top of one another and wrapped the stack around a mercury thermometer. This experiment, crude by today's standards, may have been the beginning of research concerning plant temperatures in relation to their environment. The literature on this topic is rather sparse until the 1920s. By this time thermocouples had become standard equipment, making detailed leaf temperature measurements possible. Miller and Saunders (1923) used thermocouples to measure the temperature of leaves of crop plants in Kansas under natural conditions. Their results showed that turgid leaves of most crops were essentially at air temperature, but alfalfa was consistently below air temperature by about 1°C. This was one of the first reports of leaf temperatures being less than air temperature. Further evidence of this effect was reported by Eaton and Belden (1929), who measured cotton leaf temperatures to be 2 to 4°C less than air temperature in the hot, dry climate of Arizona. Later, Wallace and Clum (1938) reported leaf temperatures as much as 7°C less than air. However, these results were not universally accepted. Curtis (1936, 1938) argued that transpirational cooling could not explain the results, but that erroneous air temperature measurements, radiative cooling, and other factors were the cause. Curtis's arguments were strong, and the view that leaves were always warmer than air prevailed for several decades, in spite of a number of reports to the contrary.

The work of Ehrler (1973) conclusively demonstrated that leaf temperatures could be cooler than the air temperature, and, in fact, were a function of the vapour pressure deficit of the air. In field experiments with cotton, he placed fine wire thermocouples in leaves and measured air temperature and vapour pressure at 1 m above the crop. His results showed a leaf-air temperature difference that ranged from −3° to +2°C, depending on the amount of soil water depletion. He also demonstrated experimentally that the leaf-air temperature difference was a linear function of the air vapour pressure deficit.

Remotely Sensed Temperatures

A decade before Ehrler's work with thermocouples, instruments for measuring emitted infrared radiation were adapted for field use. This development would foreshadow a revolution in the measurement of surface temperatures. Monteith and Szeicz (1962) and Tanner (1963) were among the first researchers to use ground-based infrared radiometers to measure plant temperatures. Infrared radiometers were also developed for mounting in aircraft (Dutton 1962; Lenschow and Dutton 1963). These devices provided spatial averages of surface temperature by recording numerous temperature measurements along a flightline. Later they would be replaced by scanning radiometers that produced images of surface temperature for large areas. Scanning radiometers were mounted in the early weather satellites and the feasibility of estimating the surface temperature of the earth from space was soon demonstrated (Rao and Winston 1963). The first of the Landsat series of earth observing satellites was launched in 1972; but it was not until 1984 that the capability for measuring surface temperature from these platforms was available. In 1977, a polar orbiting satellite was launched specifically for surface temperature measurement. This satellite, called the Heat Capacity Mapping Mission (HCMM) (Short and Stuart 1982), provided temperature data at 0.5 km resolution for about 2.5 years. Surface temperatures derived from the HCMM have been related to temperatures from standard meteorological sites (Byrne et al. 1984), and were used in frost mapping (Kalma et al. 1983).

Kinematic, Radiometric, and Aerodynamic Temperatures

Kinetic (absolute) temperature is a measure of the average kinetic energy of atomic and molecular units in motion within bodies above absolute zero. Radiometric (apparent) temperature is an estimate of the kinetic temperature of the surface of a body, as evaluated from a measurement of radiation emitted from the body. Radiometric temperatures are always less than kinetic temperatures by a factor ϵ, the emissivity, which differs for each material of which the body is composed. Kinetic temperatures and radiometric temperatures are related by the Stefan-Boltzmann Law, which states that the total radiation emittance integrated over all wavelengths depends on the fourth power of the absolute temperature of the body (Short and Stuart 1982). The relation is usually expressed as

$$R = \epsilon \sigma T_s^4 \ , \tag{3}$$

where R is the radiant exitance of the surface (W m^{-2}), ϵ the emissivity of the surface (unitless), σ the Stefan-Boltzmann constant (5.674×10^{-8} W m^{-2} K^{-4}), and T_s the surface (kinetic) temperature (°K). Equation (3) is extensively used in remote sensing of surface temperature. However, it is seldom noted that this equation applies *only if the radiant exitance is measured over the wavelength range of 0 to* ∞ (Slater 1980, p 241). Such a measurement is usually not practical and numerical methods must be used to find the temperature over a wavelength interval λ_1 to λ_2. The result is both a function of wavelength and wavelength interval. For a 2 μm wavelength interval centered on 6 μm, the exponent of T_s is about 7, instead of 4 as shown in Eq. (3), and the coefficient of T_s is no longer the Stefan-Boltzmann constant σ. The exponent is 4 only at about 11 or 12 μm. Thus, Eq. (3) holds reasonably well for the thermal channel on Landsat-5 and the HCMM (10.4-12.5 μm), but not as well for small wavelength intervals at the shorter wavelengths.

Aerodynamic temperature refers to a temperature within a vegetated surface corresponding to a particular height such that, when subtracted from a temperature at some height above the top of the vegetation, will yield the proper temperature difference from which the sensible heat transfer from the surface can be calculated (Monteith 1973, Huband and Monteith 1986). For vegetated surfaces, the aerodynamic temperature may be estimated from profiles of temperature and windspeed measured in the boundary layer. Huband and Monteith (1986) stated that the aerodynamic temperature and the radiometric temperature may not be the same for a crop canopy. Choudhury et al. (1986) found that the two temperatures were nearly the same for near-neutral conditions, but radiometric temperatures were higher than aerodynamic temperatures for stable conditions and lower for unstable conditions. The non-equivalence of these temperatures may cause errors in estimating the surface energy balance when remotely sensed radiometric temperatures are used.

Factors Affecting Radiometric Surface Temperature Measurements

Emissivity and sky radiant emittance: Surface temperature measurements are affected by the emissivity of the surface and by radiation emitted by the surroundings and then reflected from the target to the radiometer. The flux of radiation R reaching the sensor is (Fuchs and Tanner 1966)

$$R = \epsilon \sigma T_s^4 + (1 - \epsilon) B^* \ , \tag{4}$$

where B^* is the reflected radiation from the surroundings (W m^{-2}), with the other terms previously defined. Huband and Monteith (1986) reported B^* to range from 190 to 370 W m^{-2} depending mainly on cloud cover.

If $\epsilon = 1$ then B^* has no effect on R. If $\epsilon < 1$ (but correctly specified) and B^* is neglected (as is frequently done by assuming that $R = \epsilon \sigma T_s^4$), then T_s will be overestimated. If ϵ is erroneously assumed to be 1 when it is less than 1, then T_s will be underestimated. The error involved by assuming $\epsilon = 1$ is (Perrier 1971)

$$(T_s - T)/T = 1 - [\epsilon + (1 - \epsilon)B^*/\sigma T^4]^{\frac{1}{4}} , \qquad (5)$$

where T is the calculated apparent temperature obtained by assuming $\epsilon = 1$ and using $R = \sigma T^4$. If $B^* = R$ then $T = T_s$, regardless of the value of ϵ. The value of B^* is generally less than R and varies with sky conditions. The maximum error occurs when $B^* = 0$. The maximum difference between the true surface temperature T_s and the apparent temperature T is

$$T_s - T = T(1 - \epsilon^{\frac{1}{4}}) . \qquad (6)$$

Values of $T_s - T$ versus emissivity over the range from 0.9 to 1 are plotted in Fig. 1 for values of $B^* = 0$, R/2, and R. The abscissa forms the line for $B^* = R$. Values of B^* can be estimated using the relation

$$B^* = \epsilon_a \sigma T_a{}^4 , \qquad (7)$$

where ϵ_a is the sky emittance. Assuming that cloudless sky conditions prevail and that vegetation temperatures are nearly the same as air temperatures (i.e., $T_s \approx T_a$), then

$$B^*/R = \epsilon_a \sigma_a{}^4/\sigma T_s{}^4 \approx \epsilon_a . \qquad (8)$$

The wide-band sky emittance ϵ_a can be approximated by the relation (Brutsaert 1975)

$$\epsilon_a = 1.24(e_0/T_a)^{1/7} , \qquad (9)$$

where e_0 is the vapour pressure of air (in mb). However, infrared radiometers are filtered to receive radiation in restricted wavebands. Hand-held IR radiometers are usually filtered to the 8 to 14 μm bands. For example, the bandwidth of Thematic Mapper band 6 is 10.5 to 12.5 μm. Idso (1981) developed equations for the restricted wave bands from experimental data. He presented the equations

$$\epsilon_a(8\text{–}14\ \mu m) = 0.24 + 2.98 \times 10^{-8}\ e_0{}^2 \exp(3000/T_a) , \qquad (10)$$

and

$$\epsilon_a(10.5\text{–}12.5\ \mu m) = 0.10 + 3.53 \times 10^{-8}\ e_0{}^2 \exp(3000/T_a) , \qquad (11)$$

for two commonly used wavebands.

Evaluation of Eqs. (9)-(11) for 15 mb vapour pressure at $T_a = 303°K$ yields ϵ_a values of 0.807, 0.373, and 0.258, respectively. Using these values for B^*/R in Eq. (5) and calculating the percent reduction in error incurred by assuming $\epsilon = 1$ shows that the error is about 30% of the maximum shown in Fig. 1 for 10.5-12.5 μm, and about 40% of the maximum for 8-14 μm wavebands (dashed lines). All plant surfaces have a long-wave emittance of 0.95 or more, with most plant leaves being 0.97-0.98 (Gates 1964). At $\epsilon = 0.97$, the maximum error would be 2.4°C. By accounting for sky radiant emittance, the error would be about 1.7 and 1.4°C for the 10.5-12.5 μm and 8-14 μm wavebands respectively.

Atmospheric absorption: Absorption of thermal-IR radiation by atmospheric water vapour can cause serious errors in radiometric determination of surface temperature from aircraft or satellites. Kiang (1982) calculated the effect of increasing atmospheric water on the radiance received by the thermal channel of the Thematic Mapper (TM6) on Landsat-5 and showed that the radiance decreased by more than 20 percent as the precipitable water increased from 1 to 8 g/cm².

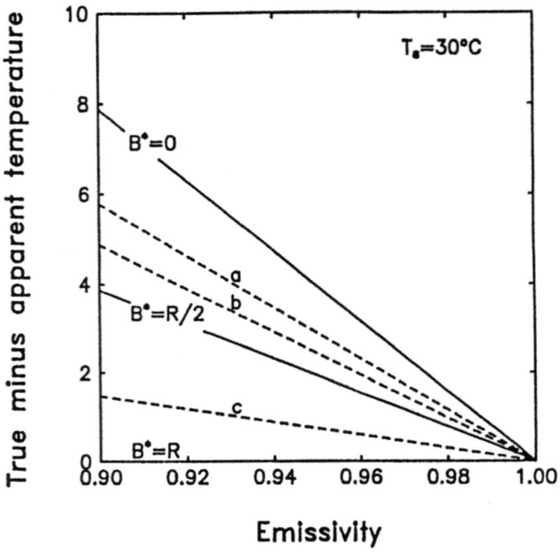

Fig. 1. True minus apparent temperature versus emissivity for several values of the sky radiance B*. Dashed lines indicate the error for a typical sky radiance and a surface temperature measured in (a) the 10.5-12.5 μm waveband; (b) the 8-14 μm waveband; and (c) for the full spectrum.

His results indicate that curves for different ground temperatures tend to converge as precipitable water increases, making temperature discrimination increasingly difficult.

Several radiative transfer models are available that provide corrections for the absorption effect if atmospheric temperature and moisture are known as a function of height, information usually obtained from radiosonde data. The LOWTRAN codes, developed at the Air Force Geophysics Laboratory, Bedford, MA (e.g. Kneizys et al. 1983) are widely used. However, a formulation recently presented by Price (1983) is computationally simpler. Price stated that estimates to within 2 to 3°C of the actual surface temperature could be expected for areas on the order of 100-300 km² if these corrections were made.

Lorenz (1968) stated that the influence of water vapour in the 8-14 μm waveband on surface temperature measurements was negligible if the distance to the target was less than 154 m. His results suggest that ground-based thermometric measurements will generally not need to be corrected for water-vapour absorption.

Radiometer calibration: Inadequate calibration of radiometers, whether used at ground level or from space, can cause serious errors in surface temperature estimations. Radiometers that measure reflected and emitted radiation within the 0.4-15 μm wavelength interval use detectors whose voltage response can be altered by the temperature of the detector (Slater 1980), thereby causing an erroneous estimate of target temperatures. To circumvent this problem, temperature compensation circuits may be included as an integral part of sensor configurations. However, adequate ambient temperature insensitivity is not always achieved. If the detector output is not adequately compensated for ambient temperature effects, radiometer calibration factors are of little value.

The ambient temperature problem is of most concern to small portable radiometers that rely on electronic compensation circuits. Most airborne scanners and satellite instruments use devices to control the detector at a specified temperature (frequently the temperature of liquid nitrogen).

However, calibration problems remain. Calibration factors may change with time, due to detector deterioration, contamination, or damage.

In-flight calibration factors for satellite sensors may be different from those determined prior to launch. Studies by Schott and Volchok (1985) indicated that the radiometric calibration of Landsat-5 TM band 6 appeared to have a systematic error in the gain. If this error were not accounted for, their data indicated that the predicted surface temperatures would have a root mean square error of 6°K. For the HCMM satellite, Barnes and Price (1980) determined that the calibration factors had changed after launch causing the satellite-derived surface temperature to be 5.2°C higher than the first calibration. The decision was made to offset the prelaunch calibration by –5.2°C. Later, Vukovich (1984) determined that the correction was no longer necessary and that 5.2°C should now be added to the HCMM surface temperatures to obtain the correct value. These reports underscore the need for adequate in-flight calibration procedures for updating of calibration factors.

EVALUATING THE SURFACE ENERGY BALANCE AT REGIONAL SCALES

Satellites provide the only viable means of obtaining measured data from which the surface energy balance can be evaluated at regional scales. However, satellite data alone are not sufficient because not all of the necessary parameters can be measured from space. Becker et al. (1987) listed a number of parameters that could and could not be evaluated in this manner. They suggested that surface temperature, global solar flux, surface albedo, and downward longwave flux can be inferred from satellite radiances, although some problems in their determination remain. Air temperature, windspeed, and water vapour pressure at a reference altitude cannot, as yet, be obtained remotely. Thus, the surface energy balance must be evaluated by combining remotely sensed data with models that predict the surface fluxes, used in conjunction with ground-based measurements.

Methods for assessing λE at regional scales include the analytical approach of Price (1980, 1982), a modelling approach used by a number of authors (e.g. Carlson and Boland 1978; Rosema et al. 1978; Soer 1980; Carlson et al. 1981; Gurney and Hall 1983; Camillo et al. 1983; Gurney and Camillo 1984; Sellers et al. 1986), and use of instantaneous midday surface temperatures with daily totals of net radiation in an empirical relation to estimate daily totals of evaporation (Jackson et al. 1977; Seguin and Itier 1983).

Analytical Approach

Price (1980, 1982) assumed that soil heat flux (G) was negligible and formed 24 h averages of the remaining terms in Eq. (1) to express the latent heat flux as

$$\begin{aligned}\langle\lambda E\rangle = SV(1-\alpha)c_0 + \epsilon_a\sigma\langle T^4\rangle - \epsilon_s\sigma[\langle T_s{}^4\rangle + 0.75\langle T_s{}^2\rangle(T_{max} - T_{min})^2] \\ - \rho c_p(\langle T_s\rangle - \langle T_a\rangle)/r_a\ ,\end{aligned} \tag{12}$$

where S is the solar constant (1375 W m^{-2}), V the atmospheric transmittance, α the surface albedo, c_0 the Fourier expansion of the angle of solar incidence, and T_{max} and T_{min} are the maximum and minimum air temperatures. In all cases, the angle brackets < > represented a 24 h average. Price used $\epsilon_a = 0.66 + 0.039e_0^{\frac{1}{2}}$ as given by Brunt (1932) with the 24 h average of e_0 to calculate incoming longwave radiation.

In Eq. (12), the quantities $<T_s>$, T_{max}, T_{min}, and α may be estimated from remote observations, with at least two measurements of surface temperature being required during the diurnal cycle. The quantities $<u>$, $<e_0>$, $<T_a>$, and V may be obtained from meteorological data from a nearby weather station, and ϵ_s and z_0 must be estimated from some knowledge of surface characteristics. The meteorological variables must be assumed constant over the area of interest, which may be on the order of 200 km².

Price (1982) calculated evapotranspiration (ET) for an HCMM scene of southeastern Washington and northern Oregon. His technique produce ET values of 10 mm/day for the irrigated areas, which were clearly distinguished from the drier surroundings. The validity of satellite-derived ET values was tested by calculating potential ET with Penman's equation, using local meteorological data as inputs. The calculated potential ET of 8.8 mm/day and some pan evaporation data compared favourably with the satellite data. Price (1982) concluded that satellite data may not provide highly accurate values of ET but such information is useful for assessing ET over large areas and for interpolation between points where surface values are known. He stressed that coordinated ground/satellite experiments are needed to evaluate the ultimate benefits and the limitations of the analytical approach.

Modelling Approach

The modelling approach uses techniques that simulate the simultaneous flow of heat and water in soil, and the surface energy balance relation is given by Eq. (1). A numerical solution of the soil heat flow equation

$$\rho c\frac{\partial T}{\partial t} = \frac{\partial}{\partial z}\left[K\frac{\partial T}{\partial z}\right] \tag{13}$$

is required. Here ρc and K are the volumetric heat capacity and thermal conductivity of soil, respectively. The thermal conductivity is frequently assumed constant but it may vary with z and/or with soil-water content.

Despite some serious difficulties when dealing with vegetated surfaces (see later), a common approach is to solve Eqs. (1) and (13) repeatedly for various sets of values of parameters such as albedo, surface roughness, soil thermal conductivity, and surface humidity. Calculated values of the surface temperature at the times of satellite observations are saved in a multidimensional array. Satellite-measured surface temperatures are entered in the array and the appropriate values of albedo, surface roughness, etc., are obtained, from which cloudiness, which affects the net radiation, are obtained from local weather station data. The thermal conductivity is frequently computed using a method such as described by de Vries (1963).

An advantage of models of this type is that, since remotely sensed data are likely to be available only once or twice a day, model parameters fitted at these times can then be used to simulate fluxes at all times of day. Thus daily means and totals, which are of most use, can be easily obtained. A disadvantage is that some soil and aerodynamic parameters are difficult to estimate. Furthermore, they may vary in time and space, thereby limiting the utility of model results.

With adequately approximated parameters and a reliable time series of meteorological data, models can provide reasonable estimates of daily ET. Gurney and Hall (1983) estimated ET rates of 2 to 5 mm/day in the Alaskan sub-Arctic using a model calibrated against HCMM surface temperatures. They reported good agreement with ground measurements of ET. Gurney and Camillo (1984) reported model estimates of ET using a once daily estimate of surface temperature to be within the 95-percent error bounds of eddy correlation measurements for day-time periods near Hanover, Federal Republic of Germany.

Gurney and Camillo (1984) pointed out that a possible source of error in modelling surface fluxes is an inappropriate definition of surface characteristics. Many models define the surface to be that of the soil, essentially ignoring any vegetation present. However, remotely sensed temperatures are affected by the relative proportions of soil and vegetation, and the assumption is made that fitting the estimated surface temperature to this composite value gives accurate ET estimates. Soil, when not completely covered by vegetation, may be considerably warmer than transpiring vegetation. The resulting composite temperature measured by the remote sensor would be higher than for the evaporating surfaces, causing an inappropriate fit between modelled and measured temperatures. In such cases, erroneous values of ET may result.

General Circulation Models and the Surface Energy Balance

The energy balance at the earth's surface is a primary process of global atmospheric ciculation. General circulation models (GCM's) require determinations of the fluxes of radiation, water vapour, sensible heat and momentum across the lower boundary of the atmosphere. Evapotranspiration models based on the surface energy balance utilize boundary conditions at a distance above the surface within the atmospheric boundary layer. Sellers et al. (1986) developed a Simple Biosphere Model (SiB) to provide a bridge between the 100-400 km grid GCM's and the smaller scale surface energy balance models.

The SiB model uses morphological and physiological characteristics of the vegetation community within a GCM grid area to derive coefficients and resistances which govern the momentum, radiation, and sensible and latent heat fluxes between the surface and the atmosphere. The world's vegetation is divided into two morphological groups: perennial trees and shrubs which constitute the upper story or canopy vegetation, and the ground cover consisting of annual grasses and other herbaceous plants. Varying amounts of trees, shrubs, and ground cover may co-exist within a given grid area. Physical aspects of the canopy are described by a temperature for the upper story canopy and another temperature for the ground cover. Two terms represent water intercepted by the canopy and by the ground cover. Soil moisture is stored in three layers. Radiation fluxes are modelled using a two-stream approximation which allows for the multiple reflection of light by leaves. The transfer of heat and water vapour from the surface to the atmospheric boundary layer is characterized by three surface resistances and three aerodynamic resistances. The model produces area-averaged sensible and latent heat fluxes.

It was anticipated that the SiB could be used within GCM's for sensitivity studies, and for climate simulations when the distribution of vegetation type and soil moisture were known (Sellers et al. 1986). However, the value of this model is limited because of the large number of parameters that must be specified. It cannot be used in a predictive mode.

Estimating Evapotranspiration using Midday Surface Temperatures

Jackson et al. (1977) calculated daily values of evapotranspiration using one-time-of-day measurements of $T_c - T_a$, and daily values of net radiation. They combined Eqs. (1) and (2), assumed $G = 0$, and wrote

$$(\lambda E - R_n)_d = -B(T_c - T_a)_i \;, \tag{14}$$

where the subscripts d and i signify daily and instantaneous values, respectively, and B is a constant. Considering B a constant implies that the aerodynamic resistance is constant, and that Eq. (14) is not affected by wind.

Differences between daily values of λE (measured with weighing lysimeters) and R_n were plotted versus $T_c - T_a$ (Fig. 2). Surface temperatures were measured between 1340 and 1400 h using hand-held infrared thermometers. Air temperatures, measured at 1.5 m above the surface, were averaged over the 20 min period. The data in Fig. 2 indicate that the sensible heat H increases as $T_c - T_a$ increases. The scatter of data points can be attributed, at least in part, to the implicit assumption of a constant aerodynamic resistance.

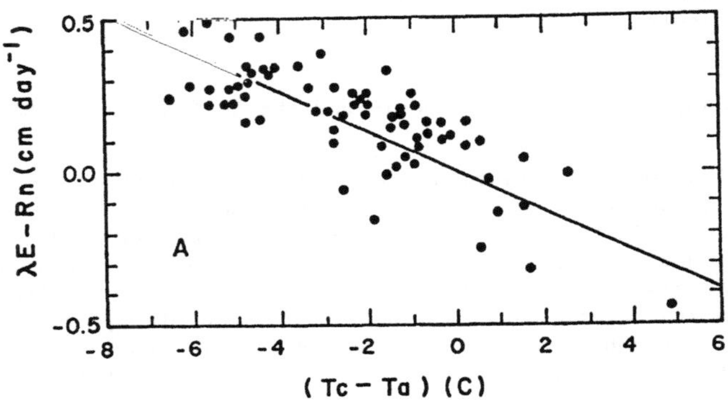

Fig. 2. The difference between evapotranspiration and net radiation as a function of canopy-air temperature difference.

Seguin and Itier (1983) applied Eq. (14) to satellite-derived values of T_s. They found B to be different from the value reported by Jackson et al. (1977). The difference can be attributed in part to the fact that the ground-based measurements of T_c were made at an angle of about 30° from horizontal, whereas the satellite measurements were essentially vertical. Seguin and Itier (1983) discussed the theoretical aspects of assuming the exchange coefficient as a statistically-derived constant B. They concluded that modest precision in λE estimations could be obtained if the daily values were summed over two- to four-week periods, and that the different values of B be used for stable and unstable cases.

Seguin and coworkers (e.g., Assad et al. 1986; Seguin et al. 1986) have used this method to estimate λE in southern France and Senegal using satellite data. Their results show essentially the same degree of scatter as the ground-based data (Fig. 2). The simplicity of the method is appealing but further validation is necessary.

EVALUATING THE SURFACE ENERGY BALANCE AT LOCAL SCALES

At sub-regional, or local, scales (a few square kilometres or less), the energy balance can be evaluated by combining remotely sensed and ground-based data. Parameters such as air temperature and windspeed need not be modelled but can be directly measured. Values of the surface fluxes measured in this manner can be directly compared with fluxes measured by standard ground-based techniques. A disadvantage, however, is that the remote measurements are made at an instant in time, and ways must be developed to extrapolate the instantaneous data to daily values. This method is best described by examining the R_n, G, and H terms of Eq. (1).

Net Radiation

The net radiation can be expressed as the sum of incoming and outgoing radiant fluxes, i.e.,

$$R_n = R_{S\downarrow} - R_{S\uparrow} + R_{L\downarrow} - R_{L\uparrow} , \tag{15}$$

where the subscripts S and L signify solar (shortwave) radiation (0.15 to 4 μm) and longwave radiation (>4 μm) respectively. The arrows indicate the flux direction.

Jackson et al. (1985) argued that the incoming terms ($R_{S\downarrow}$ and $R_{L\downarrow}$) could be measured with traditional ground-based instruments and the data extrapolated radially for some distance from the point of measurement, and that the outgoing terms ($R_{S\uparrow}$ and $R_{L\uparrow}$) could be obtained from multispectral data. They proposed that $R_{S\downarrow}$ be measured with a calibrated pyranometer sensitive to radiation over most of the solar spectrum. The area over which a point measurement of $R_{S\downarrow}$ could be extrapolated would be governed largely by the areal uniformity of the scattering and absorbing properties of the atmosphere.

Only part of the reflected solar term, $R_{S\uparrow}$, is measured with a multispectral radiometer. Because each radiometer measures a different fraction of the total reflected radiation, the total must be estimated from the partial. Jackson (1984) used a radiative transfer model to calculate the irradiance at the earth's surface for a number of atmospheric scattering and absorption conditions. Measured spectral reflectance distributions for 14 different surface conditions (bare soil to full green canopy) were used along with calculated irradiance data to determine the ratio of the radiation measured by a multispectral radiometer to the total reflected solar radiation (the P/T ratio).

The incoming longwave radiation, $R_{L\downarrow}$, can be estimated from ground-based measurements of air temperature and vapour pressure using the relation $R_{L\downarrow} = \epsilon_a \sigma T_a^4$, where the sky emittance $\epsilon_a = 1.24(e_0/T_a)^{1/7}$ (Eq. 9), and σ is the Stefan-Boltzmann constant. The outgoing longwave radiation, $R_{L\uparrow}$, is obtained from the remotely measured surface temperature, $R_{L\uparrow} = \epsilon_s \sigma T_s^4$, where ϵ_s is surface emissivity, and T_s is surface temperature (^oK). (We have ignored the small contribution to $R_{L\uparrow}$ by reflected sky radiation.) At this point, it is not necessary to evaluate ϵ_s. If the thermal radiometer is calibrated with reference to a blackbody whose emissivity is near unity, then the surface emissivities would be implicit in the apparent surface temperatures and would cancel when the emitted longwave energy is calculated (Jackson et al. 1985).

Soil Heat Flux

Soil heat flux is traditionally measured with sensors buried just beneath the soil surface. This flux is dependent on whether the surface is wet or dry, bare or vegetated, and therefore, a ground-based measurement should not be extrapolated areally. G may be as high as $0.5R_n$ for bare soil, decreasing to about $0.1R_n$ for full vegetative cover. A remote measurement of G is not possible, but several relationships between G and R_n have been proposed (Clothier et al. 1986). Jackson et al. (1987) used the relation

$$G/R_n = 0.583e^{-2.13ND} , \tag{16}$$

where ND is the normalized difference [(IR-red)/(IR+red)], a spectral index that estimates the amount of vegetation present.

Sensible Heat Flux

The final term to be evaluated is the sensible heat flux H, which is given by Eq. (2), i.e. $H = \rho c_p(T_s - T_a)/r_a$, where r_a is a stability-corrected aerodynamic resistance (adapted from Mahrt and Ek 1984)

$$r_a = \{\ell n[(z-d+z_0)/z_0]/k\}^2(1+15Ri)(1+5Ri)^{\frac{1}{2}}/u \, , \tag{17}$$

for the stable case $[(T_s - T_a) < 0]$, and

$$r_a = \{\ell n[(z-d+z_0)/z_0]/k\}^2\{1-15Ri/[1+C(-Ri)^{\frac{1}{2}}]\}^{-1}/u \, , \tag{18}$$

for the unstable case $[(T_s - T_a) > 0]$. In (17) and (18), Ri is the Richardson number $[Ri = g(T_a - T_s)(z - d)/T_a u^2]$, g the acceleration due to gravity ($m\ s^{-2}$), z the height (m) at which the windspeed and air temperature are measured, d the displacement height (m), z_0 the roughness length (m), k von Karman's constant (0.4), and $C = 75k^2[(z-d+z_0)/z_0]^{\frac{1}{2}}/\{\ell n[(z-d+z_0)/z_0]\}^2$.

Estimation of Daily Totals of λE from Instanteous Values

Jackson et al. (1983) assumed that the daily course of λE would generally follow the trend of solar radiation throughout the daylight period. They showed that, for cloudless skies, the ratio of total daily solar radiation R_d to an instantaneous value R_i could be approximated by the relation

$$R_d/R_i = 2N[\pi \sin(\pi t/N)] \, , \tag{19}$$

where t is the time (starting at sunrise) and N is the daylength (in hours). With the assumption that $\lambda E_d/\lambda E_i = R_d/R_i$, Eq. (19) allows a factor to be calculated to convert to daily totals from instantaneous values of λE for the time of day that the instantaneous measurement was made. An additional assumption was that environmental factors such as windspeed be relatively constant over the daily period.

Comparison of "Remote" Evaporation Estimates and Bowen Ratio Measurements

A weakness inherent in most remote sensing methods for evaluating the surface energy balance is the difficulty of adequately validating the results. On a local scale, validation is somewhat easier in that it is possible to make both remote and ground-based measurements applicable to the same area at the same time. A validation experiment recently reported by Jackson et al. (1987) demonstrated the potential and some problems of remote measurements for evaluating the surface energy balance. The following discussion is based on their results.

Jackson et al. (1987) mounted a four-band multispectral radiometer, a single band thermal infrared thermometer, and a video camera in a small aircraft and obtained multispectral data over 11 fields of a 770 ha farm on several days during 1985 and 1986. Data reported here are limited to the dates when Bowen ratio data (Gay and Greenberg 1985) were taken and to fields in which the Bowen ratio equipment was located. These dates were day of year (DOY) 85204 (23 July 1985), 85236 (24 August 1985), 86095 (5 April 1986), 86111 (21 April 1986), and 86175 (24 June 1986). The fields were planted to cotton in 1985 and to wheat for the first two dates in 1986. On DOY 86175, the Bowen ratio equipment was located in an alfalfa field whose dimensions were one half (0.14 by 0.8 km) those of the fields in which the other crops were grown.

Net radiation, soil heat flux, and sensible heat flux were calculated using a combination of remotely-sensed data and ground-based data from a portable micrometeorological station located in the same field as the Bowen ratio equipment. Instantaneous latent heat values were calculated using Eqs. (1), (2), (17), and (18). Daily evaporation rates λE_d were measured with the Bowen ratio system by summing the 12-min values, and estimated from the remote instantaneous values using the method described by Jackson et al. (1983).

Instantaneous values of evaporation λE_i, net radiation R_n, sensible heat flux H, and soil heat flux G for three crops are shown in Fig. 3. The flights to collect the remote data were made near 1030 MST on five days. However, on day 86175, flights were made at 6 times within a 4 hour period during the same day. These data are shown in Fig. 4. The solid square symbol in Figs. 3 and 4 refers to the same data point (1037 h on DOY 86175).

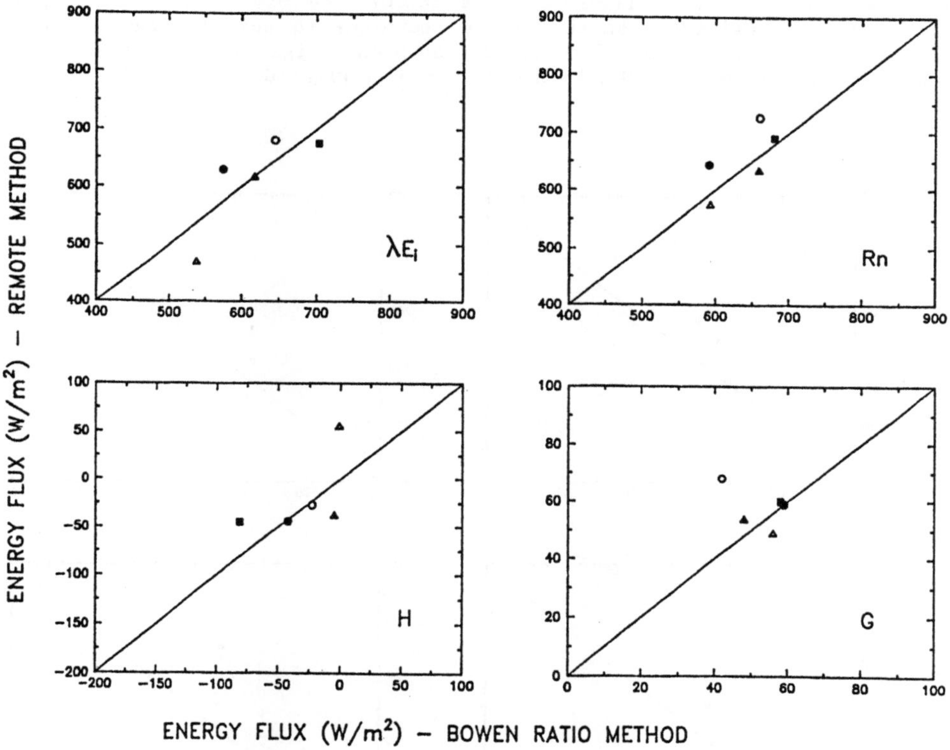

Fig. 3. Instantaneous latent heat flux λE_i, net radiation R_n, sensible heat flux H, and soil heat flux G, measured with the remote method and compared with the Bowen ratio method. The symbols represent: o, cotton, DOY 85204; ●, cotton, DOY 85236; △, wheat, DOY 86095; ▲, wheat, DOY 86111; ■, alfalfa, DOY 86175.

The difference between the remote and Bowen ratio methods for measuring λE_i was less than 10% for 4 of the 5 days, and was 12.5% for DOY 86095. The maximum difference between the two measurements for the six times on DOY 86175 was 11.4%, occurring at 0925 h. The net radiation differed by 10% or less for all

5 days and for all 6 times on DOY 86175. The soil heat flux values were quite similar for times near 1030. At other times the remote estimate was higher than the heat flux plate measurement. This result could be expected because Eq. (3) was developed from R_n and G data taken near 1030 and does not account for the diurnal trend of G. The calculations of sensible heat flux, a difficult term to evaluate with the remote method, were reasonably similar, with the greatest difference occurring on DOY 86095. A measurement error of 1.1°C in either air temperature or surface temperature would have reduced H from 56 to 0 W m^{-2} (the value from the Bowen ratio was −1) and increased the value of λE_i to 527 W m^{-2}, reducing the difference between the remote λE_i and the Bowen ratio λE_i from 12.5 to 2%.

A major advantage of using a remote method for evaluating evaporation is demonstrated in Fig. 5. Values of λE_i calculated from data acquired while flying the length of the fields are plotted to simulate their approximate location in the field. The Bowen ratio values are shown as diamonds (◆). The data presented in Fig. 5 illustrate the concurrence of the two methods, and also show the sensitivity of the remote technique to soil moisture. On day 85204, the field area from 0.3 to 1.6 km had been irrigated during the previous 24 h. The area from 0 to 0.3 km had not been irrigated for more than 2 weeks.

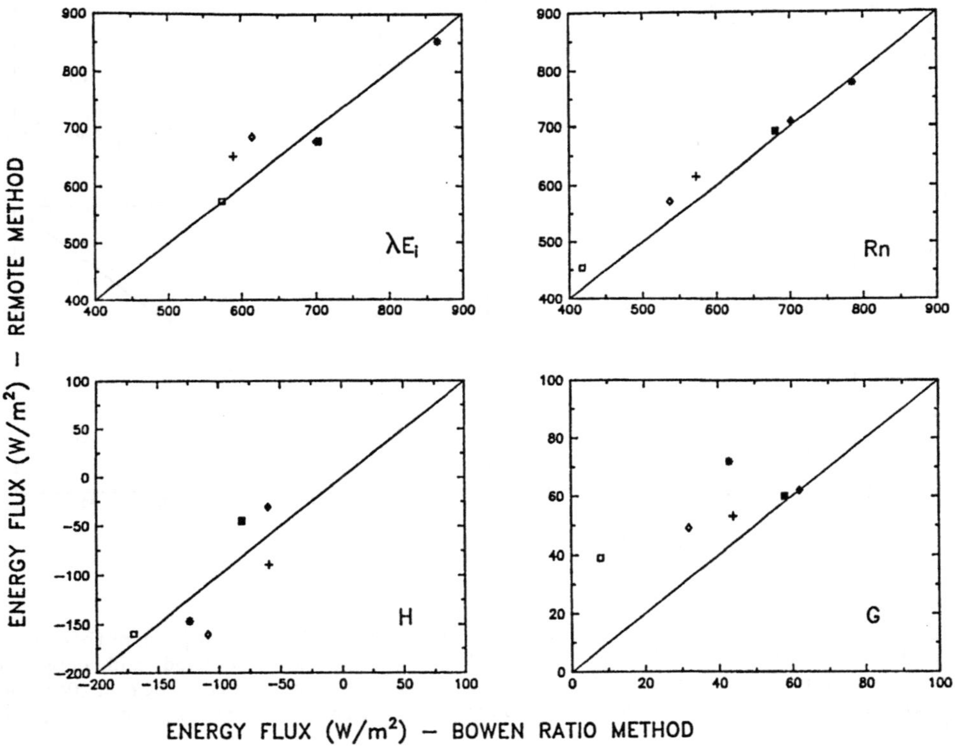

Fig. 4. Instantaneous heat flux λE_i, net radiation R_n, sensible heat flux H, and soil heat flux G, measured with the remote method and compared with the Bowen ratio method over alfalfa for six times on DOY 86175. The symbols represent: □, 0838 h; ◇, 0925 h; +, 0944 h; ■, 1037 h; ◆, 1056 h; *, 1147 h.

The difference in evaporation in the two areas, estimated from remote measurements, is nearly 25%. The same situation is evident for DOY 86095, where the highest values of evaporation correspond to the area most recently irrigated (between 0.4 and 0.8 km). On days when irrigation had been completed on the entire field, or several weeks had elapsed since irrigation, differences in evaporation across the fields were barely discernable.

Instantaneous values are excellent for showing the spatial distribution of evaporation, but daily totals are more useful for evaluating water use and for scheduling irrigations. A comparison of daily latent heat flux λE_d estimated from the instantaneous values and the measured totals from the Bowen ratio method are shown in Fig. 6. On three of the days the difference was less than 7.5%, whereas on DOY 86095 and 86175 the errors were 24.6% and 18.9%, respectively. An explanation for the difference for DOY 86095 would be the same as that for the instantaneous values discussed in an earlier paragraph. The difference for DOY 86175 may be attributed to changes in windspeed.

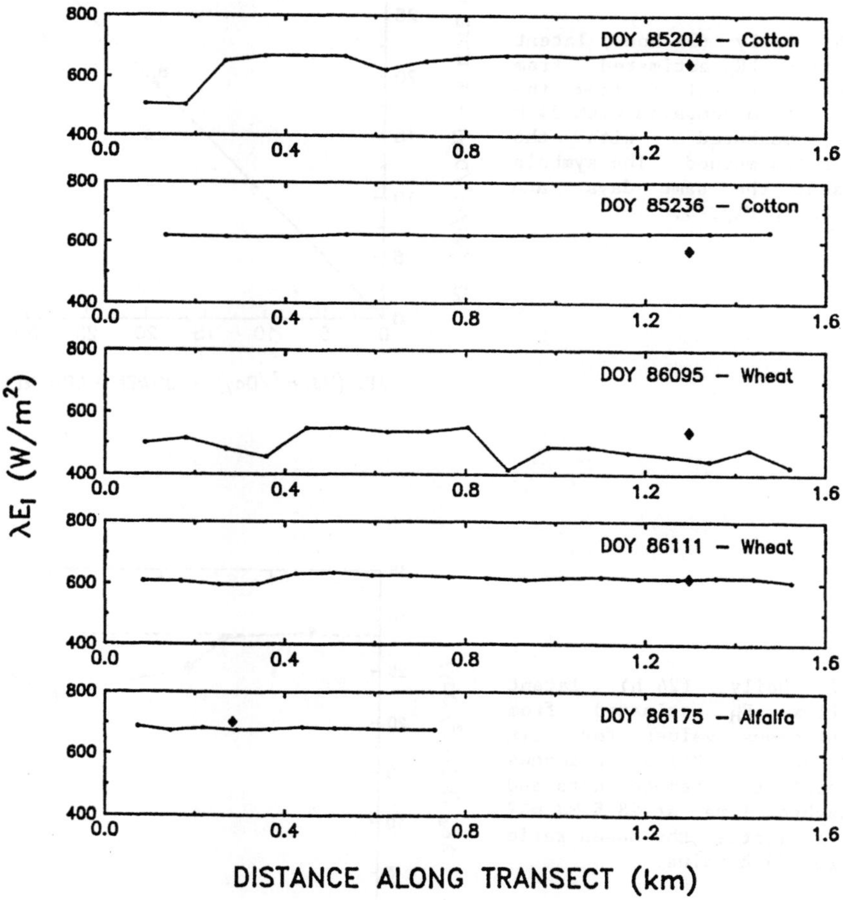

Fig. 5. Spatial distribution of instantaneous evaporation λE_i over the 1.6 km length of fields (0.8 km for alfalfa). The dots connected by lines indicate a remote measurement. The diamond (◆) indicates the point measurements of λE_i (average of 12 min) with the Bowen ratio apparatus.

Estimated λE_d values for the six measurement periods on DOY 86175 are shown in Fig. 7. The first two data points fall very close to the line representing the daily total as measured with the Bowen ratio apparatus. Subsequently, the remote estimates fall below the line, reaching a minimum at 1056 h. Windspeed values for the measurement period are shown in Fig. 8. The diamond symbols represent windspeed values at the time of the remote measurements. The average windspeed for the 24-hour period is represented by the dashed line. For the first two periods, the windspeed was near the average and subsequently dropped below the line in a manner similar to the remote λE_d. Lowest values of the windspeed were observed from about 1030 to 1130 h.

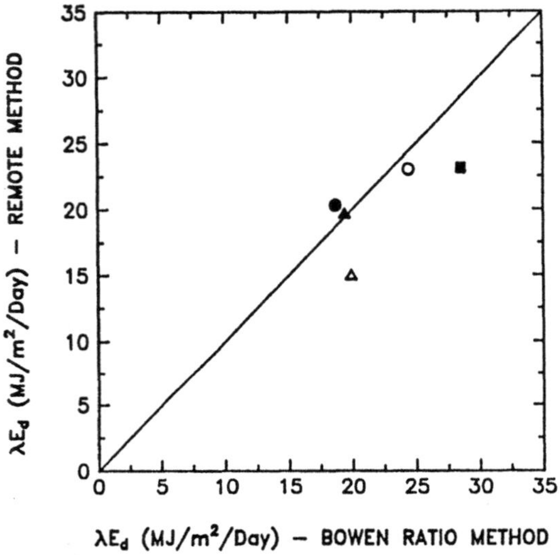

Fig. 6. Daily (24 hour) latent heat flux λE_d estimated from instantaneous values from the remote method compared with 24 h values measured with the Bowen ratio method. The symbols represent the same days and crops as in Fig. 3.

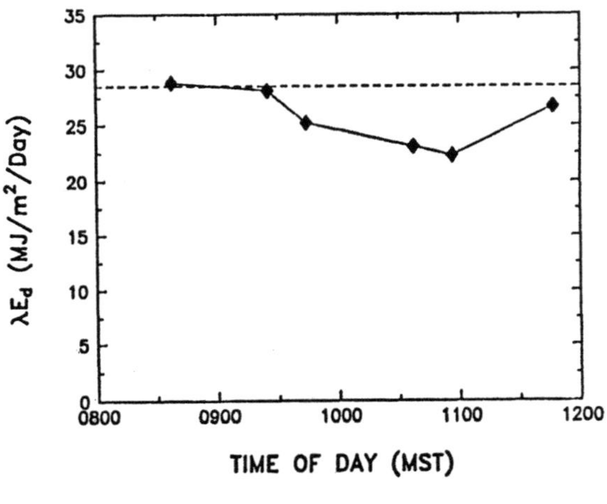

Fig. 7. Daily (24 h) latent heat flux λE_d estimated from instantaneous values for six times on DOY 86175. Diamonds represent the remote data and the dashed line at 28.5 MJ m^{-2} day^{-1} indicates the Bowen ratio measured 24 h value.

These results demonstrate a weakness in the estimation of daily totals from instantaneous measurements. When drastic changes in windspeed occur over prolonged periods, the diurnal evaporation may not be adequately described by a sine function. The data in Figs. 7 and 8 imply that, if 24-hour totals of windspeed and solar radiation are recorded, it may be possible to develop correction factors to adjust the values if the instantaneous measurements were taken during a time when the environmental conditions were not representative of the entire 24-hour period.

Assessment of Validation Experiments

Latent heat flux values obtained using an airborne remote sensing technique compared favourably with values from a Bowen ratio apparatus for relatively uniform cropped fields. The largest difference, as expected, was between the measured daily Bowen ratio values and the extrapolated remote values. This difference (25%) was within the range of differences found when several ground-based Bowen ratio and eddy correlation methods were compared[1]. Preliminary results using satellite-based data over the same fields also compare favourably with the Bowen ratio data (Raymond et al. 1987; Moran and Jackson 1987).

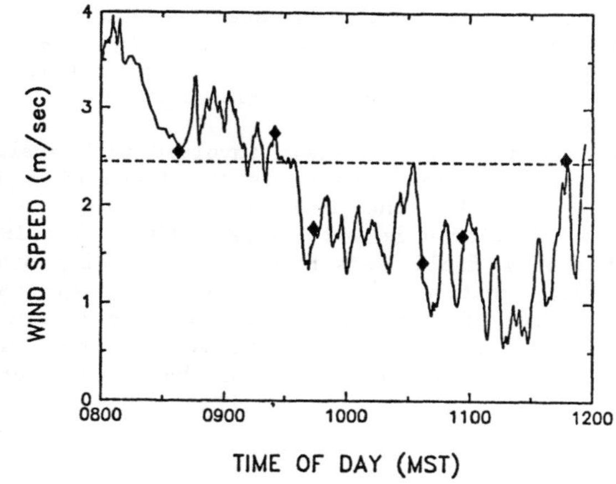

Fig. 8. Windspeed ($m\ s^{-1}$) during the measurement period (solid line) and the 24 h average windspeed (dashed line at 2.45 $m\ s^{-1}$). Diamonds indicate the time for which λE_i was calculated and λE_d estimated.

An advantage of remote sensing methods is that the spatial distribution of evaporation can be estimated over entire fields. The areal limitation of the method is determined by the distance that the ground-based micrometeorological measurements can be extrapolated. With clear sky conditions, the areal extrapolation of incoming solar radiation could extend many kilometres with little error. Under these conditions, the spatial variations in windspeed and air temperature would probably be the limiting factors.

A disadvantage of the remote method is that evaporation is evaluated at an instant in time. This is sufficient when only spatial distribution is of interest, but daily values are needed for estimating water use and for scheduling irrigations. Thus, it is necessary to extrapolate to daily values

[1]ET Kanemasu, G Asrar, D Nie, D Watts, L Fritschen, L Gay, H Weaver, B Tanner, M Tanner, J Green (1987) Inter- and intra-sensor comparisons among Bowen ratio and eddy correlation instruments. Symposium poster paper

from the instantaneous values. The simple procedure used here yielded reasonable estimates of the daily values, and provided insight into how this estimate may be improved.

For nonuniform surface conditions, it may not be possible to specify the factors required to compute sensible flux with sufficient accuracy to yield acceptable values of evaporation. For partial canopies, vegetation temperatures and exposed soil temperatures may be considerably different. The composite temperature measured by a remote sensor may not be the proper temperature to use in calculating the sensible heat flux (Eq. (2)). The situation is compounded by the fact that the aerodynamic resistance becomes more complicated under these conditions. This problem is currently under investigation (Kustas et al. 1987).

CONDLUDING REMARKS

Interest in evaluating the surface energy balance on a regional scale began prior to the launch of the first earth observation satellite. The reports reviewed here indicate that this interest continues, and some progress has been made. On a local scale, with full cover crops, surface fluxes determined using remotely sensed data compared favourably with values measured with standard ground-based techniques. At present, the remote method does not adequately account for fluxes over areas with partial canopy cover. This problem must also be addressed on the regional scale.

Solving the many problems involved with evaluating the surface energy balance from space-based sensors is the focus of a multi-national, multi-disciplinary program called the International Satellite Land-Surface Climatology Project (ISLSCP) (Schmugge and Sellers 1986). The ISLSCP program includes a series of intensive field measurements, one being the First ISLSCP Field Experiment (FIFE), that was conducted on the Konza Prairie during 1987. Results from intensive, multidisciplinary experiments such as FIFE may produce the necessary knowledge from which operational systems could be devised that would provide routine information concerning the earth's surface energy balance.

REFERENCES

Assad E, Freteaud JP, Kerr Y, Lagouarde JP, Seguin B (1986) Utilisation de la thermographie infrarouge dans l'estimation de l'évaporation a l'échelle régionale. Application au Sénégal. L'Agronomie Tropicale 40-4: 279-285

Barnes WL, Price JC (1980) Calibration of a satellite radiometer. Appl Opt 19: 2153-2161

Becker F, Camillo PJ, Choudhury BJ (1987) Review of ET estimation by means of satellite data. Proc ISLSCP Workshop on Algorithms, Jet Propulsion Laboratory, Pasadena, CA, in press

Brunt D (1932) Notes on radiation in the atmosphere. Q J R Meteorol Soc 58: 389-418

Brutsaert WH (1975) On a derivable formula for long-wave radiation from clear skies. Water Resour Res 11: 742-744

Byrne GF, Kalma JD, Streten NA (1984) On the relation between HCMM satellite data and temperatures from standard meteorological sites in complex terrain. Int J Remote Sens 5: 65-77

Camillo PJ, Gurney RJ, Schmugge TJ (1983) A soil and atmospheric boundary layer model for evapotranspiration and soil moisture studies. Water Resour Res 19: 371-380

Carlson TN, Boland FE (1978) Analysis of urban-rural canopy using a surface heat flux/temperature model. J Appl Meteorol 17: 998-1013
Carlson TN, Dodd JK, Benjamin SG, Cooper JN (1981) Satellite estimation of the surface energy balance, moisture availability and thermal inertia. J Appl Meteorol 37: 67-87
Choudhury BJ, Reginato RJ, Idso SB (1986) An analysis of infrared temperature observations over wheat and calculation of latent heat flux. Agric For Meteorol 37: 75-88
Clothier BE, Clawson KL, Pinter PJ Jr, Moran MS, Reginato RJ, Jackson RD (1986) Estimation of soil heat flux from net radiation during the growth of alfalfa. Agric For Meteorol 37: 319-329
Curtis OF (1936) Leaf temperatures and the cooling of leaves by radiation. Plant Physiol 11: 343-364
Curtis OF (1938) Wallace and Clum, "Leaf temperatures"; A critical analysis with additional data. Am J Bot 25: 761-771
Denmead OT, Bradley EF (1987) On scalar transport in plant canopies. Irrig Sci 8: 131-149
De Vries DA (1963) Thermal properties of soils. In: Van Wijk WR (ed) Physics of the plant environment. Wiley, New York, pp 210-235
Dutton JA (1962) Space and time response of airborne radiation sensors for the measurement of ground variables. J Geophys Res 67: 195-205
Eaton FM, Belden GO (1929) Leaf temperatures of cotton and their relation to transpiration, varietal differences, and yields. US Dep Agric Tech Bull 91
Ehlers JH (1915) The temperature of leaves of pinus in winter. Am J Bot 2: 32-70
Ehrler WL (1973) Cotton leaf temperatures as related to soil water depletion and meteorological factors. Agron J 65: 404-409
Fuchs M, Tanner CB (1966) Infrared thermometry of vegetation. Agron J 58: 597-601
Garratt JR (1984) The measurement of evaporation by meteorological methods. Agric Water Manage 8: 99-117
Gates DM (1964) Characteristics of soil and vegetated surfaces to reflected and emitted radiation. Proc 3rd Int Symp Remote Sens Environ, Ann Arbor, MI, pp 573-600
Gay LW, Greenberg RJ (1985) The AZET battery-powered Bowen ratio system. Proc 17th Conf on Agric Forest Meteorol, Am Meteorol Soc, Boston, MA, pp 181-182
Gurney RJ, Hall DK (1983) Satellite-derived surface energy balance estimates in the Alaskan Sub-Arctic. J Climate Appl Meteorol 22: 115-125
Gurney R, Camillo PJ (1984) Modelling daily evapotranspiration using remotely sensed data. J Hydrol 69: 305-324
Huband NDS, Monteith JL (1986) Radiative surface temperature and energy balance of a wheat canopy I. Comparison of radiative and aerodynamic canopy temperature. Boundary-Layer Meteorol 36: 1-17
Idso SB (1981) A set of equations for full spectrum and 8- to 14-μm and 10.5- to 12.5-μm thermal radiation from cloudless skies. Water Resour Res 17: 295-304
Jackson RD (1984) Total reflected solar radiation calculated from multi-band sensor data. Agric For Meteorol 33: 163-175
Jackson RD, Reginato RJ, Idso SB (1977) Wheat canopy temperature: A practical tool for evaluating water requirements. Water Resour Res 13: 651-656
Jackson RD, Hatfield JL, Reginato RJ, Idso SB, Pinter PJ Jr. (1983) Estimation of daily evapotranspiration from one-time-of-day measurements. Agric Water Manage 7: 351-362
Jackson RD, Pinter PJ Jr, Reginato RJ (1985) Net radiation calculated from remote multispectral and ground station meteorological data. Agric For Meteorol 35: 153-164
Jackson RD, Moran MS, Gay LW, Raymond LH (1987) Evaluating evaporation from field crops using airborne radiometry and ground-based meteorological data. Irrig Sci 8: 81-90
Kalma JD, Byrne GF, Johnson ME, Laughlin GP (1983) Frost mapping in Southern Victoria: an assessment of HCMM thermal imagery. J Climatol 3: 1-19

Kanemasu ET, Wesely ML, Hicks BB, Heilman JL (1979) Techniques for calculating energy and mass fluxes. In: Barfield BJ, Gerber JF (eds) Modification of the aerial environment of plants. ASAE Monograph #2, Am Soc Agric Eng, St Joseph, MI, pp 156-182

Kiang RK (1982) Atmospheric effects on TM measurements: Characterization and comparison with the effects on MSS. IEEE Trans Geosci Remote Sens GE-20: 365-370

Kneizys FX, Shettle EP, Gallery WO, Chetwynd JH Jr, Abreu LW, Selby JEA, Clough SA, Fenn RW (1983) Atmospheric transmittance/radiance: Computer code LOWTRAN 6. Air Force Geophysics Lab Rpt AFGL-TR-83-0187 Hanscom AFB Mass, 200 pp

Kustas WP, Choudhury BJ, Moran MS, Reginato RJ, Jackson RD, Gay LW (1987) Problems in the estimation of sensible heat flux over incomplete canopy cover with thermal infrared data. Am Meteorol Soc 18th Conf on Agric For Meteorol, pp 88-90

Lenschow DH, Dutton JA (1963) Surface temperature variations measured from an airplane over several surface types. J Appl Meteorol 3: 65-69

Lorenz D (1968) Temperature measurements of natural surfaces using infrared radiometers. Appl Opt 7: 1705-1710

Mahart L, Ek M (1984) The influence of atmospheric stability on potential evaporation. J. Climate Appl Meteorol 23: 222-234

Miller EC, Saunders AR (1923) Some observations on the temperature of the leaves of crop plants. J Agric Res 26: 15-43

Monteith JL (1973) Principles of environmental physics. Edward Arnold, London, pp 241

Monteith JL (1981) Evaporation and surface temperature. Q J R Meteorol Soc 107: 1-27

Monteith JL, Szeics G (1962) Radiative temperature in the heat balance of natural surfaces. Q J R Meteorol Soc 88: 496-507

Moran MS, Jackson RD (1987) Estimation of evapotranspiration and net radiation over crops using TM-satellite spectra and ground-based meteorological data. Agron Abs, p 15

Perrier A (1971) Leaf temperature measurement. In: Sestak Z, Gatsky J, Jarvis PG (eds) Plant photosynthetic production manual and methods. Junk, The Hague, pp 632-671

Price JC (1980) The potential of remotely sensed thermal infrared data to infer surface soil moisture and evaporation. Water Resour Res 16: 787-795

Price JC (1982) Estimation of regional scale evapotranspiration through analysis of satellite thermal-infrared data. IEEE Trans Geosci Remote Sens GE-20: 286-292

Price JC (1983) Estimating surface temperatures from satellite thermal infrared data - A simple formulation for the atmospheric effect. Remote Sens Environ 13: 353-361

Rameaux (1843) Des temperatures vegetables. Ann Sci Nat, Bot Biol Veg 19[2], 10

Rao PK, Winston JS (1963) An investigation of some synoptic capabilities of atmospheric "window" measurements from satellite TIROS II. J Appl Meteorol 2: 12-23

Raymond LH, Moran MS, Jackson RD (1987) Mapping latent heat energy from remotely-sensed data using ARC/INFO software. Proc Conf Spatail Data Systems for Management, Tucson, AZ, pp 38-45

Rosema A, Bijleveld JH, Reiniger P, Tassone G, Gurney RJ, Blyth K (1978) Tellus, a combined surface temperature, soil moisture and evaporation mapping approach. Proc 12th Int Symp Remote Sens Environ, Ann Arbor, MI, pp 2267-2276

Schmugge TJ, Sellers PJ (1986) The First International Satellite Land-Surface Climatology Project (ISLSCP) Field Experiment (FIFE). Proc 3rd Int Colloq Spectral Signatures of Objects in Remote Sensing, Les Arcs, France, pp 321-325

Schott JR, Volchok WJ (1985) Thematic Mapper thermal infrared calibration. Photogramm Eng Remote Sen 51: 1351-1357

Seguin B, Itier B (1983) Using midday surface temperature to estimate daily evaporation from satellite thermal IR data. Int J Remote Sens 4: 371-383

Seguin B, Lagouarde JP, Kerr Y (1986) Estimation of regional evaporation using midday surface temperature from satellite thermal IR data. Proc ISLSCP Conf. Rome, Italy, pp 339-344

Sellers PJ, Mintz Y, Sud YC, Dalcher A (1986) A simple biosphere model (SiB) for use within general circulation models. J Atmos Sci 43: 505-531

Short NM, Stuart Jr LM (1982) The Heat Capacity Mapping Mission (HCMM) anthology. NASA SP-465, National Aeronautics and Space Administration, Washington, DC, 263 pp

Slater PN (1980) Remote sensing: Optics and optical systems. Addison-Wesley, Reading, MA, 575 pp

Soer GJR (1980) Estimation of regional evapotranspiration and soil moisture conditions using remotely sensed crop surface temperatures. Remote Sens Environ 9: 27-45

Spittlehouse DL, Black TA (1980) Evaluation of the Bowen ratio/energy balance method for determining forest evapotranspiration. Atmos & Oceans 18: 98-116

Tanner CB (1963) Plant temperatures. Agron J 55: 210-211

Vukovich FM (1984) A comparison of surface temperature derived from HCMM infrared measurements with field data. Remote Sens Environ 15: 63-76

Wallace RH, Clum HH (1938) Leaf temperatures. Am J Bot 25: 83-97

SURFACE TEMPERATURE AND THE SURFACE ENERGY BALANCE: COMMENTARY

K.G. McNaughton

DSIR Plant Physiology Division, Private Bag, Palmerston North, New Zealand

Summary. The development of a method for measuring evaporation from a remote platform is desirable because it provides perhaps the only way of surveying remote areas and for providing data at the scale and frequency required in many applications. However, the difficulties should not be underestimated. This summary of the discussion at the Symposium following Dr Jackson's paper emphasizes some of those difficulties.

INTRODUCTION

At the start of his oral presentation Ray Jackson made the comment that if one wanted to know the evaporation rate from some area, then perhaps the easiest way to find out would be to ring Champ Tanner* and to ask him. Though this was intended as a tribute to one of the great men of evaporation research, it is also a valid comment on the state of our science: notwithstanding our present understanding of the physics of evaporation from vegetation, empirical knowledge and semi-empirical formulae are still very valid guides in making estimates of evaporation rates. If any new method for measuring or estimating evaporation is to have any real value, it must give better results than a telephone call to Professor Tanner.

Dr Jackson indicated that the reason for the current interest in using surface temperatures to measure evaporation is that surface temperature can now be measured very easily using sensors mounted on remote platforms. From some of these platforms large areas of land can be scanned. This is of great potential value to irrigationists, hydrologists and climatologists who need to know evaporation rates from large areas; current techniques using surface-mounted instruments are expensive and are restricted to single fields. However, Dr Jackson's point was not so much that the hydrologic practitioners have chosen to develop remote-sensing techniques to solve their problems as that the promoters of remote-sensing have chosen the measurement of evaporation as a possible application for their new-found abilities. It is largely a case of a technique in search of an application.

This is not necessarily a bad thing, but it could lead to pressures to persist with the development of these methods to beyond the point where most practitioners might judge progress to be unpromising. We might therefore ask whether there is a sufficient prospect for success that the present emphasis should be maintained. Is it likely that remotely-sensed surface temperatures can be used in some way to give measurements of evaporation that will pass the Tanner test?

Dr Jackson has given a very good account of the current state of development of the subject and a very fair evaluation of the difficulties. His account complements and supplements the review by Reiniger and Seguin (1986). I would

*Prof. C.B. Tanner, Department of Soil Science, University of Wisconsin

like to discuss further some of the difficulties that Dr Jackson has raised, which stand between remotely-sensed surface temperatures and good measurements of the evaporation rate. In doing this I will try to address most of the points raised in discussion at the meeting. I will also elaborate on a couple of matters that Dr Jackson has passed over briefly.

THE BASIC EQUATIONS

Dr Jackson discussed an approach in which remotely-sensed surface temperature, T_s, and local measurements of various meteorological elements are used together. This line of attack is by way of the energy balance equation,

$$\lambda E = R_n - G - H \, , \tag{1}$$

where H is given by

$$H = \rho c_p (T_s - T_a)/r_a \tag{2}$$

and the symbols are as defined in Dr Jackson's paper.

Surface temperature, T_s, is to be measured by a radiometer mounted on a satellite, an aircraft or an instrument mast standing on the ground. By their nature the first two of these platforms are most useful for surveying large areas, but they give less continuous coverage in time. However, I will ignore that problem and ask how well (1) and (2) can be used to measure the evaporation rate at a uniform site when a continuous record of radiometrically-measured surface temperature is available.

Equation (1) is a very robust equation: it is necessarily true and its form is a simple sum of terms. This means that we have no problems in applying it as an average over any area or time interval that we like, whether the vegetation is homogeneous or not and whether the fluxes are steady or not. There might be errors in measuring or estimating R_n or G, but these errors are common to other measurement techniques, such as Bowen ratio or eddy correlation, where λE is found by $\lambda E = R_n - G - \rho c_p \overline{w'\theta'}$, and to estimation equations such as those of Penman (1948) or Priestly and Taylor (1972).

The difficulties that are peculiar to remote sensing begin with (2). This equation has, in effect, two unknowns: T_s and r_a. 'Surface' temperature can be defined operationally as the temperature that the radiometer sees, but how then should we calculate the aerodynamic resistance, since the latter is related to the distribution of the sources of sensible heat within the canopy?

RADIOMETRIC AND AERODYNAMIC SURFACE TEMPERATURES

In discussion following Dr Jackson's paper, Mike Raupach suggested that surface temperature should be defined as the average temperature of all the canopy elements weighted by the relative contribution of each element to the overall aerodynamic conductance. With such a definition r_a would have a natural definition in terms of the turbulent transport processes in the canopy. However, use of this definition of r_a would raise the alternative problem of how the 'radiometric' temperature of a plant canopy is related to the 'aerodynamic' temperature.

Tom Denmead reinforced the point about the differences between radiometric and aerodynamic definitions of T_s. Some 80% of the sensible heat flux in a forest that he had studied came from the top half of the canopy early and late in the day, but only about half of it came from the top half at midday (Denmead,

1984). Evaporation from the forest floor ranged from 10 to 40% of total water loss from the forest. Energy for this was supplied partly in a two-stage process, with solar radiation absorbed in the canopy heating the air there, and this energy then being transported down in gusts to the forest floor to supply much of the energy consumed in the evaporation of moisture from the soil. The relationship between aerodynamic and radiometric temperature is therefore expected to be complex, and to vary diurnally with the penetration of solar radiation and the changing distribution of heat sources and sinks within the canopy.

Richard Smith, who has used radiometers to monitor crop temperatures, was more optimistic. He endorsed Dr Denmead's concern, especially in the case of sparse canopies where a significant proportion of the incident radiation penetrates to the soil surface, but thought that some kind of canopy model could be used to solve the problem even though some simplifying assumptions might be necessary. H.L. Penman had demonstrated 40 years ago that it is possible to make good practical progress with the aid of simple models (Penman 1948).

The magnitude of the errors that can arise from this difficulty are indicated in some recent papers. Huband and Monteith (1986) defined r_a using the equation

$$r_a = \ell n((z - d)/z_0)/ku_* \, , \tag{3}$$

as Dr Jackson has done, albeit with a stability correction in Dr Jackson's work, and then found that the radiometrically-measured surface temperatures of a uniform wheat crop were a full degree cooler than the T_s values needed to satisfy (2). Choudhury et al. (1986) performed a similar analysis but used z_0' $(= z_0/7)$ in place of z_0 in (3) and included a stability correction. They found differences of up to 4°C in stable conditions. An error in T_s of 1°C would lead to a typical error of 50 W m^{-2} in H from crops, and so to an equal error in λE. If the temperature error is more than 1°C, then the error in λE becomes comparable in size to that arising from the use of empirical estimation formulae in many cases.

ADVECTION

In his presentation Dr Jackson introduced the problems that remote sensing has been asked to solve by showing a photograph taken from an aircraft looking down at the patchwork quilt of fields in an irrigation area. The crops in adjacent fields showed strong visual contrasts. The photograph demonstrated nicely the potential value of a successful remote-sensing technique for calculating evaporation, since coverage of such a varied collection of fields is scarcely practical using surface-mounted instruments. The slide also illustrated one of the difficulties.

A comment at the Symposium was that Dr Jackson's analysis was one-dimensional, but advective effects must surely be taken into account where the plot sizes are relatively small. In discussion it was agreed that advection was a topic that is poorly developed in the literature, and that present models have limited predictive ability. Possible advective effects were not taken into account in remote-sensing evaluations of the evaporation.

The difficulties posed by advection were illustrated in a poster paper accompanying the session. Dunin and co-workers[1] presented the results of a field experiment on evaporation from a wheat field. The upwind field did not

[1]FX Dunin, HD Barrs, WS Meyer, ACF Trevitt (1987) Foliage temperature and latent heat flux of irrigated wheat. Symposium poster paper

contrast markedly with the experimental crop, so problems of advection and fetch/height ratios for the instrument exposures were not judged to be of major concern. Energy balance/Bowen ratio measurements agreed well with lysimeter measurements in unstable conditions, but not when conditions were stable, when evaporation rates calculated from (1) and (2) gave overestimates, with errors caused by differences in aerodynamic and radiometric surface temperatures. However, the differences changed sign and increased to about 150 W m^{-2} as the crop boundary layer became more stable. Bowen ratio measurements also diverged from the lysimeter results in a way that indicated flux divergence above the crop. Apparently advection was responsible for these unexpected results.

A few years ago I participated in an experiment on transport processes within an advective inversion over paddy rice bordered by dry land (Lang et al. 1983a,b). On one occasion we observed an internal boundary layer only a few metres deep after wind had traversed a fetch over the rice of about 600 m. The air within this remarkably shallow boundary layer was extremely stable, with nocturnal, radiative cooling at the surface being reinforced by evaporative cooling. The point is that the very strong effects of stability on the depth of internal boundary layers, and so on vertical flux divergence close to the ground, are not generally known, have never been properly described in the scientific literature, and are rarely properly considered in experimental design. The effects of local advection on evaporation rates can not easily be allowed for, even by researchers who make detailed measurements in only one field. Any accurate allowance for the effects of advection in the context of remote sensing would seem to be beyond present capabilities.

SPATIAL HETEROGENEITY

A related point is that (2) supposes that air temperature is measured within the constant flux layer over each area of interest. If thermometers are not to be placed above each distinct element of a patchwork landscape, then where is air temperature to be measured? If the temperature is taken at a nearby climate station, then how can such an air temperature be related to T_a in Eq. (2)?

A similar point could have been raised concerning wind speed measurements. In (3) aerodynamic resistance is calculated from local values of roughness lengths and wind speeds close above each field; both will vary from field to field (Hicks and Wesley 1981) or from field to forest. In (2) the error in H is porportional to the error in wind speed, so significant errors are likely if uncorrected data from a base station are used.

Aerodynamic resistance might also be affected by boundary shelter belts, or by wind accelerations caused by the local topography, or by roughness changes. Micrometeorologists have recognized for a long time that aerodynamic methods for measuring fluxes are sensitive to such disturbances. Equation (2) represents a rather unreliable method for measuring the sensible heat flux, even if air temperature and wind speed are measured in the right places.

Thus, substantial new difficulties arise as soon as we attempt to apply (2) beyond the accessible, flat, uniform fields for which it was developed. Data for testing remote-sensing methods are few and they have usually been collected at very uniform sites; the difficulties of more complex terrain have not been addressed in experiments so far; this despite the fact that the great attraction of remote-sensing methods is that they can be used to survey large areas of naturally varying terrain.

A PERSPECTIVE

In discussion, Peter Taylor expressed the opinion that numerical models of flow over complex terrain and above surfaces of varying heat and moisture properties will have to provide the link between site-specific surface measurements and spatial averages derived from satellite observations or from large-scale meteorological models. Some studies of this type are being conducted in connection with the HAPEX experiment. Aircraft-based measurements can also provide spatially-averaged fluxes and possibly provide the best tool for calibrating satellite observations.

Perhaps the best way to summarize this discussion is to ask again whether in remote sensing we have, or potentially have, a method for estimating evaporation that can fill a niche in our range of useful techniques for measuring evaporation.

I think that is fair to say that measurements of radiometric surface temperature have been made over uniform crops chiefly to develop the remote-sensing methods for wider areas. The method is not accurate enough to replace lysimeters, Bowen ratio or eddy correlation techniques in making measurements of evaporation from uniform fields. There is some prospect for improvement here if a model can be developed to relate the radiometric and aerodynamic surface temperatures. Such a model would probably need to be quite detailed, and would probably involve detailed measurements of canopy structure and condition, if it were to give results comparable to the established micrometeorological methods. I don't think there is a need to develop such a method for its own sake.

Justification for further development would have to be based on improvements in the wider-scale measurements. However, for this application the model for radiation and heat exchange in the canopy would need to be very simple since detailed information on canopy structure would not be available.

The natural role of remote sensing is in making observations over large areas, but the evaporation estimates have never been tested on this scale. Experiments such as HAPEX and FIFE are very important here, as is the use of aircraft to make flux measurements over areas that cannot be surveyed using surface-mounted flux instruments. However, the results are not yet in, and the list of difficulties is long. We have no direct evidence that evaporation can be estimated better by these means than by empirical equations using data from local climate stations - the metaphorical telephone call to Champ Tanner.

Perhaps it is still too early to say that the problems cannot be solved or that the research effort should be scaled down. Certainly remote sensing has advantages in large-scale surveys that are hard to match, and the technique can detect the progress of water stress. However, it does seem that the enthusiasm shown by some proponents of remote sensing as a means for determining evaporation rates is naive, and that perhaps some of the objectives should be reviewed. I specifically exclude Dr Jackson from this suggestion, and commend him for the useful and balanced review that he has given of this topic.

REFERENCES

Choudhury BJ, Reginato RJ, Idso SB (1986) An analysis of infrared temperature observations over wheat and calculation of latent heat flux. Agric For Meteorol 37: 75-88

Denmead OT (1984) Plant physiological methods for studying evapotranspiration: Problems in telling the forest from the trees. Agric Water Manage 8: 167-189

Hicks BB, Wesley ML (1981) Heat and momentum transfer characteristics of adjacent fields of soybeans and maize. Boundary-Layer Meteorol 20: 175-186
Huband NDS, Monteith JL (1986) Radiative surface temperatures and energy balance of a wheat crop. Boundary-Layer Meteorol 36: 1-17
Lang ARG, McNaughton KG, Chen Fazu, Bradley EF, Ohtaki E (1983a) Inequality of eddy transfer coefficients for vertical transport of sensible and latent heats during advective inversions. Boundary-Layer Meteorol 25: 25-41
Lang ARG, McNaughton KG, Chen Fazu, Bradley EF, Ohtaki E (1983b) An experimental appraisal of the terms in the heat and moisture flux equations for local advection. Boundary-Layer Meteorol 25: 89-102
Penman HL (1948) Natural evaporation from open water, bare soil and grass. Proc Roy Soc London A193: 120-146
Priestley CHB, Taylor RJ (1972) On the assesment of surface heat flux and evaporation using large-scale parameters. Mon Weather Rev 100: 81-92
Reiniger P, Seguin B (1986) Surface temperature as an indicator of evapotranspiration and soil moisture. Remote Sens Rev 1: 277-310

STOMATAL PHYSIOLOGY AND GAS EXCHANGE IN THE FIELD

I.R. Cowan

Research School of Biological Sciences, The Australian National University, GPO Box 475, Canberra, ACT, 2601, Australia

Summary. The sensitivities of rate of assimilation of CO_2 and rate of transpiration of water vapour to change in stomatal aperture are examined. We suggest that stomatal aperture is adapted to equalise the marginal benefit of assimilation and the marginal cost of transpiration. We show how the temporal and spatial variation of stomatal aperture is determined by the variation in external environment and leaf internal physiological properties. We indicate how the increase in cost of transpiration relative to the benefit of assimilation may determine the decline in stomatal aperture with decrease in the amount of water available to the roots of a plant, even though the plant may suffer no immediate lesion due to shortage of water.

The interaction of plants through their influence on their common environment is examined. We show that rate of transpiration in communities may be much less sensitive to change in stomatal aperture than in the single plant. The implications of that are discussed.

INTRODUCTION

In this paper, we shall discuss two aspects of the influence of stomata on gas exchange. One aspect highlights the function of stomata as an adaptation enhancing individual survival. The other emphasizes the influence of stomata on gas exchange from land surfaces, account being taken of the interaction of plants with each other through their corporate influence on their common physical environment.

STOMATA

Stomata are the pores in plant leaves which allow the ingress of carbon dioxide used for growth, but inevitably also allow the escape of water vapour. The aperture of each stomate is controlled by the function of the guard cells, the pair of cells adjacent to the pore. The configuration of these cells is determined by their internal osmotic pressure, and is sensitive to many factors. Figure 1, a modified version of a diagram by Raschke (1975), shows that these factors are of two kinds. Some of them are generated internally. For example, the guard cells are sensitive to the state of water within the plant, perhaps acting through the levels of various hormones. They are also sensitive to the carbon dioxide concentration within the leaf. Insofar as these factors are dependent upon the functioning of the stomata themselves, the response of the stomata can be described as feedback control of water vapour and carbon dioxide exchange. The stomata are also sensitive to signals from the external environment of the plant, such as light intensity and ambient humidity. As these signals are independent of the functioning of the leaf, but

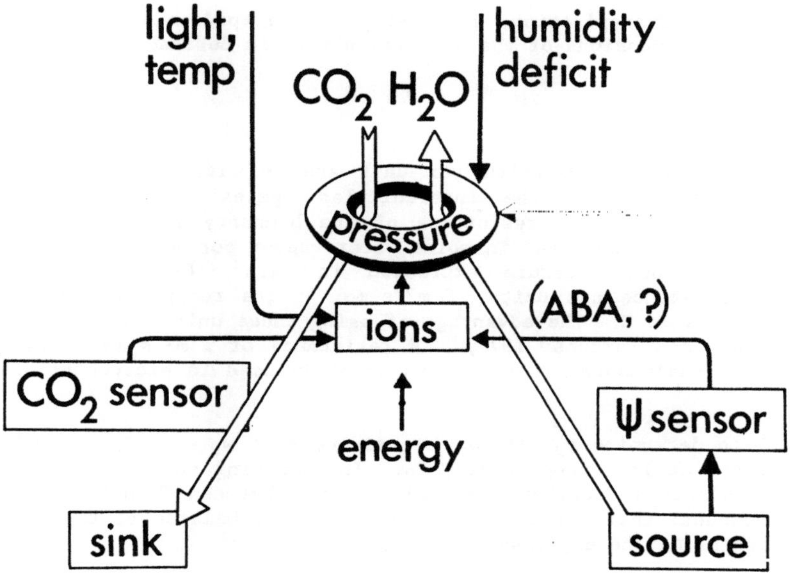

Fig. 1. Schematic diagram of stomatal control system. (adapted from Raschke 1975).

nevertheless influence water vapour and CO_2 exchange, the response of the stomata can be described as feed-forward control. The responses of guard cells to these internal and external factors have been the subject of study by plant physiologists for some 200 years. The fascination is understandable. Not only are stomata a vital adaptation enabling plants to live in the terrestrial environment, they have the virtue that the intricacies of cellular metabolism are translated mechanically into observable changes in cell shape.

However, it is not the metabolism of the guard cells that will be discussed here; rather, it is the operation of the stomata in terms of their control function. The distinction can be illustrated by a simple analogy. Suppose we needed to understand the operation of a control valve in a hydraulic pipeline, the design and specifications of the valve having been lost. We might attempt to determine the function of the device by studying the mechanism controlling the position of the diaphragm within the pipeline. But our objective might be achieved more quickly if we were to observe the flow of fluid through the valve in relation to the fluctuations of fluid pressure upstream. We might, for example, discover that the flow remained relatively constant despite large variations in the fluid pressure. The function of the valve would become immediately clear. This is the kind of approach we shall make to the function of the stomata. After all, natural selection has presumably not acted on the mechanism of the guard cells *per se*, but has operated in such a way as to provide an increasingly efficient means of controlling CO_2 uptake and water vapour loss.

GAS EXCHANGE

To develop this theme, it is first necessary to describe how water vapour loss and carbon dioxide exchange in the leaf are influenced by stomatal aperture.

Leaf resistance to transpiration, a property of the epidermis depending mainly on the number and dimensions of the stomatal pores, is defined as r_ℓ in

$$E = \frac{w_\ell - w}{r_\ell + r_b} , \tag{1}$$

where E is rate of transpiration per unit area of leaf, w_ℓ and w are the humidities of the air in the intercellular spaces and of ambient air, respectively, and r_b is the resistance of the boundary layer to transfer of water vapour. With w expressed in moles water vapour per mole air (numerically approximate to vapour pressure expressed in bars (10^5 N m^{-2})) and E in mol m^{-2} s^{-1}, resistance has units of m^2 s mol^{-1}; its reciprocal, conductance, has the same units as E. The advantage of using these units has been discussed elsewhere (Cowan and Farquhar 1977). A resistance of 1 m^2 s mol^{-1} is roughly equivalent to a resistance, in the units commonly used in micrometeorology, of 0.4 s cm^{-1}.

In using (1) to determine r_ℓ, it may be assumed that $w_\ell = w'(T_\ell)$, the humidity of air saturated at leaf temperature, T_ℓ. In computing the influence of r_ℓ on rate of transpiration, another interrelationship between E and T_ℓ has to be taken into account: that which relates to the energy balance of the leaf. The two relationships may be expressed as

$$E = \frac{w'(T_\ell) - w}{r_\ell + r_b} = \frac{1}{L}\left\{\varphi - \frac{C_p(T_\ell - T)}{r_b}\right\} , \tag{2}$$

where L is the molar heat of vapourisation of water, φ net flux of radiation received per unit leaf area, C_p molar heat capacity of air and T ambient air temperature. Frequent use is made of an approximate solution of (2) for E, first derived by Penman (1952). Here we shall be content with examining the relative sensitivity of rate of transpiration to change in leaf resistance. By differentiation we find that

$$\frac{r_\ell}{E}.\frac{\partial E}{\partial r_\ell} = \frac{-r_\ell}{r_\ell + (\varepsilon + 1)r_b} = \sigma_E, \quad \text{say,} \tag{3}$$

in which ε ($=(L/C_p)dw'(T_\ell)/dT_\ell$) is a dimensionless form of the saturation humidity - temperature relationship. The partial differential emphasises that the ambient environment, defined by w and T, is constant. It should also be noted that the influence of leaf temperature on the emission of thermal radiation from the leaf, and therefore on φ, has been ignored.

We see in (3) that the sensitivity of rate of transpiration is affected not only by the boundary layer resistance to vapour transfer, but also by the term εr_b, deriving from the coupling of vapour and energy transfer. The effect is substantial. For the purposes of numerical example, we shall take $r_\ell = 2.5$ m^2 s mol^{-1} (typical of well-watered plants in the light), $r_b = 0.5$ m^2 s mol^{-1} (an estimate appropriate for a leaf 20 mm wide with wind speed 1 m s^{-1} - see Cowan 1972), and $T_\ell = 25^0$C. Then $\varepsilon = 2.5$, yielding $\sigma_E = -0.59$.

Rate of uptake of CO_2 is determined by the simultaneous equations

$$A = \frac{c - c_i}{1.37r_b + 1.6r_\ell} = A(c_i, \text{light, temperature}) , \tag{4}$$

where c and c_i are respectively the ambient and intercellular concentrations of CO_2. The first of (4) relates to the transfer of carbon dioxide across the leaf boundary layer and through the stomatal pores. The numerical factors account for the difference in the diffusion coefficients of water vapour and

CO_2 in air. What is usually only a small effect of vapour efflux on the diffusion of CO_2 has been ignored. The second of the equations bypasses all the intricacies of photosynthetic metabolism by describing CO_2 fixation in the cells of the mesophyll tissue as a general function, A, of intercellular CO_2 concentration, light intensity, and temperature. Figure 2 illustrates some characteristics of fixation in a snow gum leaf (Wong et al. 1978). Such characteristics are obtained by placing the leaf in a cuvette in which the environment can be controlled, and measuring the rates of CO_2 and water vapour exchange, and leaf temperature. Intercellular CO_2 concentration is estimated by making use of the analogy between water vapour transfer and CO_2 transfer across the leaf epidermis and boundary layer. The concentration is varied by varying the ambient concentration of CO_2.

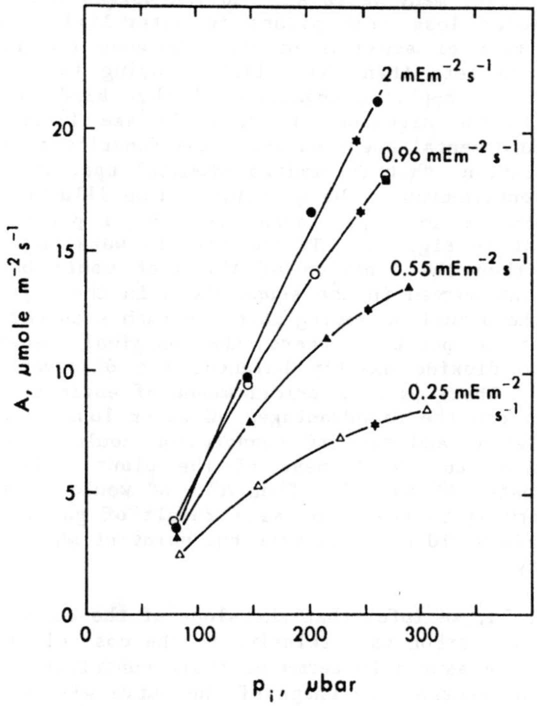

Fig. 2. Rate of assimilation of CO_2, A, as a function of internal partial pressure of CO_2, p_i, and irradiance in a leaf of snow gum. (Wong et al. 1978).

As with the transfer of water vapour, we see from (4) that rate of CO_2 uptake is limited not only by the boundary layer resistance and leaf resistance acting in series, but by an additional limitation, the requirement that intercellular CO_2 concentration is finite, such that the biochemical process of fixation is driven at the rate at which CO_2 enters the leaf. Assuming that the temperature of the leaf is close to the optimum for photosynthesis, and therefore rate of photosynthesis is not affected by small changes in temperature, then sensitivity of rate of assimilation to change in leaf resistance is

$$\frac{r_\ell}{A} \cdot \frac{\partial A}{\partial r_\ell} = \frac{-r_\ell}{0.86r_b + r_\ell + 0.63/(\partial A/\partial c_i)} = \sigma_A, \text{ say .} \qquad (5)$$

A typical magnitude of the apparent biochemical resistance to CO_2 fixation, $1/(\partial A/\partial c_i)$, for annual crop plants other than tropical cereals, is 10 m² s mol⁻¹. Thus $\sigma_A = -0.27$.

OPTIMISATION

In our numerical example, the ratio of the sensitivities

$$\frac{\sigma_E}{\sigma_A} = \frac{A}{E} \cdot \frac{\partial E}{\partial A} = 2.2 \quad .$$

While the magnitude would differ with different conditions of temperature, light intensity, and wind speed, it is generally true that it exceeds unity, except perhaps in those plants (such as tropical grasses) having a low magnitude of the biochemical "resistance", $1/(\partial A/\partial c_i)$, associated with their possession of the C_4 pathway of photosynthesis (Hatch 1976). It was this observation that gave rise, many years ago, to the suggestion that it might be useful to diminish water loss from plants in water-limited environments by using chemical inhibitors of stomatal opening, because the relative loss in carbon dioxide would be less than the relative saving in water use. Quite apart from the naivete of applying economics of that kind to plant function, there is a fallacy in the argument, as we shall see later. However, the suggestion raises a fundamental question about the function of stomata: what is it, in terms of adaptation, that determines stomatal aperture in a particular leaf in a particular environment? The question can be illustrated by reference to Fig. 3. Gas exchange in a particular leaf at a particular instant is represented by a point in Fig. 3a. If the stomata were to be wider open or less widely opened, then the function of the leaf would be represented by another point on a line curved in the sense shown in the figure. What is it that has determined the actual operating point on such a curve? Presumably the point is located at a position where the marginal benefit of further acquisition of carbon dioxide exactly balances the disadvantages of further loss of water vapour. There is no *a priori* means of estimating the advantages of carbon acquisition and the disadvantages of water loss, but let us imagine that rate of assimilation and rate of evaporation could be converted into a common currency related to the fitness of the plant. Let the fluxes so transformed be designated A^* and E^*. Then $A^* - E^*$ would be a measure of the net rate of gain accruing to the plant as a result of gas exchange (see Fig. 3b). The maximum gain would coincide with the point at which the slope of the curve $\partial E^*/\partial A^*$ is unity.

Reverting now to Fig. 3a, we infer that the slope of the curve at the operating point is the benefit of carbon gain relative to the cost of water loss, λ say, benefit and cost being measured in terms of their contributions to the fitness of the plant. But, of course, the shape of the curve will differ in time and will differ from leaf to leaf in a plant according to the variation of the environmental conditions that determine assimilation and transpiration. The argument suggests that at each instant of time and for each leaf, stomatal aperture should be such that $\partial E/\partial A$ is constant and uniform and equal to λ. What this implies for a single leaf is demonstrated in Fig. 4. The surface is the domain within which the leaf functions. Its shape is determined by the diurnal variation of the physical microenvironment of the leaf, and by leaf internal properties (of the kind shown in Fig. 2) relating to photosynthetic carbon fixation in the leaf mesophyll tissue. If stomatal conductance remained constant, then the leaf would traverse the surface, so to speak, on one of the broken curves. If stomatal aperture were to vary so that rate of transpiration remained constant, then the trajectory across the surface would be a contour. An optimal trajectory is one, such as the dotted curve, along which the slope of the surface is uniform. Evidently there is an infinity of such trajectories, each resulting in a different total loss of water during the course of the day. Each is optimal in the following restricted sense: the amount of carbon gained in the course of the day is the maximum possible corresponding to the amount of water lost.

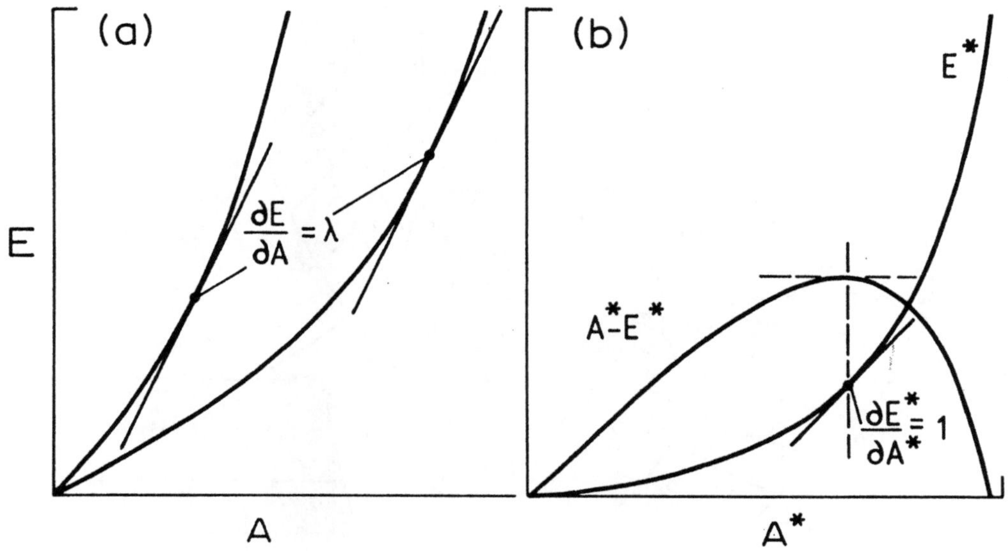

Fig. 3. (a) Relationships between rate of transpiration, E, and rate of assimilation, A. The two points represent actual rates that might be obtained in a single leaf at two instants of time (the differences being due to differences in external environment), or in two leaves in the same plant at the same instant of time (the differences being due to differences in external environment and/or differences in leaf properties). The curves represent the variations that would occur if stomatal aperture were to vary without change in external environment. (b) A hypothetical linear transformation of one of these curves, E^* and A^* being respectively the cost and benefit of water loss and carbon gain. Maximisation of net benefit would require that $dE^*/\partial A^* = 1$, with the corollary that $\partial E/\partial A$ is the benefit of carbon gain relative to the cost of water loss, λ say, and S is maintained constant and uniform.

Presumably the trajectory that is optimal in the broader sense also - that the benefit of a small virtual change in carbon gained exactly balances the cost of the corresponding change in water loss - would be determined, *inter alia*, by the amount of water available to the plant, for the cost of water loss would be greater the more imminent the likelihood of physiological drought. The matter has been taken up (Cowan 1986) by identifying the cost of water loss with the risk that the plant will become desiccated as a result of its use of available soil water. The risk depends on the pattern of water use adopted by the plant, the volume of soil tapped by the roots, the statistics of rainfall, and the rate at which water is lost by processes outside the control of the plant. Application of optimisation theory yields results of the kind shown in Fig. 5. The variations of daily carbon fixation with soil water deficit are optimal in that no other variation would yield a greater average rate of fixation for the same risk of desiccation. The full curves relate to plants having root systems which do not intermingle with those of their neighbours; the broken curves to plants with root systems occupying twice the area uniquely available to each plant, and therefore competing for water. A premise of the theory is that physiological lesions due to water shortage take place only when all of the available soil water is exhausted. Nevertheless the avoidance of the risk involves loss of growth, the loss becoming greater the more imminent the risk. There is a loose analogy with the penalty suffered by a grazing animal in danger from attack by a carnivore; the more imminent the risk the greater is the grazing time lost in sensing and taking evasive action against the danger.

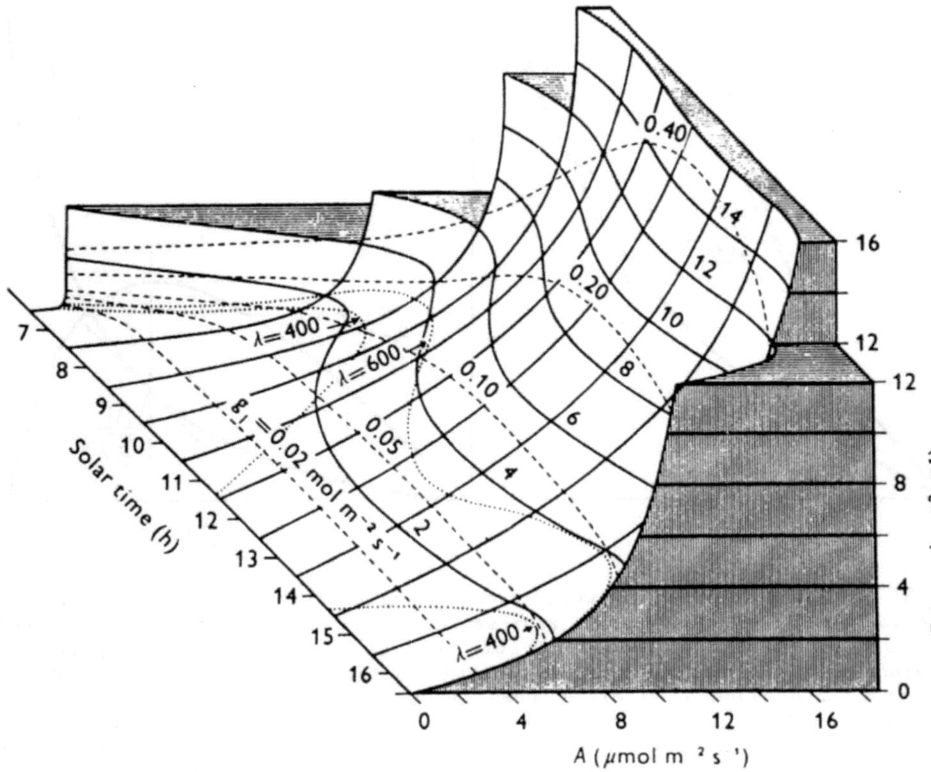

Fig. 4. Rate of transpiration, E, as a function of rate of assimilation, A, and time of day for certain assumed characteristics of leaf metabolism and environment. The magnitudes of E are indicated on the contours of the surface. The broken lines are trajectories on the surface, representing the diurnal variation of E and A for particular constant magnitudes of leaf conductance, $g_\ell (= 1/r_\ell)$. The dotted lines are trajectories for which $\partial E/\partial A = \lambda$ is constant. (Cowan and Farquhar 1977).

The supposition that stomata are adapted to maintain $\partial E/\partial A$ constant, at a magnitude dependent on the amount of water available to the plant, carries with it the implication that they must be suitably sensitive to the many factors that influence transpiration and assimilation. The climate to which plants are subjected is not as regular as Fig. 3 suggests, and is not predictable. Therefore a pre-ordained diurnal rhythm in stomatal aperture, modulated with respect to the amount of water available to the roots, would contribute only partly to the optimisation of carbon gain in relation to water loss. We have seen that stomata indeed respond to both internal and external factors that affect E and A. The extent to which these responses are consistent with the hypothesis that $\partial E/\partial A$ is constant is a matter for continuing experiment and debate. The extent to which $\partial E/\partial A$ being constant is, in fact, an ideal solution of the compromise between the water and carbon economies of plants is also open to question. For example, the criterion does not take account of possibilities that the energy involved in stomatal movements may be significant, and the characteristics relative to carbon fixation (i.e. characteristics of the kind shown in Fig. 2) may be influenced by the amount of carbon already fixed.

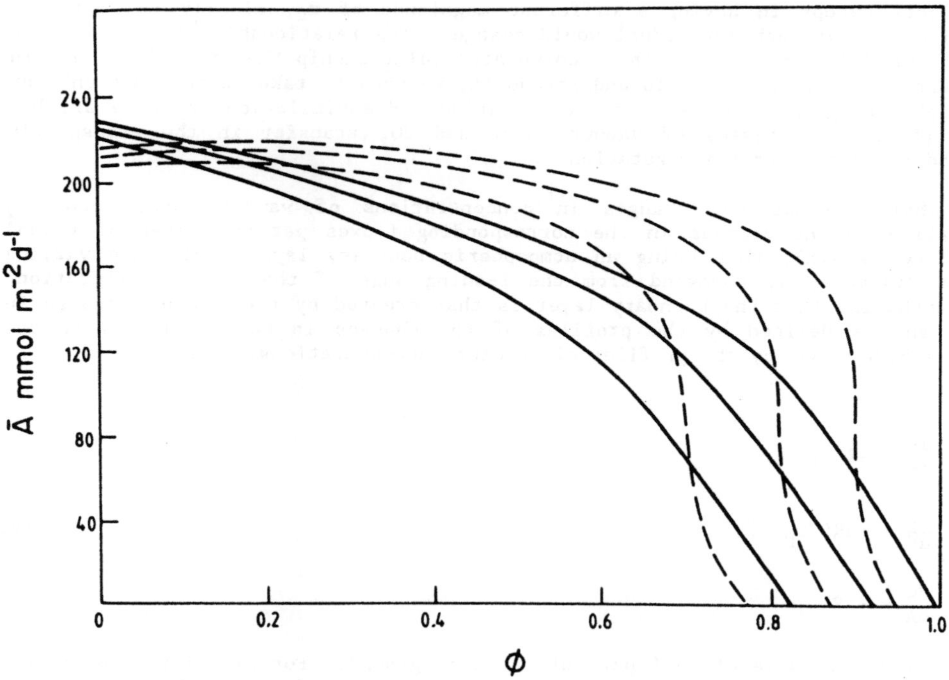

Fig. 5. Optimal variation of rate of assimilation per day, $\bar{A}$, with soil water deficit, φ. (Cowan 1986). The temporal functioning of a plant on such a curve is distributed in a way dependent on the statistics of rainfall, the variation of rate of transpiration with φ (this being related to the variation of A with φ), and other processes, such as evaporation directly from the soil, which remove water from the soil root zone. The distribution defines two important parameters: the probable average rate of assimilation, $\hat{A}$ say, and the probability, K say, the plant will become dessicated ($\varphi = 1$) in a given period of time. Each of the variations shown in the figure are optimal in the sense that no other variation would yield greater $\hat{A}$ without increase in K, or smaller K without decrease in $\hat{A}$. The full curves relate to plants in a community in which root systems do not overlap; the broken curves to a community that is similar except that root systems overlap and therefore compete for water.

The theory that has been outlined here is perhaps best regarded as an attempt to identify some of the selection pressures that may have influenced the behaviour of stomata. Given that we are unlikely to be able to identify all such pressures (let alone quantify them), or account for all the physiological constraints on the adaptation of the stomatal mechanism, it is hardly possible that we shall succeed in describing stomatal behaviour with precision.

VEGETATION AND THE ATMOSPHERIC BOUNDARY LAYER

The relationship between E and A that we have discussed is that which would occur if the stomata in an individual plant were caused to open and close without change in the external environment. But if the stomata in all the individuals in an extensive area of vegetation were to be opened or closed somewhat, or that area were to be replanted with a cultivar identical in all respects except in having a different magnitude of r_ℓ, clearly the external environment of each individual would change. The relationship between E and A would be different. It is this "corporate" relationship that is of interest in an agricultural context. To understand it, we need to take account not only of the physiological processes of transpiration and assimilation in the leaf, but the physical processes of vapour, heat and CO_2 transfer in the atmospheric boundary layer over the vegetation.

We shall express the changes in concentrations of vapour, heat, and CO_2 relative to the changes in the corresponding fluxes per unit area of ground surface as R(x), this being an atmospheric boundary layer resistance varying with distance, x, downwind from the leading edge of the altered vegetation. We emphasise that the boundary layer is that created by the supposed change in r_ℓ, and is defined by the profiles of the changes in concentrations in the atmosphere - not by the profiles of absolute concentrations. Thus

$$\frac{dw}{dE} = aR \quad ,$$

$$\frac{dT}{dE} = -aRL/C_p \quad , \tag{6}$$

$$\frac{dc}{dA} = -aR \quad ,$$

in which a is area of leaf per unit area of ground. For simplicity, we assume in what is to follow that all leaves are exposed to radiation and the atmospheric environment in the same way. The second of (6) derives from the assumption that the transfer of thermal radiation between the vegetation and the atmosphere, and within the atmosphere, is unchanged, and therefore the quantity $Lw + C_pT$ is conserved. Differentiating (2) and (4) as before, but now substituting (6) for the changes in the environment previously taken as constant, we find that

$$\frac{r_\ell}{E} \cdot \frac{dE}{dr_\ell} = \frac{-r_\ell}{(\varepsilon+1)(aR(x)+r_b) + r_\ell} = \Sigma_E, \text{ say,} \tag{7}$$

$$\frac{r_\ell}{A} \cdot \frac{dA}{dr_\ell} = \frac{-r_\ell}{0.63aR(x) + 0.86r_b + r_\ell + 0.63/(\partial A/\partial c_i)} = \Sigma_A, \text{ say.} \tag{8}$$

In evaluating (7) and (8), we have used

$$R = 0.416x^{1/9} \text{ m}^2 \text{ s mol}^{-1} \tag{9}$$

with x in m, this being a particular example of the solution of the advection problem due to Philip (1959), but see also Philip (1987), and one shown in the Appendix to be appropriate (with some reservations about accuracy close to the leading edge) for vegetation having a roughness length of 0.1 m and sustaining a drag equivalent to a friction velocity of 0.4 m s^{-1}. (Note that R, as used here, has units m^2 s mol^{-1}, whereas R_1 in the Appendix has the meteorological units s m^{-1}.) With a taken as 2, the sensitivities Σ_E, Σ_A and their ratio are as shown in Table 1.

At the leading edge, the sensitivities Σ_E and Σ_A are the same, for reasons that must be obvious, as the sensitiviites σ_E and σ_A for a single leaf discussed earlier in this article. The rate of change in sensitivity with distance close to the leading edge is probably over estimated because of the artificiality of the profile of eddy diffusivity close to the vegetation assumed by Philip, and discussed in the Appendix. However, the calculations of Table 2 in the Appendix suggest that R(x) of the Philip analysis is likely to converge quite closely within a few hundred metres of the leading edge towards that which would emerge from a more realistic description of the K-profile. Caution is again required in interpreting the rate of change of R(x) with x when x is greater than some 20 km, for the equivalent boundary layer height is comparable with what is conventionally taken as the height of the planetary boundary layer. Neither the Philip nor the linear representation of K(z) is satisfactory on this scale, even in neutral conditions. In any case the resistance of the boundary layer cannot be allowed to increase indefinitely. To discuss transfer processes at very much greater distances, one would need to take account of condensation in the atmosphere, divergence of terrestrial radiation, diurnal variation in radiation, large scale vertical movements of air, and many other factors, all of which influence the absolute magnitude of E, and which are here avoided, intentionally, by dealing only with relative changes in E over limited distances. For a discussion of some of these factors, see de Bruin (1983).

Table 1. Influences of change in stomatal resistance, r_ℓ, extending over fetch x. R is atmospheric boundary layer resistance; Z is equivalent boundary layer height given by $R = \int_0^Z dz/K$, where K is turbulent diffusivity expressed in consistent units (molar flux/gradient in mole fraction); Σ_E, Σ_A and Σ_T, defined in text, are the sensitivities of rate of transpiration, E, rate of assimilation, A, and temperature of air close to vegetation, T, to the change in resistance.

x m	R(x) $m^2\ s\ mol^{-1}$	Z height m	Σ_E	Σ_A	Σ_T $10^3\ ^oC$	Σ_E/Σ_A
0	0	0	-0.59	-0.27	0	2.17
1	0.42	0.2	-0.35	-0.26	0.22	1.36
2	0.45	0.3	-0.34	-0.26	0.23	1.33
5	0.50	0.5	-0.32	-0.25	0.24	1.28
10	0.54	0.9	-0.31	-0.25	0.25	1.24
20	0.58	1.5	-0.30	-0.25	0.26	1.20
50	0.64	3.1	-0.29	-0.25	0.28	1.15
100	0.69	5.3	-0.27	-0.25	0.29	1.11
200	0.75	9.0	-0.26	-0.25	0.30	1.07
500	0.83	18.4	-0.25	-0.24	0.31	1.02
1000	0.90	31.5	-0.24	-0.24	0.32	0.98
2000	0.97	54.0	-0.23	-0.24	0.33	0.95
5000	1.07	110.2	-0.21	-0.24	0.34	0.90
10000	1.16	188.9	-0.20	-0.23	0.35	0.87
20000	1.25	323.9	-0.19	-0.23	0.36	0.82
50000	1.38	660.5	-0.18	-0.23	0.37	0.79
100000	1.50	1132.4	-0.17	-0.23	0.38	0.76

Within downwind distances of 200 m to 20 km, the sensitivity Σ_E is 1/2 to 1/3 of what it is at the leading edge. The reduction in the sensitivity Σ_A is very much smaller, with the result that Σ_E/Σ_A ($=(A/E)dE/dA$) is less than unity. That is to say, the relative reduction in rate of assimilation in a new cultivar, differing only from its parent in having an increased stomatal resistance, would exceed the relative reduction in rate of transpiration. This does not necessarily imply that such a cultivar is deficient for agricultural purposes. Reduction in rate of transpiration might be of paramount importance in water-limited environments (although reduced density of planting may be a more efficaceous way of eking out the water that is available). The discussion does show, however, the importance of taking account of boundary layer effects in any attempt to engineer the water-use characteristics of crop plants. It is worth noting that enhancement of the biochemical capacity of plants to fix carbon at a given stomatal resistance is not subject to as stringent a qualification. This is consistent with the observation that the concentration of CO_2 in the lower atmosphere does not change much as air moves across surfaces having different rates of CO_2 exchange.

In contrast to the influences of a change in r_ℓ on transpiration and assimilation, which diminish with distance downwind, the influences on temperature, humidity, and CO_2 concentration increase, as combination of (6), (7) and (8) shows. The quantity $(1/E)dT/dr_\ell$, calculated with the same numerical assumptions used in calculating Σ_E and designated Σ_T, is shown in Table 1. It cannot be translated into numerical values without also assuming a magnitude for E. Taking $E = 3$ mmol m^{-2} s^{-1}, the increase in T at $x = 20$ km is 1.1°C for an increase in r_ℓ of 1 m^2 s mol^{-1}. Several parameters that have been taken as constant in this article, ϵ, φ, and $\partial A/\partial c_i$, are, in fact, sensitive to temperature and a rigorous treatment of the advection problem for large differences in plant surface properties would have to take that into account. The change in CO_2 concentration is, as observations show, small, because of the predominant influence on assimilation rate of the leaf internal biochemical resistance.

The effects of the variables in (7) and (8) on dE/dA are readily discerned, and it is unnecessary to provide further numerical examples. If r_ℓ is small, a is large, and R is large (as it tends to be with vegetation having small aerodynamic roughness), then dE/dA diminishes rapidly with the fetch x. That will generally obtain with rapidly growing, densely planted, annual crops. If r_ℓ is large, as it is when soil water is in short supply, then dE/dA decreases slowly, individual plants functioning more nearly as plants in isolation. This is also so with vegetation, particularly open forest, that is aerodynamically very rough. A recent discussion of the properties of different kinds of vegetation in relation to stomatal control of transpiration has been provided by Jarvis and McNaughton (1986).

DISCUSSION

Evidently there are two different types of relationship between E and A in vegetation (see Fig. 5). One, that discussed first in this article, pertains to the changes that would take place in an individual plant if leaf resistance in that plant alone were to change. It is this type of relationship, it has been suggested, that is peculiarly relevant to the adaptation of the stomatal mechanism, for presumably adaptation has come about through selection of aberrant individual stomatal behaviour. The other relationship is that which actually takes place when stomatal aperture changes not only in an individual plant, but in neighbouring plants also as a result, for example, of a change in the availability of soil water. It describes the function of plant communities, including agricultural crops.

On the face of it, our analysis suggests that, contrary to the notion of improving the economics of carbon gain relative to water loss by artificial closure of stomata, there may be gains to be made in decreasing leaf resistance. However, it is probable that artificial selection has brought this about, albeit inadvertantly, and that rate of assimilation in crop plants cannot be much enhanced by further opening of the stomata at present levels of CO_2 concentration. The matter is somewhat academic though, because atmospheric CO_2 concentration is increasing about 1.4 ppm per annum due, primarily, to the burning of fossil fuels. Stomata are responding to this. It has long been known that stomatal aperture diminishes with increase in ambient CO_2 concentration. There is recent evidence, gathered from herbarium specimens and confirmed with experiments in controlled environments, that the numbers of stomata, relative to the numbers of epidermal cells, is decreased in plants grown with enhanced CO_2 concentration (Woodward 1987). However, although these effects diminish the increase in availability of CO_2 at the sites of photosynthesis, they do not obviate it. Breeding and selection should be oriented to ensure that the future increase in availability is translated into the maximum possible increase in economic growth. The ways in which the increase in stomatal resistance, and possible changes in patterns of plant growth and distribution, will contribute to future trends in climate are interesting problems, but ones beyond the scope of this article.

APPENDIX

In his treatment of advection, Philip (1957, 1987) analyses the simultaneous equations

$$\frac{\partial \Phi}{\partial z} = -u\frac{\partial \theta}{\partial x}$$

and

$$\Phi = -K\frac{\partial \theta}{\partial z} \quad ,$$

in which Φ is the vertical flux of an entity, the concentration of which is θ, u is wind speed, K is eddy diffusivity, z is height above the land surface, and x is distance downwind. Philip takes $u = u_1 z^m$ and $K = K_1 z^n$ with the restrictions, $n < 1$, $m > n-2$. He obtains analytical solutions of the equations for two sets of boundary conditions. In both, $\theta = 0$ for $x = 0$, $z = 0$. The additional condition for the first is $\theta = 1$ for $x > 0$, $z = 0$. For the second, it is $\Phi = 1$ for $x > 0$, $z = 0$. In each case the boundary layer resistance, $\theta(0,x)/\Phi(0,x)$, is of the form $R_1 x^p$ where $p = (1-n)/(2+m-n)$. Provided p is small, the magnitudes of R_1 for the constant surface concentration and constant surface flux conditions differ very little. They may both be expressed as

$$R_1 = \frac{1}{(1-n)K_1}\left[\frac{(2+m-n)^2 K_1}{u_1}\right]^p \left[\frac{\Gamma(1+p)}{\Gamma(1-p)}\right]^{\frac{1}{2}} \quad .$$

In a numerical example, Philip (1987) takes $m = 1/7$, $n = 6/7$, $u_1 = 5.19\ m^{6/7}\ s^{-1}$, $K_1 = 0.15\ m^{8/7}\ s^{-1}$. We have adopted the same magnitudes of m and n, but have preferred to use $u_1 = 3.5\ m^{6/7}\ s^{-1}$ and $K_1 = 0.32\ m^{8/7}\ s^{-1}$, yielding $R_1 = 16.62\ x^{1/9}\ s\ m^{-1}$. In doing so, we have wished to represent, as closely as possible within the restrictions imposed by the adoption of Philip's analysis, transfer from vegetation having a roughness length of 0.1 m, and sustaining a drag equivalent to a friction velocity of $0.4\ m\ s^{-1}$, these being assumed to typify the condition of a wheat crop on a day of light to moderate wind.

Caution is required in the use of Philip's analysis. The profiles of K and u differ from the linear and logarithmic expressions more commonly taken to represent the profiles in near-neutral conditions. More importantly, the analysis assumes $K = 0$ at $z = 0$ (indeed it is this fact that requires $n < 1$ in the power law representation of K, for if $n > 1$ then the boundary layer resistance is infinite). Table 2 shows a comparison of the boundary layer structure as represented by Philip (with the particular parameter values we have adopted) and that corresponding to the equations $K = ku_*(z+z_0)$ and $u = (u_*/k)\ln(1+z/z_0)$, with von Karman's constant k being 0.4, the friction velocity u_* being $0.4\ m\ s^{-1}$, and the roughness length z_0 being 0.1 m.

The differences are not great within the region 5 < z < 500 m. The parameter $\int_0^z dz'/K$, which is the "resistance" of the slab of atmosphere between the vegetation and height z, is particularly well matched. One may infer that, at distances downwind sufficiently large for the fluxes of vapour, heat, and CO_2 to be substantially invariant with height in the layer below z = 5 m, the expression for R(x) due to Philip will not differ much from that which would emerge from a numerical solution using the linear profile for K and the logarithmic profile for u.

Table 2. Profiles of windspeed u, diffusivity K, and "resistance" $\int_0^z dz'/K$, used in this article and those (distinguished by a superscript) corresponding to a linear profile of K and logarithmic profile of u with roughness length 0.1 m and friction velocity 0.4 m s^{-1}.

z m	u m s^{-1}	K m^2 s^{-1}	$\int_0^z dz'/K$ s m^{-1}	u^+ m s^{-1}	K^+ m^2 s^{-1}	$\int_0^z dz'/K^+$ s m^{-1}
0	0	0	0	0	0.02	0
0.1	2.52	0.04	15.7	0.69	0.03	4.33
0.2	2.78	0.08	17.4	1.10	0.05	6.87
0.5	3.17	0.18	19.8	1.79	0.10	11.20
1	3.50	0.32	21.9	2.40	0.18	15.0
2	3.86	0.58	24.2	3.04	0.34	19.0
5	4.40	1.27	27.5	3.93	0.82	24.6
10	4.86	2.30	30.4	4.61	1.62	28.8
20	5.36	4.17	33.6	5.30	3.22	33.1
50	6.12	9.15	38.3	6.20	8.02	38.9
100	6.76	16.57	42.2	6.91	16.02	43.2
200	7.46	30.02	46.6	7.60	32.02	47.5
500	8.50	65.84	53.2	8.51	80.02	53.2
1000	9.39	119.28	58.7	9.21	160.02	57.6

REFERENCES

Cowan IR (1972) Mass and heat transfer in laminar boundary layers with particular reference to assimilation and transpiration in leaves. Agric Meteorol 10: 311-319

Cowan IR (1986) Economics of carbon fixation. In: Givnish TJ (ed) On the economy of plant form and function. Cambridge University Press, Cambridge, pp 133-170

Cowan IR, Farquhar GD (1977) Stomatal function in relation to leaf metabolism and environment. Symp Soc Exp Biol 31: 471-505

de Bruin HAR (1983) A model for the Priestley-Taylor parameter α. J Climate Appl Meteorol 22: 572-578

Hatch MD (1976) The C_4 pathway of photosynthesis: mechanism and function. In: Burris RH, Black CC (eds) CO_2 metabolism and plant productivity. University Park Press, Baltimore, pp 59-81

Jarvis PG, McNaughton KG (1986) Stomatal control of transpiration: scaling up from leaf to region. Adv Ecol Res 15: 1-49

Penman HL (1952) The physical bases of irrigation control. Rep 13th Int Hort Congr, London 2: 913-924

Philip JR (1959) The theory of local advection: I. J Meteorol 16: 535-547

Philip JR (1987) Advection, evaporation, and surface resistance. Irrig Sci 8: 101-114

Raschke K (1975) Stomatal action. Ann Rev Plant Physiol 26: 309-340

Wong SC, Cowan IR, Farquhar GD (1978) Leaf conductance in relation to assimilation in *Eucalyptus pauciflora*. Sieb ex Spreng. Plant Physiol 62: 670-674

Woodward FL (1987) Stomatal numbers are sensitive to increases in CO_2 from pre-industrial levels. Nature 327: 617-618

STOMATAL PHYSIOLOGY AND GAS EXCHANGE IN THE FIELD: COMMENTARY

B.J. Legg

Institute of Engineering Research, Wrest Park, Silsoe, Bedford MK45 4HS, UK

Summary. An optimisation theory of stomatal movement is used to show the type of survival strategy that plants might adopt during drought. The theory cannot readily be used to explain stomatal movement in cultivated crops, as they are now grown in an alien environment, but it could be used in future to determine the ideal response in any given climate.

There are still difficulties with all methods for measuring gas exchange in the field. Micrometeorological methods require a large uniform fetch and we cannot yet be sure that all the theoretical problems have been resolved. The use of field enclosures changes the environment around the vegetation: in recent studies this has not affected transpiration but has had a major effect on CO_2 uptake rates. Combined micrometeorological-physiological methods are still the best in many situations, but we are not yet able to predict the surface resistance to water vapour loss for vegetation growing in dry soils.

INTRODUCTION

It is very appropriate that the subjects of stomatal physiology and field gas exchange should be linked together. In many early studies they were not, and it was assumed in Penman's early work (1952), for example, that the transpiration rate of all well watered crops was the same. Similarly, some workers assumed that photosynthesis-light response curves obtained from dissected leaves in the laboratory could be applied directly to leaves in the field. It has now been shown that stomata do influence gas exchange, whether one is considering water vapour, carbon dioxide, or even sulphur dioxide. But for water vapour there are feedback loops whereby a change in stomatal resistance changes the water potential of the plant and the vapour pressure deficit of the air close to the leaves, and both of these have some further influence on the stomata. The control systems are still not fully understood, especially in advective conditions, but over the last 21 years there have been advances in our understanding of stomata, reviewed here by Cowan (1988), and equally great advances in our ability to measure gas exchange in the field, with the CSIRO Division of Environmental Mechanics taking the lead.

STOMATAL PHYSIOLOGY

Almost all higher plants have stomata that vary their aperture throughout the day, and this observation prompts two very obvious questions. What external or internal stimuli are they responding to? And why do they do it?

Most stomatal physiologists have concentrated on the first question, and accepted a rather vague answer to the second: they open to allow carbon dioxide in, but as this inevitably causes a loss of water, they must close again if the plant water potential becomes dangerously low. Cowan (1977) and Cowan and Farquhar (1977) have taken the question "Why?" more seriously, and have used their answer to develop an elegant and powerful theory of stomatal behaviour. The question is, of course, teleological, but if we once phrase the question of "what strategy for stomatal movement would we expect to give a plant a competitive advantage over its neighbours?", it is then quite acceptable and much simpler to ask the question "why do stomata move?" thereafter.

Cowan's fundamental hypothesis is that stomata move in order to optimise the ratio of carbon dioxide uptake to water use, and it is important to recognise that this does not lead to one unique strategy, but to a whole family of strategies. For a given daily cycle of environmental conditions one plant, by moving its stomata, may assimilate A_1 units of carbon and limit water loss to the minimum amount possible, E_1. Another plant, by optimising its stomatal movements, may assimilate A_2 units of carbon, and restrict water loss to E_2. It can be shown, however, that if A_2 is greater than A_1, then the ratio of E_2/A_2 must be larger than E_1/A_1. In other words: if a plant chooses to grow fast, it can do so only by using water profligately; if a plant chooses to use water conservatively, it must necessarily grow slowly.

At present only carbon uptake and water use are included in the theory, and there are some fairly subtle effects. For example, a decrease in evaporation rate causes a rise in leaf temperature which in turn affects the rate of photosynthesis. But time-delayed effects are not allowed for, nor is the possibility that evaporation could prevent permanent damage due to high temperatures.

There is an intriguing analogy with the strategy that a herbivore may choose in trying to survive predators (Cowan 1988). Some may devote all their time and energy to growing and multiplying quickly, and accept large losses due to predation (some insects exemplify this); others may choose to spend a lot of time and energy avoiding predators, and thereby grow slowly (grazing mammals). At first sight the analogy seems weak. The strike from a predator is always sudden, and usually fatal, whereas many plants spend a considerable part of their lives under water stress, but manage to complete their life cycle without dying of drought (N Collis-George, personal communication 1987). But it depends whether one views plant response to stress "as derangements that arise from stress-induced lesions at vulnerable sites in metabolism, or as potentially adaptive changes that reflect ordered operation of metabolic regulatory mechanisms and favour the plant as a whole during and after stress" (Hanson and Hitz 1982). Cowan inclines towards the second view, and so the plant sacrifices present growth for future survival in the same way that some herbivores grow more slowly in exchange for a longer life expectancy. Death, by drought or by the teeth of a carnivore, may come to only a few individuals, but the risk is real enough that an evolutionary advantage has been gained.

To pursue the analogy further we might expect herbivores to spend more time and energy avoiding predators when the threat is greatest, and we might also expect stomata to become more sensitive when the soil around part of the root system becomes dry, giving portents of impending doom (BE Clothier, personal communication 1987). Recent experiments by Gollan et al. (1986) and Blackman and Davies (1985) tend to support this hypothesis, and it seems possible that a signal is sent to the stomata from the roots by means of a reduction in cytokinin export. But does this mean that plants may stop using water when part of the root system is in dry soil, despite the fact that water is freely available elsewhere? This is related to the question of what average soil water potential the roots sense, and whether roots form channels for water to move freely from wet soil to dry. Although the latter has been observed, it is probably not a very large effect, and we can be reassured that the small amount of water redistribution via the roots can be explained without the need for complex valves and switches. The fact that the hydraulic conductivity of wet soil is far higher than that of dry soil means that water can enter roots from wet soil far more easily that it can escape into dry zones (PAC Raats, personal communication 1987). Similarly, plants are unlikely to panic when only a small part of the soil is dry, but will respond when the dry zones are large enough to indicate more serious depletion.

As we have discussed, there are many survival strategies that a plant may follow. The winning strategy will depend on the environmental conditions, including rainfall, that are normally encountered, and it will also depend on what other plants are competing for the water. For example, there is no point conserving water and growing slowly if rainfall is abundant, nor if your neighbour has decided to grow fast and use it all up. It would be interesting to know whether some plants, such as water lillies or marsh plants, have adapted to an abundant supply of fresh water (ARG Lang, personal communication 1987). Although some measurements of the water use of marsh plants have been made (Linacre et al. 1970), they were not sufficiently detailed to show whether the stomata close in the dark or whether they respond to the vapour pressure deficit of the air. The stomata of mangroves have been studied (Clough et al. 1982) and show a similar response to other plants, but there may be a severe restraint on the availability of water because of its salinity.

Before we can apply the optimisation theory to agricultural crops, we must be very careful. Many are grown in circumstances that are radically different from those in which they evolved. For example, wheat is cultivated as an annual, whereas wild wheat is a perennial. The competition from other plants is also very different, and may even be eliminated by the use of herbicides. An extreme example of odd behaviour is found in kiwifruit, whose stomata stay open at night, contributing up to 20% of the total transpiration, and under severe water stress the stomatal resistance does not rise above 500 s m^{-1} even when the consequence is that leaves may fall off (KJ McAneney, personal communication 1987). Did the ancestor of the kiwifruit evolve in a very wet environment? Or has selective breeding enhanced some desirable qualities and destroyed others that are not essential for a cultivated plant?

It is quite in order, however, to ask whether the optimisation theory can indicate what stomatal behaviour is most desirable in cultivated crops. If they are to be irrigated, the answer is simple: they should have a low stomatal resistance and adopt the high growth and very high transpiration stategy. There is some evidence that plant breeding for high yield has unwittingly achieved this, and the surface resistance of well managed and well watered arable crops, including pasture, is in the range of 25-50 s m^{-1}. The surface resistance of wild vegetation, including forests and rangelands, seldom if ever falls this low. Prairie grassland and heather have values nearer 120 s m^{-1}; forests on poorer soils in the humid tropics, or on chalk soils in temperate climates, are higher than this; and even the most productive Douglas-fir forests in British Columbia have a surface resistance of 60 s m^{-1} (KG McNaughton, personal communication 1987). It is not so clear what strategy should be adopted by dryland crops. It will depend on the likely rainfall during the growing season. But optimisation theory could be used to provide an answer, and this is a very worthy objective for future research.

MEASUREMENT OF GAS EXCHANGE IN THE FIELD

The fact that gas exchange in the field has been measured by such a wide diversity of methods (Table 1) is perhaps an indication that none are totally satisfactory. In general, the micrometeorological methods do not interfere with the environment, but demand very large uniform plots. Also, there have been fairly regular discoveries that earlier theories were wrong (more of this

Table 1. Measurement techniques for field gas exchange

	Gases other than water vapour	Plot size	Interference to environment
1. Micrometeorological			
1.1 Eddy correlation	Yes	>100h*	No
1.2 Bowen ratio plus gas conc. gradient	Yes	>100h	No
1.3 Advection budget	Yes	~10 m	No
2. Physiological			
2.1 Chambers for individual leaves/organs	Yes	~0.1 m	Yes
2.2 Chambers for whole plants or small communities	Yes	1-10 m	Yes
2.3 Stem water flow			
- heat pulse	No	1-10 m	No
- heat budget	No	0.1-1 m	No
3. Combined physiological-micrometeorological			
3.1 Diffusion through surface and aerodynamic resistances	Yes	>0.1 m	No
3.2 Combination equation			
- Penman-Monteith	No	>0.1 m	No
4. Hydrological	No	>1000 m	No

*h = crop height

later), and it would be audacious to assume that the equations in present use are completely correct. Physiological methods can be used on a much smaller scale, and give precise results; but there is still fierce debate about how much enclosures change the environment and how much they change the magnitude of gas fluxes being measured. Futhermore, because of point-to-point variability, the sampling problem is large. Combined physiology and micrometeorological methods try to give the best of both worlds, but it has often proved difficult to choose the appropriate surface concentration, and time consuming to derive the appropriate average value for surface resistance.

This paper does not attempt to review the whole subject, but outlines a few recent advances.

Micrometeorological Methods

The measurement of gas exchange by eddy covariance is the most fundamental of all methods, but there have been problems both with instrumentation and with theory.

The ideal gas concentration sensor has a very small sampling volume and high frequency response, and it must be used with an anemometer that gives velocity fluctuations, preferably in three dimensions, at the same point in space. Open-path, optical-absorption instruments seem best suited for this, and several experimental infrared absorption analysers have been developed. One, by Ohtaki and Matsui (1982), uses rotating optical filters so that the concentration of water vapour and carbon dioxide can be measured in rapid succession in the same sampling volume. This instrument has been used to measure fluxes above vegetation, and more recently to study the annual cycle of carbon dioxide absorption and release by the oceans[1].

The need for high frequency response can be relaxed by making measurements at greater heights, though at the expense of needing a larger uniform fetch. But large eddies also dominate the transport process within and above forest canopies, and OT Denmead and EF Bradley[2] have recently shown that an adequate time response can be obtained by drawing an air sample from the forest canopy and passing it through a standard laboratory infrared gas analyser (IRGA). They sampled air from between 3 and 25 m height in a 20 m high plantation of Pinus ponderosa, and passed it through 150 m of 5 mm i.d. polypropylene tube to two identical infrared gas analysers connected in series, one fitted with a water vapour detector and one with a CO_2 detector. The output signals were electronically filtered and showed a fairly flat frequency response between 0.01 and 0.3 Hz. The smoothing that occurred within the sampling tubes (Philip 1963) was at higher frequencies and was probably unimportant. For comparison they also measured the water vapour fluctuations *in situ* using a Lyman-α detector, and the carbon dioxide flux above the canopy by using the Bowen ratio and vertical gradient of mean carbon dioxide concentration. Results show that while the IRGA underestimated the water vapour flux by a constant proportion of 40%, the average CO_2 flux using the IRGA for eddy correlation was only 5%, on average, below estimates from the mean concentration profile.

Theoretical problems with the eddy covariance method have revolved around the interference caused by fluxes of heat and water vapour on the relatively small fluxes of minor constituents such as carbon dixoide. A definitive paper was written by Webb et al. (1980), and has not been seriously challenged. But E Ohtaki and colleagues[1] measured 'raw' values of CO_2 flux over oceans of -0.05 to -0.017 mg m^{-2} s^{-1} and correction terms of $+0.10$ mg m^{-2} s^{-1} and $+0.08$ mg m^{-2} s^{-1} due to sensible heat and water vapour flux, respectively. It is worrying when correction terms exceed the measured value, but the same effect can be found when measuring fluxes of other trace gases over vegetation, such as ozone (GW Thurtell, personal communication 1987) and nitrogen oxides (Denmead 1983). EK Webb (personal communication 1987) observes that for CO_2 fluxes over vegetated surfaces the correction terms may be as large as 50%; they should never be ignored on the assumption that they are probably negligible. Furthermore, it can be estimated that over oceans, where the CO_2 flux is generally very small, the correction terms may exceed the true value of the flux by a factor of 20 or more. Thus, errors in the correction terms are considerable, and this may preclude the use of the eddy covariance method. It is possible to eliminate the need for the heat flux correction term by sampling the air and smoothing out temperature fluctuations[2], but even this seems

[1]E Ohtaki, O Tsukamoto, Y Iwatani (1987) Turbulent transport of carbon dioxide in the atmospheric boundary layer over the ocean. Symposium poster paper

[2]OT Denmead, EF Bradley (1987) Eddy-correlation measurement of the CO_2 flux in plant canopies. Symposium poster paper

unlikely to overcome the problem over oceans where temperature fluctuations would need to be eliminated to a very high degree of accuracy (EK Webb, personal communication 1987).

Micrometeorologists have for many years attempted to measure gas fluxes above vegetation from the mean concentration gradients. Initially it was assumed that the eddy diffusivity for momentum (K_M), heat (K_H), water (K_W) and other gases (K_G) would be identical, but fairly early work showed that K_H and K_W are considerably larger than K_M in unstable conditions. More recent work has provided evidence that even in thermally neutral boundary layers K_H exceeds K_M close to vegetation by as much as a factor of two (Raupach and Legg 1984). This effect is related to the fact that diffusion theory is not valid close to sources or sinks (Raupach 1988), and the source and sink heights are not the same for momentum and heat. At present we know of no *a priori* reason to assume that K_G is the same as K_H and K_W, especially where the sources are different, but there is indirect evidence from work in which the CO_2 flux above crops has been measured for long enough to compare with the dry matter increase of the crop (e.g., Biscoe and Gallagher 1978). It would certainly seem safer to estimate K_G from the Bowen ratio (e.g., Denmead and Bradley 1985) than to derive it from the wind profile.

We have emphasised the need for a long and uniform fetch in applying all micrometeorology methods, but it is worth underlining the errors that occur if they are used over complex terrain (JJ Finnigan, personal communication 1987). The Bowen ratio, for example, relies on measurements made at two different heights on a mast and a measurement of net radiation at the same point. Each sensor on the mast "sees" concentrations of CO_2 or water vapour, or temperatures, that are determined by air particles diffusing from a patch of ground upwind. The higher sensor sees concentrations representative of a larger area further upwind than does the lower sensor, while the net radiometer views a third area directly beneath. If these areas are not uniform, either through change in vegetation type, effective sun angle, or source strength caused by wind and turbulence changes in complex terrain, systematic errors are possible. Related problems apply equally to measurement by eddy covariance.

The one micrometeorological method that can be used on small areas is the advection budget, used, for example, by Denmead et al. (1977) to measure ammonia losses from small fertilised areas. Unfortunately, it can only be used readily for gases that are not ubiquitous; but we may hope for a reduction in fetch requirement for other methods as our understanding of advection improves (JR Philip, personal communication 1987).

Physiological Methods

Physiologists, plant breeders and agronomists all wish to measure the performance of small groups of plants in the field. Traditional growth analysis is destructive and very labour intensive, and can measure only the integral response of plants to the variable weather over several weeks. The heat pulse technique for measuring the water flow in individual stems is now well established, though Green and Clothier (1987) have shown that while Swanson's (1983) analysis gives very accurate answers for apple trees, it underestimates the flow in kiwifruit vines by 37%. The error is caused by very heterogeneous flow across the stem due to large xylem vessels. It is clearly important that independent calibration methods are used for each new situation in which the method is used.

Perhaps less well known is the steady state heat budget method for small stems that was proposed initially by Vieweg and Ziegler, and developed and used by Sakuratani (1981). A budget equation relates the electrical heating applied to a short insulated length of stem to the losses by conduction through the insulation, conduction in the plant stem, and convection by the water flowing through the stem. For soybeans and sunflowers (and presumably many other species) the heat loss by convection in the water stream accounts for more than 75% of the total for much of the day, and as the conduction terms can be estimated fairly accurately, this provides a useful absolute method of measuring sap flow. The method gives a continuous output, and can be used in the field.

For carbon dioxide exchange, the use of chambers is still by far the most popular method, and it is important to establish whether results obtained are likely to relate closely to the undisturbed world outside. A most thorough study of the transpiration of 10 m tall trees inside and outside a ventilated chamber has been made by Foster and Leuning (1987). They obtained temperatures and humidities within the chamber that were close to ambient, and showed that transpiration in the chambers was within 10% of that outside. The air speed was different inside the chamber, but the aerodynamic resistance of the leaves was not changed by enough to have a major influence on transpiration. This result is consistent with the view that forest transpiration is controlled predominantly by stomatal resistance, and this is unchanged in the chamber provided the radiation and vapour pressure deficit are the same as outside. However, this is not so when the foliage is wet, and further work is needed to show whether the same conclusions apply to a more dense canopy.

In similar experiments, Dunin and Greenwood (1986) also showed transpiration of Eucalyptus maculata to be unchanged by the presence of a chamber, but OT Denmead and colleagues[3] found the rate of CO_2 uptake within the chamber to exceed that outside by as much as a factor of two. The ratio was near to unity under cloudy skies and around midday, but close to two in the mid-morning and mid-afternoon when the sun was shining. The explanation offered is that the chamber acts as a giant diffuser, and so illuminates the leaves more uniformly and causes a higher rate of photosynthesis. At first sight this appears to conflict with other experimental evidence that changing the angle of grass leaves to improve the utilisation of sunlight has only a small effect on the rate of net photosynthesis (Sheehy et al. 1979), and that greenhouses clad with diffusing glass were not found to produce dramatically better crops than those with clear glass. But there may be an important difference for E. maculata, whose leaves are mainly vertical (CJ Bell, personal communication 1987). At low sun angle the leaves at the top of the canopy are light saturated, and a diffusing chamber distributes the light more evenly. At high sun angle, however, the vertical leaves cause the light to penetrate deeply into the canopy whether the enclosure is present or not. It would be interesting to know whether a full analysis of light profiles linked to the individual leaf photosynthetic response can account for all of the difference observed, or whether some difference was also caused by the trees in the chamber not being representative of the rest of the canopy, or by systematic error in the external measurement using the Bowen ratio and the gradient of mean CO_2 concentration. It would also be interesting to repeat the experiment with a chamber clad with clear, nondiffusing plastic film.

Until these issues have been resolved, we cannot assume that the photosynthetic rate measured inside canopies is representative of undisturbed vegetation.

CONCLUDING REMARKS

At a time when research in many Western countries is being challenged to prove that it is relevant to practical problems, it is useful to consider what

[3]OT Denmead, FX Dunin, SC Wong, EAN Greenwood (1987) Photosynthesis by trees in tents and forests. Symposium poster paper

benefits have arisen, or are likely to arise, from our studies of stomatal physiology and gas exchange in the field.

The need to measure the water use of natural vegetation and of agricultural crops is perhaps self evident. Research has already had a major impact on hydrology, and been used in planning water supplies, siting reservoirs, forecasting floods, and estimating in advance the effects of changing land utilisation on river flow and aquifer recharge. The estimation of evaporation has also become the basis for calculating irrigation need worldwide. The measurement of carbon dioxide exchange is used more for research than as an operational tool. It is used to compare the photosynthetic capacity of new breeds of agricultural crop, to measure the reduction in growth caused by disease, and the economic benefit that may be obtained from chemical application. If recent attempts to use instrumentation on aircraft to measure CO_2 fluxes are successful, it may also be possible to improve crop yield forecasts on a routine basis.

But what of the studies of stomatal physiology? The movement of stomata has proved so fascinating and enigmatic that many would consider this research to be fully justified simply to satisfy human curiosity. But it has also led to a deeper understanding of the interaction between vegetation and the atmosphere, and forms the basis for the combined physiological and micrometeorological methods of measuring gas exchange (Table 1). Thus, laboratory measurements of photosynthesis can be extrapolated to field conditions; the uptake rate of sulphur dioxide can be related to aerial concentrations and the surface resistance of the crop (though the role of stomata is less important when the foliage is wet); and the Penman-Monteith equation shows how stomata can affect the partitioning of radiant energy between sensible heat and water vapour.

To return to the question of how much stomata affect evaporation, it can be shown directly from the Penman-Monteith equation that the relative sensitivity of evaporation, E, to surface resistance, r_s, is given by

$$\frac{\delta E/E}{\delta r_s/r_s} = \frac{r_s/r_a}{1 + \Delta/\gamma + r_s/r_a}, \qquad (1)$$

where Δ is the slope of the saturated vapour pressure curve against temperature, γ is the psychrometric constant, and r_a is the boundary layer resistance.

In temperate climates Δ/γ is approximately 2 (correct at a temperature of 18°C), so for short vegetation and arable crops where r_s/r_a is close to 1.0 the ratio $(\delta E/E)/(\delta r_s/r_s)$ is equal to 0.25 and evaporation rate is relatively insensitive to surface resistance; but for forests where r_s/r_a may exceed 10.0 this ratio exceeds 0.77. It also follows that the evaporation from wet foliage is only 1.3 of that from dry foliage for short vegetation, but may exceed 4.0 for forests (Monteith 1981). For both of these reasons, the idea of "potential evaporation" has been useful in agricultural meteorology, but is never used by forest meteorologists (KG McNaughton, personal communication 1987).

The next challenge is to see whether our understanding of stomatal response to the environment and soil water availability has advanced to the point where evaporation from drying soils can be estimated reliably from the Penman-Monteith equation. The alternative which is commonly used is to calculate potential evaporation and assume that actual evaporation is less than this by a proportion that decreases in an arbitrary way as soil water deficit increases. The former approach is certainly more elegant (Monteith 1981), and some progress has been made by Russell (1980), but it has yet to be shown that the method is more accurate as well as being more elegant.

ACKNOWLEDGEMENTS

The author is very grateful to Ian Cowan, whose talk stimulated the dicussion that led to this commentary, and to all those who contributed verbally or through posters. Ian Cowan also provided answers to the difficult questions.

REFERENCES

Biscoe PV, Gallagher JN (1978) A physical analysis of cereal yield. I Production of dry matter. Agric Prog 53: 34-50

Blackman PG, Davies WJ (1985) Root to shoot communication in maize plants of the effects of soil drying. J Exp Bot 36: 39-48

Clough BG, Andrews TJ, Cowan IR (1982) Physiological processes in mangroves. In: Clough BF (ed) Mangrove ecosystems in Australia. Structure, function and management. Australian National University Press, Canberra, pp 193-210

Cowan IR (1977) Stomatal behaviour and environment. Adv Bot Res 4: 117-228

Cowan IR (1988) Stomatal physiology and gas exchange in the field. This volume, pp

Cowan IR, Farquhar GD (1977) Stomatal function in relation to leaf metabolism and environment. In: Jenkins DH (ed) Integration of activity in the higher plant, The Soc for Exp Biol Symp No XXXI, University Press, Cambridge

Denmead OT (1983) Micrometeorological methods for measuring gaseous losses of nitrogen in the field. In: Freney JR, Simpson JR (eds) Gaseous loss of nitrogen from plant-soil systems. Martinus Nijhoff/Dr W Junk, The Hague, p 133

Denmead OT, Simpson JR, Freney JR (1977) A direct field measurement of ammonia emission after injection of anhydrous ammonia. Soil Sci Soc Am J 41: 1001-1004

Denmead OT, Bradley EF (1985) Flux-gradient relationships in a forest canopy. In: Hutchison BA, Hicks BB (eds) The forest-atmosphere interaction. D Reidel, Dordrecht, pp 421-442

Dunin FX, Greenwood EAN (1986) Evaluation of the ventilated chamber for measuring evaporation from a forest. Hydrol Proc 1: 47-62

Foster IJ, Leuning R (1987) Comparison of three methods for estimating transpiration by single trees: ventilated chamber, leaf energy budget, and Penman-Monteith equation. CSIRO Div For Res User Series 7, 34 pp

Gollan T, Passioura JB, Munns R (1986) Soil water status affects the stomatal conductance of fully turgid wheat and sunflower leaves. Aust J Plant Physiol 13: 459-464

Green SR, Clothier BE (1987) Water use of kiwifruit vines and apple trees by the heat-pulse technique. J Exp Bot, in press

Hanson AD, Hitz WD (1982) Metabolic responses of mesophytes to plant water deficits. Ann Rev Plant Physiol 33: 162-203

Linacre, ET, Hicks BB, Sainty GR, Grauze G (1970) The evaporation from a swamp. Agric Meteorol 7: 375-386

Monteith JL (1981) Evaporation and surface temperature. Q J R Meteorol Soc 107: 1-27

Ohtaki E, Matsui T (1982) Infrared device for simultaneous measurement of fluctuations of atmospheric carbon dioxide and water vapour. Boundary-Layer Meteorol 24: 109-119

Penman HL (1952) The physical bases of irrigation control. Rep 13th Int Hort Congr 2: 913

Philip JR (1963) The damping of a fluctuating concentration by continuous sampling through a tube. Aust J Phys 16: 454-463

Raupach MR (1988) Canopy transport processes. This volume, pp

Raupach MR, Legg BJ (1984) The uses and limitations of flux-gradient relationships in micrometeorology. Agric Water Manage 8: 119-131

Russell G (1980) Crop evaporation, surface resistance, and soil water status.

Agric Meteorol 21: 213-226
Sakuratani T (1981) A heat balance method for measuring water flux in the stem of intact plants. J Agric Meteorol 37: 9-17
Sheehy JE, Cobby JM, Ryle GJA (1979) The growth of perennial ryegrass: a model. Ann Bot 43: 335-354
Swanson RG (1983) Numerical and experimental analysis of implanted-probe heat pulse theory. Unpublished PhD thesis, University of Alberta
Webb EK, Pearman GI, Leuning R (1980) Correction of flux measurements for density effects due to heat and water vapour transfer. Q J R Meteorol Soc 106: 85-100

AIR FLOW OVER COMPLEX TERRAIN

J.J. Finnigan

CSIRO Division of Environmental Mechanics, GPO Box 821, Canberra, ACT, 2601, Australia

Summary. Understanding of boundary layer flow in complex terrain has proceeded through an interplay between field experiment and mathematical modelling. Wind-tunnel simulation has had surprisingly little influence to date although the situation is changing now. In this review changes to the mean flow are considered first and it is shown that over low hills a reasonable qualitative understanding is provided by simple linear theory but, when hills are steep enough to cause separation, a basic physical description of the flow is still elusive. The changes to turbulence over most of the flow regime can best be described as a response to changes in the mean rates of strain with the turbulence 'memory' determining how much of the upwind flow is reflected in the measured stresses. Mathematical models range from finite-difference numerical schemes with sophisticated turbulence closures to linearized analytical models. These latter have dominated the conceptual approach to both modelling and data analysis for the last decade.

INTRODUCTION

Despite its all-embracing title this review will treat only one aspect of flow in complex terrain: the nature of the turbulent boundary layer over isolated hills. In fact, the compass of the paper is restricted still further to conditions of near-neutral stratification. These bounds are dictated to a large degree by what information is available. Our firmest conclusions are based on the interplay of wind-tunnel and field experiment, both of which for different reasons are biased towards neutral flow.

When the geometric scale is reduced in a wind-tunnel model, various devices must be used to maintain the essential characteristics of the turbulence. One criterion that must be observed is that the Reynolds Number (Re) remain above some minimum value, where $Re = UL/\nu$. U and L are typical velocity and length scales and ν is the kinematic viscosity, so that maintaining high Re implies that as L shrinks, U must increase. In flows where buoyancy plays a role, it is the Froude Number, Fr ($= U/NL$) that must be kept the same in prototype and model. The Brunt Vaisala frequency, N ($= [(g/T)(\partial T/\partial z)]^{\frac{1}{2}}$) is a measure of the stratification, with T, a reference temperature, g the acceleration of gravity and $\partial T/\partial z$ the vertical temperature gradient. The limitations on changing N in a wind tunnel are severe so to maintain Fr constant as the scale is reduced, U must decrease. These conflicting requirements make it almost impossible in practice to model turbulence and buoyancy effects simultaneously, at least in most present wind tunnels.

It is in field experiments, therefore, that the effects of buoyancy and stratification on the turbulence over a hill are to be observed. Indeed they have been observed, most notably in the expeditions led by Bradley in Australia, but with one exception (Bradley 1983) these data have not yet been published. In the northern hemisphere in contrast, a constant motivation for studying wind over hills has been the siting of wind turbines, and hills have been chosen for their propensity for strong winds with a consequent bias towards the neutral case.

The Richardson Number, Ri ($= N^2/(\partial U/\partial z)^2$) is a more usual measure of the local effects of stratification than Fr and is a more useful parameter in turbulent flow. If we approximate local $\partial U/\partial z$ by U/L, we can set up a correspondence between Fr and Ri:

$$Ri \sim 1/Fr^2 \ .$$

When buoyancy effects dominate the flow, Fr approaches zero so $|Ri|$ approaches infinity. For Ri negative this denotes the free convection limit or in the case of hills, anabatic flow. Although this regime is certainly important, we know so little about it that we will abandon further discussion forthwith. For Ri positive we know that turbulent production ceases when Ri exceeds about 1/4, $Fr < 2$. Buoyancy, however, dominates the flow dynamics when $Fr \lesssim 1$ so that we can consider the regime where turbulent exchange and boundary layer processes control the flow separate from that where stability effects like lee waves and blocking dominate. This does not mean there is no overlap. Nonlinear processes in the low Fr range such as lee wave breaking can reduce the local density gradient and allow Ri to fall below 1/4 but these effects are neither well studied nor understood at present. For further information the reader is referred to the recent review by Baines (1987) on blocking and the paper by Hunt and Snyder (1980) on modelling buoyant flows.

In this review we first discuss the coordinate system in which most of the data are presented and define some essential variables. There follows a brief chronological overview to give a feeling for the steps by which understanding has advanced over the last fifteen years. Next we attempt to draw together and make some sense of the mass of published experimental work from field and wind tunnel, both over low hills and those with extensive separation regions. We then discuss mathematical models; these have been the touchstone of understanding in the subject and in tracing their development we can see which ideas have held centre stage at different times since 1974. A brief summary with suggestions for further research concludes the review.

COORDINATE SYSTEMS

When studying complex shear flows, we can choose between analyzing the system in an externally imposed coordinate system, which may be quite arbitrary as in the case of rectangular Cartesian coordinates, or related to the geometry of the underlying surface, such as in terrain-following systems, or we can allow the wind field itself to fix the reference frame. This second choice finds its simplest expression in 'physical streamline coordinates'. In this system all physical quantities are precisely those that would be measured in a rectangular Cartesian frame locally aligned with the streamlines, while distances are measured in units of length.

Such a coordinate system was described (for two-dimensional flows) by Finnigan (1983). The details of the coordinate frame are shown in Fig. 1 and the streamwise (x direction) and cross-stream (z direction) momentum equations for steady adiabatic mean flow are, respectively:

$$U\partial_x U = -\frac{1}{\rho}\,\partial_x P - \partial_x\overline{u^2} + \frac{\overline{u^2} - \overline{w^2}}{L_a} - \partial_z\overline{uw} + \frac{2\overline{uw}}{R} - \nu\partial_z\Omega \ , \qquad (1)$$

$$\frac{U^2}{R} = -\frac{1}{\rho}\,\partial_z P - \partial_x\overline{uw} + \frac{2\overline{uw}}{L_a} - \partial_z\overline{w^2} + \frac{\overline{u^2} - \overline{w^2}}{R} + \nu\partial_x\Omega\ , \tag{2}$$

where U and u are the mean and fluctuating components of velocity in the x direction and w is the velocity fluctuation in the z direction. By definition there are no mean wind components in the y and z directions. ρ is the mean density, ν the kinematic viscosity, P the mean pressure and Ω the y component of mean vorticity. ∂_x denotes differentiation in the x coordinate direction, ∂_y in the y direction and so on.

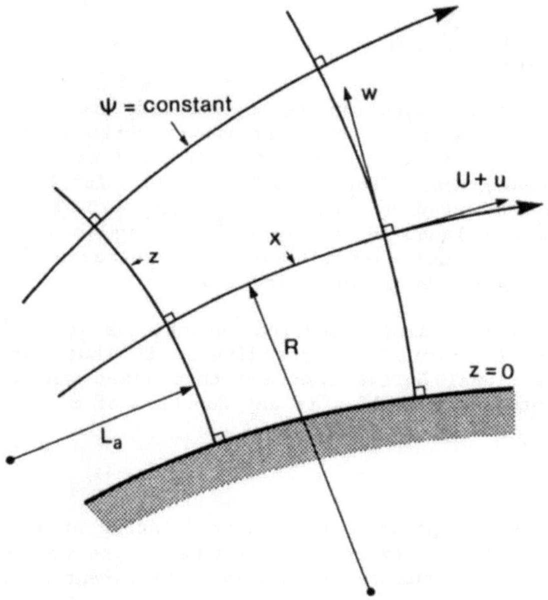

Fig. 1. The 2-D streamline coordinate system. The curves Ψ = constant, Ψ being the stream function, are the x coordinate lines. Their orthogonal trajectories are the z lines. The streamline at z = 0 is the surface. R and L_a are the local radii of curvature of the x and z lines.

The unit base vectors in the x, y and z directions are, respectively, the tangent, binormal and principal normal to the streamline. This means that the axes cannot be arbitrarily relabelled. The flow equations were originally derived and written in the aerodynamic convention (Finnigan 1983) with the y axis and v velocity component identified with the principal normal direction, z and w perpendicular to the plane of symmetry and the three base vectors taken in the order x,y,z forming a right-handed triplet. To retain the same equations under the micrometeorological convention we adopt here, that is with z and w replacing y and v, and vice versa, we must stipulate that the unit vectors taken in the same order form a *left-handed* triplet.

Two length scales, L_a and R, appear in the equations. As shown in Fig. 1, these are the local radii of curvature of the z and x coordinate lines. They are also properties of the velocity field:

$$\frac{1}{U}\,\partial_x U = \frac{1}{L_a} \tag{3}$$

and

$$\Omega = \left[\frac{U}{R} - \partial_z U\right] . \qquad (4)$$

Both R and L_a are signed quantities, being positive if their local centres of curvature lie in the positive z and x directions, respectively. For example, on a hill top R is negative since the centre of curvature of streamlines there lies in the negative z direction.

In the streamline system the mean continuity equation is trivial and is replaced by the geometric identity (Finnigan 1988a):

$$\partial_x\left[\frac{1}{L_a}\right] - \frac{1}{L_a^2} + \partial_z\left[\frac{1}{R}\right] - \frac{1}{R^2} = 0 . \qquad (5)$$

In (1) and (2) the mean advection terms simplify to one term, denoting advection along a streamline rather than the two or three required in Cartesian coordinates; this is at the expense of extra "metric" terms in the expression for momentum flux divergence. For example, in (1) familiar gradient terms $-\partial_x\overline{u^2}$ and $-\partial_z\overline{uw}$ are augmented by $+(\overline{u^2} - \overline{w^2})/L_a$ and $+2\overline{uw}/R$. If we view the momentum equation as a balance of stresses on the walls of an infinitesimal control volume, $dx\wedge dy\wedge dz$, these extra terms appear because the control volume changes its shape from place to place in the flow.

The price to be paid for retaining physical quantities in their familiar form in equations referred to curvilinear coordinates is that partial derivatives are replaced by directional derivatives and that mixed second derivatives no longer commute. In other words, if a is any function of x,y,z (scalar, vector, tensor), then

$$\partial_x\partial_z a \neq \partial_z\partial_x a .$$

This point must always be kept in mind when calculating or modelling in such a system. In fact, "physical" streamline coordinates are no longer coordinates in the strict sense at all but constitute what differential geometers term a "moving frame" (Spivak 1979), in this case a moving frame derived from a coordinate system. Given an arbitrary vector field in space, such as the velocity field of a fluid flow, one cannot in general obtain a true orthogonal coordinate system aligned with the vector lines of the field, the streamlines in this example. It is possible only if the flow field is "complex-lamellar", that is, if the vorticity vector is everywhere normal to the velocity vector. Two-dimensional (2-D) and axially symmetric flows are obvious examples of complex lamellar behaviour. Examples of flows which are not complex lamellar are the atmospheric boundary layer with an Ekman spiral or most flows over three-dimensional (3-D) topography.

Although one cannot have an orthogonal streamline coordinate system in most 3-D flows, it is possible to obtain equations in a moving frame that is a natural extension of the 2-D "physical streamline coordinates" (Finnigan 1988a). The 3-D equations, however, contain an extra length scale, σ, the "torsion" or twist of the streamlines. σ is related to the x component of the vorticity but not in a straightforward way. The derivation and ramifications of the 3-D extensions to (1), (2) and (5) are explored by Finnigan (1988a).

The 3-D streamwise momentum equation, for example, in the limit of small acceleration, curvature and torsion, i.e. neglecting terms of $O\left[\frac{1}{L_a^2}, \frac{1}{R^2}, \frac{1}{\sigma^2}, \frac{1}{RL_a} \ldots\right]$ cf $O\left[\frac{1}{L_a}, \frac{1}{R}, \frac{1}{\sigma} \ldots\right]$ is:

$$U\partial_x U = -\frac{1}{\rho}\partial_x P - \partial_x\overline{u^2} + \frac{\overline{u^2} - \overline{w^2}}{L_a} - \partial_z\overline{uw} + \left[\frac{2}{R} + \frac{R}{\sigma^2}\right]\overline{uw}$$

$$- \partial_y\overline{uv} + \frac{R}{\sigma L_a}\overline{uv} + \frac{\overline{wv}}{\sigma} + \text{viscous terms} . \qquad (6)$$

v is the fluctuating component of velocity in the y direction. The z and y directions are still defined by the principal normal and binormal to the streamline but streamlines are no longer plane curves. Notice also that the torsion, $1/\sigma$, usually appears coupled with R and L_a, a point we shall return to briefly when we discuss 3-D flow response. When curvature, acceleration and torsion are zero, streamline coordinates coincide with rectangular Cartesian coordinates aligned with the mean flow, and ∂_x, ∂_y, ∂_z become partial derivatives.

In the pages to come we will see that the available information on hill flows has been gathered primarily over 2-D ridges or on the centrelines of axially symmetric hills or, in more complex field situations, has been interpreted in this way. Where this assumption cannot be made, we find it very difficult to fit turbulence data in particular into the pattern that emerges from the simpler configurations. In most of our considerations we will not therefore need to be concerned with the problem of orientating the axes. In practical situations where streamline coordinates are used, however, this problem must be faced.

There is an obvious attraction to using streamline coordinates in field experiments as the x axis can be made parallel to the velocity vector a posteriori by rotating from the arbitrary initial measurement orientation until V = W = 0, V and W being the mean velocity components in the y and z directions. This still leaves the y and z axes free to rotate in the plane normal to the streamline. In 2-D flows or close approximations to them, z can be located in the plane spanned by U and the vector normal to the underlying terrain but in strongly 3-D flows this procedure will not work. In fact, the problem of choosing the principal normal and binormal directions from a single point measurement of mean wind and turbulence is not well-posed and its solution takes us into a facet of turbulence modelling which we cannot pursue here.

This is the main disadvantage of streamline coordinates. One great advantage is that the system permits the natural extension of knowledge gained in horizontally homogeneous flow to more complex situations. Our view of a turbulent fluid as a continuum in which stresses are functionals of the mean velocity and pressure field lies at the root of most practical models of turbulent flow and provides a useful shorthand for discussing measurements. It tacitly assumes that the turbulence characteristics measured at some point have been acquired as a population of eddies have advected to the point along mean streamlines, being distorted by the mean strain field, interacting with other eddies and smoothed out by viscosity along their journey. In horizontally homogeneous flow described in Cartesian coordinates the streamlines and x axis coincide and strain rates measured relative to the coordinate axes are relative to the streamlines also. Physical streamline coordinates is the only system that retains this convenient property in complex flows and, if we expect to extend assumptions and parameterisations from the simplest configuration to more general ones, it is in this system that we should attempt it.

Throughout this review we shall use streamline coordinates wherever possible. Practically all published data are available in this form or in a close approximation to it. When it is not and departs severely from the convention, we will say so.

CHRONOLOGICAL OVERVIEW

By the early 1970s understanding of the horizontally homogeneous atmospheric surface layer rested on a solid foundation. The Kansas and Minnesota boundary layer expeditions (Haugen et al. 1971; Kaimal et al. 1976) had confirmed the utility of Monin-Obukhov similarity theory and a good start had been made in applying mixed layer scaling to the convective layer (Wyngaard 1988). Although many details and refinements remained to be explored, the time was clearly ripe for a move to more complicated situations and in 1974 and 1975 a bridgehead was made in the study of boundary layer flow over hills by the appearance of five papers. Although not the first to address the subject, they mark the beginnings of the modern approach and display many of the features that were to become familiar over the next decade. Two of them (Frost et al. 1974 and 1975) made little lasting impact but the papers of Taylor and Gent (1974) and Deaves (1975) displayed all of the essential features of their particular approach – a numerical nonlinear formulation with mixing length turbulence closure – in a way that was not really improved upon for ten years. In 1977 Taylor extended his computations to allow for real terrain input and to cope with scales over which Geostrophic forcing and the Ekman boundary layer had to be considered (Taylor 1977a,b), but these efforts were to be the high water mark of the all-out numerical approach until the mid-1980s thanks to the influence of the fifth model published in 1975, that of Jackson and Hunt, which not only signalled a change in theoretical direction but set the agenda for field experimentation and analysis of data for the next decade.

They described a two-layer model comprising an inviscid outer flow region distant from the surface married to an inner, surface layer where turbulent stress divergence affected the momentum balance (Jackson and Hunt 1975, henceforth referred to as JH75). Their model was linearized, which restricted it in principle to low hills, and it was analytic. This second point was perhaps the main reason for its great influence. It could be applied readily to a variety of terrain and its predictions could serve as guidelines to estimate mean velocities in many different circumstances (Hunt 1980; Taylor and Lee 1987).

Once the implications of the JH75 approach had been absorbed, a string of developments followed, each one attempting to remove the restrictions of the original JH75 model and, as more field data became available, to make it match experiment. Mason and Sykes (1979) produced a 3-D version and, though Sykes (1980) published a careful analysis that sounded some notes of caution about internal inconsistencies in the basic formulation, Walmsley et al. (1982), Taylor et al. (1983b) and Mason and King (1985) continued to improve the correspondence with measurement by adjustments to the model structure. The most recent efforts by Walmsley et al. (1986) and Beljaars et al. (1987) have extended the formulation to cope with simultaneous changes in roughness and elevation, and the requirements this imposes for a more sophisticated turbulence closure have pushed model computations back into the numerical domain. While still much faster than comparable finite difference approaches, the schemes thus lose much of the appeal of analytic solutions.

Even more recently the arrival of reliable turbulence data from the major field experiments of the early 1980s has rekindled interest in numerical models that can accept sophisticated turbulence closures. Prominent amongst these is the scheme of Zeman and Jensen (1987), who use a second order closure and are the first to take proper account of the turbulence's response to the extra rates of strain imposed by the hill.

Although they took some time to get going, field experiments dedicated to probing wind structure over low hills were being published by 1979. Mason and Sykes (1979) produced mean flow results from Brent Knoll, an isolated hill in

Somerset, while Sacré (1979) reported mean wind and turbulence measurements over Pouzauges hill in Western France. Sacré was in advance of his contemporaries by ascribing turbulent stress changes to streamline curvature but, possibly because his site was less than ideal, his conjectures had little impact. Bradley's (1980) results on a steep wooded hill in Canberra appeared to match the JH75 predictions very well, a surprising result since his hill was so much steeper than the model allowed. Bradley next went to Bungendore ridge (Bradley 1983), a very gentle hill where measurements should have fitted the theory perfectly but, perversely, didn't. This paradox prompted a search for a better field site that would contain a hill not so abrupt as to do violence to the tenets of linear theory but where significant changes could be expected in measurable quantities.

The same search had already begun elsewhere and the group at Atmospheric Environment Service (AES), Canada, had settled on Kettle's Hill in Alberta (Taylor et al. 1983a). Ultimately, the search for the perfect hill site was to drive both AES and the U.K. Meteorological Office yet further afield, to the Outer Hebrides in fact, where the AES group led an international team in a two season study of Askervein hill in S. Uist (Taylor and Teunissen 1983, 1985) while the Met. Office group followed a less ambitious program at Blasheval on N. Uist (Mason and King 1985). The Askervein results are now out in report form and are starting to appear in the refereed literature. Askervein seems certain to become one of the most significant boundary layer experiments of recent times.

Bradley's group meanwhile had set up shop on Cooper's Ridge, an excellent site near Goulburn, N.S.W. In the early stages of the experiment, only changes between an upwind reference mast and the summit were measured but a concurrent wind tunnel study (Finnigan et al. 1988a) revealed that a more interesting story would emerge if changes in mean wind and turbulence were followed along streamlines up the hill. The interplay between field, wind tunnel and theoretical developments prompted by the derivation of an appropriate coordinate system soon had the Canberra group heading in a significantly different direction to other centres, emphasizing the response of turbulence to extra strain rates in a streamline frame of reference as well as the role of turbulence memory (Coppin et al. 1986; Finnigan and Raupach 1985).

Where was the wind tunnel in all this? Early experiments at Colorado State University financed by the U.S. Department of the Environment in their search for the ideal site for a wind turbine (Meroney et al. 1978) contained a great deal of data as well as good and bad ideas on interpretation, but their failure to publish in refereed journals doomed the work to have less influence than it deserved. Hunt and Snyder (1980) did wind-tunnel measurements on a 3-D hill to provide a neutral flow reference for their main study, stratified flow and dispersion, but didn't simulate the atmospheric boundary layer in the approach flow. The only wind-tunnel study to date to have exerted great influence upon the direction of modelling and analysis has been that of Britter et al. (1981). This has been less because of the experiment itself than through their introduction of rapid distortion theory to calculate turbulence changes in the outer inviscid flow region.

Without having universal impact, the simulations of Kettle's Hill in the AES's Toronto wind tunnel (Teunissen et al. 1982) doubtless guided their thinking while the early analysis of their own wind-tunnel data by Bradley's group in Canberra seemed to emphasize the importance of streamline curvature (Finnigan and Raupach 1985). More recently, Teunissen et al. (1987) have published comparisons between wind tunnel simulations of Askervein Hill at different scales, and these should provide a valuable adjunct to the field data and mathematical modelling of that site. We should also mention the series of papers that have emanated from the U.S. Environmental Protection Agency's North Carolina wind tunnels with Snyder and Arya prominent amongst the authors.

These have concentrated on steep hills and the fate of scalar pollutants in the wake.

The relatively slight influence exerted by these simulations on the theory of flow over low hills calls for some explanation. A possible cause is the hold that the JH75 theory had from 1975-85. Model surface roughness needs to be exaggerated in a wind tunnel to maintain acceptable turbulence levels. The asymptotic scaling laws of the JH75 theory predicted that the region of turbulence influence on a rough hill would be vanishingly thin, down among the roughness elements in fact. This certainly was cause for concern to Britter et al. (1981) as well as to later wind tunnellers who in consequence have tried to relax standard wind tunnel criteria for modelling atmospheric flows (Teunissen et al. 1987; Teunissen and Shokr 1985), or placed a smooth hill behind a rough upwind surface. The result is that many of the wind-tunnel measurements mix topographical effects with those of roughness change and are correspondingly difficult to interpret.

THE STRUCTURE OF THE WIND FIELD

Most of what we know about turbulent flow over hills comes from the field and wind-tunnel experiments that we have skimmed over in the last section. The problems of field experimentation are well known: relatively few points can be sampled even at the expense of great logistical effort, the approaching wind speed and direction is outside the investigator's control and ideal sites are difficult to find. The problems of wind-tunnel simulation are of a different kind. In order to achieve Reynolds No. independence, the tunnel floor and model hill surface must be made 'aerodynamically rough', that is, the transfer of momentum to the surface must be almost entirely through the form drag on individual roughness elements rather than through viscous stress. For randomly distributed roughness beneath a horizontally homogeneous boundary layer this is generally taken to occur at

$$z_0^+ = \frac{u_* z_0}{\nu} > 3$$

(Schlichting 1960) where z_0 is the surface roughness length parameter (see (25) below). Consider now the problem of modelling a hill of height $h = 100$ m at 1/1000 scale in a large boundary layer wind tunnel. A typical operating wind speed is ~ 12 m s^{-1} at which speed the interdependent quantities z_0 and u_* might be 0.7 m s^{-1} and 0.12 mm (data from Finnigan et al. 1988a) giving $z_0^+ = 5.6$. Any substantial reduction in z_0 will cause a comensurate drop in u_* and the flow will no longer be fully rough.

At a scale of 1/1000, however, the equivalent full scale z_0 is 0.12 m, characteristic of scattered bushes or tall crops. A grass or heather covered hill like Cooper's Ridge or Askervein would have a full scale $z_0 \sim 3$ cm, which at 1/1000 scale gives a model z_0 value of 3.10^{-5} m, and aerodynamically smooth flow in the wind-tunnel. Teunissen et al. (1987) have investigated whether the aerodynamically rough criterion can be relaxed by testing four wind tunnel models of Askervein hill with different geometric scales and surface roughness in three separate wind tunnels and comparing the results with full scale. Their roughest model had a value of $z_0^+ = 1.7$ and also had the largest scale, 1/800, while the smallest z_0^+, 0.093, went with the smallest model at 1/2500 scale. So the models varied from less than fully rough to aerodynamically smooth, which occurs for $z_0^+ < 0.2$. We shall return to their results below.

The Mean Flow

Vertical profiles of mean velocity typical of a low isolated hill are shown in Fig. 2. These data are taken from Finnigan et al. (1988a), an extensive data set from a rough ($h/z_0 = 412$) wind-tunnel model hill that will be used to illustrate many of the points in this section. We will refer to this experiment henceforth as WTHILL. The most striking features of Fig. 2 are the pronounced increase in velocity near the surface on top of the hill, the slight deceleration at the lowest levels just upwind of the hill and the velocity deficit in the wake.

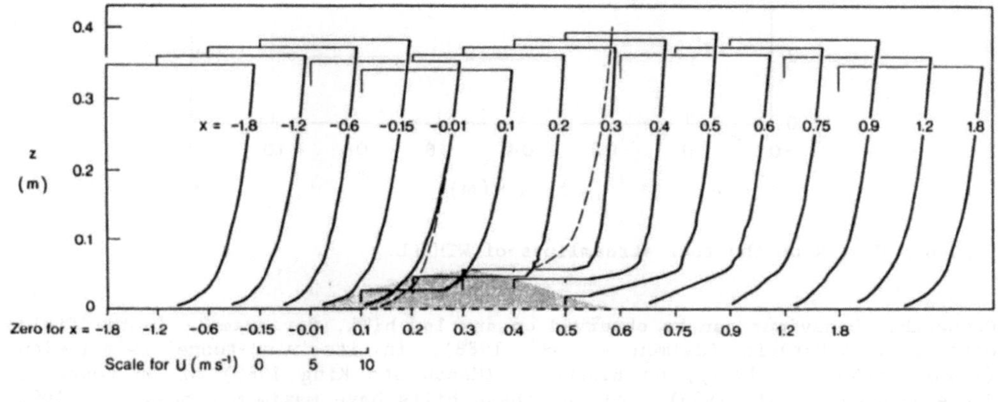

Fig. 2. Profiles of U measured at various x stations on WTHILL. The upstream profile (x = −1.8 m) is superimposed as a chained line on the local profiles at x = 0.01 m and x = 0.3 m, the foot and the crest of the hill which extends from x = 0 to x = 0.6 m.

The evolution of U along streamlines at different heights can be seen in Fig. 4, the streamlines themselves being plotted to scale over WTHILL in Fig. 3. On streamline 1 the deceleration in front of the hill is pronounced while the speedup is close to its maximum value. Deceleration in the wake region behind the hill is also exaggerated on the lowest streamline and decreases as we move further from the surface, the velocity recovery becoming monotonic by streamline 3.

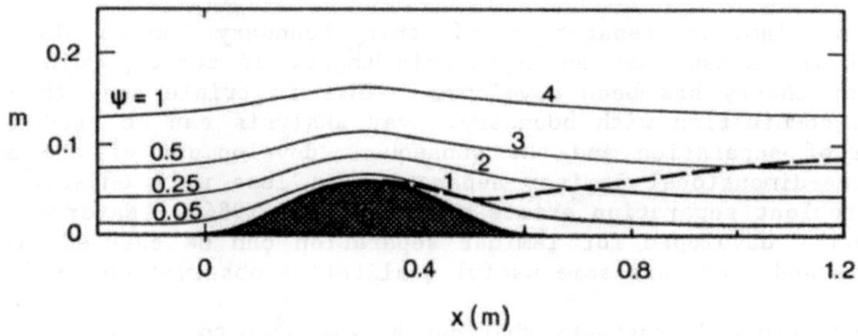

Fig. 3. The four streamlines of WTHILL drawn to scale over the hill profile. The stream function values are $0.05\ m^2s^{-1}$ on streamline 1 and so on. The shaded area behind the hill shows the wake, defined as the region of mean velocity deficit and turbulence increase.

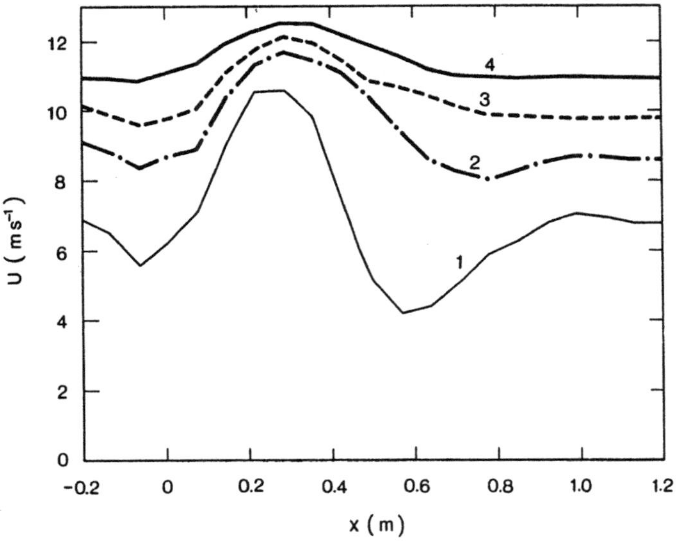

Fig. 4. U vs x on the four streamlines of WTHILL.

Comparable behaviour can be observed on any low hill, for example in the field data from Askervein (Salmon et al. 1988), in its wind-tunnel simulation (Teunissen et al. 1987), on Blasheval (Mason and King 1985) or on Cooper's Ridge (Coppin et al. 1986). All of these hills have maximum slopes of ~ 10°. On hills with steeper slopes we begin to encounter regions of separation and the near surface flow patterns are changed substantially.

<u>The separation bubble and wake region; general considerations</u>: When separation occurs, it changes the wind field not only in the separation region itself but over the entire hill so that defining its limits becomes a prerequisite of establishing the overall flow field. We can define the separation point as the place where a hitherto grazing streamline takes off from the surface. On a smooth wall this coincides with the point where both the skin friction and the normal gradient of mean velocity at the surface, $\partial_z U|_{z=0}$ vanish. In turbulent flow, however, particularly on a rough wall, these points need not be coincident and vanishing of the velocity gradient and the surface shear stress seems a necessary but not sufficient requirement for separation (Batchelor 1976).

Two-dimensional laminar separation of thin boundary layers is now well understood in the sense that an asymptotic theory of corresponding rigour to boundary layer theory has been developed. This is triple deck theory (Meyer 1983) and in combination with boundary layer analysis can be used to predict the position of separation and the subsequent development of the separation bubble. Three-dimensional laminar separation is less well understood and no theory of turbulent separation exists at all (Smith 1986). Nevertheless, some of the arguments developed for laminar separation can be carried over to the turbulent case and there are some useful qualitative observations to be made.

One feature of triple deck theory that can be extended to turbulent flow is the contraction of the horizontal length scale near the point of separation. In other words, the usual boundary layer situation where normal gradients of mean quantities are much larger than tangential gradients and its consequence, the description of the flow by parabolic equations of the boundary layer type, disappear near the separation point. There, the scale of tangential gradients becomes of the same order as that of the normal gradients and the flow description reverts to the full elliptic situation. (In the laminar case, normal length scales are of $O(Re^{-\frac{1}{2}})$ and tangential scales are of $O(Re^{-3/8})$, Smith 1986.) The obvious consequence is that the separation point cannot be predicted by following the flow development downwind but must result from a global analysis of the entire flow field.

Separation is inevitably preceded by deceleration of the tangential flow near the wall and, when the momentum of the fluid moving parallel to the surface is sufficiently reduced by the decelerating pressure gradient, it is diverted away from the wall. A crucial factor in deciding when this will occur is the turbulent stress divergence as we see from the form of (1) close to the surface:

$$U\partial_x U \simeq -\partial_x P/\rho + \partial_z \tau \ , \qquad (7)$$

where τ $(= -\overline{uw} + \nu\partial_z U)$ is the kinematic shearing stress. The decrease in horizontal length scales means that very close to the separation point (7) is not a useful description of the flow, even on a grazing streamline, and we must revert to the full Eq. (1); nevertheless, it does help to clarify the processes leading to separation. If momentum is transferred to a grazing streamline from the faster moving fluid above more rapidly than it is removed by friction below, then the flow remains attached. It follows that any agent that acts to suppress turbulent momentum transfer to the surface in a region of adverse pressure gradient will tend to promote separation. A particularly pervasive agency of shear stress supression is the convex streamline curvature that occurs over a hill top. We will treat this mechanism in greater detail when we consider the turbulent structure; for the moment we simply point out that the extremely high curvature and accompanying adverse pressure gradient at a salient edge ensures the separation of flow there. This is the one situation where the separation point is easy to predict. At a salient edge also all of the deceleration necessary for separation occurs at the edge itself and the measurable surface flow may appear to accelerate right up to the point where it separates.

If the mechanism of flow separation is poorly understood, then the recirculating flow in the 'bubble' region between separation and reattachment is even less well known. Some progress has been made in predicting the extent of the separation bubble behind a backward facing 2-D step using turbulence closure models (see, for example, Speziale 1987) but turbulent 3-D separation is still innaccessible to computation. However, some increase in qualitative understanding has been gained by Hunt et al. (1978) and Tobak and Peake (1982), who have applied the principles of topology to the vector velocity and stress fields in the separation region. Although useful in interpreting measurements, their approach provides no predictive finding except in one important respect. They found that a 3-D separation bubble, closed in the mean, is kinematically impossible, a result of considerable importance in the study of scalar transport and pollutant trapping behind a hill.

The only unequivocal data on separation bubble behaviour come from wind-tunnel tests as the bubble height is characteristically of O(h), too high to be explored by the meteorological towers of field experiments. The tunnel data are, unfortunately, bedevilled by the fact that many of the studies are of smooth hills in rough wall boundary layers, which enables measurements to be made very close to the surface but introduces the complication of smooth to rough transition.

Separation on 2-D ridges: The first question we can ask is how steep must a hill be before separation occurs? Bouwmeester et al. (1978) seem to have found the answer for smooth symmetrical triangular ridges, where they found complete separation behind a ridge at 18° slope (1:3) and no separation when the slope was reduced to 14° (1:4). Given our earlier remarks on salient edges, we must interpret this second result as meaning that the inevitable separation bubble behind the sharp crest reattached almost immediately. Snyder and Britter (1987) also measured the flow field over a smooth symmetrical triangular ridge of 30° slope. Unfortunately, they made no direct measurement of the separation bubble extent but it was certainly well established and the reattachment point was behind the downwind base of the hill, the bubble length being about 3-4h. On the smooth 60° ridge studied by Arya and Shipman (1981) the bubble extended for 13h and reached a maximum depth of 2.5h.

The differences when the hill surface is rough are considerable, producing on the very rough (h/z_0 = 120), 30° triangular ridge of Castro and Snyder (1982) a deep bubble of depth ~ h that reattached about 9h behind the downwind foot of the ridge. The difference between smooth and rough separation is related to the increased near surface momentum that is possible on a smooth wall and, we may speculate, to the tendancy of the concentrated vortex sheet in smooth separation to roll up faster than the thicker, more diffuse sheet accompanying turbulent flow detachment.

On curved hills with profiles roughly describable as Gaussian or sinusoidal, that is symmetrical with zero slope at crest and foot, the position of separation is less well defined. On three rough curved ridges ($h/z_0 \simeq 120$) with maximum slopes 26°, 16° and 10° respectively, Arya et al. (1987) reported complete, steady separation on the 26° hill but only intermittent, unsteady separation on the two shallower ridges. Britter et al. (1981) on a rough ridge with maximum slope 12.5°, (h/z_0 = 500) and Finnigan et al. (1988a) at 9°, (h/z_0 = 412), also saw only intermittent separation behind the crest. In a more recent study of a slightly steeper (16°) but much rougher (h/z_0 = 50) hill Finnigan et al. (1988b) observed complete separation in a region about 2h in horizontal extent.

<u>Separation on 3-D hills:</u> Solid data on the effects of three dimensionality come primarily from two papers: Castro and Snyder (1982), who studied a succession of very rough (h/z_0 = 120) triangular hills with aspect ratio increasing from cones to 2-D ridges, and Arya and Gadiyaram (1986), who looked at two slightly rough (h/z_0 = 3360) cones of differing height to diameter ratio. The slope angle of Castro and Snyder's hills was fixed at 30° and a consistent picture emerged of increasing bubble length with aspect ratio. A similar trend of increasing proportion of downwind hill surface covered by the bubble as aspect ratio increased was also observed. Note that behind their rough 30° cone the bubble is about 1h in depth and reattaches behind the base of the hill whereas they report in the same paper that on a smooth cone of the same geometry the bubble is much shallower, barely extends to the foot of the cone and covers less than half the area.

On their 26.5° cone Arya and Gadiyaram (1986) measured almost exactly the same behaviour as on Castro and Snyder's rough 30° hill but on reducing the angle to 17.5° detected only localized intermittent separation behind the crest. We note that the rather low roughness used in Arya and Gadiyaram's experiment resulted in a value of $z_0^+ \sim 1.07$, less than fully rough.

Although field experiments do not provide such unequivocal data on the depth of the recirculation region, reversed flow below the tower height can certainly be detected. Hence with the wind normal to the major axis of the roughly elliptical Askervein, whose maximum slope was ~ 12° and $h/z_0 \sim 4000$, a region of intermittent reversed flow was seen behind the crest (Mickle et al. 1988). On Blasheval, steeper and almost axisymmetric with a maximum slope of 24° and $h/z_0 \sim 10{,}000$, Mason and King (1985) observed a well established separation region. Periodic vortex shedding and a reversed flow region behind the almost axisymmetric Black Mountain (slope = 20°, h/z_0 = 170) is a phenomenon well known to sailors on the nearby lake but no separation was detected when the wind flowed over the crest of Cooper's Ridge at 8.5° with h/z_0 ~ 3000.

Table 1. Separation over two-dimensional ridges and three-dimensional and axisymmetric hills. WT = wind tunnel.

Hill type	Surface	Max Slope	Separation	Reference
		Two-dimensional ridges		
triangular (WT)	smooth	14°	no	Meroney et al. (1978)
" (WT)	"	18°	yes	" " " "
" (WT)	"	30°	yes	Snyder & Britter (1987)
" (WT)	"	60°	yes	Arya and Shipman (1981)
Cooper's Ridge	(h/z_0 = 3000)	8.5°	no	Coppin et al. (1986)
WTHILL (WT)	(h/z_0 = 412)	9°	intermittent	Finnigan et al. (1988a)
curved (WT)	(h/z_0 = 4000)	10°	no	Arya et al. (1987)
Askervein	(h/z_0 = 120)	12.5°	intermittent	
curved (WT)	(h/z_0 = 500	12.5°	intermittent	Britter et al. (1981)
curved (WT)	(h/z_0 = 120)	16°	intermittent	Arya et al. (1987)
curved (WT)	(h/z_0 = 50)	16°	yes	Finnigan et al. (1988b)
curved (WT)	(h/z_0 = 120)	26°	yes	Arya et al. (1987)
triangular (WT)	(h/z_0 = 120)	30°	yes	Castro and Snyder (1982)
		Three-dimensional and axisymmetric hills		
Cone (WT)	smooth	30°	yes	Castro and Snyder (1982)
Cone (WT)	(h/z_0 = 3360)	17.5°	intermittent	Arya and Gadiyaram (1986)
Black Mountain	(h/z_0 = 170)	20°	yes	Bradley (1980)
Blasheval	(h/z_0 = 10000)	24°	yes	Mason and King (1985)
Cone (WT)	(h/z_0 = 3360)	26.5°	yes	Arya and Gadiyaram (1986)
Cone (WT)	(h/z_0 = 120)	30°	yes	Castro and Snyder (1982)

All these results are summarized in Table 1 in a form that makes it easier to define the quantitative limits to steady separation. It is clear that on 2-D ridges, whether curved or triangular, the critical angle for steady separation is around 18° but that this figure can be reduced by increasing surface roughness. As the aspect ratio of the hill decreases, the extent and width of the separation region also shrinks so that on a smooth cone, 30° seems to be about the minimum angle to ensure steady separation but adding surface roughness drops this angle to somewhere between 20.0° and 17.5°. Without a well established theory of turbulent separation or much more data, we can do little more than establish these broad limits. A striking conclusion, however, is the sensitivity to surface roughness of separation even on very abrupt hills, where one would expect topographical effects to totally dominate. This ought to make us very reluctant to relax the criterion of aerodynamic roughness in wind-tunnel models, particularly around the critical slope values of ~ 10°-20°. Hence analysis of field data from Askervein (Mickle et al. 1988) suggests that for some wind directions (~235°) a small region of intermittent separation exists behind the crest but this seems to have disappeared in the simulation by Teunissen et al. (1987), where the fully rough criterion was not met.

Flow in the separation region: We have little information about the details of flow in the bubble region; however, both Castro and Snyder (1982) and Arya and

Gadiyaram (1986) have noted the establishment and persistence of axial vorticity from the trailing arms of the horseshoe vortex that develops around a cone. These vortices cause a marked downwash field on the centreline of the near wake, a conclusion of some interest in estimating plume trajectories. With just a small amount of asymmetry in the shape of a 3-D hill a strong single trailing vortex may develop, as was observed by aircraft measurements in the field study of Ailsa Craig (Jenkins et al. 1981). The persistance of streamwise vorticity in a 2-D boundary layer is a well known phenomenon and can be inferred from the equation for Ω (see Tennekes and Lumley 1972), which displays no term coupling the dominant deformation rate, $\partial_z U$, and the streamwise vorticity, so that turbulent diffusion alone is available to smear out the rotation.

Whether or not a region of true separation exists behind a hill, a pronounced wake region is always established. This region is characterized by reduced mean velocity and increased turbulence and is remarkably persistent. One notable generic feature of the 2-D wake region is an elevated layer of high shear that appears behind the hill crest and continues downwind at about height h giving the upper part of the wake region some of the characteristics of a turbulent mixing layer. This elevated shear bounds the region of maximum velocity defect and its appearance is independent of whether true flow separation has occurred. Behind 2-D ridges complete recovery of the mean velocity profile does not occur before travelling about 60h downwind (Arya and Shipman 1981). Behind 3-D hills recovery is usually more rapid although, as we have just noted, axial vortices can be very persistent. The whole subject of atmospheric wake flows, their downstream development, character and parameterization, is treated by Taylor (1988) in this volume so we will not pursue it further here.

Upstream separation: We drew attention to the near-surface deceleration upwind of the hill as one of the characteristic features of the mean velocity field. When this deceleration is large enough, it can induce separation. As in the case of a salient edge, separation inevitably occurs if there is a discontinuity in upwind surface slope but we have even less information on the values of slope and surface roughness that delimit the occurrence and size of the bubble than we had for the downwind case. Snyder and Britter (1987) observed no upstream separation (evidently a vanishingly small region) before their 30° ridges and cones while Arya and Shipman (1981) with a smooth 60° triangular ridge seem to observe a well established upwind bubble although they accord it little attention in their paper. It seems likely that because the region of upwind adverse pressure gradient is confined to the lowest streamlines, substantially larger slopes are required to produce separation upwind than downwind. We note that when the wind approaches the hill obliquely, the upwind deceleration may disappear even at the lowest streamlines. This was observed on Askervein (Salmon et al. 1988).

Speedup: The mean velocity speedup on the hill top holds an obvious interest both for wind power enthusiasts and for engineers interested in predicting wind loading on buildings, and many early investigations of hill flows aimed at quantifying just this feature. Thus, the series of reports from the Fort Collins wind tunnels summarized by Meroney et al. (1978) looked at the influence of hill shape on wind turbine siting while the Askervein hill study was also motivated by wind energy interest. We follow convention and define the speedup Δs as the fractional change in speed at a fixed height above the surface so

$$\Delta s(x,z) = \frac{U(x,z) - U_0(z)}{U_0(z)} ,$$

where $U_0(z)$ is the upstream velocity profile, although working in streamline coordinates it would make some sense to define Δs as the fractional change in velocity along a streamline. We further define Δs_{max} as the maximum value of Δs, which usually occurs at some height above the hill top. On hills low enough to avoid separation, the general features of the speedup can be understood through the linearized theory of Jackson and Hunt. They divided the flow field into an inner region of depth ℓ, where turbulent stress divergence, mean flow acceleration and pressure gradient are of comparable size, and an outer region where inviscid dynamics prevail. JH75 is linear in the velocity perturbations and assumes that within the inner layer the rules governing turbulence behaviour in horizontal equilibrium wall flows continue to hold. As such its quantitative predictions are restricted to very low hills (JH75 will be discussed in more detail later); nevertheless, it forms a useful qualitative guide in many cases, indeed it is often the only guide.

The theory assumes that the perturbation pressure field around the hill is determined primarily by the inviscid, outer region dynamics and that this perturbation can be approximated by potential flow around the hill shape, the velocity scale being taken as $U_0(L)$. L is the horizontal scale of the hill defined as the horizontal distance from crest to half-height point. Within the outer region, consequently, Δs takes the values given by classical theory. In the inner layer, in contrast, approach velocities are of $O(U_0(\ell))$ but the velocity perturbation there is still driven by the pressure gradient derived from the outer flow solution. If the approach flow, $U_0(z)$, is taken as logarithmic with z_0 typical of real hills, then $U_0(L) >> U_0(\ell)$ and Δs in the inner region is much larger than in the outer flow. In fact, on the crest of 2-D ridges JH75 predicts that $\Delta s(\ell)$ is roughly twice as large as the value given by potential flow with a velocity scale $U_0(\ell)$.

Taylor et al. (1987) have comprehensively reviewed the application of JH75 to the flow over low hills and recommend the formula

$$\Delta s_{max} \sim \left[\frac{U_0{}^2(L)}{U_0{}^2(\ell)}\right] \cdot \left[\frac{h}{L}\right] \zeta ,$$

where $\zeta(x,z)$ is a function that factors the precise shape of the hill into the potential flow calculation of the pressure field. For values of L, h and z_0 typical of hills like Askervein, Blasheval or Cooper's Ridge, $\zeta \sim 1$ and $(U_0{}^2(L)/U_0{}^2(\ell)) \sim 2$ so that

$$\Delta s_{max} \sim 2h/L , \tag{8}$$

the value recommended for 2-D ridges in JH75.

Taylor and Lee (1987) used these ideas in formulating simple guidelines for estimating wind changes over small scale topographic features. Over 3-D, axially symmetric hills they recommend $\Delta s_{max} \sim 1.6h/L$ and for 2-D escarpments $\Delta s_{max} \sim 0.8h/L$. These rules-of-thumb seem to be accurate within ± 25% on low hills. For example, the wind-tunnel results of Finnigan et al. (1988a) gave a maximum speed up of ~ 0.5 (after correcting for wind tunnel blockage) compared with an estimate of $\Delta s = 0.6$ from (8), while in the series of rough curved hills studied by Arya et al. (1987) those without separation bubbles also fitted (8) within 10%. On other low hills also reasonable agreement has been measured, thus Δs values on Kettle's Hill averaged about 15% higher than the estimates, Askervein values were about 10% lower and there was almost exact agreement between 2 m and 15 m height above Cooper's Ridge. Most of these hills fall within the limits of surface roughness ($10^5 > L/z_0 > 10^3$) within which the above formulae were derived; all were relatively low hills with maximum slope no more than ~ 10° and none had significant changes in surface roughness.

When hills became steep enough to establish steady separation regions, many of the basic tenets of JH75 break down. The theory assumes that the pressure field is given approximately by irrotational inviscid flow around the hill shape. With separation, however, the far-field solution is given by flow around the body plus separation bubbles; in the separation regions the pressure field is dominated by the bubbles themselves whilst the attached flow responds to both influences. Since the flow to the crest is generally attached, speedup on steeper hills still correlates with increasing h/L but with L now determined by the shape of hill plus separation bubbles. The appearance, and then growth, of bubble length with increasing hill steepness therefore puts an upper limit on Δs of about 1.25, and this value is obtained on axisymmetric hills where, as we have already seen, the separation bubble is minimized.

For example, on the steep 2-D ridge studied by Arya and Shipman (1981), h/L was 4 but the measured Δs_{max} was only 1.2. The separation bubble length was 13h in that case, while Snyder and Britter (1987) on their high aspect ratio ridges with h/L = 1 found Δs_{max} values clustering around 0.7. On a cone, however, with the same h/L but a much smaller separation region Δs_{max} was 0.9. On 3-D hills in general the picture that emerges is far from clear. At Blasheval, Mason and King (1985) compared a measured value of $\Delta s \sim 0.7$ at 8 m on the hill top with estimates of about 1.2 from more sophisticated developments of JH75. There was a well established separation region behind Blasheval so this result is not unexpected. In contrast, on a rough 30° cone in the wind tunnel with definite separation, Arya and Gadiyaram (1986) measured $\Delta s = 1.5$ compared with a value of $1.6h/L = 1.6$ but on a shallower cone, where $1.6h/L = 0.96$, obtained $\Delta s = 1.65$. On Brent Knoll, an almost axisymmetric hill with intermittent separation, Mason and Sykes (1979) had $1.6\ h/L = 0.9$ but measured $\Delta s = 1.3$ while on Black Mountain, a very rough 3-D hill, Bradley (1980) had $1.6\ h/L = 1.0$ and measured $\Delta s_{max} = 1.07$.

Clearly we are on shaky ground in applying the JH75 theory or its descendents in these circumstances. The precise value of ζ for a cone is different from that on a more gentle hill as is the ratio $U_0{}^2(L)/U_0{}^2(\ell)$, but these are not the main problems. If the hills are steep enough to cause separation, they also contradict the linearity, equilibrium turbulence formulation and weak interaction (between inner and outer flow fields) of the theory.

JH75 places the maximum in $\Delta s(x,z)$ at the height ℓ above the hill top. A maximum at some height is characteristic of all measurements and may be so pronounced as to produce a maximum or jet in the total velocity profile if the hill is steep enough (Arya and Gadiyaram 1986; Bradley 1980). A good deal of effort has been expended therefore in determining the most appropriate value for ℓ; two suggestions, those from JH75 and Jensen (Taylor et al. 1987), are compared with measured positions of Δs_{max} by Taylor et al. (1987); however, they observe only approximate agreement with the Jensen formula and find the height predicted by the original JH75 choice to be much too large. After linearizing the momentum equation and adopting as a turbulence closure the flat land form of the surface log law, JH75 arrived at the following relationship for ℓ:

$$\frac{\ell}{L}\,\ell n\left[\frac{\ell}{z_0}\right] = 2k^2 \ , \tag{9}$$

where k (= 0.4) is the von Karman constant, while Jensen, using comparable assumptions but slightly different reasoning, obtained

$$\frac{\ell}{L}\,\ell n^2\left[\frac{\ell}{z_0}\right] = 2k^2 \ . \tag{10}$$

Jensen's formula results in ℓ values roughly four times smaller than JH75's over most of the range of interest and seems to predict the position of Δs_{max} a good deal better than the original JH75 formula. Nevertheless, the additional assumptions required to obtain (10) rather than (9) seem somewhat ad hoc and over real hills, where typical values of Δs (~ 0.75) already obviate the linearization of JH75 and where, as we shall see below, the equilibrium form of the log law is a poor description of the turbulent stress, even approximate agreement is surprising.

The Turbulence

The changes to the turbulence as the flow goes over a hill occur in response to the changing mean strain rates, that is, to changes in the mean velocity field. The formal description of the Reynolds stresses in 2-D, adiabatic steady flow is given by the rate equations (Finnigan 1983):

Advection(A) Production(P) Pressure-strain(PS)

$$U\partial_x\overline{u^2} = -2[\overline{u^2}\,\partial_x U - \overline{uw}\Omega] + \frac{2}{\rho}\,\overline{p\partial_x u}$$

Pressure Transport(PT) Turbulent Diffusion (TD)

$$-\frac{2}{\rho}\,\partial_x\overline{pu} - \partial_x\overline{u^3} + \frac{1}{La}(\overline{u^3} - 2\overline{uw^2}) - \partial_z\overline{u^2w} + \frac{3}{R}\,\overline{u^2w}$$

Viscous dissipation (V) (VT)

$$-\epsilon_u + (\text{viscous transport})\ . \tag{11}$$

A P PS

$$U\partial_x\overline{w^2} = -\left[4\overline{uw}\,\frac{U}{R} - 2\overline{w^2}\partial_x U\right] + \frac{2}{\rho}\,\overline{p\partial_z w}$$

PT TD

$$-\frac{2}{\rho}\,\partial_z\overline{pw} - \partial_x\overline{uw^2} + \frac{3\overline{uw^2}}{La} - \partial_z\overline{w^3} + \frac{1}{R}\,(\overline{w^3} - 2\overline{u^2w})$$

V VT

$$-\,\epsilon_w + (\text{viscous transport}) \tag{12}$$

A PS PT TD V VT

$$U\partial_x\overline{v^2} = \frac{2}{\rho}\,\overline{p\partial_y v} - \frac{2}{\rho}\,\partial_y\overline{pv} - \left[\partial_x - \frac{1}{La}\right]\overline{uv^2} - \left[\partial_z - \frac{1}{R}\right]\overline{wv^2} - \epsilon_v + (vt) \tag{13}$$

(Note that in 2-D flow there is no direct production of y direction turbulent energy.)

A P PS

$$U\partial_x\overline{uw} = -\left[2\overline{u^2}\,\frac{U}{R} - \overline{w^2}\Omega\right] + \frac{1}{\rho}\left[\overline{p(\partial_x w + \partial_z u)}\right]$$

PT TD

$$-\frac{1}{\rho}\left[\partial_x\overline{pw} + \partial_z\overline{pu}\right] - \partial_x\overline{u^2w} + \frac{1}{La}(2\overline{u^2w} - \overline{w^3}) - \partial_z\overline{uw^2} + \frac{1}{R}(2\overline{uw^2} - \overline{w^3}) \tag{14}$$

$$+\ (\text{viscous terms})$$

A P

$$U\partial_x\tfrac{1}{2}\overline{q^2} = -(\overline{u^2} - \overline{w^2})\partial_x U - \overline{uw}\left[\partial_z U + \frac{U}{R}\right]$$

PT + TD V VT

$$-\left[\partial_x - \frac{1}{La}\right]\left[\frac{1}{\rho}\overline{pu} + \tfrac{1}{2}\overline{uq^2}\right] - \left[\partial_z - \frac{1}{R}\right]\left[\frac{1}{\rho}\overline{pw} + \tfrac{1}{2}\overline{wq^2}\right] - \epsilon + (vt)\ , \tag{15}$$

where $\frac{1}{2}\overline{q^2} = \frac{1}{2}(\overline{u^2} + \overline{v^2} + \overline{w^2})$.

Equation (15) for turbulent kinetic energy (tke) is the sum of (11), (12) and (13).

The terms in the tke equation group into three kinds: production terms that describe the creation of turbulent energy through the work done by existing turbulent stresses against the mean rates of strain; destruction terms that represent the dissipative effect of viscosity on the turbulent motion, and transport terms that describe the movement of turbulence from place to place in the flow to redress local imbalances between production and destruction. Transport can be effected by the mean flow, when it is called advection, by turbulent velocity and pressure fluctuations, usually called turbulent diffusion and pressure transport respectively, and by molecular motion. This last effect is always negligible at the Reynolds numbers we are concerned with. Moving to the component stress equations, a further effect must be considered: the fluctuating turbulent pressure, as well as being an agent of transport, also forces a flow of energy from one component of normal stress to another, tending to make the turbulence more isotropic and acting as the main destruction term for shear stress $\overline{uw}$. The two effects are usually called pressure-strain and pressure destruction, respectively. Each term in (11) to (15) is labelled appropriately. Note that the term on the left-hand side in each case is the mean transport or advection term, which takes a particularly simple form in streamline coordinates.

The level of a turbulent stress at some point in the flow is the result of a balance between production and destruction on the spot and turbulence transported in from elsewhere in the flow. If the contribution of the transport terms is negligible, the flow is said to be in 'local equilibrium', a state of affairs characteristic of the horizontally homogeneous atmospheric surface layer. In such a flow advection is zero since the mean velocity is normal to the only direction of inhomogeneity. If the flow is also adiabatic, then we observe that turbulent transport (pressure transport plus turbulent diffusion) is also negligible except very close to a rough surface. There, much of the turbulence produced within two or three times the height of the roughness elements is transported downwards to maintain high turbulence levels between the elements themselves (Raupach 1988). In this region, the roughness sublayer, the flow is not in local equilibrium even though mean flow advection is negligible.

When the dominant transport term is advection, then a more picturesque view of the turbulent state is provided by the concept of 'turbulence memory'. The idea is that although we are treating the turbulent fluid as a continuum, it is a continuum obtained by ensemble averaging a large number of quasi-deterministic large structures or eddies which are responsible for most of the diffusing action of the turbulence, which owe their particular character to the pattern of mean flow straining in which they evolved and which have a characteristic lifetime of growth and decay. For a recent review of this picture of turbulence see Landahl and Mollo-Christensen (1986). At a point in the flow, therefore, the observable turbulence consists of a population of such eddies advecting past, each at a different stage in its cycle of growth and decay and each characteristic of the mean strain field it has passed through to that point.

The terms in (11)-(15) can also be identified in a qualitative way with stages in the growth and decay of eddies. It is the largest turbulent motions that interact most efficiently with the mean flow and so it is into those scales that mean flow kinetic energy is first transferred. This process is described by the production terms. Since the mean strain rates, U/R, U/L_a $\partial_z U$, appear explicitly in these terms, changes in strain rate are reflected instantaneously in the rate of production. This is why the large eddies reflect the strain fields they have encountered during their lifetimes. This lifetime in turn depends on the rate at which energy is passed to smaller and smaller eddies and eventually dissipated as heat. This process of eddy-eddy interaction is embodied in the triple product terms, that is, turbulent diffusion and the terms involving pressure[1], while the ultimate conversion of eddy motion to heat, which occurs on the very smallest scales, is contained in the viscous dissipation. The dissipation process itself occurs at length scales too small for changes in the mean strain rates to affect it but the *rate* of dissipation is controlled by the rate at which energy can be passed from larger to smaller eddies in the "eddy cascade". The time scale characterizing both this process and the related pressure destruction of shear stress is called the eddy "lifetime" or "turnover time", T_ϵ, and it does respond to changes in the mean strain rates but more slowly than does production.

We can define T_ϵ in various ways, the most appropriate one in nonequilibrium flow being

$$T_\epsilon = \tfrac{1}{2}\overline{q^2}/\epsilon \ . \tag{16}$$

The corresponding length scale,

$$\ell_\epsilon = T_\epsilon U \ , \tag{17}$$

represents the typical distance that an eddy will be advected along a streamline during its lifetime. If ℓ_ϵ is small compared with the length scale for streamwise inhomogeneity, then we can expect the flow to be approximately in local equilibrium because the local destruction rate will have had time to adjust to the local rate of production. As Wyngaard (1982) demonstrates, this kind of local balance is implicit in turbulence models based on eddy diffusivities.

It is natural, therefore, to ask how much of the hill flow regime we might expect to exhibit local equilibrium since this will define the region where we can expect sensible answers from eddy diffusion models. Over low hills an order-of-magnitude estimate can be obtained by adopting the methods of JH75 to compare advection and production of turbulent energy. To do this we simply assume the local equilibrium relationships that obtain in the homogeneous surface layer and write:

$$\text{production} = -\overline{uw}\ \partial_z U \sim \frac{u_*^3}{kz} \ ,$$

$$\text{advection} = U\partial_x \tfrac{1}{2}\overline{q^2} \sim U\,\frac{\Delta\tfrac{1}{2}\overline{q^2}}{L} \sim U\,\frac{u_*^2}{L} \ .$$

Here we have considered only shear production in the expectation that any local equilibrium region will be close enough to the wall for this to be the dominant production term and have taken u_*^2 to be a typical measure of the change in tke, $\Delta\tfrac{1}{2}\overline{q^2}$, over the horizontal hill scale, L. If we now let $z = \bar{\ell}$ be the height at which the two processes are comparable, we have

[1]Strictly the terms containing fluctuating pressure must be regarded as having two effects depending on the source of the pressure. Turbulent pressure fluctuations can arise from coupling between the turbulence and the mean flow or from turbulent-turbulent interaction. Pressure fluctuations from the first source provide the so-called "rapid" part of the pressure-strain or pressure destruction terms and should be interpreted as causing a distortion of the turbulent eddies under the influence of the mean flow. Pressure fluctuations arising from turbulent-turbulent interactions, in contrast, are responsible for the part of the pressure-strain or destruction terms corresponding to eddy-eddy interaction.

$$\frac{u_*^3}{k\overline{\ell}} = \frac{u_*}{k}\ \ell n\left[\frac{\overline{\ell}}{z_0}\right]\frac{u_*^2}{L} ,$$

whence $\frac{\overline{\ell}}{L}\ \ell n\left[\frac{\overline{\ell}}{z_0}\right] = 1$. (18)

Hence $\overline{\ell}$ turns out to be the same (practically) as the JH75 inner layer scale ℓ, the depth over which turbulent stress divergence is important in the momentum balance. Since for local equilibrium we require $z \ll \overline{\ell}$, that is, advection << production (or dissipation), we infer that only over the lowest part of the region of turbulence influence is local equilibrium a valid concept.

The converse of the local equilibrium region is the rapid distortion region where eddies suffer negligible dissipation in their passage over the hill, although they still respond to the mean strain rates they encounter. We can estimate the bounds of this region through the same JH75 assumptions by comparing the time T for an eddy to be advected a distance L with the eddy turnover time T_ϵ. We write:

$$T = L/U; \quad T_\epsilon = u_*^2/\frac{u_*^3}{kz} ,$$

where we have taken $\frac{1}{2}\overline{q^2} \sim u_*^2$ and used the usual surface layer expression for the dissipation rate. Then if $\overline{\overline{\ell}}$ is the height at which $T = T_\epsilon$, we obtain:

$$\frac{\overline{\overline{\ell}}}{L}\ \ell n\left[\frac{\overline{\overline{\ell}}}{z_0}\right] \sim 1 ,$$

whence $\overline{\overline{\ell}} \sim \overline{\ell} \sim \ell$. At heights much larger than ℓ, $T \ll T_\epsilon$ and we are in the region where changes to the turbulence can be computed by linearized turbulence or "rapid distortion" theory (Hunt 1978). Rapid distortion theory calculates stresses by integrating production rates over a predetermined eddy lifetime. It also allows intercomponent transfer by the rapid part of the pressure-strain term and transport by advection but its conclusions reflect most strongly upwind strain history.

Estimates of the depth of these regions obtained as we have done are obviously flawed because we have assumed, a priori, local equilibrium forms to calculate the extent of nonequilibrium behaviour; nevertheless, they provide useful order-of-magnitude figures and it is instructive to see how large they are in some practical situations. For example, on Askervein hill ($L = 280$ m, $h = 116$ m, $z_0 = 0.03$ m, $\ell = 15$ m) local equilibrium will apply in the lowest 2 m or so while above 100 m, say, rapid distortion calculations should give an accurate estimate of changes to the turbulence. Over the rough WTHILL of Finnigan et al. (1988a) ($L = 0.15$ m, $h = 0.06$ m, $z_0 = 0.12$ mm, $\ell = 7$ mm) the physical roughness height z_r was 7 mm (gravel) and, as we have already mentioned, over such rough surfaces there exists a roughness sublayer of depth $\sim 2.z_r$ where local equilibrium is denied by turbulent diffusion. Hence, there is no local equilibrium region on this model hill, while by about one hill height above the surface we have entered the rapid distortion regime.

Changes to the mean strain rates: Far upwind of the hill $\partial_z U$ is the only mean strain rate and very close to the surface we expect it to remain the dominant one since the other strains that appear in the production terms of (11)-(15), U/L_a (representing plane straining) and U/R, (rotation) must vanish with U at the wall. On the crest of the hill the existence of a peak value in Δs implies an increase in $\partial_z U$ below this point and a decrease above. The peak in Δs occurs typically at about $\ell/4$. This means that on the hilltop the shear is enhanced in the region where we might expect local equilibrium but reduced over most of the flow regime. At the upwind foot of the hill, where the flow decelerates, $\partial_z U$ is reduced at all heights. This is clearly displayed in Fig. 2 as well as in the mean flow data from Askervein (Mickle et al. 1988), or in most of the experiments discussed earlier. Downwind of the hill crest is the elevated region of strongly increased $\partial_z U$ bounding the wake, which can also be seen in most of the wind-tunnel simulations canvassed in the preceeding sections.

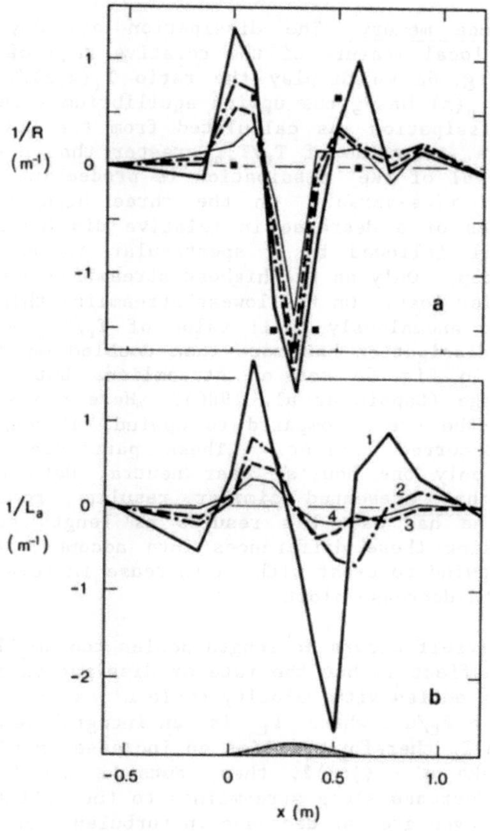

Fig. 5. (a) 1/R vs x on the four streamlines of WTHILL. (■ hill surface). (b) $1/L_a$ vs x on the 4 streamlines. Curves identified as in (a).

The parameters $1/L_a$ and $1/R$, the inverse length scales associated with U/L_a and U/R, are plotted in Fig. 5 for the four streamlines of WTHILL. Although their peak magnitudes are similar, a striking difference is the symmetry of $1/R$ compared with the skew symmetry of $1/L_a$. We can infer from (5), the kinematic condition linking $1/R$ and $1/L_a$, that over hills with h/L ratios typical of WTHILL or Askervein, U/L_a and U/R are of the same order. It is instructive to use the equilibrium assumptions of JH75 to compare them to $\partial_z U$. Writing

$$\frac{U}{L_a} \sim \frac{\Delta U}{L} = \frac{2hU}{L^2}$$

and

$$\partial_z U \sim \frac{u_*}{kz} ,$$

if $\hat{\ell}$ is the height where $\partial_z U \sim U/L_a$, then

$$\frac{\hat{\ell}}{L} \ell n\left[\frac{\hat{\ell}}{z_0}\right] = \frac{L}{2h} , \tag{19}$$

and for hills like Askervein (h = 116 m, L ~ 300) L/2h ~ 1 so that $\hat{\ell} \sim \ell$. In other words, shear dominates the 'extra' strain rates only at heights $z << \ell$ while over most of the flow regime, U/L_a, U/R and $\partial_z U$ are of the same order.

Dissipation and turbulence memory: The dissipation or eddy turnover time scale, $T_\epsilon = \frac{1}{2}q^2/\epsilon$, is a local measure of the relative rate of dissipation of turbulent energy. In Fig. 6a we display the ratio $T_\epsilon(x,z)/T_{\epsilon 0}$ on the four streamlines of WTHILL, $T_{\epsilon 0}(z)$ being the upwind equilibrium value of T_ϵ on the same streamline. The dissipation was calculated from the Kolmogorov formula for the inertial subrange. A value of $T_\epsilon/T_{\epsilon 0}$ greater than one implies that, relative to the local level of tke, dissipation is proceeding less vigorously than at equilibrium and vice-versa. On the three highest streamlines a consistent pattern emerges of a decrease in relative dissipation rate to the upwind foot of the hill followed by a spectacular increase as the flow accelerates to the hill top. Only on the highest streamline does this increase fail to redress the earlier loss. On the lowest streamline things are somewhat more complicated with an anomalously small value of $T_\epsilon/T_{\epsilon 0}$ at X/L = -4, but there too the relative dissipation has more than doubled on the crest. The same ratio is plotted in Fig. 6b not on streamlines but for z = 2 m and z = 16 m on Cooper's Ridge (Coppin et al. 1986). Here too we see increased relative dissipation at the crest compared to upwind, although here an even greater increase has occurred earlier. These particular data are only tentative, however, as only one hour's, near neutral data are included in Fig. 6b. Mason (1986) has presented similar results from Nyland's hill (his Fig. 10) although he has cast his results as length scales Λ^W, where $\Lambda^W = [(\overline{(w^2)}^{3/2})/\epsilon]$. Taking these differences into account, we see the same pattern of change from upwind to crest with an increase in relative dissipation closer to the ground but a decrease aloft.

The dissipation process itself occurs at length scales too small for changes in the mean strain rate to affect it but the rate of dissipation is controlled by the rate at which large eddies with velocity scale u' can interact with their neighbours. Hence $T_\epsilon \propto \ell_t/u'$, where ℓ_t is an integral length scale of the turbulence. Decrease in T_ϵ therefore implies an increase in u' or decrease in ℓ_t or both. If we take $u' \sim (\frac{1}{2}\overline{q^2})^{\frac{1}{2}}$, then, outside the local equilibrium region, u' may actually decrease along streamlines to the hill top, as we shall see shortly, implying an even greater decrease in turbulent length scale.

Two related causes are to be found for this decrease: the stabilizing effect of streamline curvature near the hill top and streamline convergence. The analogy between buoyancy and curvature in shear flows has been known for a long time. Bradshaw (1969) formalized the similarities in the 'curvature Richardson No', Rc, which, in analogy with its diabatic namesake, expresses the balance between shear and curvature forces on the fluid. In streamline coordinates it takes the form $Rc = 2U/R\Omega$, which means it is also a 'flux curvature Richardson number', the ratio of tke production by shear and curvature, as is readily seen from (15). Since Ω (= $[U/R - \partial_z U]$) is usually negative in the boundary layer, Rc is negative (denoting instability) when R is positive and vice versa. Hence we expect curvature effects to stabilize the flow on the hill top and make it unstable at the upwind foot of the hill. R_c can also be interpreted as the squared ratio of two time scales, the vorticity timescale Ω^{-1} and Nc^{-1}, where N_c (= $[2U\Omega/R)]^{\frac{1}{2}}$) is the frequency of small adiabatic inviscid oscillations normal to the curved streamlines. N_c is the curved flow counterpart of the Brunt-Vaisälä frequency. In a diabatic flow we can form a length scale ℓ_c which we define as the typical distance a fluid element may move in the z-direction before it has exchanged all its vertical

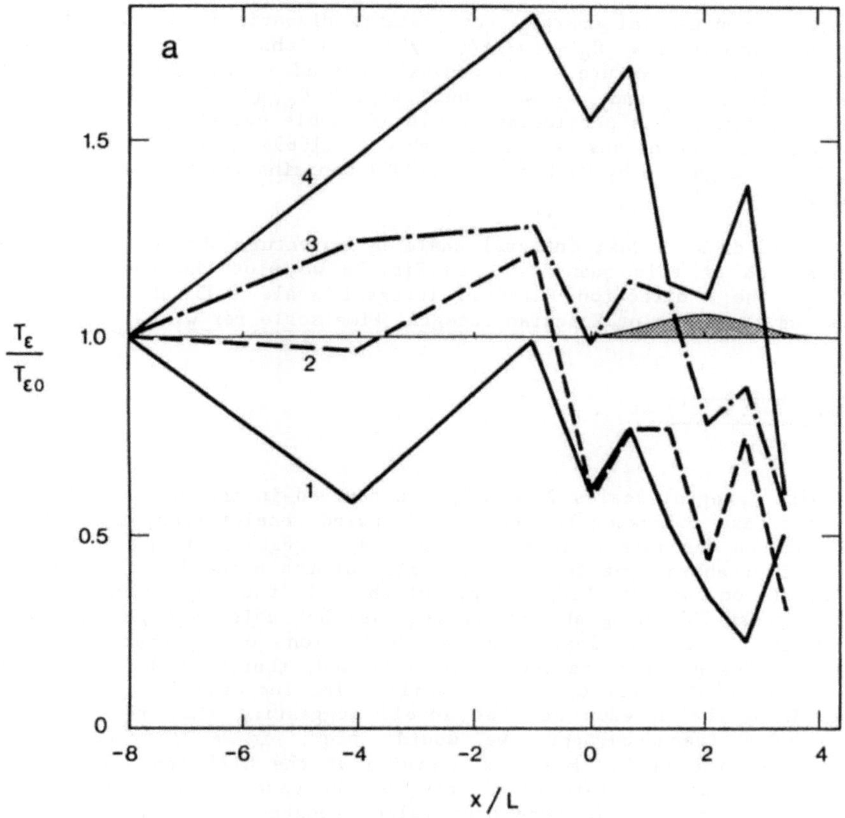

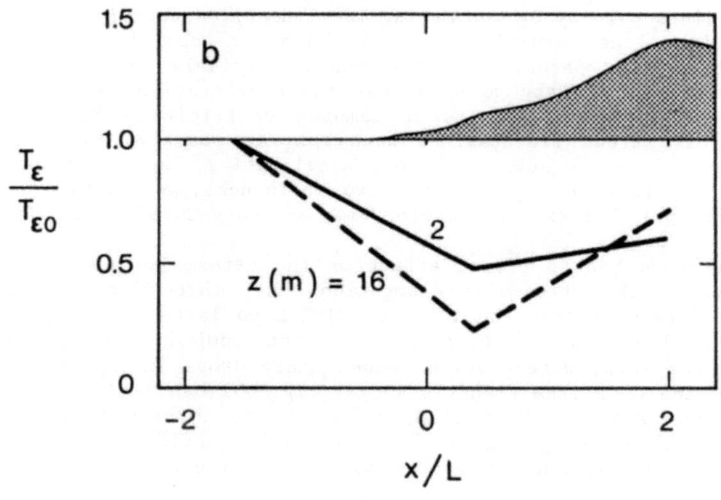

Fig. 6. (a) $T_\epsilon/T_{\epsilon 0}$ vs x/L on the four streamlines of WTHILL. (b) $T_\epsilon/T_{\epsilon 0}$ vs x/L at z = 2 m and z = 16 m on Cooper's Ridge.

kinetic energy for potential energy. In a stable diabatic flow, $\ell_c = (\overline{w^2})^{\frac{1}{2}}/N$. In a stably curved flow $\ell_c = (\overline{w^2})^{\frac{1}{2}}/(2U\Omega/R)^{\frac{1}{2}}$ so that ℓ_c decreases with increasing (negative) curvature. The convexly curved streamlines on the hill top therefore reduce ℓ_c and, as we should expect ℓ_c and ℓ_t to be related, increase dissipation. This particular result of stable curvature has been well documented on many occasions since Bradshaw's (1969) paper. The recent experimental investigation by Muck et al. (1985) contains references to much of the earlier work.

The reduction of cross stream integral scale by curvature should be reflected in other measures of this quantity. In Fig. 7a we plot the evolution over WTHILL of L_w, the z-direction Eulerian integral scale defined as $L_w = UT_w$, where T_w is the single point Eulerian integral time scale for w:

$$T_w = \int_0^\infty \frac{\overline{w(t)w(t + t')}}{\overline{w^2}}\, dt' \quad . \tag{20}$$

(The streamwise integral scales T_u and L_u are derived in the same way.) After an initial decrease following the region of upwind deceleration, L_w increases rapidly behind the unstable (concave streamlines) curvature at the foot of the hill only to fall sharply at the crest on all but the highest streamline. At the lower levels on Cooper's Ridge (Fig. 7b) there is the suggestion of a fall in L_w at the crest following an earlier increase but this is emphatically not the case at 16 m. It is clear from the definition of L_w that unless the frequency of eddies passing the sensor is increased, thereby reducing T_w, then L_w must increase as U speeds up over the hill. The increase in L_w at 16 m is roughly equal to the speedup at that level, suggesting that no change has occured in eddy frequency, or, we would infer, in eddy production or interaction rate, the train of eddies arriving at the hill foot simply being convected past the crest. This is clearly not the case on WTHILL where, even on the highest streamline, T_w must have fallen substantially just before the hill top.

The second factor reducing integral scales is streamline convergence. If Rc is derived in the classical way by converting the linearized equation for velocity perturbations into the equation for a harmonic oscillation with the displacement of a fluid parcel as the dependent variable (see, for example, Turner 1973, p 11), it is easy to show that the acceleration term U/L_a enters into the oscillator formulation as a damping coefficient, decreasing the amplitude of cross stream fluctuations when U/L_a is positive, and increasing them in a region of deceleration. In our accelerating, curved, hill flow in short, it is impossible to separate the two influences, a conclusion already suggested by (5). To this extent, acceleration and curvature are coupled.

Stabilizing curvature has no direct effect on the streamwise fluctuations but acceleration does. The streamwise elongation that, through continuity, must accompany cross-stream contraction leads on WTHILL to large increases in L_u at the hill top following an initial fall in the upwind deceleration zone. (Fig. 8a.) In the wake, deceleration causes sharp drops in L_u and a rise in L_w. Once again the data from Cooper's Ridge (Fig. 8b) behave in the same way. An immediate conclusion is that changes in L_w are a better guide to the dissipation integral scale L_ϵ than are changes in L_u. Evidently much of the u variance is caused by motions on scales too large to contribute to eddy-eddy interactions.

The memory length scale, $\ell_\epsilon = UT_\epsilon$, is a measure of how much upwind strain 'history' the ensemble of eddies measured at some point will reflect in its structure. We recall that for local equilibrium we require $\ell_\epsilon \ll L$ say, where we can employ L as a typical length scale over which the mean strain rates

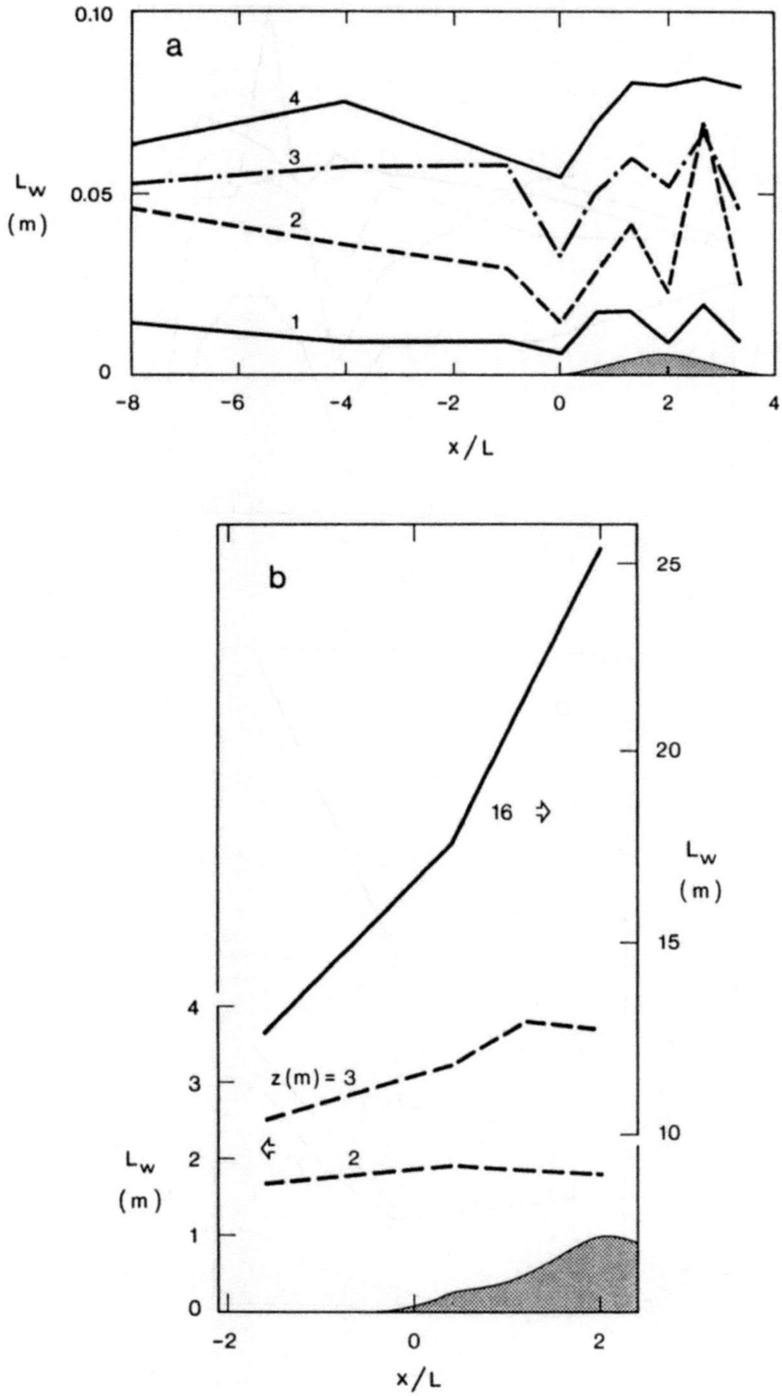

Fig. 7. (a) L_w vs x/L on the four streamlines of WTHILL. (b) L_w vs x/L at z = 2 m and z = 3 m (left-hand scale) and at z = 16 m (right-hand scale) on Cooper's Ridge.

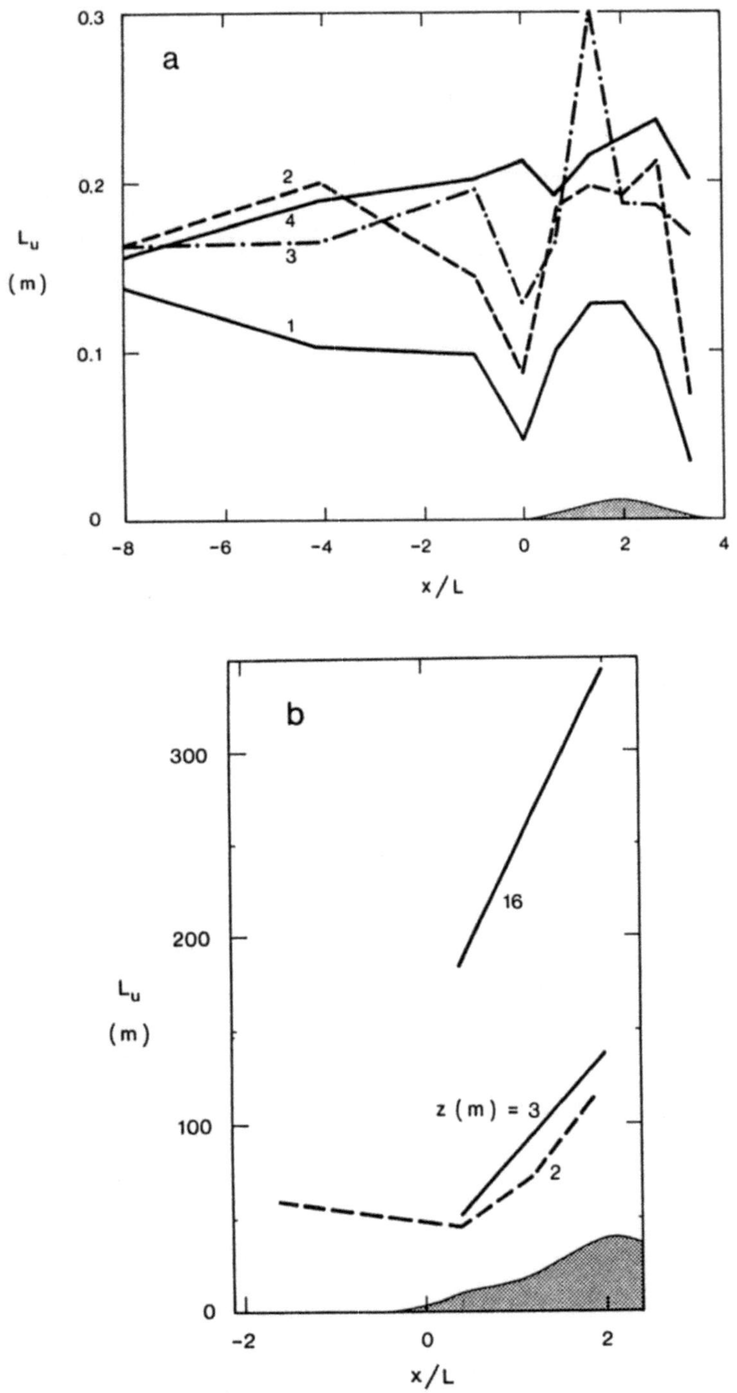

Fig. 8. (a) L_u vs x/L on the four streamlines of WTHILL. (b) L_u vs x/L at z = 2 m, z = 3 m and z = 16 m on Cooper's Ridge.

change. ℓ_ϵ on the four streamlines of WTHILL is plotted in Fig. 9a. On WTHILL L = 0.15 m so that up to the point that the wake is encountered, ℓ_ϵ is always $\gtrsim$ L. This is good support of our earlier rough estimate that there is no local equilibrium region on this rough hill. Perhaps more surprising is the result shown in Fig. 9b from z = 2 m on Cooper's Ridge. There L ≃ 250 m and ℓ_ϵ increases from L/4 to about L/2 at the hill top, suggesting that local equilibrium is confined to a very shallow layer indeed, to lower than 0.1ℓ, in fact, since ℓ = 20 m on this hill. At 16 m, ℓ_ϵ goes from about 1.5 L upwind to 5 L on the crest.

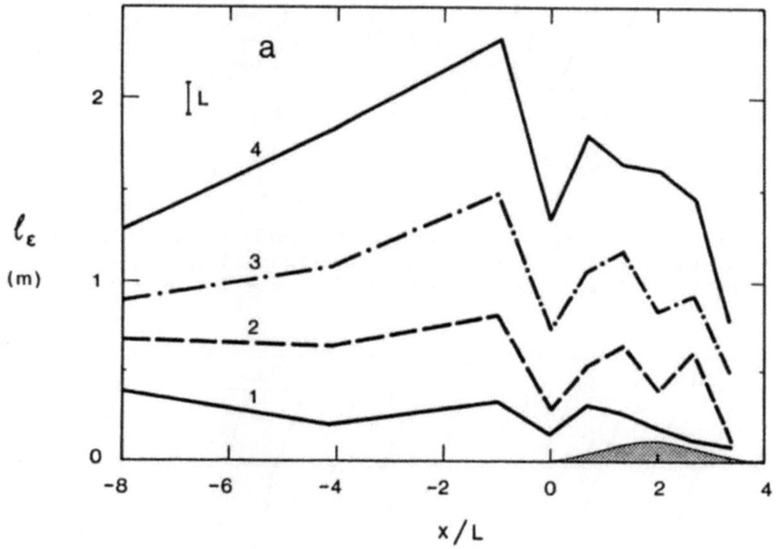

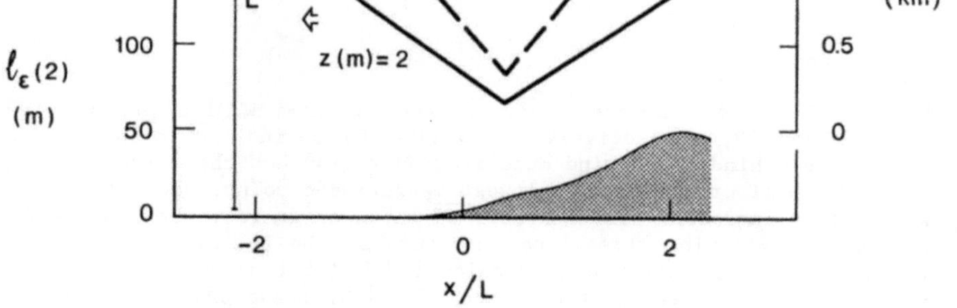

Fig. 9. (a) ℓ_ϵ vs x/L on the four streamlines of WTHILL. The horizontal scale L is shown for comparison. (b) ℓ_ϵ vs x/L at z = 2 m (left-hand scale) and z = 16 m (right-hand scale) on Cooper's Ridge. For comparison L is shown relative to both ordinate scales.

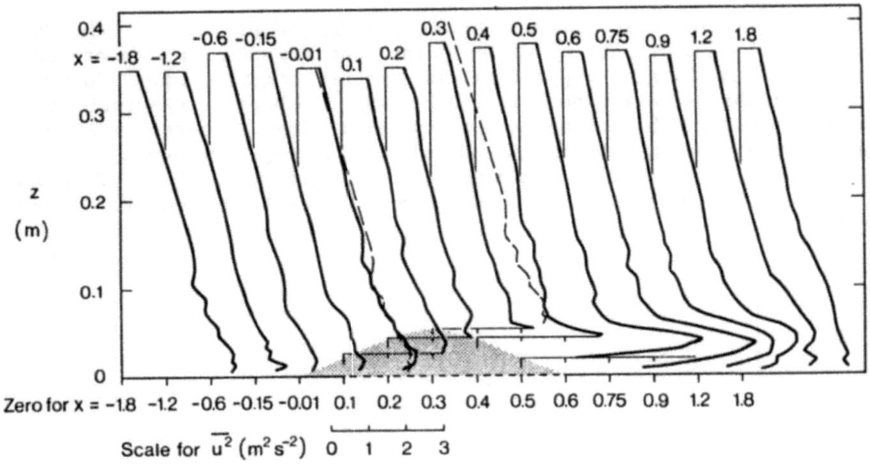

Fig. 10. Profiles of $\overline{u^2}$ measured at various x stations on WTHILL with the upstream (x = −1.8 m) profile (chained line) superimposed on the local profile at x = −0.01 m and x = 0.3 m. The hill extends from x = 0 to x = 0.6 m.

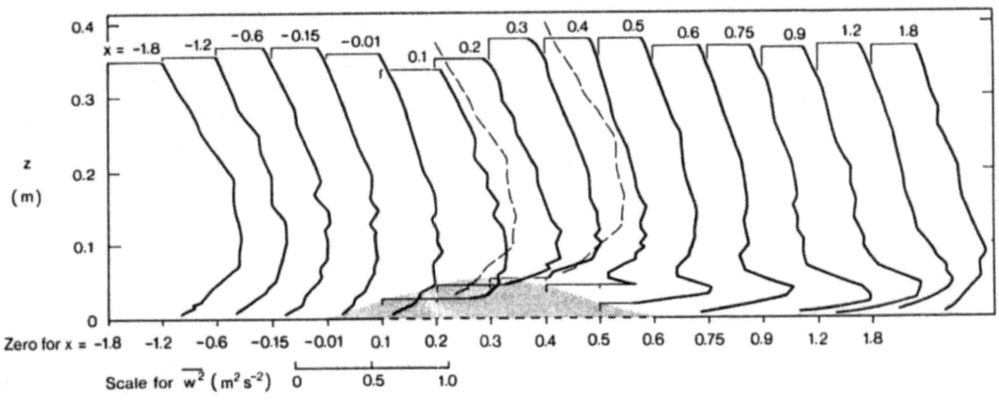

Fig. 11. As for Fig. 10 but for $\overline{w^2}$. Upstream profile is superimposed at x = 0.1 and x = 0.3.

Turbulence Moments

Profiles of the second moments, $\overline{u^2}$, $\overline{w^2}$ and $\overline{uw}$ from WTHILL are plotted in Figs. 10, 11 and 12, respectively. Relative to upwind, $\overline{u^2}$ displays a very slight increase behind the upwind deceleration region but then drops rapidly to the top of the hill at all but the lowest measurement point, where there is the suggestion of an increase. In contrast, $\overline{w^2}$ shows an increase behind the region of concave, destabilizing streamline curvature at the bottom of the hill. On the hill top it shows little change below z = h/2 but is much reduced above. Also, $-\overline{uw}$ shows an increase at the bottom of the hill and pronounced reductions down to the lowest measurement level on the hill top. In the wake all three moments show strong peaks, clearly delineating the wake as the region of most active turbulent exchange. The evolution of the stresses can be seen particularly clearly as we follow the four streamlines over the hill (Figs. 13,

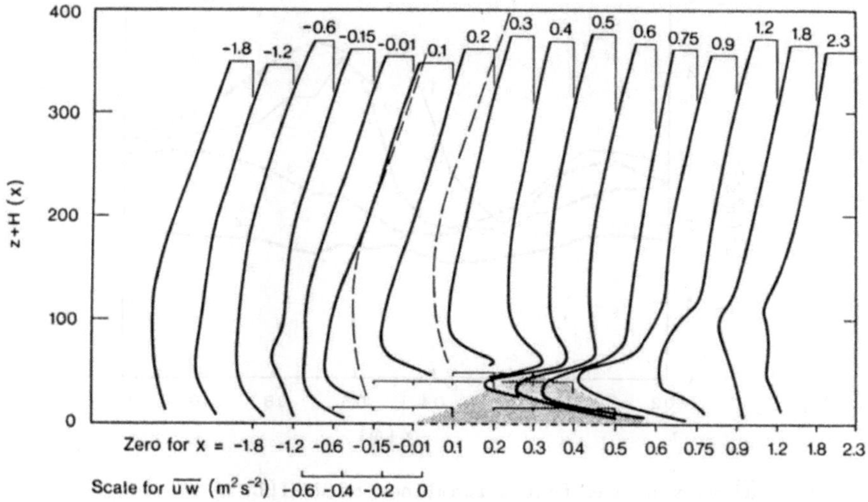

Fig. 12. As for Fig. 10 but for $\overline{uw}$. Note that $\overline{uw}$ is negative so that profiles extend to the left of their zero marks.

14 and 15), the large increases in stress behind the hill corresponding closely to the point where a particular streamline encounters the wake. Comparing the streamwise behaviour of these moments with that of the 'extra' strain rates, U/L_a and U/R, from Fig. 5, it is immediately obvious that they follow very closely the relevant production terms over a length of the upwind streamline roughly equal to ℓ_ϵ. This direct link between production rate and stress level is not surprising given that we have already deduced that over WTHILL, the turbulence response will be biased towards the rapid distortion regime. We recall from (11), (12) and (13) that:

$$\text{Production } \overline{u^2} = -2\left[\overline{u^2}\,\frac{U}{L_a} - \overline{uw}\,\frac{U}{R} + \overline{uw}\,\partial_z U\right], \tag{21}$$

$$\text{Production } \overline{w^2} = -2\left[2\overline{uw}\,\frac{U}{R} - \overline{w^2}\,\frac{U}{L_a}\right], \tag{22}$$

$$\text{Production } \overline{uw} = -\left[(2\overline{u^2} - \overline{w^2}) + \overline{w^2}\,\partial_z U\right]. \tag{23}$$

Remembering that the ratio $\overline{u^2}:\overline{w^2}:-\overline{uw}$ is approximately 5.0:1.25:1.0 and neglecting $\partial_z U$, we can see that *changes* in $\overline{u^2}$ production will be dominated by normal strains while $\overline{uw}$ *changes* will be linked primarily to curvature. Production of $\overline{w^2}$ responds to both influences with a bias towards a curvature response.

Changes in $\overline{u^2}$, $\overline{w^2}$ and $-\overline{uw}$, essentially the same as those of Figs. 10, 11 and 12, have been observed in practically every case where turbulence measurements have been performed over 2-D ridges or on the centrelines of 3-D hills with one important proviso. On smoother hills it has proved possible to make measurements closer to the surface, in a region where $\partial_z U$ is the dominant strain rate. Hence, if measurements are made on the hill top below the peak in Δs, the increased production of $\overline{u^2}$ from the term $-2\overline{uw}\partial_z U$ may exceed the decrease due to $-2\overline{u^2}U/L_a$ and very close to the surface a region of increased $\overline{u^2}$ may be observed. Similarly, in the upwind deceleration region, the loss of

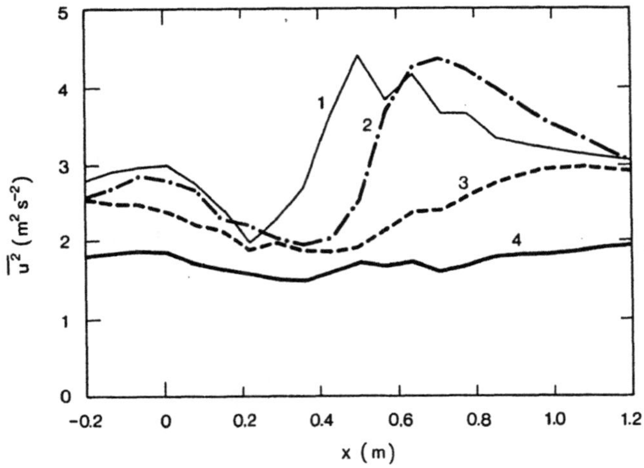

Fig. 13. $\overline{u^2}$ vs x on the four streamlines of WTHILL.

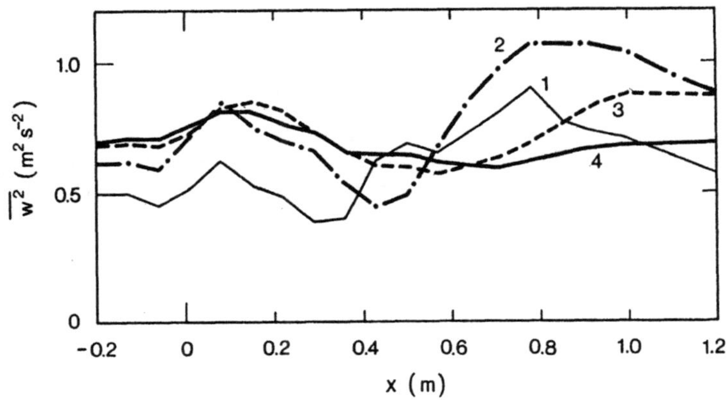

Fig. 14. $\overline{w^2}$ vs x on the four streamlines of WTHILL.

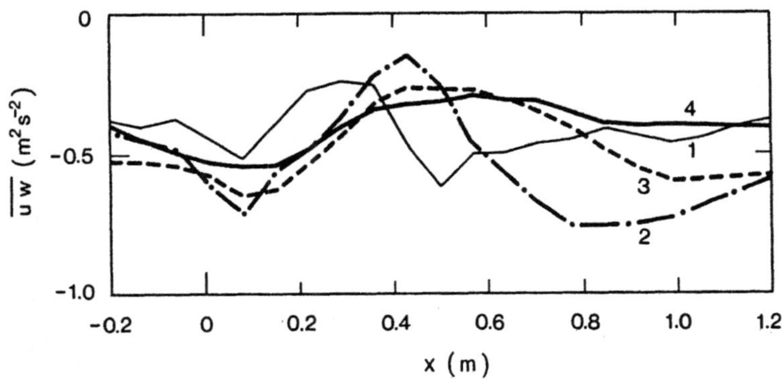

Fig. 15. $\overline{uw}$ vs x on the four streamlines of WTHILL.

production at the lowest level caused by reduced $\partial_z U$ cannot be redressed by the increase in $-2\overline{u^2}U/L_a$ and $\overline{u^2}$ falls. These effects are clearly seen in Fig. 16, which displays data from Askervein. The same effect occurs in the case of $\overline{w^2}$ and $\overline{uw}$ on the hill top, with the difference that $\overline{w^2}$, having no shear production term, obtains its increase by 'feeding' on increased $\overline{u^2}$ via the pressure-strain term. Since $\overline{uw}$ relies on this weaker increase in $\overline{w^2}$ to couple it to shear in the term $-\overline{w^2}\partial_z U$, the increases in both $-\overline{uw}$ and $\overline{w^2}$ are confined to a much thinner region than the increase in $\overline{w^2}$. Both Fig. 16 from Askervein and Fig. 17 from Cooper's Ridge demonstrate this clearly with $-\overline{uw}$ and $\overline{w^2}$ exceeding their upwind values at Askervein only at the lowest measuring point, while at 2 m on Cooper's ridge they are both less than upwind although $\overline{u^2}$ exceeds its upwind value below 5 m.

This pattern of near-surface response has been observed whenever measurements can be made in the region of increased shear on the hill top, decreased shear upwind, that is, on Askervein and Cooper's Ridge, on Blasheval (Mason and King 1985) and in the wind-tunnel simulations of Kettle's Hill (Teunissen et al. 1982), of Askervein (Teunissen et al. 1987), on the rough 2-D triangular ridge studied by Snyder and Britter (1987) and on the three rough curved ridges tested by Arya et al. (1987).

Where smooth hills have been placed in a rough wall boundary layer, however, precisely to enable measurements to be made within the inner layer, the expected increase is often masked by the effect of the rough-smooth transition. For example, on all of the smooth models tested by Snyder and Britter (except the cone), $-\overline{uw}$ on the hill top was reduced at all levels relative to upwind, and similar results were noted in $\overline{u^2}$ by Meroney et al. (1978) on smooth 2-D ridges.

It is impossible to deduce from the ratios of stresses alone the depth of the region of true local equilibrium. Increases in $\overline{u^2}$, $\overline{w^2}$ and $-\overline{uw}$ in the inner layer can be accounted for as successfully by rapid distortion theory as by a local equilibrium hypothesis, although we do not, of course, suggest that rapid

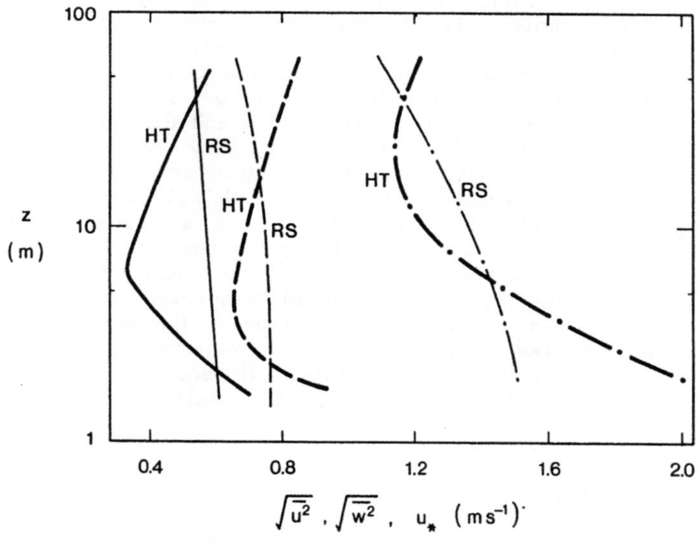

Fig. 16. Vertical profiles of $(\overline{u^2})^{\frac{1}{2}}$ (—·—); $(\overline{w^2})^{\frac{1}{2}}$ (— —) and u_* (= $(-\overline{uw})^{\frac{1}{2}}$) (——) from two positions on Askervein Hill with the wind perpendicular to the long axis of the hill (235°). HT denotes hill top and RS the upwind reference station. (Mickle et al. 1988).

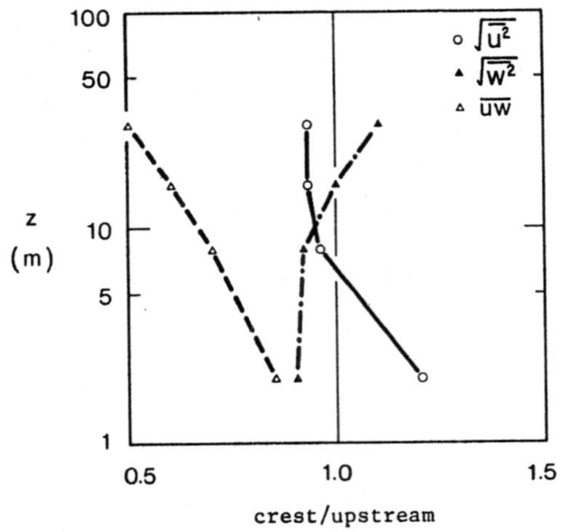

Fig. 17. Vertical profiles of ratios of r.m.s. normal stresses and of $\overline{uw}$ between crest and upstream on Cooper's Ridge. (Coppin et al. 1986).

distortion prevails so close to the surface. It has become common, nevertheless, to refer to the region where changes in $\partial_z U$ dominate as the 'local equilibrium region', a misleading nomenclature.

Another misunderstanding that has crept into the literature concerns the application of rapid-distortion theory. Britter et al. (1981) seem to have been the first to apply results of classical rapid distortion theory to hill flow. There are, in fact, only three analytical solutions to the rapid distortion equations appropriate to this problem. They describe the changes in initially specified turbulence following irrotational straining, simple shearing and solid body rotation (Townsend 1980). Analytical solutions for combinations of these simple strains, such as a curved shear flow or an accelerated shear flow, have not proved possible. Britter et al. (1981) chose the first of these, plane strain without shear or rotation as their model. The solution predicts a decrease in $\overline{u^2}$ and increases in $\overline{w^2}$ and $\overline{v^2}$ on the hill top. However, this choice is equivalent to ignoring U/R and $\partial_z U$ compared with U/L_a in the production terms shown in (11)-(14). In reality these three are of the same order over most of the rapid distortion regime and experiments in both field and wind tunnel, for example Finnigan et al. (1988a), Mickle et al. (1988) or Coppin et al. (1986), where both u and w components have been measured, show decreases in $\overline{w^2}$ on the hill top as the stabilizing effect of curvature overrides the amplification of $\overline{w^2}$ by acceleration. It is unfortunate, therefore, that reference to "rapid distortion" in this context is usually taken to mean this plane-strain solution. The misunderstanding is exacerbated by the repeated publication of $\overline{u^2}$ measurements, which show a decrease at the crest, without the related $\overline{w^2}$ measurements (which *also* decrease); most recently, for example, in the simulations of Askervein by Teunissen et al. (1987).

Three-Dimensional Effects

On the centreline of axially symmetric hills we see second-moment responses essentially similar to those over ridges with the proviso that the reduction in $\overline{w^2}$ aloft may be somewhat less than in the 2-D case. The data of Mason and King

(1985) from Blasheval, Mason (1986) from Nyland's Hill or Arya and Gadiyaram (1986) in the wind tunnel are good examples.

The extreme near surface speedup possible on smooth cones may also lead to $\partial_z U$ values enhanced to the point where the rough-smooth transition effect may be cancelled out. Hence, on the smooth triangular ridges of Snyder and Britter (1987) $-\overline{uw}$ on the crest was less than $-\overline{uw}$ upstream but on their smooth cone, this ratio was reversed. Similarly Pearse (1982) recorded low level decreases in $\overline{u^2}$ on top of smooth cones with h/L values of 0.3 and 0.5 but an increase for h/L = 1.0. In contrast, on a rough cone with h/L = 1.0, and correspondingly lower speedup, Arya and Gadiyaram (1986) saw no such increase.

As soon as the three dimensionality becomes more complex, however, the data are more reluctant to fit into this simple framework. Bradley (1980), for example, on the steep, rough Black Mountain measured increases in both $\overline{u^2}$ and $\overline{w^2}$ up to z = h/2 on the hill top while $-\overline{uw}$ increased below z/h = 0.1 and decreased above.

One extra effect that has not been considered so far but which may have a significant role to play is that of the third extra strain rate, streamline torsion. We have worked primarily with 2-D streamline coordinates where the extra strain rates, curvature U/R and acceleration U/L_a, are familiar. In 3-D flows in general, the price of aligning the local x axis parallel to streamlines is the appearance in the flow equations of the twist or torsion of the streamline (Finnigan 1988a, and see the discussion of coordinates above). There are to date very few data on the sensitivity of flow to twist. What there are, mainly from studies of flow in helically coiled pipes (Kao 1987), indicate that when torsion and curvature are of similar magnitude, their combined effect on the flow is both profound and quite different to that of curvature alone. A glance at (6) reveals that σ generally appears in combinations with R and L_a in 3-D flow equations.

Streamline twist is automatically induced when shear flow approaches a 2-D ridge obliquely. If we make the crude assumption that only the normal component of the approach flow is accelerated over the hill (Bradley 1983), we can form a good estimate of the induced torsion. Using data from Cooper's Ridge with a typical measured logarithmic approach flow at 45° to the ridge axis, we conclude that on the hill top at 10 m, $1/\sigma \sim 5.0 \times 10^{-3}\ m^{-1}$ while $1/R \sim 1.5 \times 10^{-3}\ m^{-1}$.

Streamline torsion also accompanies flow around the sides of isolated hills and is present, of course, in the trailing vortices of the wakes. We have as yet virtually no turbulence measurements from hillsides to confuse us but it seems unlikely that we would be able to fit them into the emerging pattern of a response to mean strain without including this important factor.

Turbulence in the Wake

As in the case of the mean flow, Taylor (1988) comprehensively reviews the downstream development of the wake turbulence. I confine myself, therefore, to remarks about the near wake or separation region. As before, what we do know here comes primarily from a few wind-tunnel studies. Of these only the studies of Castro and Snyder (1982) and Arya et al. (1987) employed pulsed wire anemometers so that we can have confidence in their data in the separation region. Behind hills with only intermittent separation, Britter et al. (1981) employed a pulsed wire while Finnigan et al. (1988a,b) used a coplanar triple wire (Legg et al. 1984). Although this cannot cope with reversed flow, it allows reasonable measurements in turbulent intensities up to ~ 70%, much more than with an X-wire.

These two quantities are plotted in Fig. 18 as ratios of their values far upwind.

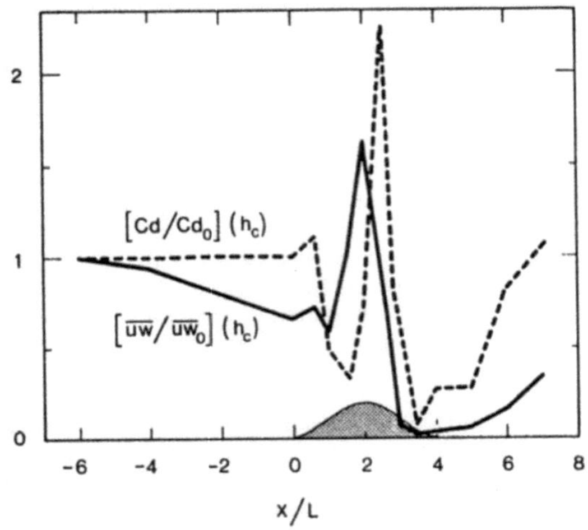

Fig. 18. Data from a wind-tunnel model hill covered with a simulated crop canopy of height h_c. Ratios of drag coefficient at (x,h_c) relative to its value upwind $[Cd/Cd_0](h_c)$ vs x/L (– – –) and of shear stress $[\overline{uw}/\overline{uw}_0](hc)$ vs x/L (——).

Clearly $-\overline{uw}(h_c)$ shows the behaviour we now expect in a region where $\partial_z U$ dominates the strain rate; $-\overline{uw}$ falls in the upwind acceleration region, climbs rapidly to exceed its upwind value by the crest and plummets in what was a definite separation bubble. ($\overline{uw}$ measurements in this region are tentative, of course, even with a triple wire.) The behaviour of the drag coefficient is perhaps more unexpected. A slight rise at the foot of the hill corresponds to the increased mixing accompanying unstable flow curvature. Cd then drops rapidly and has only recovered to 85% of its upwind value by the crest. Interestingly it continues to rise around the separation point, indicating that U is dropping faster than $-\overline{uw}$ as separation is approached from upstream and then falls to very low values in the bubble itself.

These variations result from profound changes in the eddy structure and mean flow penetration of the canopy (Finnigan et al. 1988b) and though they represent a somewhat extreme case of a rough surface, the mechanisms responsible can be expected to operate to some degree over all rough hills.

MATHEMATICAL MODELS

Mathematical models are the shop windows in which the advancing theory of hill flow has been displayed for the last 15 years. They predated published field experiments by five years and major experiments have, more often than not, been first announced in company with the latest model that they inspired. See, for example, Mason and Sykes (1979), Mason and King (1985) or Zeman and Jensen (1987).

An ideal mathematical model of turbulent hill flow would be capable of reproducing the flow structure that we have described with minimum time and effort. There are two possible candidates: closure models implemented in Eulerian coordinates and large eddy simulations. At the present time we must

All of the 2-D data show much the same order of increase in the turbulence moments as are displayed in Figs. 10, 11 and 12. These can be explained in the same way as the upstream results as primarily a direct response to change in strain rate, the dominant strain being provided by the region of enhanced shear bounding the wake. Hence, we see immediate large peaks in $\overline{u^2}$ and $-\overline{uw}$ whose production is directly coupled to shear and a slight delay until the maximum in $\overline{w^2}$ occurs, presumably after pressure-strain transfer from $\overline{u^2}$. The main difference behind axially symmetric hills is the pronounced bimodal distribution of shearing stress in the lateral direction (Arya and Gadiyaram 1986), a not unexpected result in view of the existence of two counter-rotating axial vortices behind such obstructions. An effect noted in the rough WTHILL studied by Finnigan et al. (1988a) was a pronounced shift to lower frequency of the peak of the $\overline{w^2}$ spectrum just behind the hill crest. This shift corresponds to the large upsurge in $\overline{w^2}$ energy behind the hill top on the two lower streamlines (Fig. 14) and is clearly coupled to the intermittent separation bubble there.

We recall from Fig. 8a the sudden contraction in L_u behind the hill, which suggests that the physical scale of the bubble or wake is reflected immediately in the turbulence. It would be incorrect to imagine all the turbulence within the near wake region to be of bubble scale though; the experiments of Pronchik and Kline (1983) on flow behind a backward facing step show that much of the turbulence within the separation bubble is produced in the reattachment zone and then is advected into the bubble by the reversed flow. It is therefore of much smaller scale than the bubble dimensions.

Surface Shear Stress

The shearing stress at the surface, τ_0, is physically interesting in its own right, for example, as the direct cause of particle dislodgment during wind erosion, but equally importantly, it forms the preferred boundary condition for mathematical models. In this role it is usually linked to the velocity profile through some version of the logarithmic 'law of the wall'. In the usual derivation of the 'log law' (Yaglom 1979), dimensional analysis is used to express the velocity shear near the surface as:

$$\partial_z U = \frac{u_*}{kz} \varphi_M \left[\frac{z}{z_s}, \frac{z}{R}, \frac{z}{L_a}, \frac{z}{\sigma}\right] , \qquad (24)$$

where k (= 0.4) is von Karman's constant and φ_M is a function of the dimensionless combinations of all the length scales that might affect $\partial_z U$. $u_* = \sqrt{\tau_0/\rho}$ is the friction velocity and z_s is a length scale characterizing the surface roughness. Although we have included z/σ for completeness in (24), we concentrate on the 2-D case from now on. If a range of z can be found where φ_M is independent of z/z_s, z/R, and z/L_a, and where the complementary relationship describing flow far from the wall is simultaneously valid, then it can be shown that φ_m must be a constant and the familiar log law follows upon integrating (24) (Yaglom 1979):

$$U(z) = \frac{u_*}{k} \ln(z/z_0) . \qquad (25)$$

The roughness length z_0 now emerges naturally as the surface length scale, viz $z_s = z_0$. Since $(\nu\, \partial_z U - \overline{uw}) = \tau/\rho$, it is clear that (24) is equivalent to an eddy diffusivity formulation for the turbulent shear stress with diffusivity $K_m = u_* kz$.

Wyngaard (1982) has shown that using an eddy diffusivity is equivalent to an assumption of local equilibrium in the rate equation for the relevant turbulent stress, hence use of the log law in the form of (25) strictly implies local equilibrium. However, mild departures from this ideal state can be tolerated without losing the log law formulation entirely, the classic example, of course, being Monin-Obukhov similarity in the diabatic surface layer when dependence of φ_M on the stability dependent ratio z/L_{MO} is retained. (L_{MO} is the Monin-Obukhov length.)

For φ_M to be independent of z/z_0, z/L_a and z/R (and numerically equal to 1) we require $z/z_0 \to \infty$ while simultaneously z/R, $z/L_a \to 0$. Over smooth walls this is possible because there, $z_0 = \nu/u_*$, a very small number indeed. A large number of measurements have been made of curved or accelerating flow over smooth walls where φ_M is seen to approach 1 as $z \to 0$. Over rough walls the situation is not so simple. There it may not be possible to make z so small that z/R and z/L_a approach zero without entering the roughness sublayer. Put another way, local equilibrium is seriously violated within the roughness sublayer because of turbulent diffusion and so the eddy diffusivity formulation of (24) cannot hold. Further from the surface we enter a region where advection becomes comparable to production or destruction; local equilibrium disappears and (25) is no longer valid.

If the surface is sufficiently rough, these two regions may overlap completely; no region of local equilibrium and no useful region of log law dependence then exists. This is the case on the rough WTHILL of Finnigan et al. (1988a). Over many less rough hills, however, we can find a region where approximate local equilibrium obtains and where a modified form of (25) can be used. The modifications come from allowing φ_M to retain some weak dependance on z/L_a and z/R (see Yaglom (1979) for examples of z/L_a dependence in engineering flows).

We saw in (23) that modifications to $\overline{uw}$ production over the hill are more strongly coupled to curvature than to acceleration so it is reasonable to expect a formal application of Bradshaw's buoyancy - curvature analogy to work. Bradshaw (1969) proposed to replace z/L_{MO} by Rc in the empirically known formulae for $\varphi_M(z/L_{MO})$. Hence, in the case of stably curved flow the Monin-Obukhov stable form $\varphi_M = (1 + 4.7\, z/L_{MO})$ (Tennekes 1982) would be replaced by $\varphi_M = (1 + 4.7Rc)$. Finnigan (1988b) reviews the application of the analogy to various field experiments where independent measurements of $-\overline{uw}$ and $\partial_z U$ are available, that is, to Nyland's Hill, Cooper's Ridge, Askervein and Black Mountain. He finds hilltop φ_M values ranging from around 1.5 (Nyland's and Cooper's) to 1.1 (Askervein and Black Mountain) and confirms that the WTHILL data did not fit the modified log law. The observed values of φ_M are clearly indicative of streamline curvature induced stability on the hill top. These changes in φ_m are accompanied also by a reduction in z_0. Unfortunately, we do not yet have upstream measurements available to see whether the unstably curved region admits the same parameterization.

The largest cause of uncertainty in the φ_M estimates comes in choosing a value of u_* since direct measurement of τ_0 is very difficult on a rough wall and extrapolation of the lowest measured value of $-\overline{uw}$ to the ground begs the question of the interdependence of $\tau(z)$ and $\partial_z U$. Finnigan and Raupach (1985) proposed that the curvature effect on WTHILL was so strong that τ_0 on the crest would actually be less than far upwind while adherents of strict JH75 thinking have matched their lowest measurements with a τ_0 derived by assuming the unmodified form of (25) (e.g., Zeman and Jensen 1987). Recent measurements by Finnigan et al. (1988b) suggest that the answer lies somewhere in between. They made wind-tunnel measurements over a model hill covered with a simulated crop canopy ($h/z_0 = 50$). This allowed a coplanar triple wire to be traversed through the canopy so that measurement of $-\overline{uw}$ and pitot pressure *through* the roughness were obtained. Since practically all of the tangential stress on the plane defined by the top of the roughness elements is absorbed on the model crop stalks, this stress, τ_0, could be determined unequivocally. As the mean velocity U does not vanish at the top of the crop either, it was possible to form the drag coefficient Cd ($= [-\overline{uw}(h_c)/U^2(h_c)]$), where h_c is the crop height.

rule out large eddy simulations because the capabilities of available computers limit the method to flows with only one direction of inhomogeneity.

Eulerian turbulence closure models can be further divided into two groups: schemes that retain the nonlinear form of the flow equations and linearized models. However crude the turbulence closure, nonlinearity forces us to adopt numerical finite difference methods while linear models, in principle at least, admit analytical solutions. Both approaches have been used extensively with linearized models holding centre stage from about 1975, although the modelling effort started with and has now returned to nonlinear formulations.

We can say immediately that a model that is to reproduce the mean flow and turbulent behaviour described in the previous section must have certain features. First of all, the equations to be solved must reflect all the important physical processes at work. This is a particularly contentious point with models that discard certain terms in the flow equations as negligible; linear models form an obvious case. To take a concrete example, most linearized analytical models like JH75 adopt terrain-following coordinates near the surface and use a scale analysis to discard the vertical advection term, $\hat{W}\partial\hat{U}/\partial\hat{z}$ (where ^ is used to emphasize that these are not streamline coordinates), so that the remaining advection term, $\hat{U}\partial\hat{U}/\partial\hat{x}$, is identical with that of streamline coordinates. On the other side of the equation, the same assumptions lead to the retention of only one term in the divergence of stress, $-\partial\widehat{\overline{uw}}/\partial\hat{z}$. Measurements on a grazing streamline over WTHILL, however, (Finnigan et al. 1988a) reveal that $-\partial_x\overline{u^2}$, $(\overline{u^2}-\overline{w^2})/L_a$ and $-\partial_z\overline{uw}$ are all of comparable size. The equation set solved by JH75, in other words, misses out an important component of the momentum balance on this rough hill. This problem is not confined to linearized models, as we shall see below.

We also saw that for hills of only moderate h/L, the influence of the wake region on the pressure field is strong. The pressure distribution around the hill is only asymmetric because of the wake and only on the shallowest hills would we expect to be able to calculate the flow downwind of the crest without allowing interaction between the velocity and pressure fields. On hills of greater steepness, the flow field is influenced both by separation and the wake region. This means that we should solve for the dependent variables of the problem, U, R, P, and the turbulent stresses simultaneously, that is, we should treat the flow equations as an elliptic system. In fact, this is absolutely essential if we hope to model the separation region itself with any fidelity. Desirable as this may be, elliptic problems are notoriously hard to treat by finite difference methods and we will usually see models assuming that the pressure field is known a priori, thereby reducing the equation set to a parabolic system equivalent to Prandtl's boundary layer equations.

The next problem to be faced is that of choosing the level of turbulent closure. To reproduce accurately the behaviour of the turbulent stresses, we must cope with the differing responses of the component variances to straining by the mean flow, to changes in dissipation rate, to the effect of turbulence memory and to the sudden change in turbulent length scales as we pass into the wake region. The obvious candidate to do this is second order closure. Indeed, Sykes (1980) points out that even over an asymptotically small hill changes in turbulent stresses cannot be modelled by any less sophisticated scheme. Nevertheless, as we shall see, by far the most popular closure involves an eddy diffusivity.

Turbulent closures that depend on an eddy diffusivity K to relate the turbulent stresses to the mean rates of strain fall into three classes, called in the jargon, mixing length ℓ, k-ℓ and k-ϵ methods (Launder and Spalding 1972). Noting that dimensionally K must be the product of a velocity scale V_t and a length scale ℓ_M, the simple mixing length model obtains both by specifying only the length scale ℓ_M. Hence $V_t = \ell_M|\partial_z U|$ and $K = \ell_M V_t = \ell_M{}^2.|\partial_z U|$. This method is only useful when $\partial_z U$ is the dominant strain rate, such as in equilibrium boundary layer flows. Then ℓ_M is usually taken to be proportional to z. In k-ℓ models the velocity scale V_t is obtained by solving an auxiliary equation, that for turbulent kinetic energy, $\frac{1}{2}\overline{q^2}$, whence $V_t = (\frac{1}{2}\overline{q^2})^{\frac{1}{2}}$, (tke is often called k in aerodynamic literature, hence k-ℓ). Adopting a separate differential equation for V_t introduces some relaxation into the specification of K. The last method, k-ϵ closure, solves both the tke equation to obtain V_t and the equation for dissipation ϵ to obtain a length scale.

The third essential feature of the model structure is the adoption of sensible boundary conditions. As a lower boundary condition, most models either explicitly or implicitly force the normal gradient of shear stress at the surface to be zero and term this an equilibrium layer (e.g., Zeman and Jensen 1987). In fact, the limit as $z \to 0$ of the streamwise momentum equation (1) requires that:

$$\partial_z \tau \Big|_{z=0} = \frac{1}{\rho}\, \partial_x P \Big|_{z=0} \tag{26}$$

and momentum transfer in the lowest layers of the flow can be significantly distorted by setting $\partial_z \tau = 0$ at $z = 0$. In particular, as we saw earlier, close to separation the decelerating effect of the pressure gradient $1/\rho\ \partial_x P$ on the momentum gradient $\partial_x \frac{1}{2} U^2$ is modulated by the shear stress gradient. At the surface the positive $1/\rho\ \partial_x P$ is balanced entirely by $\partial_z \tau$. In a model calculation, removing this term by setting $\partial_z \tau = 0$ allows the full decelerating effect of the pressure gradient to be exerted on the mean velocity. Other things being equal, the consequence of this would be to predict premature separation.

Finally, we would like our ideal model to be easy, rapid and cheap to compute. To summarize, the yardsticks against which we will measure the models are: the approximations embodied in the set of equations to be solved; the structure of the model, elliptic or parabolic; the turbulence closure used; the boundary conditions satisfied, and, last but not least, the simplicity of solution.

The first serious attempts to model turbulent hill flows appeared in 1974 and 1975. Of the five models published then, four – Taylor and Gent (1974), Frost et al. (1974, 1975) and Deaves (1975) – fell into the first of our categories, nonlinear models with numerical, finite difference solutions. To the fifth, the linear model of Jackson and Hunt (1975), we will return shortly.

In its conception, Taylor and Gent's model (TG74) was scarcely improved upon in the next decade. They solved the full nonlinear flow equations in a curvilinear coordinate system obtained by conformally mapping the flow over the hill, and retained the important extra 'metric' terms that appear in the equations as a result of the transformation. They comment that the equations were solved iteratively, which I take to be the same as an implicit solution or, equivalently, that the equation set was treated as an elliptic system. The equations were closed by two different methods: a simple mixing length and a k-ℓ closure. The shearing stress was then written as

$$\tau = K \frac{\partial U}{\partial z} ,$$

with proper allowance made for the coordinate frame. TG74 thus displays two important features that we have alluded to in earlier sections. The first of these is the use of a coordinate system that closely parallels the flow streamlines in the expectation that a horizontally homogeneous parameterization of stress, in which shear stress dominates, will continue to hold. The second is the realization that in a strongly distorted flow, a single scalar eddy diffusivity cannot be multiplied into the mean strain rate, $(\partial_j U_i + \partial_i U_j)$ to obtain the deviatoric stress tensor. This was a fact already well known in 1974, see, for example, Launder and Spalding (1972). Although TG74 avoided this trap by making $\overline{u^2}$ and $\overline{w^2}$ constant ratios of the evolving $\frac{1}{2}\overline{q^2}$, with hindsight we can see their solution to be flawed as these ratios are not fixed but change over the hill as $\overline{u^2}$ responds primarily to acceleration and $\overline{w^2}$ to curvature. The lower boundary condition TG74 used was equivalent to forcing a region of constant stress at the surface. Most probably as a result of this the model predicts an exaggerated peak in $-\overline{uw}$ at the hill top compared to the few reliable data available, e.g., Mason (1986), Finnigan et al. (1988b).

In 1977 Taylor (Taylor 1977a,b) produced two developments of TG74 that allowed it to be applied to arbitrary 2-D hills, the main departure from the earlier paper being to end the reliance on conformal mapping. Meanwhile Deaves (1975) had taken a slightly different approach. While retaining mixing length closure for the turbulent stresses, he chose to solve the mean vorticity equation. Although ensuring that the problem was treated elliptically, this technique restricted it to two dimensions. Deaves was also the first person to acknowledge that the effect of extra strain rates like curvature and acceleration limited the application of his model to low hills.

Measured against our stated yardsticks, therefore, the three models in which Taylor had a hand and that of Deaves were elliptic, which should have allowed reasonable fidelity in describing the wake region, but their constant stress boundary conditions would make predictions of the point of separation suspect. Taylor's models retained all the terms in the flow equations but Deaves discarded enough that his model was effectively restricted to low hills. In each case the turbulence closure was insufficient to reproduce the Reynolds stress behaviour and their nonlinearity dictated finite difference computer solution.

In 1975 the direction of hill flow modelling set off along a different path which it was to follow for the next decade. The change was prompted by the paper of Jackson and Hunt (1975), who attacked the problem with the techniques of linear, small perturbation analysis and assymptotic matching. Their model was parabolic, had the simplest turbulence closure, used the incorrect constant stress boundary condition and was framed in linearized equations. Its overwhelming strength was its simplicity. The prize for ignoring these desiderata was an analytic solution, ample justification in the opinion of most users as its subsequent popularity showed.

The ubiquity of the underlying framework of the Jackson and Hunt theory is well attested by the number of times we have already called upon it to bring order to a mass of data. Its efficacy as a quantitative theory we have touched on only in passing. Before we can go further we need to explain the theory in more detail than we have done hitherto. JH75 divided the flow field into two regions with different dynamics: an outer layer, inviscid and irrotational, where the velocity and pressure perturbations caused by the hill are solutions of the Euler equations, and an inner layer of depth ℓ, where turbulent stress divergence affects the mean flow. In this inner region the turbulence is supposed to be in local equilibrium with the mean shear.

To obtain analytic solutions to this system, simplifying assumptions are needed. First of all the equations of motion in both layers are made linear in Δu, the perturbation to the mean velocity, thereby restricting the model to low hills. Second a simple mixing length closure is used for the turbulence in the inner layer. Third and most far reaching, the background advection velocity in each layer is made constant with height. In the inner layer it is $U_0(\ell)$, where $U_0(z)$ is the undisturbed upwind profile, and in the outer layer $U_0(L)$. Without the assumption of constant advection velocity, the inner layer solution would present the classical problem of the small perturbation of a parallel viscous shear layer (Drazin and Reid 1981). If the eddy diffusivity K were constant, then the stationary normal mode amplitudes of the disturbance Δu would be governed by an Orr-Sommerfield equation, while retaining z dependence in K leads to a rather more complicated problem where the normal modes can sometimes be expressed in terms of confluent hypergeometric functions (Jones 1968) but where recovery of the physical velocity perturbation inevitably requires numerical techniques. In short, the mathematical structure of the problem is changed at a fundamental level by adopting a constant advection velocity.

Matching of the flow between the two layers was done in the limit z_0/L, $\Delta u/u_* \to 0$, a much more restrictive condition than simply linearizing the momentum equations would impose. In this limit the shear within the inner layer virtually disappears and $U_0(\ell) \to U_0(L)$. In this limit also, the pressure perturbation that drives the inner layer solution is given by the limit as $z \to 0$ of the outer layer potential flow solution. Here we see a strong link with classical Prandtl boundary layer theory in that the interaction between inner and outer layer is weak and the inner layer equations, with P specified, can be solved as a parabolic system.

The depth ℓ of the inner layer was determined by matching the linearized advection term, $U_0\ \partial\Delta U/\partial x$, with the gradient of perturbation shear stress, $\partial\Delta\tau/\partial z$, to arrive at the result we saw in (9), or equivalently,

$$\frac{\ell}{L} = 2k\,\frac{u_*}{U_0(\ell)}\ . \qquad (27)$$

In other words, ℓ is the depth of the region where $\partial\Delta\tau/\partial z$ is of the same order as either component of the perturbation Bernoulli term, $\partial_x[\Delta P/\rho + U_0\Delta U]$. This turns out to be a surprisingly thin region in many cases. For example, in the case of Finnigan et al.'s (1988a) WTHILL, ℓ is 7 mm, equal to the physical roughness height although on Askervein hill, ℓ = 15 m. Certainly on the rough WTHILL, the depth over which the turbulent stress divergence actually plays a significant role in the momentum budget is ~10ℓ, a sign perhaps that over hills of finite size, the equilibrium formulations invoked in deriving (27) can break down decisively.

The main conclusion of the JH75 paper was that the speedup Δs in the inner layer would be substantially greater than inviscid theory predicted. The reason for this and its predicted magnitude we have already discussed. The pressure perturbation is set by an inviscid irrotational calculation with a velocity scale $U_0(L)$. Within the inner layer this perturbation acts on a region where the background velocity $U_0(z)$ is substantially smaller than $U_0(L)$ so that the velocity perturbations there are proportionately larger. The theory predicts the height of the maximum speedup Δs_{max} to occur at $z = \ell$ above the hill crest and that for a 2-D ridge Δs_{max} will be of order 2h/L.

The division of the flow region into two layers is a useful clarification of the physics but the assumption that within the region where the turbulence affects the mean flow it is in equilibrium with the local strain rates, although essential for analytical solutions, is, as we have already seen, an oversimplification. Making this assumption allowed Jackson and Hunt to use the equilibrium form of the law of the wall, (25), not only as a boundary condition, as the earlier numerical models had done, but as the turbulence closure also, thereby exacerbating and deepening the layer where cross-stream momentum transfer is distorted. We recall that both curvature and acceleration lead to effective φ_M values as high as 1.5 on the hill top while Wyngaard (1982) has also pointed out that advection may lead to effective φ_M values different from unity if the flow in a region far from local equilibrium is forced into the log law mould.

Although JH75 was applauded for its analytical form, it was soon realized that it predicted speedups close to the ground a good deal larger than were observed

(Britter et al. 1981). A series of developments followed which were aimed at removing some of the restrictive assumptions of the model.

The first of these was by Mason and Sykes (1979). They extended JH75 to three dimensions without altering any of its essential features. Their 3-D version gave reasonable agreement with their measurements of Δs at $z = 2$ m over Brent Knoll which were reported in the same paper.

In 1980 Sykes published a more rigorous asymptotic analysis, similar to JH75 but owing much to triple deck theory (Meyer 1983). He showed that for proper matching of velocity perturbations between the layers, a third layer must be interposed between the outer inviscid and the wall layer. In this layer the turbulence is not in local equilibrium and Sykes concluded that even over low hills where a linearized solution was possible, second-order closure was necessary to model the turbulent stresses adequately in this middle layer. The main reason for the exaggerated speedup generated by the original JH75 theory was seen to be the use of an excessively large advection velocity $U_0(\ell)$ in the inner layer. Another problem was the fact that the driving perturbation pressure is not modified by the inner layer, the weak interaction or boundary layer assumption. Walmsley et al. (1982), Taylor et al. (1983b) and, independently, Mason and King (1985) modified the theory to cope with these problems.

Finnigan et al. (1988a) have shown that the near surface pressure over WTHILL is approximated very closely by a rotational inviscid solution obtained by assuming that the vorticity of the upwind logarithmic profile is conserved along streamlines. It is, in fact, this difference between JH75's irrotational inviscid solution for the outer layer and the *rotational* inviscid solution that is addressed by the pressure field modifications of Walmsley et al., Taylor et al. and Mason and King. Their models remain weak interaction formulations.

Their main improvements were to provide different advection velocities in the inner and outer layers, essentially decoupling the choice of advection velocity from the definition of the layer depths. Taylor et al. (1983b), for example, instead of using $U_0(L)$ and $U_0(\ell)$ found that $U_0(c_1L)$ and $U_0(c_2\ell)$, with $c_1 = c_2 = 0.135$, fitted their test case results better. The problem with using arbitrary advection velocities lies in finding a rational way of choosing them. Taylor et al. (1983b) chose velocities that gave the best match between the linearized theory and Taylor's (1977a) nonlinear numerical model with mixing length closure. For hills of finite size, however, two features of the numerical model differ from the linear case: the inclusion of nonlinear advection terms and the height dependence of the advection velocity. In fitting the linear model to the nonlinear results it is not clear which feature has a greater effect. Sensibly Taylor et al. confined the comparison to h/L values where they expected nonlinearity to have relatively little influence.

A further refinement of Taylor et al. (1983b) was to exploit the linearity of the model and compute a different velocity field for each Fourier component of the hill profile. The solution for the complete hill was the sum of the solutions for each Fourier component.

Mason and King (1985) arrived at virtually the same result as Taylor et al. by a different method. If a hilltop solution was required at a height Δz above the ground, then $U_0(\Delta z)$ was taken as the advection velocity for that solution. This means in effect that the model must be rerun for each height. The two models TWS83 and MK85 reproduce the perturbations of mean velocity upwind of Kettle's Hill and Blasheval, respectively, but consistently overpredict velocities in the lee of the hill. The most recent development of the linearized, two layer form is that of Beljaars et al. (1987). They compared the standard mixing length closure with a k-ℓ scheme and also incorporated changes in surface roughness in the model. Comparisons with the data from the

Askervein experiment showed that the k-ℓ closure version predicted Δs very well upwind of the crest but as usual overpredicted in the lee. There were substantial differences, however, between the measured and predicted shear stresses.

Although these developments add many complications to the original JH75 formulation, the main advantage of the linearized, two-layer approach remains its modest demands on computing time, especially when implemented in the efficient spectral form used by the later versions. However, because of the intrinsic limitations imposed by its parabolic structure and linearity, the agreement found by Beljaars et al. is probably as good as it is going to get. The approach is adequate for the upwind velocity field on low hills but cannot cope with the wake or the turbulence.

By 1985 enough turbulence data was available from Askervein, Blasheval, Cooper's Ridge, Nyland's Hill and the wind tunnel simulations at AES of Kettle's Hill and Askervein to prompt a revival of interest in finite difference models capable of improved turbulence closure or elliptic treatment. The first of these (Raithby et al. 1987) was only a slight improvement in conception on Taylor's (1977a,b) papers. It used k-ϵ closure, was elliptic in conception and retained the incorrect, constant stress lower boundary condition. Its main advantage seems to be a more efficient computation scheme. It does somewhat better than most other models at matching the mean velocity field in the lee of Askervein, presumably because it is elliptic, but its reproduction of turbulence quantities is poor; it fails to get the profile of $\frac{1}{2}\overline{q^2}$ correct even at the equilibrium upstream reference point. Like Beljaars et al. in the linear approach, this model may mark the limit of performance of eddy diffusivity closures, that is, a reasonable prediction of the mean velocity field over a hill of moderate steepness but poor description of the turbulence. The main improvement still to be made lies in removing the constant stress boundary condition that inevitably distorts momentum transport behind the hill crest, possibly promoting separation and changing the momentum deficit in the wake.

The paper of Zeman and Jensen (1987) (henceforth ZJ87) took quite a different tack. It was the first model to come to grips with the complications of the turbulent response. ZJ87 made no attempt at an elliptic formulation, concentrating their computing effort on a sophisticated second-order closure scheme with the pressure field determined a priori by a potential flow calculation. A major innovation was to implement their scheme in streamline coordinates, albeit in a rather complex way involving the Von-Mises transform. A disadvantage of this approach is that not all the extra terms appearing in the equations have obvious physical meaning and some simplifications valid for low hills only had to be made. Nevertheless, they confirmed the influence of streamline curvature on the development of the turbulence except very close to the wall. This last conclusion may be flawed, however, since once again they used as a lower boundary condition the flat, constant stress relationship, $\partial_z\tau = 0$, which automatically denies curvature influence. Their model produced a good match with profiles of $\overline{uw}$, and $\overline{w^2}$ down to 2 m on the crest of Askervein, but did rather less well at simulating $\overline{u^2}$. It is not clear why this should be so although several inconsistencies remain in the turbulence closure ZJ87 used, particularly in the incorporation of curvature and acceleration effects in the rapid pressure term.

An even more recent model has been reported (Sykes 1987) that combines elliptic formulation with second-order closure. Curiously Sykes uses a simplified form of closure with no rapid terms because he believes that the role of the rapid distortion region has been over emphasized. An interesting new direction in this paper is the calculation of scalar fluxes, for many applications the real reason for the interest in wind structure!

Two things seem striking in this summary of effort in modelling hill flows. The first, of course, is the shift in emphasis to linear asymptotic models between 1975 and 1985. Despite the fact that measurements were being made in situations that clearly violated assumptions of the theories, great efforts were made to interpret field data in terms of the depth of the inner layer ℓ and the equilibrium assumptions for turbulence near the ground. Undoubtedly it was the attractiveness of the analytic solution and the generality that went with it that was behind the influence of linear modelling, an apt comment on our infatuation with computer simulation. The second thing is a pervasive sense of reinventing the wheel. Already by 1972 considerable experience had been gained in modelling turbulent flows by all levels of closure up to third order. The monograph by Launder and Spalding (1972) gives an excellent summary of when one might expect the various approaches to give reliable results but little hint that this knowledge existed comes from reading any of the papers I have cited until Zeman and Jensen (1987).

CONCLUSIONS

This field of study is presently in a phase of rapid development. The relatively simple paradigm of the JH75 linear theory provided a period of stability which enabled experiments to be designed over low hills and analysis to proceed. Its descriptive power is exceeded by the more detailed data sets now becoming available, however, and although some of its basic principles will continue to be useful, others must be discarded or augmented.

In the case of the mean flow the most serious gap in our understanding is the prediction and description of separation. This is fundamentally an elliptic, strong interaction problem and so lies outside the compass of JH75, although once the extent of separation is established, JH75 then provides a reasonable estimate of the upwind attached flow. In understanding separation itself we are in an awkward position. In the main, micrometeorological theory has relied heavily upon prior knowledge gained in engineering flows. (Flow in plant canopies provides a notable exception to this statement.) Unfortunately no useful body of work exists on turbulent separation from rounded bluff bodies with rough walls. It is necessary, therefore, to embark upon a fundamental study of this situation if understanding of flow over hills of even moderate steepness is to be advanced. The obvious way to do this, at the outset at any rate, is through physical modelling.

When we consider the turbulence, things are somewhat different. There exists no quantum barrier like the separation question to describing its evolution once the mean flow is known. The two limits of turbulent response, the equilibrium and rapid-distortion regimes, can, in fact, be described quite well although in the rapid-distortion regime all the relevant strain rates must be included, which may prescribe a numerical solution. The problem lies in that over most of the interesting region turbulence behaviour falls between these two stools. There, turbulence remembers a significant strain history but also a significant dissipation history. We might call this the second-order closure regime except that most second-order closure models are, in fact, strongly biased towards the near-equilibrium situation. For example, most extant second-order models use expressions for the rapid pressure-strain terms which are precisely those given by rapid-distortion theory in the limit of zero strain (Maxey 1982). The formulae that emerge from rapid-distortion theory are strongly dependent on both the total strain absorbed by a typical eddy in its lifetime and the kind of strain, be it shear, acceleration or curvature. These, as we have seen, vary considerably with position over the hill. It is possible that the simplest useful theory for turbulent reponse, therefore, might follow the lines proposed by Townsend (1980) in his model of flow over water waves: a rapid-distortion calculation to determine the ratio $\overline{u_i u_j}/\overline{u_i u_i}$

together with a solution of the tke equation to fix the absolute stress levels. This is certainly indicated over rough hills where the entire flow regime is pushed towards the rapid-distortion limit.

REFERENCES

Arya SPS, Capuano ME, Fagen LC (1987) Some fluid modelling studies of flow and dispersion over two-dimensional low hills. Atmos Environ 21: 753-764

Arya SPS, Gadiyaram PS (1986) An experimental study of flow and dispersion in the wakes of three-dimensional low hills. Atmos Environ 20: 729-740

Arya SPS, Shipman MS (1981) An experimental investigation of flow and diffusion in the disturbed boundary layer over a ridge - 1. Mean flow and turbulence structure. Atmos Environ 15: 1173-1184

Baines PG (1987) Upstream blocking and flow over mountains. Ann Rev Fluid Mech 19: 75-98

Batchelor GK (1976) An introduction to fluid dynamics. Cambridge University Press

Beljaars ACM, Walmsley JL, Taylor PA (1987) A mixed spectral finite-difference model for neutrally stratified boundary-layer flow over roughness changes and topography. Boundary-Layer Meteorol 38: 273-303

Bouwmeester RJB, Meroney RN, Sandborn VA (1978) Sites for wind power installations: Wind characteristics over ridges. Colo State Univ Fl Mech and Wind Eng Prog Rep No RLO/2438-78/1 UC 60

Bradley EF (1980) An experimental study of the profile of wind speed, shearing stress and turbulence at the crest of a large hill. Q J R Meteorol Soc 106: 101-124

Bradley EF (1983) The influence of thermal stability and angle of incidence on the acceleration of wind up a slope. J Wind Eng Indust Aerodyn 15: 231-242

Bradshaw P (1969) The analogy between streamline curvature and buoyancy in turbulent shear flow. J Fluid Mech 36: 177-191

Britter RE, Hunt JCR, Richards KJ (1981) Airflow over a two-dimensional hill: Studies of velocity speed-up, roughness effects and turbulence. Q J R Meteorol Soc 107: 91-110

Castro IP, Snyder WH (1982) A wind tunnel study of dispersion from sources downwind of three-dimensional hills. Atmos Environ 16: 1869-1887

Coppin PA, Bradley EF, Katen PC (1986) The evolution of atmospheric turbulence over a two-dimensional ridge. In: Proc. 9th Australasian Fluid Mechanics Conference, Auckland, NZ, pp 569-572

Deaves DM (1975) Wind over hills: A numerical approach. J Indust Aerodyn 1: 371-391

Drazin PG, Reid WH (1981) Hydrodynamic stability. Cambridge University Press

Finnigan JJ (1983) A streamline coordinate system for distorted turbulent shear flows. J Fluid Mech 130: 241-258

Finnigan JJ (1988a) Streamline coordinates and moving frames in 3-D fluid flows. J Fluid Mech, to be submitted

Finnigan JJ (1988b) The law of the wall in flow over hills. Boundary-Layer Meteorol, to be submitted

Finnigan JJ, Raupach MR (1985) Turbulent flow over two-dimensional ridges. In: Proc 2nd Workshop on Wind Engineering and Industrial Aerodynamics, Melbourne, Vic, CSIRO Div of Building Research

Finnigan JJ, Raupach MR, Bradley EF, Aldis GK (1988a) A wind tunnel study of turbulent flow over a two-dimensional ridge. Q J R Meteorol Soc, to be submitted

Finnigan JJ, Brunet Y, Raupach MR (1988b) Turbulent flow in a model plant canopy on a two-dimensional ridge. Boundary-Layer Meteorol, to be submitted

Frost W, Mans JR, Fichtl GH (1974) A boundary-layer analysis of atmospheric motion over a semi-elliptical surface obstruction. Boundary-Layer Meteorol 7: 165-184

Frost W, Harper WL, Fichtl GH (1975) Analysis of atmospheric flow over a surface protrusion using the turbulence kinetic energy equation. Boundary-Layer Meteorol 8: 401-417
Haugen DA, Kaimal JC, Bradley EF (1971) An experimental study of Reynolds stress and heat flux in the atmospheric boundary layer. Q J R Meteorol Soc 97: 168-180
Hunt JCR (1978) A review of the theory of rapidly distorted turbulent flow and its applications. Fluid Dyn Trans 9: 121-152
Hunt JCR (1980) Wind over hills. In: Wyngaard JC (ed), Workshop on the Planetary Boundary Layer, Am Meteorol Soc, Boston, pp 107-144
Hunt JCR, Snyder WH (1980) Experiments on stably and neutrally stratified flow over a model three-dimensional hill. J Fluid Mech 96: 671-704
Hunt JCR, Abell CJ, Peterka JA, Woo H (1978) Kinematical studies of the flow around free or surface-mounted obstacles: applying topology to flow visualization. J Fluid Mech 86: 179-200
Jackson PS, Hunt JCR (1975) Turbulent wind flow over a low hill. Q J R Meteorol Soc 101: 929-955
Jenkins GH, Mason PJ, Moores WH, Sykes RI (1981) Measurements of the flow structure around Ailsa Craig, a steep, three-dimensional, isolated hill. Q J R Meteorol Soc 107: 833-851
Jones WL (1968) Reflexion and stability of waves in stably stratified fluids with shear flow: a numerical study. J Fluid Mech 34: 609-624
Kaimal JC, Wyngaard JC, Haugen DA, Coté OR, Izumi Y, Caughey SJ, Readings CJ (1976) Turbulence structure in the convective boundary layer. J Atmos Sci 33: 2152-2169
Kao HC (1987) Torsion effect on fully developed flow in a helical pipe. J Fluid Mech 184: 335-356
Landahl MT, Mollo-Christensen E (1986) Turbulence and random processes in fluid mechanics. Cambridge University Press
Lander BE, Spalding DB (1972) Lectures in mathematical models of turbulence. Acaemic Press, London
Legg BJ, Coppin PC, Raupach MR (1984) A three-hot-wire anemometer for measuring two velocity components in high intensity turbulent boundary layers. J Phys E 17: 970-976
Mason PJ (1986) Flow over the summit of an isolated hill. Boundary-Layer Meteorol 37: 385-405
Mason PJ, King JC (1985) Measurements and predictions of flow and turbulence over an isolated hill of moderate slope. Q J R Meteorol Soc 111: 617-640
Mason PJ, Sykes RI (1979) Flow over an isolated hill of moderate slope. Q J R Meteorol Soc 105: 383-395
Maxey MR (1982) Distortion of turbulence in flow with parallel streamlines. J Fluid Mech 124: 261-282
Meroney RN, Sandborn VA, Bouwmeester RJB, Chien HC, Rider M (1978) Sites for wind power installations: physical modeling of the influence of hills, ridges and complex terrain on wind speed and turbulence. Colo State Univ Fl Mech and Wind Eng Prog Rep No RLO/2438-77/3 UC 60
Meyer, RE, 1983: A view of the triple deck. SIAM J Appl Math 43: 639-663
Mickle RE, Cook NJ, Hoff AM, Jensen NO, Salmon JR, Taylor PA, Tetzlaff G, Teunissen HW (1988) The Askervein Hill project: Vertical profiles of wind and turbulence. Boundary-Layer Meteorol, submitted
Muck KC, Hoffmann PH, Bradshaw P (1985) The effect of convex surface curvature on turbulent boundary layers. J Fluid Mech 161: 347-369
Pearse JR (1982) Wind flow over conical hills in a simulated atmospheric boundary-layer. J Wind Eng Indust Aerodyn 10: 303-313
Pronchik SW, Kline SJ (1983) An experimental investigation of the structure of a turbulent reattaching flow behind a backward facing step. Report MD-42 Thermo Science Div Dept of Mech Eng, Stanford University, CA
Raithby GD, Stubley GD, Taylor PA (1987) The Askervein Hill project: A finite control volume prediction of three dimensional flows over the hill. Boundary-Layer Meteorol 39: 247-267
Raupach MR (1988) Canopy transport proceses. This volume, pp

Sacré C (1979) An experimental study of the airflow over a hill in the atmospheric boundary layer. Boundary-Layer Meteorol 17: 381-401

Salmon JR, Bowen AJ, Hoff AM, Johnson R, Mickle RE, Taylor PA, Tetzlaff G, Walmsley JL (1988) The Askervein Hill project: Mean wind variations at fixed heights above the ground. Boundary-Layer Meteorol, submitted

Schlichting H (1960) Boundary-layer theory. McGraw-Hill, New York

Smith FT (1986) Steady and unsteady boundary-layer separation. Ann Rev Fluid Mech 18: 197-220

Snyder WH, Britter RE (1987) A wind tunnel of the study of the flow structure and dispersion from sources upwind of three-dimensional hills. Atmos Environ 21: 735-751

Speziale CG (1987) On nonlinear K-ℓ and K-ϵ models of turbulence. J Fluid Mech 178: 459-475

Spivak M (1979) A comprehensive introduction to differential geometry, Vol II. Publish or Perish Inc, PO Box 7108, Berkeley, CA

Sykes RI (1980) An asymptotic theory of incompressible turbulent boundary-layer flow over a small hump. J Fluid Mech 101: 647-670

Sykes RI (1987) Numerical calculation of turbulent flow over a wavy boundary. ARAP Tech Memo 86-10 ARAP Div of CRT Inc 50, Washington Rd, Princeton NJ

Taylor PA (1977a) Some numerical studies of surface boundary-layer flow above gentle topography. Boundary Layer Meteorol 11: 439-465

Taylor PA (1977b) Numerical studies of neutrally stratified planetary boundary-layer flow above gentle topography. Boundary-Layer Meteorol 12: 37-60

Taylor PA (1988) Turbulent wakes in the atmospheric boundary-layer. This volume, pp

Taylor PA, Gent PR (1974) A model of atmospheric boundary-layer flow above an isolated two-dimensional hill: An example of flow over gentle topography. Boundary-Layer Meteorol 7: 349-362

Taylor PA, Lee RJ (1987) Simple guidelines for estimating wind speed variations due to small scale topographic features. Climatol Bull 18(2): 3-32

Taylor PA, Teunissen HW (1983) Askervein '82: Report on the September/October 1982 experiment to study boundary-layer flow over Askervein, South Oist. Internal Report No MSRB-83-8, Atmos Environ Service, Downsview, Ontario, Canada

Taylor PA, Teunissen HW (1985) The Askervein Hill project; Report on the September/October 1983 main field experiment. Internal Report MSRB-84-6, Atmos Environ Service, Downsview, Ontario, Canada

Taylor PA, Mason PJ, Bradley EF (1987) Boundary-layer flow over low hills. Boundary-Layer Meteorol. 39: 107-132

Taylor PA, Mickle RE, Salmon JR, Teunissen HW (1983a) The Kettle's Hill experiment - site description and mean flow results. Report AQRB-83-002-L Atmos Environ Service, Downsview, Ontario, Canada

Taylor PA, Walmsley JL, Salmon JR (1983b) A simple model of neutrally stratified boundary-layer flow over real terrain incorporating wavenumber dependent scaling. Boundary-Layer Meteorol 26: 169-189

Tennekes H (1982) Similarity relations, scaling laws and spectral dynamics. In: Nieuwstadt FTM, van Dop H (eds), Atmospheric turbulence and air pollution modelling, D Reidel, Dordrecht, pp 37-68

Tennekes H, Lumley JL (1972) A first course in turbulence. MIT Press, Cambridge, MA

Teunissen HW, Shokr (1985) The Askervein Hill project: Wind tunnel simulation (smooth model) at length scale 1:1200. Internal Report MSRB 85-1 Atmos Environ Service, Downsview, Ontario, Canada

Teunissen HW, McLean IR, Flay RGJ (1982) Wind tunnel simulation of planetary boundary layer flow over an isolated low hill. In: Proc 4th Int Symp on Wind Energy Systems, BHRA, Cranfield, England

Teunnissen HW, Shokr ME, Bowen AJ, Wood CJ, Green DWR (1987) The Askervein Hill project: Wind-tunnel simulations at three length scales. Boundary-Layer Meteorol 40: 1-29

Tobak M, Peake DJ (1982) Topology of three-dimensional separated flows. Ann Rev Fluid Mech 14: 61-85

Townsend AA (1980) The response of sheared turbulence to additional distortion. J Fluid Mech 98: 171-191

Turner JS (1973) Buoyancy effects in fluids. Cambridge University Press

Walmsley JL, Salmon JR, Taylor PA (1982) On the application of a model of boundary layer flow over low hills to real terrain. Boundary-Layer Meteorol 23: 17-46

Walmsley JL, Taylor PA, Keith T (1986) A simple model of neutrally stratified boundary layer flow over complex terrain with surface roughness modulation (MS3DJH/3R). Boundary-Layer Meteorol 36: 157-186

Wyngaard JC (1982) Boundary-layer modelling. In: Nieuwstadt FTM, van Dop H (eds) Atmospheric turbulence and air pollution modelling, D Reidel, Dordrecht, pp 69-106

Wyngaard JC (1988) Convective processes in the lower atmosphere. This volume, pp

Yaglom AM (1979) Similarity laws for constant pressure and pressure-gradient turbulent wall flows. Ann Rev Fluid Mech 11: 505-540

Zeman O, Jensen NO (1987) Modification of turbulence characteristics in flow over hills. Q J R Meteorol Soc 113: 55-80

AIR FLOW OVER COMPLEX TERRAIN: COMMENTARY

N.O. Jensen

Department of Meteorology and Wind Energy, Risø National Laboratory, 4000 Roskilde, Denmark.

Summary. This commentary begins with a reiteration of what is known, and what has been simplified in order for general understanding to be obtained. It goes on to contrast a couple of points made by Finnigan: on the utility of Jackson-Hunt-type models; and on the limitations of wind modelling. It concludes with a short summary of poster papers presented, and with some remarks on future research needs.

INTRODUCTION

In relation to both John Finnigan's paper (Finnigan 1988) and this contribution, the term complex terrain is a little pretentious; most of what is being put forward refers to low, solitary, two-dimensional ridges. The distinctive feature that makes this more complicated than classical air-foil theory is that the dimensions of the object are of the same order, or smaller, than the height of the incipient turbulent boundary layer: the main questions are how the hill modifies the mean wind profile and how the turbulence responds to this modified flow field.

Recent research into this situation has resulted in some general statements about the flow behaviour. Some of these are outlined in Finnigan (1988), but see also Taylor et al. (1987). However, real topography is usually quite far from this ideal, and comparatively more difficult to treat. (Besides, a very large range of "generic cases" would be needed in order to be able to generalize about this.) Estimates of flow perturbations over real terrain are therefore best made by use of numerical models and digitized contour maps. General implicit $k-\varepsilon$ schemes have been published, but the closure at this level is not good enough to give a faithful representation of the turbulence. The linearized models with Fast-Fourier-Transform (FFT) solutions are vastly more effective. Full turbulence models, such as that of Zeman and Jensen (1987), have not yet been adapted to such situations.

Steeper topography involves flow separation and possibly recirculation. In the far field this can probably be dealt with as a wake phenomenon (see Taylor 1988), but in the vicinity of the hill further theoretical development is needed. Even qualitatively, it is important to distinguish between two- and three-dimensional separation; the latter not featuring a separation streamline leaving the object, and hence not a "closed" recirculation cell (see Fig. 1).

Excluding the cases of katabatic and anabatic winds, and some studies of lee waves in stably stratified situations, the effect of buoyancy forces is usually neglected in studies of flow over hills.

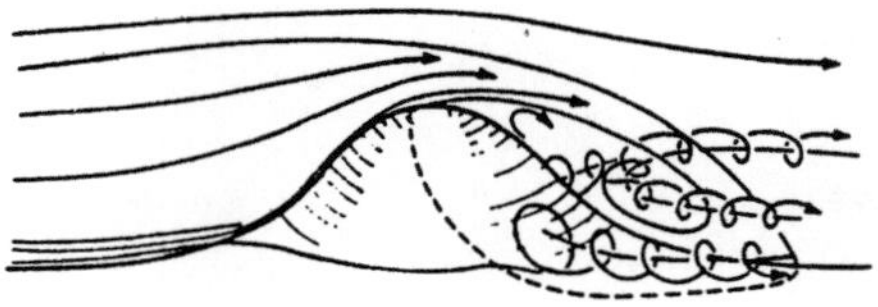

Fig. 1. Streamline diagram of three-dimensional separation. Note that the stagnation streamline does not originate on the hill. (Hunt et al. 1979).

As can be seen from simple arguments, the height to which a hill sets up a pressure perturbation in the field depends on the length of the hill (the height of the hill merely determines the amplitude of the perturbation). Since the atmosphere usually is slightly stably stratified above the boundary layer, hills of that length scale or larger will set up gravity waves. If one wishes to exclude this effect, the size of the hill is limited to sufficiently short length scales.

With this in mind, the effect of nonzero surface heat flux is (to an increasing degree of approximation and complexity in the analysis) to alter the mean flow field, the turbulence profiles, and the turbulence budget (Hunt et al. 1987). Further research is needed to clarify the effect and importance of these points.

In the remainder of this commentary, I would like to focus on two points made in Finnigan (1988): first, the usefulness of linearized flow models, and, second, the turbulence structure above a typical hill and the difference one would expect between this and a wind tunnel model of it. Finally, I present a brief summary of the papers given in the poster session, and give some remarks regarding future research needs in this field.

LINEARIZED FLOW MODELS

In fluid mechanics, and especially in turbulence and boundary layer contexts, it is customary to use simplified approaches, such as dimensional analyses and truncated equations that contain significant terms only in the equations of motion. A useful trick, valid for small perturbations, is to linearize the advection term. A rudimentary set (or one could say a classical example) is:

$$u_0 \frac{\partial u}{\partial x} = - \frac{\partial p}{\partial x} + K \frac{\partial^2 u}{\partial z^2} \tag{1}$$

$$u_0 \frac{\partial w}{\partial x} = - \frac{\partial p}{\partial z} + K \frac{\partial^2 w}{\partial z^2} \tag{2}$$

$$\frac{\partial u}{\partial x} + \frac{\partial w}{\partial z} = 0 \quad , \tag{3}$$

where u is the horizontal velocity perturbation, u_0 is a constant advection velocity, x denotes the horizontal direction of the flow, z is the vertical direction, w the vertical velocity perturbation, p the pressure perturbations, and K is the eddy viscosity. By taking the Fourier transform in the x-direction, we obtain

$$iku_0\hat{u} + ik\hat{p} - K\frac{\partial^2\hat{u}}{\partial z^2} = 0 \quad (4)$$

$$iku_0\hat{w} + \frac{\partial\hat{p}}{\partial z} - K\frac{\partial^2\hat{w}}{\partial z^2} = 0 \quad (5)$$

$$ik\hat{u} + \frac{\partial\hat{w}}{\partial z} = 0 \quad , \quad (6)$$

where the transformed quantities are denoted by a hat and k is a wavenumber. By, for instance, eliminating $\hat{w}$ and $\hat{p}$, we obtain a fourth order differential equation for $\hat{u}$ with the solution:

$$\hat{u} = a\left[\exp(-|k|z) - \exp\left[-\left[\frac{iku_0}{K}\right]^{\frac{1}{2}}z\right]\right] \quad , \quad (7)$$

where the two positive "roots" have been eliminated by the condition that $\hat{u} \to 0$ for $z \to \infty$, and where a is a constant that can be determined from a kinematic boundary condition ($a \propto u_0h/L$, where h is the height and L the length scale of the hill).

The basic message from this solution is that the flow response is dual: there is an outer layer, not depending on K (the turbulence) in which the perturbation decays on a scale $|k|^{-1}$, and there is an inner layer, the depth of which is determined by the turbulence (still in terms of K). The terms "inner" and "outer" can be justified by the following: for a simple sinusoidal hill the only wavenumber is $k = 2\pi/L$, where L is the wavelength. So the outer solution decays on that length scale. Denoting the inner scale by ℓ and estimating K as being of magnitude $u_*\ell$, where u_* is the friction velocity, we obtain by identity from the square root argument in the last term of (7) that ℓ is of order $L(u_*/u_0)$ and thus small compared to L.

Such simplified methods are always open to critisism, but nevertheless usually quite powerful. Thus, the above analysis immediately leads to a prediction of the phase lead φ of the surface stress perturbation τ. Since $\hat{\tau} = K\partial\hat{u}/\partial z|_{z=0}$, we get from (7)

$$\hat{\tau} = a\left[\left[\frac{ku_0}{2K}\right]^{\frac{1}{2}} - |k|\right] + ia\left[\frac{ku_0}{2K}\right]^{\frac{1}{2}} \quad (8)$$

or

$$\tan\varphi = \left[\frac{ku_0}{2K}\right]^{\frac{1}{2}} / \left[\left[\frac{ku_0}{2K}\right]^{\frac{1}{2}} - |k|\right] \quad , \quad (9)$$

which in view of the above scale arguments is of order unity. Thus, the maximum of the surface stress occurs 1/8L forward of the crest. This demonstrates that even this simple model contains the necessary dynamics to capture that important detail. In comparison with more detailed calculations, the phase is slightly overpredicted, however (see Fig. 2). For a more detailed description of a linearized model, see Troen and de Baas (1986). A celebrated version of these types of models is the MS3DJH/3 by Taylor et al. (1983), in which MS stands for Mason and Sykes (1979) and JH for Jackson and Hunt (1975).

Finnigan (1988) suggests that one should do away with this approach entirely (Jackson-Hunt-type models) and concentrate on methods that give a more faithful representation of the turbulence response and the effect of streamline curvature. If one is interested in the detailed turbulence response, I do not entirely disagree. Examples could be deposition and scalar transport. Also, the precise phase of the surface stress is important for dune migration. However, if one is mainly interested in the mean flow field over real (three-dimensional) terrain, linearized models using FFT techniques offer very quick calculations.

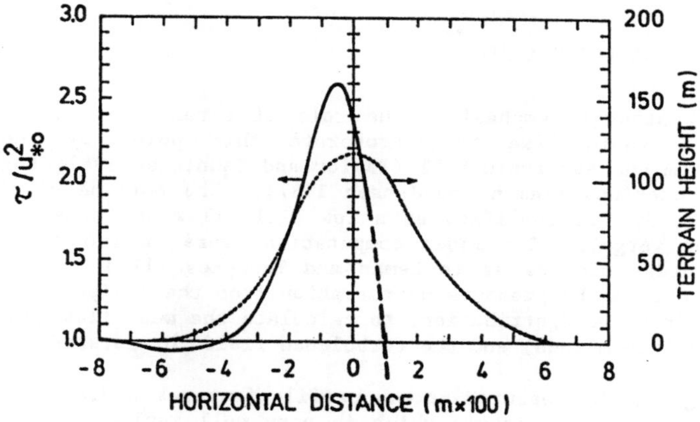

Fig. 2. Surface shear stress along a hill (dotted line) normalised with the undisturbed upstream value, computed by the model of Zeman and Jensen (1987). The position of the stress maximum estimated from the simple model described in the text would be at x ~ -125 m.

It is even possible to include a better turbulence representation in linearized models than the one given by a straight application of K-closure. For neutral conditions, the stationary turbulent kinetic energy budget can be written

$$0 = \left[\tau \frac{\partial u}{\partial z}\right]_{total} - \text{Divergence} - \varepsilon \quad , \tag{10}$$

where ε is the dissipation rate. Linearizing the production term gives

$$(\tau_0+\tau)\frac{\partial(u_0+u)}{\partial z} \approx \tau_0 \frac{\partial u_0}{\partial z} + \tau_0 \frac{\partial u}{\partial z} + \tau \frac{\partial u_0}{\partial z} \quad ,$$

where the first term on the right-hand side approximately balances the dissipation. Modelling the divergence as $\partial/\partial z(-K\partial\tau/\partial z)$ then leads to the following new set:

$$\begin{aligned}
u_0 \frac{\partial u}{\partial x} &= - \frac{\partial p}{\partial x} + \frac{\partial \tau}{\partial z} \\
u_0 \frac{\partial w}{\partial x} &= - \frac{\partial p}{\partial z} - \frac{\partial \tau}{\partial z} \\
\tau_0 \frac{\partial u}{\partial z} + \tau \frac{\partial u_0}{\partial z} + K \frac{\partial^2 \tau}{\partial z^2} &= 0 \\
\frac{\partial u}{\partial x} + \frac{\partial w}{\partial z} &= 0 \quad ,
\end{aligned} \tag{11}$$

where $K = \tau_0^{1/2} z$ has absorbed a factor of order 2.5, deriving from the ratio of turbulent kinetic energy and τ_0, and where we have assumed a ratio of unity between τ_0 and the variance of vertical velocity fluctuations.

This has not yet been pursued further, but it is certainly feasible. However, the nice two-layer structure of the solution will probably now be disturbed by a third scale.

TURBULENCE STRUCTURE ABOVE A HILL

Finnigan (1988) strongly emphasized the role of streamline curvature on the turbulence. I would like to corroborate this point by showing some measurements from the Askervein hill (Taylor and Teunissen 1987) together with some model computations (Zeman and Jensen 1987). The outline of the hill is depicted in Fig. 3. It qualifies as a low hill (flow not separated although possibly on the verge). The model computations were in two dimensions, but used a second-order closure as in Zeman and Tennekes (1975). The method of solution is to obtain the pressure perturbation from the inviscid solution, and then with this pressure distribution, to calculate the mean flow (including the position of the streamlines) and the turbulence fields by forward integration.

As shown in Fig. 4, the measured stress profile features a distinct minimum at a height of about 6 m, a feature which is very well replicated in the model. However, the approximately halved stress is still far larger than what one would estimate on the basis of an eddy viscosity model; the large mean-wind speedup over the Askervein hill causes the wind-speed gradient to be nearly zero from 4 m up to 50 m (and likely even much higher), which would lead to zero stress in this whole region.

Thus, the turbulence time scale of decay is long enough for the advection of upstream turbulence to overcome the local lack of production. In fact, if it were not for the curvature of the streamlines, the reduction would be as small

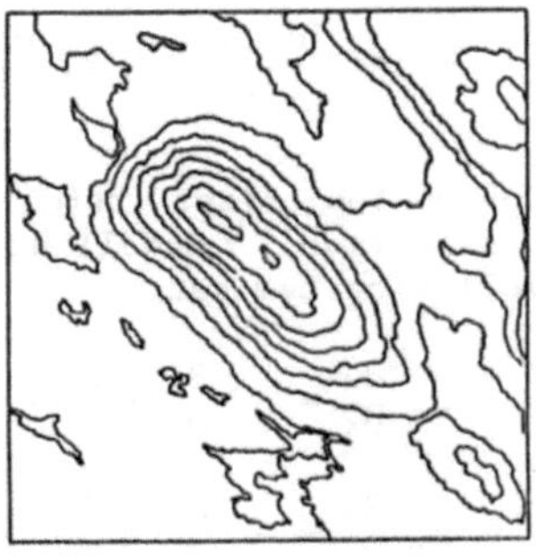

Fig. 3. The Askervein hill seen from the southwest (left), and its plan form with maximum height 126 m above mean sea level (the highest contour is 120 m). The right-hand frame measures approximately 2.55 x 2.55 km; north is up.

as that shown by the dotted line in Fig. 4. Thus, the stress minimum is caused by the stabilizing effect of the convex curvature of the streamline over the crest.

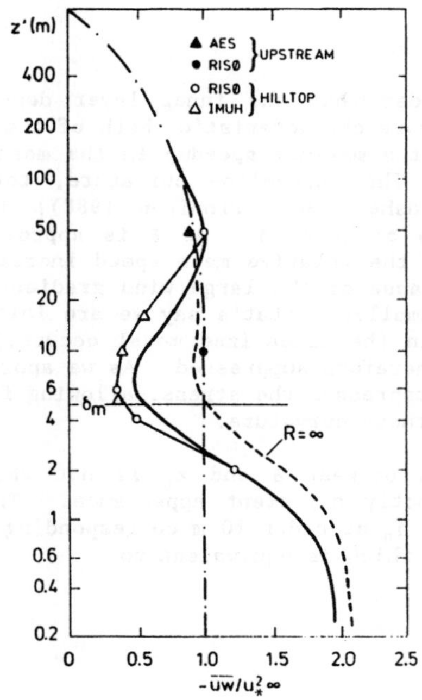

Fig. 4. Comparison of measured and modelled shear stress distribution above the crest of the Askervein hill. The dash-dotted line is the modelled upstream situation and the heavy solid line represents hilltop conditions. The dashed line is also for the hilltop, but with no curvature terms in the turbulence equations, other things being equal. (Zeman and Jensen 1987).

Towards the surface, the stress increases with a very steep gradient. Unfortunately, the measurements extended down to only 2 m (in a later supplementary measurement campaign we have tried hot-wire measurements down to 20 cm above the ground, but these data have not been analysed yet.) Using an eddy viscosity argument, which in this case seems more permissible since near the ground the time scale of turbulent eddies is small compared with the transport time over the hill, postulating equilibrium conditions near the surface, and linearizing, we obtain a relative surface stress perturbation $\Delta\tau/\tau$ twice as large as the mean flow perturbation $\Delta u/u$. At Askervein this latter quantity is about 1 for heights below 4 m and we would then expect a doubling of the stress near the surface.

The Zeman-Jensen model employs a constant stress layer assumption below the lowest streamline. Admittedly it would be more correct to use $\partial\tau/\partial z = \partial p/\partial x$ as a boundary condition, so there is room for improvement. However, this condition would have to be imposed at the top of the roughness elements ($\sim 10z_0$) and an additional "canopy model" would have to be used below that height in order to partition some of the drag into form drag. Since the scale separation between the height of the stress-minimum (let us call that δ) and z_0 (the surface roughness) is so large in the Askervein case (a factor of 100 more), these speculations are probably not of great importance.

The height δ can be estimated from the boundary condition above, which in finite difference form would be $\Delta\tau/\delta \sim \Delta p/L$. Since $\Delta\tau \sim (\Delta u/u)\tau$, $\Delta p \sim \Delta u \cdot u$ (from Bernoulli), and $\tau \equiv u_*^2$, we have

$$\frac{\delta}{L} = \left[\frac{u_*}{u}\right]^2 . \tag{12}$$

Thus, δ is of smaller order than the inner layer depth of the linearized solution, but is probably more characteristic, both of the height of the stress minimum and the height of the maximum speedup in the mean wind profile, since they are all connected. The streamline curvature, through its equivalent Richardson number Rc (Bradshaw 1969; Finnigan 1988), seems to sharpen the conditions for being above or below δ. As δ is approached from below, the stress decreases and hence the relative mean speed increases. In this region Rc is relatively small because of the large wind gradient. The wind gradient somewhat above δ is much smaller. (Let's say we are initially above ℓ, where the maximum perturbation in the linearized model occurs.) Hence Rc tends to be large, and the stress therefore suppressed. As we approach δ, the curvature increases, which further suppresses the stress, allowing for larger mean speeds near δ than in the case without curvature.

When the scale separation between δ and z_0 is not very large, the stress profile may have a distinctly different appearance. This would occur on a wooded hill (canopy height h_0 of order 10 m corresponding to $z_0 \sim 1$ m), as for Askervein. Equation (11), which is equivalent to

$$\frac{\delta}{z_0}\left[\ell n\frac{\delta}{z_0}\right]^2 \sim \frac{L}{z_0} \quad , \tag{13}$$

can be approximated by

$$\frac{\delta}{z_0} = a\left[\frac{L}{z_0}\right]^{2/3} \quad . \tag{14}$$

On Askervein $z_0 \sim 1$ cm, $\delta \sim 4$ m and $L \sim 250$ m (in this case 1/4 of its width, to comply with the standard definition of L), which results in $a \sim 0.3$. Thus with z_0 increased to 1 m, this would result in a value of δ of about 12 m or a height of the same order as the canopy height necessary to produce a z_0 of 1 m. In this case I hypothesize that the stress distribution necessary to balance $\partial p/\partial x$ is taken over by form drag in the canopy (see Fig. 5). For a wind tunnel simulation this situation would very easily arise since the surface must be kept rough for the flow to stay fully turbulent. Thus, for a scale ratio of 1/2000 and satisfying $z_0 u_*/\nu > 10$, we would have a situation close to the "rougher" of the alternatives above (with $u_* \sim 0.3$ m s^{-1} and $\nu = 1.6 \times 10^{-6}$ m^{2}s^{-1}).

For a discussion of other turbulence profiles, see Zeman and Jensen (1987).

OTHER TOPICS

In addition to the lecture by Finnigan, I should like to discuss five Symposium poster papers dealing with flow over complex terrain.

Two dealt with experimental investigations of the turbulence structure above complex terrain: the first for a simple, isolated, two-dimensional ridge called

Cooper's Ridge[1], which has z_0, h and L very similar to Askervein. The second paper[2] is also about a two-dimensional ridge (at Wyong), but very rough (z_0 ~ 1 m) and one in a succession of other similar ridges. The distance between ridges is ~ 2 km and the height ~ 150 m, i.e., a similar but slightly smaller aspect ratio than Askervein or Cooper's Ridge. Measurements were made at 30 and 60 m above a particular crest.

The observed turbulence profiles in these two experiments have some qualitative features in common, but also show differences, especially for the σ_w and τ profiles. In the Cooper's Ridge case σ_w does not increase to larger than upstream values as the surface is approached, i.e., there is very little difference between the 10 and 2 m level, see Fig. 6. However, according to Askervein one would expect that result, but with lower values in between and a minimum around 4 m. A more serious discrepancy exists between the stress profiles (see Figs. 4 and 6). Presently we have no explanation for this, but it clearly needs to be resolved.

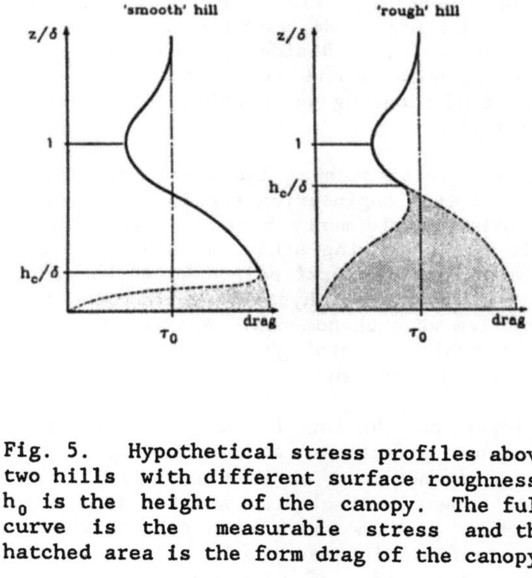

Fig. 5. Hypothetical stress profiles above two hills with different surface roughness. h_0 is the height of the canopy. The full curve is the measurable stress and the hatched area is the form drag of the canopy.

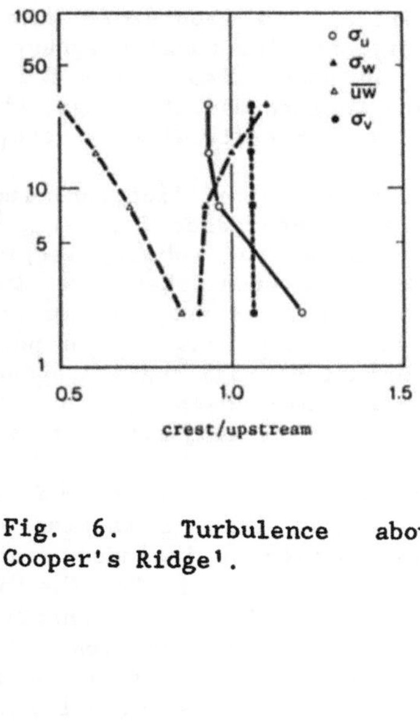

Fig. 6. Turbulence above Cooper's Ridge[1].

The main conclusion of the Wyong experiment is that $\sigma_{u,v,w}/u_*$ ratios are similar to ratios found over flat terrain. This is at variance with the results from Cooper's Ridge, but not with Askervein, where one would expect a slightly decreased value of σ_u/u_* and a slightly increased value of σ_w/u_* of order 10%), both due mostly to rapid distortion of σ_u and σ_w. More significantly, distorted ratios would occur at much lower levels (largely due to reduced values of u_* in the air layer that is most influenced by curvature). This would not be possible to measure at Wyong, however, because of the high forest canopy.

[1]PA Coppin, EF Bradley (1987) Changes in turbulence structure over a two-dimensional ridge. Symposium poster paper

[2]EF Bradley, PA Coppin (1987) Turbulence above a ridge in hilly terrain. Symposium poster paper

The third paper[3] compares predictions from the MS3DJH/3R model (Walmsley et al. 1986) with wind measurements obtained at the Fremantle wind turbine test site. The comparison was not completely successful. There are probably two main reasons for that: the azimuth sectors (quadrants) over which the data were averaged were too large, allowing flow over vastly different terrain transects to be averaged together; and the model used a very much smoothed terrain map and was thus prohibited from replicating the effect of "smaller bumps" in the terrain, which are known to have a significant local influence on the flow.

The last two papers deal with flow over topography where density effects, or stable stratification, play a dominating role in the dynamics. RO Pitts and TJ Lyons[4] modelled a downslope flow over a two-dimensional escarpment using a version of the hydrostatic Pielke (1974) model. They were able to reproduce the qualitative features of a shooting flow with a hydraulic jump, in one case, and, in another, a feature which in the jargon is called a vertically propagating hydrostatic wave. To a nonspecialist as the present author, it all looks like a stationary standing gravity wave, where the streamlines are set into oscillation with amplitudes larger than the height variation of the underlying terrain. In any case, the large wind speeds predicted downwind of the escarpment are in agreement with observations.

P Manins[5] studied a succession of two-dimensional ridges by means of both numerical and laboratory water flume modelling. He was interested in the conditions for quasi-stagnant flow in the valleys. Besides the obvious result that stagnant flow occurs for small Froude numbers and not for large ones, he found an intermediate range, where a kind of blocking was created in connection with a weak recirculating rotor in the valley.

Density (or stratification) induced flow effects such as lee waves are clearly of large importance for air pollution and wind engineering applications. Full scale data are probably lagging the model developments here, and experiments need to be undertaken. As an example of an ongoing study, I should like to mention a joint project between Risø and the Landsverkfrøðingurin at the Faroe Islands, where one of the purposes is to check the hypothesis that the most devastating winds there are caused by a lee wave phenomenon. Appropriate field stations have been set up. When a suitable climatology of these situations has been obtained, aircraft measurements are foreseen.

When we finally return to small ridges and hills, it seems to be model development that is lagging. The various effects of stratification or heat flux need to be included - and here we do not mean obvious effects like lee waves - but rather, how the turbulence budgets are altered and how this feeds back to the integral quantities. Another problem which is barely touched on yet is separation of flow. It is of practical interest to be able to predict the extent of the separation zone and the position of the reattachment point; it is of great theoretical interest to be able to predict the precise point of separation on a rounded shape; not to mention how these things depend on the flow situation (z_0 etc.). It is known that, once developed, the separation has profound upstream influences (it alters the pressure distribution, for example). For many applications, air pollution dispersion, sediment transport, and dune building, for instance, separation is of great interest. It can be modelled by brute-force, fully-implicit $k - \varepsilon$ schemes. However, the question is whether this is good enough to further the understanding of some of the above.

[3]SJ Dear, MF Steketee, TJ Lyons (1987) An analysis of microscale flow variations over the South Fremantle wind turbine test site with particular reference to wind power production. Symposium poster paper

[4]RO Pitts, TJ Lyons (1987) Validation of a hydrostatic model for downslope flow over the Darling Scarp. Symposium poster paper

[5]P Manins (1987) Wind blocking in periodic valleys. Symposium poster paper

REFERENCES

Bradshaw P (1969) The analogy between streamline curvature and buoyancy in turbulent shear flow. J Fluid Mech 36: 177-192
Finnigan JJ (1988) Air flow over complex terrain. This volume, pp
Hunt JCR, Leibovich S, Richards KJ (1987) Stratified shear flows over low hills. 1. Effects of upwind shear. Q J R Meterol Soc, in press
Hunt JCR, Puttock JS, Snyder WH (1979) Turbulent diffusion from a point source in stratified and neutral flows around a three-dimensinal hill; Part I - Diffusion equation analysis. Atmos Environ 13: 1227-1239
Jackson PS, Hunt JCR (1975) Turbulent wind flow over a low hill. Q J R Meteorol Soc 101: 929-955
Mason PJ, Sykes RI (1979) Flow over an isolated hill of moderate slope. Q J R Meteorol Soc 105: 383-395
Pielke RA (1974) A three-dimensional numerical model of the sea breezes over south Florida. Mon Wea Rev 102: 115-139
Taylor PA (1988) Turbulent wakes in the atmospheric boundary layer. This volume, pp
Taylor PA, Mason PJ, Bradley EF (1987) Boundary-layer flow over low hills - A review. Boundary-Layer Meteorol 39: 107-132
Taylor PA, Teunissen HW (1987) The Askervein Hill Project: Overview and background data. Boundary-Layer Meteorol 39: 15-39
Taylor PA, Walmsley JL, Salmon JR (1983) A simple model of neutrally stratified boundary-layer flow over real terrain incorporating wavenumber-dependent scaling. Boundary-Layer Meteorol 26: 169-189
Troen I, de Baas A (1986) A spectral diagnostic model for wind flow simulation in complex terrain. In: Proc EWEC 86, European Wind Energy Association, Conference and Exhibition, Rome, 7-9 October 1986
Walmsley JL, Taylor PA, Keith T (1986) A simple model of neutrally stratified boundary-layer flow over complex terrain with surface roughness modulations (MS3DJH/3R). Boundary-Layer Meteorol 36: 157-186
Zeman O, Jensen NO (1987) Modification of turbulence characterisitcs in flow over hills. Q J R Meteorol Soc 113: 55-80
Zeman O, Tennekes H (1975) A self-contained model for the pressure terms in the turbulent stress equations of the neutral atmospheric layer. J Atmos Sci 32: 1808-1813

CONVECTIVE PROCESSES IN THE LOWER ATMOSPHERE

John C. Wyngaard

National Center for Atmospheric Research*, Boulder, CO 80307, U.S.A.

Summary. Progress in understanding convective processes in the lower atmosphere over the past 25 years has come primarily as a result of vastly improved research tools. These tools - better instruments and numerical simulations on supercomputers - have given us a fairly complete view of the structure of boundary-layer convection, if not yet a solution to the "turbulence problem".

Having more detailed data on convective turbulence than ever before, we can now see how it differs from its mechanical counterpart in shear flows. The Reynolds equations in convection tend not to be locally balanced, which can undermine simple second-order closures for convective flows. Convection has some remarkable transport properties for scalars; I suggest these are due to its combination of persistent, coherent vertical motions of apparently large Lagrangian integral time scale and the vertical asymmetry of this velocity field.

Despite this progress, many of the problems we face in boundary-layer meteorology today severely test our experimental techniques. As we probe larger-scale problems such as its interactions with cloud convection, its behaviour over regional scales, its response to distributed inhomogeneities, and its stochastic variability, we need new tools. A promising candidate is lower tropospheric profiling with ground-based radars and lidars.

STATE OF THE FIELD - A BRIEF OVERVIEW

In their discussion of the turbulent transport of heat in the atmospheric boundary layer, Priestley and Swinbank (1947) speculated that "... an instrument might be devised to measure $\overline{wT'}$ directly ...". It soon was (Swinbank 1951). By the mid 1960s several micrometeorological research groups were routinely measuring the turbulent fluxes of both temperature and momentum. These early flux sensors were delicate, expensive, and had inadequately understood response, but they allowed strong advances in micrometeorology. For example, complete measurements of the Reynolds budgets for heat and momentum fluxes were made in the 1968 Kansas experiments (Haugen et al. 1971), even before they were made in laboratory flows.

Today, turbulent flux and momentum measurements are standard in micrometeorological research. Several three-component sonic anemometers (including one developed in CSIRO, Canberra (Coppin and Taylor 1983)) are on the market. My own institution is now developing ASTER, a portable facility that will allow users to do eddy-correlation measurements of the near-surface fluxes of heat, moisture, momentum, and some trace constituents.

---

* The National Center for Atmospheric Research is sponsored by the National Science Foundation.

In a discussion of the dynamics of the small-scale turbulent heat flux, Deardorff (1961) stated "Solution of these problems through integration of the hydrodynamic equations appears to be out of the question ...". Only a few years later he began such integrations through a technique now called large-eddy simulation (LES), in which the energy-containing range of turbulence is calculated directly in three spatial dimensions and time. The paper describing his LES study of turbulent channel flow using a 6,720 point numerical grid (Deardorff 1970a) is widely cited as the first in the field. Today, LES is providing data bases of unprecedented detail, and is considered to have great potential in research (Wyngaard 1984).

These two examples illustrate what I feel is our most important advance over the past 25 years: the development of vastly improved research tools. They have helped us to learn a good deal about the structure of the lower atmosphere, enabling us, in turn, to help our constituents with a wide variety of problems. Webb's (1984) comprehensive survey of temperature and humidity structure in the boundary layer, written for a volume on electromagnetic wave propagation through the atmosphere, demonstrates this well. Its thorough discussions of the structure and dynamics of the boundary layer, and their implications for refractive index structure, include references to over 260 published works. Its reference list reveals a sharp upward break in references/year at about 1970; nearly 85% of the references are after that date. Most of this 85% are based on experimental data of the sort that were not available earlier.

These new tools - better instruments, and numerical simulations on supercomputers - have not yet given us a solution to the "turbulence problem". We still lack a completely rational theory for the structure of high Reynolds number turbulence. In the view of some, however, we may now be closer to that goal. The renormalization group theory, for example (Dannevik et al. 1987), has had some impressive successes. Thanks to these new research tools, our improved access to the structure of turbulence now allows us to make much better assessments of the fidelity of such theoretical approaches.

THE CONVECTIVE BOUNDARY LAYER

Overall Structure and Scaling

Beginning in the early 1970s acoustic sounder returns allowed us to visualize the convective boundary layer (CBL) as a fully turbulent region with a distinct top. Fueled by the buoyancy from surface heating and/or evaporation, its large eddies fill the space available - up to the capping inversion. Figure 1 is a sketch of the CBL showing some of these features.

Deardorff and Peterson (1980) showed that the rate equation for h(x,y), the local top of the PBL, is

$$\partial h/\partial t + [v(h) + v_e].\nabla h = w(h) + w_e \quad , \tag{1}$$

where v(h) is the velocity at h, v_e is the entrainment velocity, and w(h) and w_e are their respective vertical components. They showed that Reynolds-averaging produces a slightly more complicated equation for the average height, $\bar{h}$. That equation says that $\bar{h}$ changes in response to horizontal and vertical entrainment and advection, and through the divergence of the turbulent flux of h fluctuations. In what follows I assume that the CBL depth is determined by an inversion and denote the average height by the more familiar term z_i.

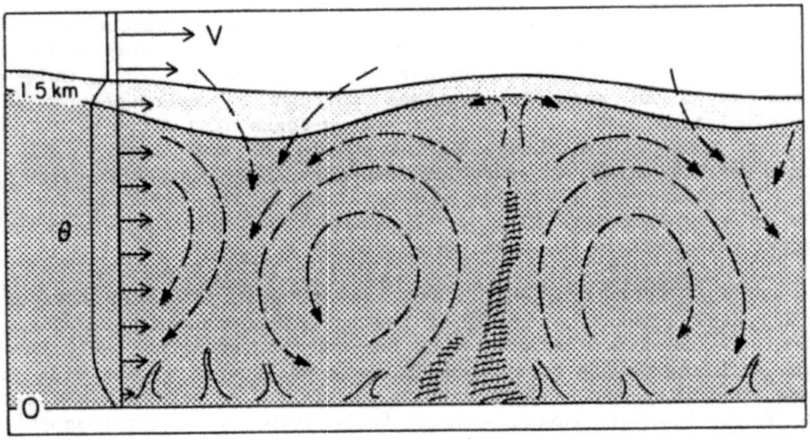

Fig. 1. An artist's representation of the CBL, showing its large, convective eddies, plumes, and the sharp capping inversion. (Wyngaard 1985).

The physics of entrainment remains elusive, in part because it is extremely difficult to observe and measure in detail; some progress has come recently through innovative laboratory experiments, however (e.g., Hannoun and List 1987). The vertical entrainment velocity,

$$w_e = \frac{dz_i}{dt} - w(z_i) \quad , \tag{2}$$

is typically of the order of 1 cm s^{-1}, very much smaller than the velocity fluctuations involved; this hints that entrainment is an intermittent process, and laboratory experiments confirm this. A similar situation exists for the dissipative eddies in a turbulent flow. Their velocity scale is $(\nu\epsilon)^{1/4}$, where ϵ and ν are the dissipation rate and kinematic viscosity, respectively. It is also typically of order 1 cm s^{-1} in the CBL, again not because the dissipative eddies are that weak, but because they are intermittent.

About the time acoustic sounders revealed the strong vertical coherence of the CBL, researchers began treating it as an entity. In recent years, certainly, the notion that the surface layer can be treated separately from the boundary layer as a whole, as assumed in Monin-Obukhov similarity, has been in decline.

The dominant scales for CBL turbulence are easily derived. Since convection fills the space available, the eddy size is of order z_i. Standard turbulent energetics then gives the velocity scale q. Following Tennekes (1970), we do a global balance, equating buoyant production and viscous dissipation:

$$\frac{g}{\Theta}\,\overline{w\theta}_0 = \frac{q^3}{z_i} \;; \quad q = \left[\frac{g}{\Theta}\,\overline{w\theta}_0 z_i\right]^{1/3} \equiv w_* \ . \tag{3}$$

Here g is the acceleration of gravity, Θ and θ are the mean and fluctuating potential temperature, and a subscript zero means surface value. On the basis of his LES studies, Deardroff (1970b) also argued that w_* is the velocity scale for CBL turbulence.

Since w_* refers only to the turbulence produced by buoyancy (i.e., to the convective turbulence), one might wonder about the role of the mechanical turbulence produced by the wind shear-Reynolds stress interaction. Assuming a wind profile having negligible shear outside the surface layer, one can estimate (Wyngaard 1983a) that the CBL-averaged rates of shear production <SP> and buoyant production <BP> are in the ratio

$$-\frac{u_*}{U}\frac{z_i}{L} = \frac{< BP >}{< SP >}, \qquad (4)$$

where U is the mean wind speed in the CBL, u_* is the friction velocity, and $L = -u_*^3\Theta/(kg\overline{w\theta}_0)$ is the Monin-Obukhov length, with $k \sim 0.4$ the von Karman constant. There is evidence (Deardorff 1973) that the shift from forced to free convection (as judged qualitatively by the emergence of well-mixed mean wind and temperature profiles) seems to begin at $-z_i/L \sim 1.5$ (when <BP> is only 10% of <SP>, according to Eq. (4) with $U = 15u_*$) and is marked at $-z_i/L \sim 4.5$ (when it is only 30%). One wonders: How can an apparently small rate of buoyant production of turbulence have such a large effect on boundary layer structure?

Since we often judge transition to free convection by the approach to a well-mixed state, we can gain insight by comparing something more directly related to eddy diffusivity K. We expect that $K \sim q\ell$, where q and ℓ are velocity and length scales of the energy-containing eddies; we take q as the square root of the turbulent kinetic energy, and ℓ as z_i. The eddy diffusivity for scalars is not always well behaved, so I'll consider that for momentum.

Based on data from the AMTEX (Lenschow et al. 1980) and Minnesota (Kaimal et al. 1976) experiments, a representative value for the momentum diffusivity in mid-CBL is (Wyngaard 1983b).

$$K_c \simeq 0.05 q_c z_i , \qquad (5)$$

where the subscript c refers to convective turbulence. Equation (5) also agrees well with data from a laboratory convection tank (Deardorff and Willis 1987).

For purely shear-driven turbulence, Brost et al. (1982) found from measurements in a neutral, cloud-topped PBL that

$$K_s \simeq \frac{0.05}{3} q_s z_i \qquad (6)$$

with s referring to shear-driven turbulence. This also agrees well with data from a neutral, laboratory layer (Townsend 1976). Equations (5) and (6) indicate that convective turbulence is three times as diffusive as mechanical turbulence. The data leading to Eqs. (5) and (6) also indicate that $q_c \simeq w_*$ and $q_s \simeq 5^{1/2}u_*$; using these and the definitions of w_* and L, we can write $-z_i/L \simeq 4.5(q_c/q_s)^3$. Thus, at $-z_i/L \simeq 4.5$ (the point of marked transition, according to Deardorff) the convective and mechanical energies are about equal; since the convective eddies are 3 times as diffusive, however, the eddy diffusivity for momentum is more than tripled.

The diffusivities for momentum and scalars differ substantially in convective turbulence. Scalars are much more diffusive - sometimes infinitely more so, as I will show later. Scorer (1985) attributes this to the properties of the horizontal vortex lines which are the initial response of the turbulence field to buoyancy forces.

Another intriguing aspect of CBL turbulence is its huge Reynolds number. Recall the familiar description of turbulence in terms of its energy spectrum

(Tennekes and Lumley 1972). Although this strictly applies to homogeneous turbulence, it is qualitatively useful for the PBL as well. The energy-containing range is characterized by wavenumbers of order $1/\ell$, or eddy sizes of order ℓ. The small-scale end of the spectrum contains the dissipative eddies of wavenumber of order $1/\eta$, where $\eta = (\nu^3/\epsilon)^{1/4}$ is the Kolmogorov microscale. Since $\epsilon = q^3/\ell$, it follows that the "scale separation" is $\ell/\eta = (q\ell/\nu)^{3/4}$. The group $(q\ell/\nu)$ is a Reynolds number characteristic of the energy-containing eddies; we will denote it simply as Re.

In a typical CBL, $q \sim w_* \sim 1$ m s^{-1}, $z_i \sim 1$ km, so that Re $\sim 10^8$, orders of magnitude larger than in laboratory turbulence. In the convection tank of Willis and Deardorff (1974), for example, Re $\sim 4 \times 10^3$. Beyond a threshold Reynolds number the structure of the energy-containing range of turbulence is usually observed to depend at most only weakly on the fluid viscosity and, hence, on Reynolds number (Townsend 1976). For that reason, this relatively low Reynolds number convection tank has been an excellent means of simulating turbulence structure and dispersion processes in the CBL.

It became clear over the years that the dissipative activity in turbulence is not uniformly distributed in space and time, but tends to be intermittent, being concentrated in spatially localized bursts. It was also clear by the late 1960s that this intermittency increases as the separation between energy-containing and dissipative ranges, i.e., Re$^{3/4}$, increases. Thus, PBL turbulence has very high intermittency. When turbulence fine-structure researchers discovered this around 1970, a sizeable number of them began to work outdoors.

One motivation for fine-structure measurements at large Re was to test Kolmogorov's hypothesis (Batchelor 1960) that the small-scale structure of turbulence is universal and depends only on ϵ and ν. For example, the skewness S of the streamwise derivative of streamwise velocity, which is related to the inertial transfer of energy by vortex stretching and can be measured with a wide-band hot-wire anemometer, would be a constant, independent of Re, under that hypothesis.

Figure 2 shows some of these early S data from the PBL. They were measured at 5.7, 11.3 and 22.6 m during the 1968 Kansas experiments (Wyngaard and Tennekes 1970). The low-Re laboratory data available until then had led researchers to conclude that S was indeed constant at about -0.4, but the PBL data at much larger Re spanned the range from about -0.6 to -1.0.

Kolmogorov (1962) had, in fact, modified his hypothesis since its original formulation in 1941, suggesting that since dissipation was quite variable in space and time, fine structure should scale with the local value of dissipation rather than its ensemble average as he originally proposed. The further hypothesis that this locally averaged dissipation rate was distributed log-normally, which was qualitatively consistent with observations, then led to predictions that fine structure statistics varied with intermittency and, hence, with Reynolds number. The $S \sim K^{3/8}$ curve in Fig. 3 was derived in that way.

This Reynolds-number dependence is thought to be confined to dissipation-range statistics, however, and many aspects of the structure of energy-containing range turbulence in the CBL can be reliably simulated in laboratory convection tanks at much lower Reynolds number (e.g., Willis and Deardorff 1974).

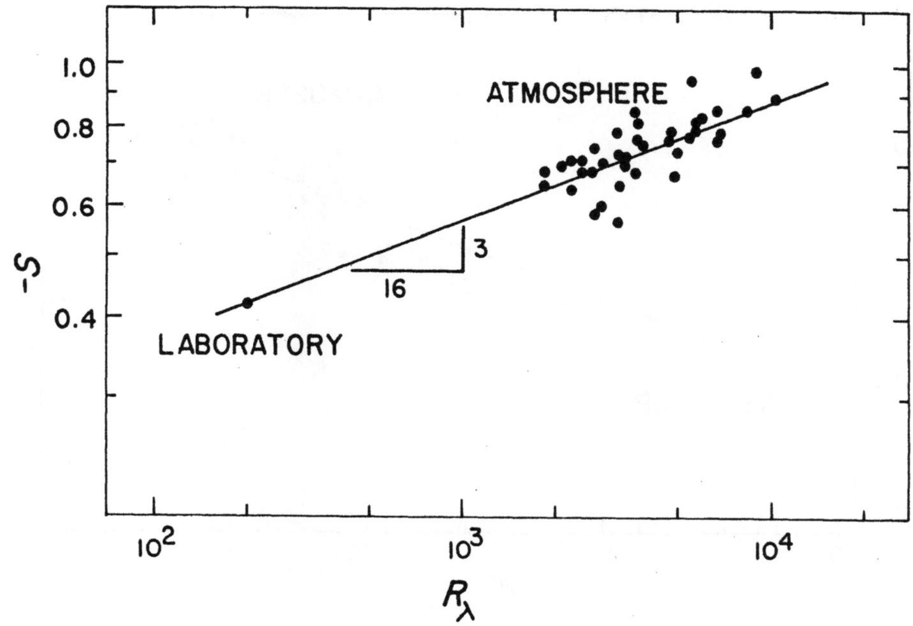

Fig. 2. Laboratory and atmospheric data on S, the skewness of the streamwise velocity derivative, versus turbulence Reynolds number $R_\lambda \sim Re^{1/2}$. The 3/16-power prediction is from an intermittency model. (Wyngaard and Tennekes 1970).

Surface-layer Similarity

Monin-Obukhov (M-O) similarity is now part of the foundation of micrometeorology. Based on the hypothesis that distance from the surface z, g/Θ, u_*, and $\overline{w\theta}_0$ are the governing parameters for surface-layer structure, it provides a convenient framework for transferring research results to applications. In recent years we have also learned more about the limits of its applicability to the CBL.

Consider, for example, that if eddies of velocity scale w_* fill the flow (Fig. 1), then horizontal wind fluctuations in the surface layer should also be sensitive to w_*. If so, however, they cannot be M-O similar because w_* depends on z_i, which is not an M-O parameter.

It took us a long time to come to grips with this; there are suggestions of this failure in the Lumley and Panofsky (1964) monograph, but papers using w_* scaling for σ_u and σ_v, the standard deviations of horizontal wind fluctuations, (e.g., Panofsky et al. 1977) were some years off.

We can illustrate both a success and a failure of M-O scaling with some results from the 1968 Kansas experiments. Figure 4 shows that the dimensionless standard deviation of vertical wind fluctuations σ_w/u_*, measured under unstable conditions at 5.7, 11.3 and 22.6 m, collapses well under M-O scaling. Under very unstable conditions it has long been argued that the behaviour of σ_w

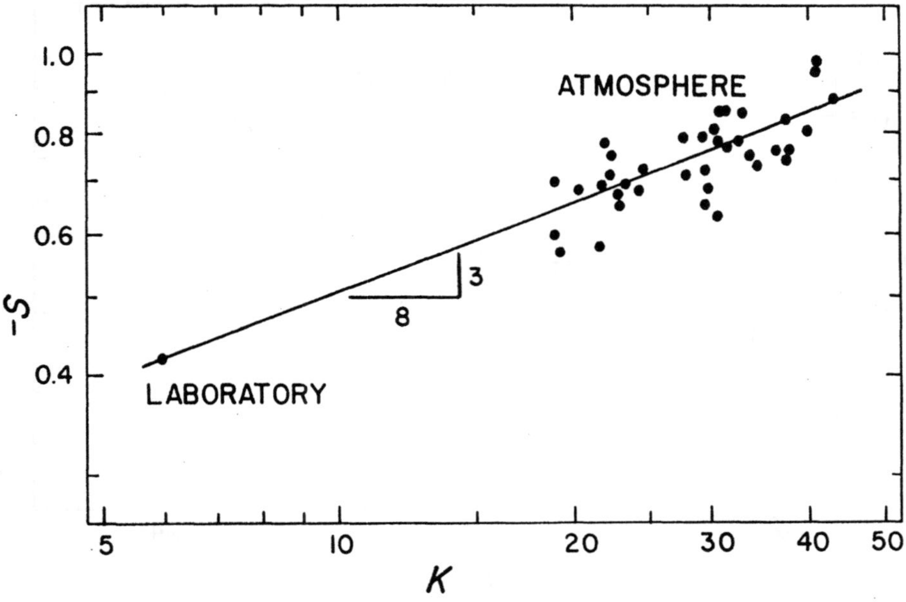

Fig. 3. Skewness S versus K, the kurtosis (the dimensionless fourth moment), for the streamwise velocity derivative. The 3/8-power prediction is based on an intermittency model. (Wyngaard and Tennekes 1970).

should closely approach that in true free convection, where u_* vanishes. The simplest scaling in that limit takes σ_w proportional to the free-convection velocity scale $u_f = (g\overline{w\theta}_0 z/\Theta)^{1/3}$, which implies $\sigma_w/u_* \sim (-z/L)^{1/3}$; this is consistent with the behaviour of the σ_w data in Fig. 4. This is called "local free convection" behaviour (Wyngaard et al. 1971).

Figure 4 shows that σ_v/u_*, by contrast, has much more scatter. Figure 5, upper panel, shows further that σ_v does not depend on z, so that the hint of the $(-z/L)^{1/3}$ trend in Fig. 4 is not due to local free convection behaviour; instead, it is due to departures from M-O similarity.

If we accept that σ_v is not M-O similar, but instead in the unstable limit scales with w_*, then from the definitions of L and w_* we have $\sigma_v/u_* \sim z_i^{1/3}(-L)^{-1/3}$. Although z_i was not measured in the Kansas experiments, it undoubtedly varied much less than L for the runs used here, so the L dependence dominates; this is consistent with the behaviour in the upper panel of Fig. 5. We can also rewrite the expression as $\sigma_v/u_* \sim (-z/L)^{1/3}(z/z_i)^{-1/3}$, which vaires only weakly about $(-z/L)^{1/3}$ behaviour at fixed z. This is consistent with the bottom panel of Fig. 5.

Unfortunately, not much is known about the extent to which other statistics may deviate from M-O similarity due to the effect of these large, convective eddies. Businger (1973) suggested that the failure of the dimensionless temperature gradient to show the predicted $(-z/L)^{-1/3}$ behaviour in local free convection was due to the influence of w_*, but I know of no attempts to test this with measurements.

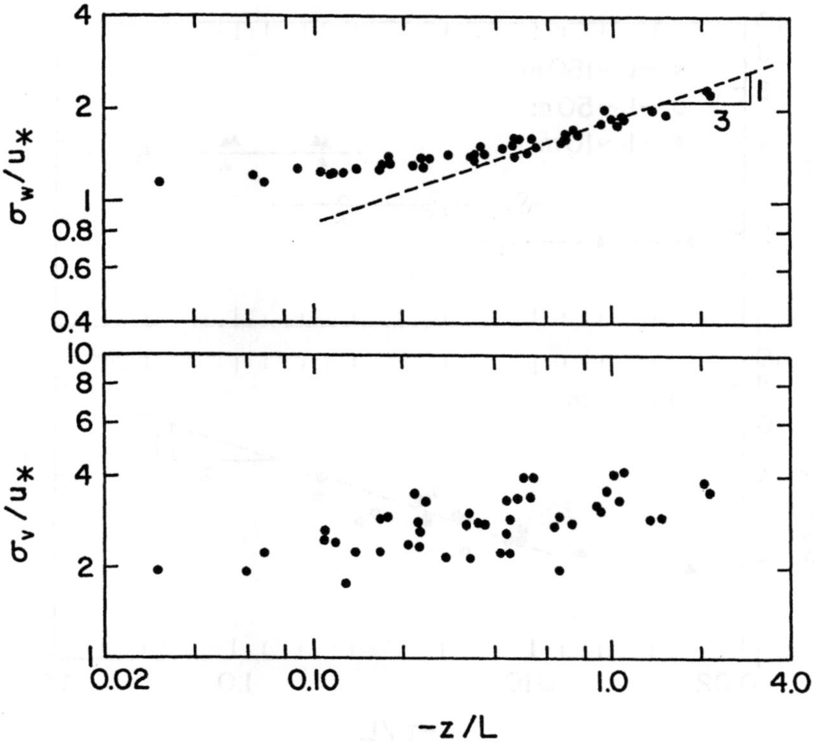

Fig. 4. Surface-layer measurements of σ_w (upper panel) and σ_v (lower panel) under unstable conditions in the 1968 Kansas experiments.

One might also expect the dissipation rate ϵ to deviate from M-O similarity on the following grounds. Since vertical velocity fluctuations in the unstable surface layer scale with the local free convection velocity scale u_f, their interactions should lead to a dissipation rate of order $\epsilon_c \sim u_f^3/z \sim g\overline{w\theta}_0/\Theta$. The large convective eddies randomly generate local mean winds of order w_* in the surface layer; their interactions with the surface roughness should lead to a dissipation rate of order $\epsilon_s \sim v_*^3/z$, where v_* is the effective friction velocity due to these large eddies. We can rewrite this as $\epsilon_s \sim C_D^{3/2} w_*^3/z \sim C_D^{3/2} g\overline{w\theta}_0 z_i/(\Theta z)$, where C_D is a drag coefficient. We expect that C_D depends on the surface roughness, so that ϵ_s should not be M-O similar. The two contributions ϵ_c and ϵ_s are of the same order at height $z \sim C_D^{3/2} z_i$, which for typical conditions is of the order of 1 m; above that height ϵ_c (which is M-O similar) should dominate. This departure from M-O similarity might not be detectable except very near the surface, therefore.

A second mechanism, entrainment at the boundary layer top, can also cause deviations from M-O similarity. Entrainment-induced turbulence can extend down through the mixed layer to the surface, particularly in the case of humidity. For example, if a moist CBL deepens rapidly by entraining overlying dry air, as after the passage of a convective storm (Fitzjarrald and Garstang 1981) or during the morning in clear conditions (Deardorff 1974), the entrainment flux of humidity at the CBL top can exceed the surface flux. The resulting flux divergence causes drying of the CBL air. One might expect that fluctuating humidity signals measured in the surface layer in such conditions would show low-humidity bursts that could have originated only from above the CBL.

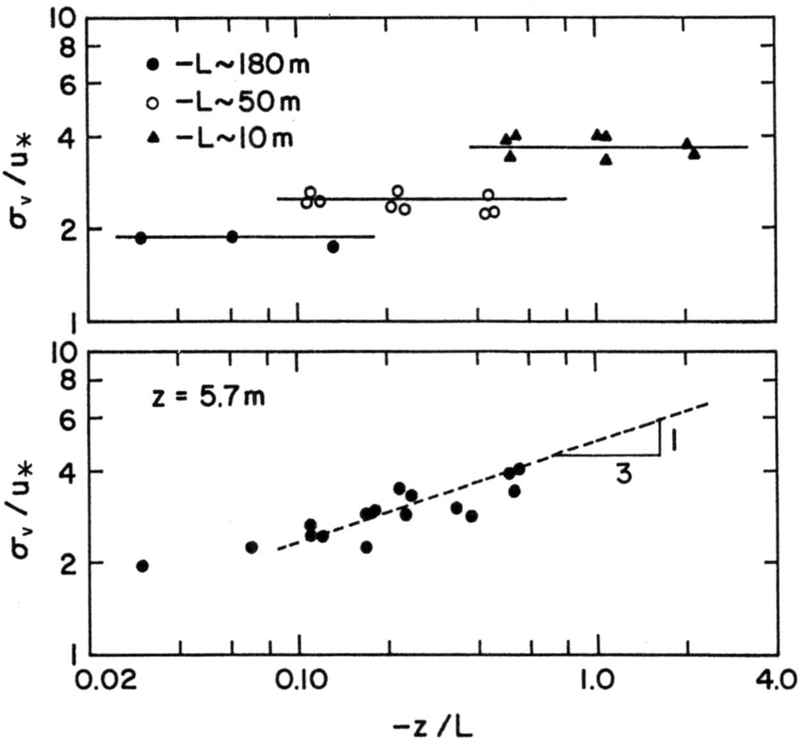

Fig. 5. Replots of the σ_V data of Fig. 4, showing the nature of the deviations from M-O similarity.

Deardorff (1974) did find such bursts in his LES study. In general, then, a fluctuating scalar field in the surface layer of an entraining CBL contains both the turbulence generated locally by shear and convection, and entrainment-induced turbulence imported from the mixed-layer top. Only the first contribution can be M-O similar, of course.

Such deviations from M-O similarity might be important in applications where measurements of certain surface-layer variables are used to infer others. There the similarity relation plays a key role. An example is obtaining the fluxes of momentum, heat, and moisture from measurements of the structure parameters of velocity, temperature, and humidity. The theory for doing so is straightforward and based on M-O similarity of the structuring parameters (Wyngaard and Clifford 1978). A recent survey (Andreas 1987) of surface-layer data on the temperature and humidity structure parameters shows some unusual behaviour that could be due to deviations from M-O similarity.

Mixed-layer Turbulence

When they identified w_* as the turbulent velocity scale in the CBL, Tennekes (1970) and Deardorff (1970b) also proposed that the temperature scale is $\theta_* = \overline{w\theta}_0/w_*$. By extension, the scale for a fluctuating scalar field c having a surface flux $\overline{wc}_0$ is $c_* = \overline{wc}_0/w_*$.

The notion that turbulence statistics in the CBL scale with z_i, w_*, θ_*, and c_* is called the "mixed-layer scaling" hypothesis. It has been widely applied over the past 15 years, and clearly works well for many properties. Deardorff (1970b) applied it to vertical velocity and temperature fluctuations from the laboratory and the atmosphere, showing that it did collapse data from different sources. Figure 6 shows another example of how effectively mixed-layer scaling can collapse data from one experiment.

Figure 7 shows a schematic of the CBL with the surface, mixed, and interfacial layers differentiated, and showing vertical profiles of the mean mixing ratio and vertical flux of a passive, conservative scalar. In general there is a scalar flux at both bottom and top of the mixed layer, the latter induced by the entrainment process. Thus, in using $\overline{wc}_0/w_*$ as the scale for scalar fluctuations in the mixed layer, one assumes that the contribution from the entrainment process is small. Figure 8 shows how poor an assumption this can be.

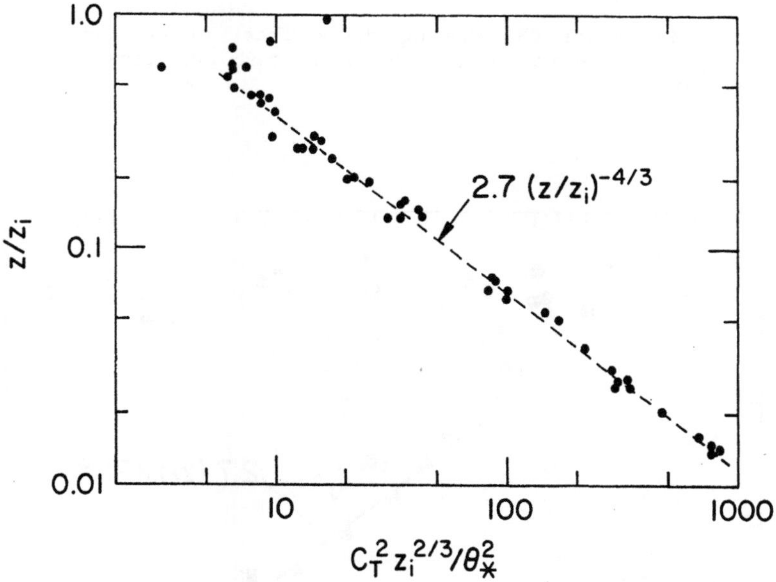

Fig. 6. Mixed-layer similarity of the temperature structure parameter $C_T{}^2$ from the 1973 Minnesota experiments. (Kaimal et al. 1976).

There are other sources of turbulence in the CBL; for example, there are large-scale, quasi-hydrostatic, "standing eddies" driven in part by horizontal temperature, variations along the lower surface. These might look like mean circulations to the tower-based experimenter but like turbulence to the aircraft observer. Cloud convection can also influence the structure of the PBL below. Its effects can be indirect, as in limiting the incoming solar radiation, changing the surface energy budget, and reducing locally the strength of boundary layer convection. Cloud convection also directly drives boundary layer ciculations by extracting moist PBL air, developing subsidence around the cloud and generating large-scale pressure perturbations. It seems that neither of these additional types of turbulence has yet been studied in any depth, but each could be important in air-quality applications, for example.

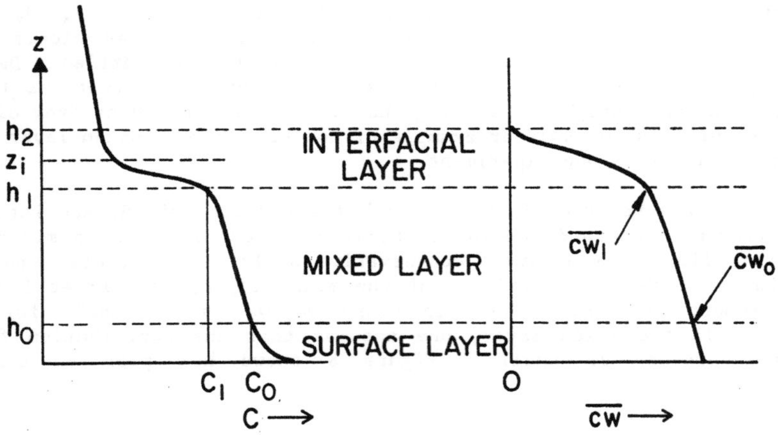

Fig. 7. A schematic of the CBL showing the surface, mixed, and interfacial layers, and the profiles of mean mixing ratio and turbulent flux of a passive, conservative scalar.

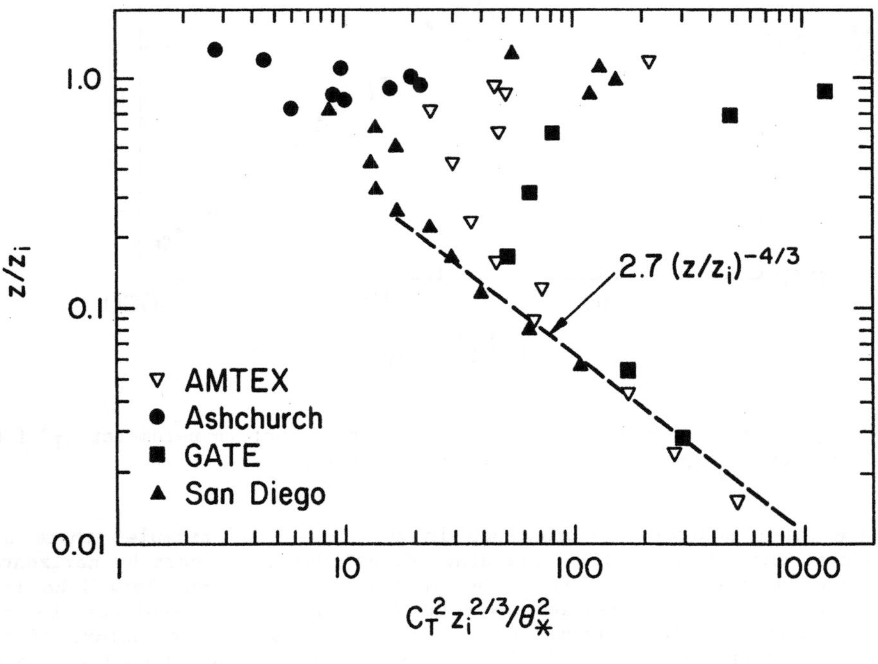

Fig. 8. A failure of mixed-layer similarity for C_T^2 caused by strong entrainment effects. (Wyngaard and LeMone 1980).

Numerical Modelling

Traditional wisdom is that one can't deal in practice with all the complications of the details of turbulence, and that to make the turbulence problem tractable one must deal instead with its average properties. This has prompted much experimental study of statistics, particularly during the past 20 years when reliable turbulence sensors have been widely available. By now researchers have examined all the usual second-moment budgets and even some for third moments. As our familiarity with these budgets has grown, so has their use in numerical models. We can illustrate some of the progress here through discussion of a Reynolds flux budget.

The budget for the turbulent flux of a conservative scalar c in large Reynolds number flow is

$$\frac{\partial}{\partial t}\overline{u_i c} = -U_j\frac{\partial}{\partial x_j}\overline{u_i c} - \overline{u_i u_j}\frac{\partial C}{\partial x_j} - \overline{u_j c}\frac{\partial U_i}{\partial x_j} - \frac{\partial}{\partial x_j}\overline{u_i u_j c} - \frac{1}{\rho_0}\overline{c\frac{\partial p}{\partial x_i}} - \frac{g_i}{\Theta}\overline{\theta c} - 2\epsilon_{ijk}\Omega_j\overline{u_k c} . \tag{7}$$

We use the Cartesian tensor convention that repeated indices are summed over 1,2, and 3. Here x_i is the spatial coordinate, u_i is the fluctuating velocity, p is fluctuating pressure, ρ_0 is mean density, g_i is the gravity vector, and Ω_i is the angular velocity of the Earth.

The terms on the right represent, in order, mean advection, gradient production, shear production, turbulent transport, pressure destruction, buoyant production, and Coriolis effects. Note that only the vertical ($i = 3$) budget has a buoyancy term. Near the surface, at least, the shear production enters only the horizontal budgets; this has the odd effect of producing horizontal scalar fluxes near the surface even when the mean scalar gradient is purely vertical.

In the vertical scalar flux budget the gradient and buoyant production and turbulent transport terms tend to be the largest, and of order flux/τ_t where τ_t is a turbulent time scale. In this CBL τ_t is of order $z_i/w_* \sim 15$ min. The mean advection or time-change terms will usually have a significantly larger time scale; this is why we can often use quasi-steady, locally homogeneous second-moment budgets, and simple scaling hypotheses, for transient, inhomogeneous CBL problems.

By measurements of the other terms in Eq. (7) we have been able to infer that the pressure covariance generally is not small, but instead represents the principal loss mechanism for scalar flux. Further, we know that when the mean scalar gradient is large, gradient production is the principal source of vertical flux. A crude but often valid truncation of the vertical scalar flux budget is then

$$\overline{wc} = -\overline{w^2}\tau_t\frac{\partial C}{\partial z} , \tag{8}$$

which implies a flux-gradient relation with a diffusivity $\overline{w^2}\tau_t$. This is a satisfying result, because it brings measurements of these rather arcane budgets into line with our expectations, experience, and practice.

The development of this level of insight into the flux budgets of both scalars and momentum coincided with emergence of second-order modelling in the early 1970s. The budgets did not indicate in general that a turbulent flux is maintained by the corresponding mean gradient; rather, they showed that while gradient production of fluxes is generally important, shear production, buoyant production, and transport can also be important. When mean gradient production does dominate, as in most mechanical turbulence, the flux tends to be down the mean gradient. In buoyant flows, however, fluxes can be supported entirely by buoyancy or transport and, hence, uncoupled from the mean gradient, as in the case of heat flux in mid-CBL (Deardorff 1972). The lure of second-order closure was the opportunity to include all these effects in one master set of differential equations.

Early on it was clear that a "full" set of second-order equations did have some turbulence-like properties, and did predict many aspects of the broad structure of a wide range of turbulent flows. That generated quite a flood of second-order modelling, although by the late 1970s some (e.g., Liepmann 1979) were outspoken about the lack of critical assessment of much of the work.

From this distance we can see some of the problems that came with the purely statistical approach represented by second-order modelling. The usual form of second-moment equations have "realizability problems"; variances do not always remain positive, for example (Schumann 1977). Lumley (1979) proposed a second-order closure approach that embraces realizability, although it is more complicated than others.

Eventually it also became clear that buoyancy-driven turbulence is quite different in some ways from shear-driven turbulence. The seeds of this can be seen in the "Priestley-Swinbank" effect: buoyancy can directly create a heat flux that has no direct connection to the mean lapse rate (Priestley and Swinbank 1947).

Subsequent LES work suggests, not surprisingly, that the pressure field in the CBL is strongly influenced by buoyancy effects which are quite coherent spatially. In view of this, one wonders about the justification for the usual practice of using purely local closures for pressure covariances. In their 1964 monograph, Lumley and Panofsky state that "... a differential equation implies that local quantities are determined by local conditions, and this is not often the case in turbulent flow". This seems particularly true for convective turbulence.

There is, indeed, evidence that pressure covariances in the CBL behave differently than in mechanical turbulence. For example, it appears that in the convective surface layer the pressure covariance is the principal source of horizontal velocity variances; a part of this pressure covariance represents the transfer of energy "uphill" from the vertical component to the more energetic horizontal ones (Wyngaard 1980). This seems odd when viewed in the purely local perspective that we often use for mechanical turbulence.

In summary, CBL turbulence is spatially coherent in the vertical, with large, energetic dominant eddies that fill the flow. Its second moment budgets have the usual (gradient) production mechanism, but also have strong transport and buoyancy effects; its pressure terms are the principal sinks, but represent a combination of effects – buoyancy, turbulent-turbulent interactions, mean shear, and even (in the upper ocean) significant Coriolis forces. The overall result tends to be complicated local constitutive equations - and a local decoupling of fluxes and mean gradients. This is what the statistical approach to the study of CBL turbulence has taught us - that its Reynolds equations can be very complicated and not as enlightening as we might have hoped.

A CLOSER LOOK AT CONVECTIVE TURBULENCE

How is it Different?

The turbulence in the CBL has some odd constitutive relations. One demonstration of this is in the recent LES results of Wyngaard and Brost (1984) and Moeng and Wyngaard (1984), who considered the horizontally homogeneous diffusion of a passive, conservative scalar through the convective PBL. Since the scalar conservation equation is linear, the general case (which, as we saw earlier, has fluxes at both top and bottom of the mixed layer) can be considered as the superposition of what they called "bottom-up" and "top-down" problems, and the total scalar field represented as the sum of bottom-up and top-down components. The bottom-up problem is defined as having a surface flux but zero flux at the mixed-layer top; the top-down problem is the reverse. It is only the bottom-up problem, of course, that is addressed by the M-O and mixed-layer scaling hypotheses.

We found through LES studies that the eddy diffusivities for top-down and bottom-up scalars are quite different. The bottom-up diffusivity showed a singularity in mid-layer and was everywhere larger than the top-down one. It has been known for some time that the diffusivity for the total field was not well-behaved.

A mixed-layer scaling hypothesis for each of the top-down and bottom-up scalar fields then follows naturally. The bottom-up scalar variance, for example, would follow

$$\overline{c_b^2} = c_b^{*2} g_b(z/z_i) \quad , \tag{9}$$

where $c_b^* = \overline{wc_0}/w_*$ and g_b is a universal function of z/z_i; this is the usual mixed-layer scaling. The top-down scalar variance would follow

$$\overline{c_t^2} = c_t^{*2} g_t(z/z_i) \quad , \tag{10}$$

where $c_t^* = \overline{wc_1}/w_*$ and g_t is another universal function of z/z_i. Moeng and Wyngaard (1984) found that $g_b(z/z_i) \neq g_t(1-z/z_i)$; the equality would hold if the vertical structure of the CBL were symmetric about the midplane. The top-down variance would seem to be a major contributor to the scatter in the data of Fig. 8.

The scalar variance in the general case is therefore

$$\overline{c^2} = \overline{(c_b+c_t)^2} = \overline{c_b^2} + 2\overline{c_b c_t} + \overline{c_t^2} \quad . \tag{11}$$

The LES results indicate that the covariance term is not zero; that is, the top-down and bottom-up scalar fields are correlated.

If we consider the component diffusion processes separately, we can see how the flux-gradient relations differ in the two cases, according to the LES results. Consider bottom-up diffusion, which satisfies

$$\frac{\partial C_b}{\partial t} + \frac{\partial}{\partial z} \overline{c_b w} = 0 \tag{12}$$

and is nonstationary. Figure 9 shows a quasi-steady idealization of the process; the mean concentration evolves continuously in response to the flux divergence, but the mean gradient does not change because the flux profile is

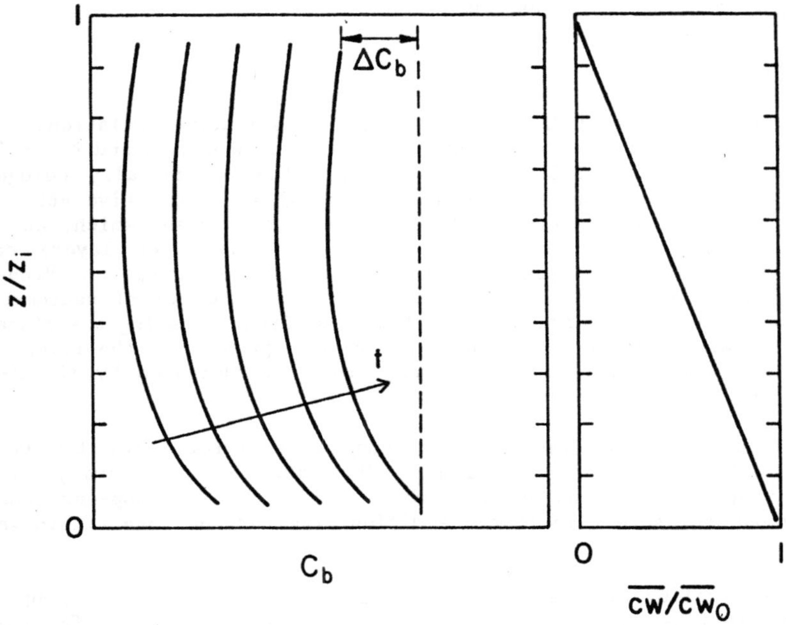

Fig. 9. Bottom-up diffusion. Note that the mean profile (left panel) is not monotonic. (Wyngaard 1987).

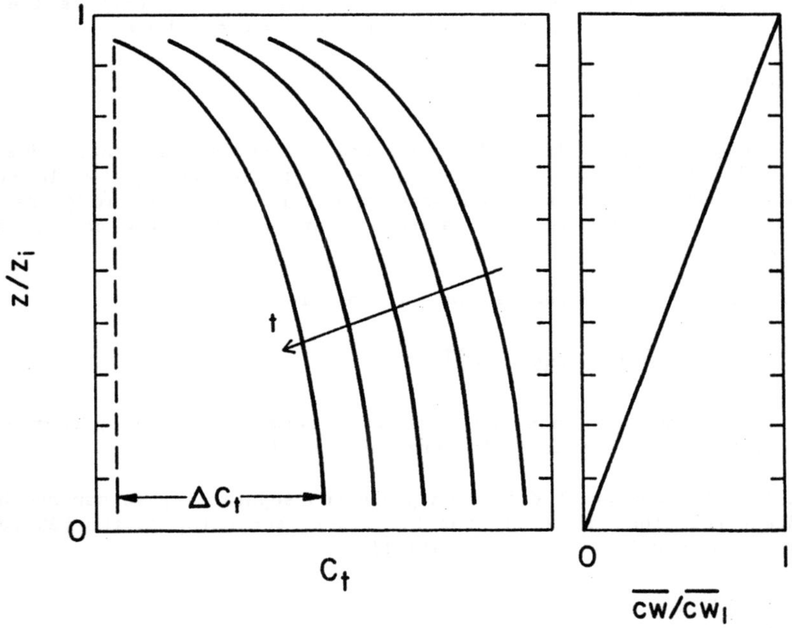

Fig. 10. Top-down diffusion. Note that the mean jump ΔC_t is larger than ΔC_b, Fig. 9. (Wyngaard 1987).

linear. Figure 10 shows the top-down process. Note that the mean profile shapes have two differences: the jump in C across the mixed layer is large in the top-down case, and the bottom-up profile is non-monotonic. The first can occur with a K profile that is not symmetric about mid-CBL, but the second requires a singularity in the bottom-up K. Thus, the top-down and bottom-up K profiles are quite different.

These LES results have not been confirmed directly through measurements in the CBL or in the convection tank. However, they are consistent with the observation that mean virtual potential temperature Θ_v is typically well-mixed in the CBL, while mean water vapour mixing ratio Q is often not. The entrainment flux of virtual potential temperature is typically of the order of 1/4 the surface flux, and of opposite sign. The LES results indicate that a given scalar flux supports about 4 times the mean gradient in top-down diffusion than in the bottom-up case, however. As a result, the contributions of the entrainment and surface fluxes to $\partial\Theta_v/\partial z$ nearly balance over much of the CBL, giving a well-mixed Θ_v profile. The humidity flux tends to be positive at both bottom and top; thus, each supports a negative mean gradient, leading to the decrease in Q with height often observed in the mixed layer (Mahrt 1976).

Since in the Boussinesq approximation buoyancy forces are proportional to the deviation in virtual potential temperature, one sees that there is a feedback mechanism operating to keep Θ_v nearly constant over most of the CBL. If $\partial\Theta_v/\partial z$ were significantly positive, for example, then upward-moving eddies would be retarded by buoyancy; this would decrease the turbulence level in the upper CBL, decrease the entrainment rate, and decrease the entrainment flux of θ_v. The top-down contribution to $\partial\Theta_v/\partial z$ would then decrease in magnitude, tending to eliminate the original perturbation in $\partial\Theta_v/\partial z$. One can trace a similar feedback loop for negative $\partial\Theta_v/\partial z$.

One can imagine generating these top-down and bottom-up diffusion results through a superposition of the results from individual sources placed at bottom or top, respectively, and arrayed downstream. Hence, the plume diffusing from a point or line source should also show unusual diffusion properties. This is indeed the case, as shown in the 1970s by Lamb, in his innovative use of Deardorff's LES results (Lamb 1982) and by Willis and Deardorff (1976) through their tank experiments on diffusion. Plumes from surface emissions "lift off", and the locus of maximum concentration rises downstream; likewise, plumes from release near the CBL top show descent. These findings are at odds with the traditional parameterizations of plume dispersion and have revolutionized short range dispersion modelling for the CBL over the past 15 years.

A general constitutive equation that accounts for these results is that of Fiedler (1984),

$$\frac{\partial}{\partial z}\,\overline{wc} = \int R\left[\frac{z}{z_i},\frac{z'}{z_i}\right]\,[C(z,t) - C(z',t)]dz' \quad . \tag{13}$$

Stull (1984) has proposed a similar closure. This is an attractive generalization of eddy diffusivity; it allows for the local flux-gradient relation to vary from the K form, and if the transfer matrix R is asymmetric, it even allows for differences between top-down and bottom-up diffusion. The challenge is to determine R. I know of no work that relates it through first principles to other, measureable flow properties, but Fiedler and Moeng (1985) have constructed an R that reproduces LES results on diffusion in the CBL.

Why is it Different?

Evidently, buoyancy-driven eddies behave differently from shear-driven ones. Correlation lengths in convective turbulence tend to be large; acoustic and lidar sounder records suggest that convective plumes are coherent through the depth of the CBL, for example. Even in the surface layer, turbulence spectra continuously expand toward lower frequencies (larger scales) as the stability changes from stable, through neutral, to unstable (Kaimal et al. 1972).

In convective turbulence the fluctuating pressure field, which we saw earlier plays a key role in the Reynolds flux equations, satisfies a Poisson equation whose leading terms represent inertial and buoyancy effects, respectively:

$$\frac{1}{\rho_0}\nabla^2 p = \overline{u_{i,j}\, u_{j,i}} - u_{i,j}\, u_{j,i} + \frac{g}{\Theta}\,\theta_{,3} \quad . \tag{14}$$

Here commas denote differentiation with respect to spatial coordinates. Since convection is the response to buoyancy forces, we hypothesize that the two types of terms in Eq. (14) are of the same order, so that

$$\frac{q^2}{\ell^2} \sim \frac{g}{\Theta}\frac{\theta}{\ell} \quad , \tag{15}$$

where q and θ are turbulent velocity and temperature scales, and ℓ is a length scale. This hypothesis is consistent with the LES results of Moeng and Wyngaard (1986). If we take q and θ as the usual mixed-layer scales w_* and θ_*, then it follows from Eq. (15) and the definitions of w_* and θ_* that $\ell \sim z_i$. This suggests the fluctuating pressure field is also very coherent spatially. Perhaps this also indicates why purely local closures tend to fail in convective turbulence.

While much of the evidence of the unusual behaviour of convective turbulence stems from detailed LES and laboratory studies, we can interpret these differences in simple, physical terms. Convection in the CBL is forced by the heating and/or evaporation at the surface; the resulting buoyancy forces on fluid elements persist until eliminated by diffusion or a stable lapse rate. Shear-driven turbulence lacks this persistent forcing, being driven instead by random pressure, viscous, and inertial forces. As a result one expects that buoyant forcing tends to give more sustained vertical velocities than found in shear turbulence. Thus, one might also expect that the Lagrangian integral time scale of vertical velocity, normalized with the Eulerian velocity and length scales, is larger in convective turbulence than in mechanical turbulence. However, since CBL turbulence is bounded in the vertical by the surface and the capping inversion, its Lagrangian vertical velocity is the derivative of a stationary function, so that its integral time scale is identically zero (Tennekes and Lumley 1972). It is clear, however, that at short times (i.e., before the boundary effects are felt) CBL turbulence does act as if it has a very large Lagrangian integral time scale (Lamb 1982; Weil 1985).

In classical Rayleigh-Benard convection between parallel plates the top surface is cooled at the same rate the bottom is heated; thus the heat flux, and hence the buoyant production rate of turbulence, is constant with height. Its vertical velocity field is therefore symmetric about the midplane. By contrast, the heat flux in the CBL is concentrated near the bottom, making its vertical velocity field strongly skewed and asymmetric about the midplane. The largest w-fluctuations are found in updrafts; the downdrafts tend to be weaker, compensating motions. Thus, another distinctive property of CBL turbulence is this asymmetry of its vertical velocity field.

I believe this combination of persistent, coherent vertical motions of apparently large Lagrangian integral time scale, and the vertical asymmetry of this velocity field, is responsible for the unusual properties of convective turbulence. Support for this comes from calculations with a simple updraft-downdraft model of the CBL (Wyngaard 1987). The model showed that differences between top-down and bottom-up diffusion can be generated by the interaction of the nonstationarity of the diffusion process with the skewness of the transporting vertical velocity field. Support also comes from Langevin model simulations of diffusion in the CBL (Sawford and Guest 1987) that show that systematic differences in the top-down and bottom-up diffusivities appear with an asymmetric vertical velocity field of sufficiently large Lagrangian integral time scale.

TWO CHALLENGES

New Experimental Techniques

I have argued that most of the progress we have enjoyed over the past 25 years stems from experiment. Many of the problems we face in boundary-layer meteorology today, however, severely test our experimental techniques. As we probe larger-scale aspects of the CBL, such as its interactions with cloud convection, its behaviour over regional scales, and its response to distributed inhomogeneities, it seems clear that our traditional tools, towers and in-situ sensors, will not be adequate. Research aircraft cannot carry this experimental burden alone; we need new tools.

One growth area here could be lower tropospheric profiling with a variety of ground-based remote sensors such as radars and lidars. They can yield profiles of wind, turbulence and possibly temperature, moisture, and some trace constituents, within the lowest few kilometres. They could be deployed in networks to achieve regional coverage. Eklund et al. (1988) have developed a boundary layer wind profiler that holds promise for such an application.

Another growth area could be simulation, both in the laboratory and on the computer. I have emphasized that much of our understanding of the CBL has come from convection tank and LES studies. It seems clear that much of the progress in short range dispersion modelling over the past 15 years has also come through that route, as Weil explains in detail elsewhere in this volume (Weil 1988).

Stochastic Variability

Turbulence theories have traditionally been posed in terms of ensemble-mean properties, but experimentalists have preferred to work with time averages. Under statistically stationary conditions the two converge, but one of the startling things we learned in the last 25 years of CBL research is how long an averaging time can be required for this convergence. In the 1968 Kansas experiments, for example, Haugen et al. (1971) found that one-hour-averaged fluxes were only constant to within 20% in the constant-flux layer. These variations are consistent with the theory of the approach of time averages to ensemble averages (Wyngaard 1973).

It follows, however, that the behaviour of the CBL over periods much shorter than this required averaging time is not well represented by its ensemble-average properties. Thus, in many problems involving convection in the lower atmosphere we need more than the "climate", or ensemble mean statistics; we also need information on the variations of a flow from realization to realization. One of the most difficult aspects of turbulence concerns what I might call this "stochastic variability" property.

As Weil discusses, this is now a pressing concern in air-quality modelling, where the most likely result (the ensemble average) is not all we want; we also want the expected variations about this mean. We don't know much about the physics here.

REFERENCES

Andreas EL (1987) On the Kolmogorov constants for the temperature-humidity cospectrum and the refractive index spectrum. J Atmos Sci 44: 2399-2406
Batchelor GK (1960) Homogeneous turbulence. Cambridge University Press
Brost RA, Wyngaard JC, Lenschow DH (1982) Marine stratocumulus layers. Part II: Turbulence budgets. J Atmos Sci 38: 818-836
Businger JA (1973) Turbulent transfer in the atmospheric surface layer. In: Haugen DA (ed) Workshop on micrometeoroloy. Am Meteorol Soc, Boston, pp 67-100
Coppin PA, Taylor KJ (1983) A three-component sonic anemometer/thermometer system for general micrometeorological research. Boundary-Layer Meteorol 27: 27-42
Dannevik WP, Yakhot V, Orszag SA (1987) Analytical theories of turbulence and the ε expansion. Phys Fluids 30: 2021-2029
Deardorff JW (1961) On the direction and divergence of the small-scale turbulent heat flux. J Meteorol 18: 540-548
Deardorff JW (1970a) A numerical study of three-dimensional turbulent channel flow at large Reynolds numbers. J Fluid Mech 41: 453-480
Deardorff JW (1970b) Convective velocity and temperature scales for the unstable planetary boundary layer and for Rayleigh convection. J Atmos Sci 27: 1211-1213
Deardorff JW (1972) Theoretical expression for the countergradient vertical heat flux. J Geophys Res 77: 5900-5904
Deardorff JW (1973) Three-dimensional numerical modeling of the planetary boundary layer. In: Haugen DA (ed) Workshop on micrometeorology. Am Meteorol Soc, Boston, pp 271-311
Deardorff JW (1974) Three-dimensional numerical study of turbulence in an entraining mixed layer. Boundary-Layer Meteorol 7: 199-226
Deardorff JW, Peterson EW (1980) The boundary-layer growth equation with Reynolds averaging. J Atmos Sci 37: 1405-1409
Deardorff JW, Willis GE (1987) Turbulence within a baroclinic laboratory mixed layer above a sloping surface. J Atmos Sci 44: 772-778
Eklund WL, Carter DA, Balsley BB (1988) A UHF wind profiler for the boundary layer: Brief description and initial results. J Atmos Ocean Tech, in press
Fiedler BH (1984) An integral closure model for the vertical turbulent flux of a scalar in a mixed layer. J Atmos Sci 41: 674-680
Fiedler BH, Moeng CH (1985) A practical integral closure model for mean vertical transport of a scalar in a convective boundary layer. J Atmos Sci 42: 359-363
Fitzjarrald DR, Garstang M (1981) Boundary-layer growth over the tropical ocean. Mon Wea Rev 109: 1762-1772
Haugen DA, Kaimal JC, Bradley EF (1971) An experimental study of Reynolds stress and heat flux in the atmospheric surface layer. Q J R Meteorol Soc 97: 168-180
Hannoun IA, List J (1987) Model for turbulent entrainment at a shear-free density interface. In: Preprints, Third International Symposium on Stratified Flows, Cal Tech, February 3-5, 1987
Kaimal JC, Wyngaard JC, Izumi Y, Coté OR (1972) Spectral characteristics of surface layer turbulence. Q J R Meteorol Soc 98: 563-589
Kaimal JC, Wyngaard JC, Haugen DA, Coté OR, Izumi Y (1976) Turbulence structure in the convective boundary layer. J Atmos Sci 33: 2152-2169
Kolmogorov AN (1962) A refinement of previous hypotheses concerning the local structure of turbulence in a viscous incompressible fluid at high Reynolds number. J Fluid Mech 13: 82-85
Lamb RG (1982) Diffusion in the convective boundary layer. In: Nieuwstadt FTM, van Dop H (eds) Atmospheric turbulence and air pollution modelling. Reidel, Dorchrecht, pp 159-229

Lenschow DH, Wyngaard JC, Pennell WT (1980) Mean-field and second-moment budgets in a baroclinic, convective boundary layer. J Atmos Sci 37: 1313-1326

Liepmann HW (1979) The rise and fall of ideas in turbulence. Am Scientist 67: 221-228

Lumley JL (1979) Computational modeling of turbulent flows. Adv Appl Mech 18: 123-176

Lumley JL, Panofsky HA (1964) The structure of atmospheric turbulence. Interscience, New York

Moeng CH, Wyngaard JC (1984) Statistics of conservative scalars in the convective boundary layer. J Atmos Sci 41: 3161-3169

Moeng CH, Wyngaard JC (1986) An analysis of closures for pressure-scalar covariances in the convective boundary layer. J Atmos Sci 43: 2499-2513

Mahrt L (1976) Mixed layer moisture structure. Mon Wea Rev 104: 1403-1418

Panofsky HA, Tennekes H, Lenschow DH, Wyngaard JC (1977) The characteristics of turbulent velocity components in the surface layer under convective conditions. Boundary-Layer Meteorol 11: 355-361

Priestly CHB, Swinbank WC (1947) Vertical transport of heat by turbulence in the atmosphere. Proc R Soc London A 189: 543-561

Sawford BL, Guest FM (1987) Lagrangian stochastic analysis of flux-gradient relationships in the convective boundary layer. J Atmos Sci 44: 1152-1165

Schumann U (1977) Realizability of Reynolds stress turbulence models. Phys Fluids 20: 721-725

Scorer RS (1985) A vivid mechanical picture of turbulence. In: Hunt JCR (ed) Turbulence and diffusion in stable environments. Clarendon Press, Oxford, pp 97-109

Stull RB (1984) Transilient turbulence theory. Part I: The concept of eddy-mixing across finite distances. J Atmos Sci 41: 3351-3367

Swinbank WC (1951) The measurement of vertical transfer of heat and water vapour and momentum in the lower atmosphere with some results. J Meteorol 8: 135-145

Tennekes H (1970) Free convection in the turbulent Ekman layer of the atmosphere. J Atmos Sci 27: 1027-1034

Tennekes H, Lumley JL (1972) A first course in turbulence. MIT Press, Cambridge, MA, USA.

Townsend AA (1976) The structure of turbulent shear flow. Cambridge University Press

Webb EK (1984) Temperature and humidity structure in the lower atmosphere. In: Brunner FK (ed) Geodetic refraction - effects of electromagnetic wave propagation through the atmosphere. Springer, Berlin, pp 85-141

Weil JC (1985) Updating applied diffusion models. J Climate Appl Meteorol 24: 1111-1130

Weil JC (1988) Atmospheric dispersion - observations and models. This volume, pp

Willis GE, Deardorff JW (1974) A laboratory model of the unstable planetary boundary layer. J Atmos Sci 31: 1297-1307

Willis GE, Deardorff JW (1976) A laboratory model of diffusion in the convective planetary boundary layer. Q J R Meteorol Soc 102: 427-445

Wyngaard JC (1973) On surface layer turbulence. In: Haugen DA (ed) Workshop on micrometeorology. Am Meteorol Soc, Boston, pp 101-149

Wyngaard JC (1980) The atmospheric boundary layer - Modeling and measurements. In: Bradbury JS, Durst F, Lauder BE, Schmidt FW, Whitelaw JH (eds) Turbulent shear flows 2. Springer-Verlag, Berlin Heidelberg, pp 352-365

Wyngaard JC (1983a) Lectures on the planetary bounday layer. In: Lilly DK, Gal-Chen T (eds) Mesoscale meteorology - Theories, observations, and models. Reidel, Dordrecht, pp 603-650

Wyngaard JC (1983b) The mean wind structure of the baroclinic, convective boundary layer. In: Reiter ER, Baozhen Z, Yongfu Q (eds) Proceedings of the First Sino-American Workshop on Mountain Meteorology. Am Meteorol Soc, Boston, pp 371-396

Wyngaard JC (1984) Large-eddy simulation: Guidelines for its application to planetary boundary layer research. Available from Defense Technical Information Center, Bldg 5, Cameron Station, Alexandria, VA 2304-6100, AD-A146381

Wyngaard JC (1985) Structure of the planetary boundary layer and implications for its modeling. J Climate Appl Meteorol 24: 1131-1142

Wyngaard JC (1987) A physical mechanism for the asymmetry in top-down and bottom-up diffusion. J Atmos Sci 44: 1083-1087

Wyngaard JC, Brost RA (1984) Top-down and bottom-up diffusion of a scalar in the convective boundary layer. J Atmos Sci 41: 102-112

Wyngaard JC, Clifford SF (1978) Estimating momentum, heat and moisture fluxes from structure parameters. J Atmos Sci 35: 1204-1211

Wyngaard JC, LeMone MA (1980) Behavior of the refractive index structure parameter in the entraining convective boundary layer. J Atmos Sci 37: 1573-1585

Wyngaard JC, Tennekes H (1970) Measurements of the small-scale structure of turbulence at moderate Reynolds numbers. Phys Fluids 13: 1962-1969

Wyngaard JC, Coté OR, Izumi Y (1971) Local free convection, similarity, and the budgets of shear stress and heat flux. J Atmos Sci 28: 1171-1182

CONVECTIVE PROCESSES IN THE LOWER ATMOSPHERE: COMMENTARY

E.K. Webb

CSIRO Division of Atmospheric Research, Private Bag No. 1, Mordialloc, Vic., 3195, Australia

Summary. The review by John Wyngaard in this Symposium (Wyngaard 1988) has presented a wide coverage of research findings for the convective boundary layer of the atmosphere. Much progress has been made in establishing relationships from experimental data. However, a satisfactory theoretical understanding of convective turbulence has not yet been achieved.

Here some further comments are made on the convective structure in the surface layer and in the convective boundary layer. Small-scale turbulence and plumes transfer heat from the ground into the surface layer. In a completely decoupled mode of convection, this heated surface layer as a whole gives rise to thermal walls arranged in an irregular polygonal net; these extend up through the CBL to heights of around 300 m where they channel the warm air into the bases of columnar thermals.

INTRODUCTION

In view of the leading part played by John Wyngaard in atmospheric boundary layer research, in particular for the convective boundary layer (CBL), we have been most fortunate that he could attend the Symposium and review progress in this field.

His extensive review has outlined the major advances in knowledge which have been achieved through well designed experimental observations and laboratory and computer simulations. Much progress has been made in establishing empirical relationships from experimental data, and this has led to considerable improvements in our ability to characterize the CBL in a variety of applications.

He has commented, however, that we have not yet reached a satisfactory theoretical understanding of convective turbulence and its associated "stochastic variability". In particular, he has referred to the process of entrainment at the top of the CBL and its effects extending down into the CBL. At present there are severe practical difficulties in the way of making sufficiently detailed measurements in the entrainment layer, and much more remains to be elucidated about the entrainment process.

For the future, he has emphasized the continuing importance of experiment, and the need for further developments in experimental techniques to attain a sufficiently comprehensive coverage. He envisages that there could be an increasing role for laboratory and computer simulation to aid the further development of our understanding of the CBL.

EVIDENCE ON CBL STRUCTURE

Progress of Research: the CBL and Entrainment

It is interesting to consider the stages of progress in CBL research by focussing on the work of John Wyngaard and his colleagues on a subject of continuing investigation, the entrainment at the top of the CBL and its effects.

Figure 8 of Wyngaard's review paper (Wyngaard 1988) reproduces observed C_T^2 with mixed-layer scaling, showing the very wide scatter caused by entrainment effects extending down through most of the CBL depth. (C_T^2 effectively represents the inertial-subrange part of the temperature variance.) For comparison a set of model results for the temperature variance $\overline{\theta^2}$ obtained by Wyngaard and colleagues is reproduced here in Fig. 1 (from Moeng and Wyngaard (1984) and Wyngaard (1985)). The curves show results of computer large-eddy simulations incorporating the top-down bottom-up formulation (Wyngaard and Brost 1984).

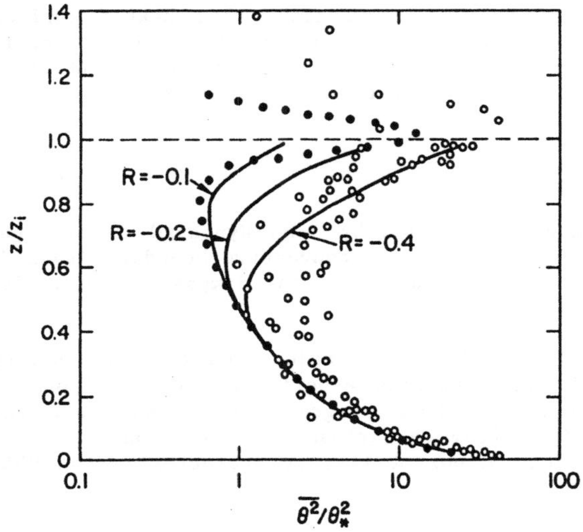

Fig. 1. Variance of temperature fluctuations normalized with mixed-layer free convection scale θ_*. R is the ratio of heat flux at the top of the CBL to that at the underlying surface. Points show laboratory data from Willis and Deardorff with R = -0.23 (filled circles) and atmospheric data from Caughey and Palmer with R between -0.1 and -0.4 (open circles). Curves show results of model calculations by Moeng and Wyngaard. (Moeng and Wyngaard 1984; Wyngaard 1985). (Note that the vertical scale is linear here but logarithmic in Fig. 8 of Wyngaard (1988)).

The diagram reproduced in Fig. 1 demonstrates in a practical way the various stages of research development:

(i) The success of the previously developed free-convection mixed layer scaling is reflected in the close fit to the data in the lowest part of the CBL, z/z_i up to about 0.1. Here z is height and z_i is the height of the inversion layer at the top of the CBL.

(ii) The main, and gratifying, point of Fig. 1 is that introduction of the top-down bottom-up formulation does successfully lead to major progress in representing the widely scattered values in the upper part of the CBL in terms of the ratio R (ratio of heat flux at the top of the CBL to that entering at the bottom).

(iii) Further research is needed to understand the entrainment process sufficiently well to express the heat flux at the top of the CBL, and thus R (as well as other characteristics of the entrainment layer), in terms of assignable quantities such as the surface heat flux H_o, and z_i, and the stability and wind speed in the overlying atmosphere.

(iv) The need for further research to formulate the detailed convection process through the whole CBL depth is apparent in some departure between calculated curves and data points in Fig. 1, especially around the middle heights of the CBL. This basic advance would necessarily be interactive with any realistic formulation of the entrainment process.

As pointed out by Wyngaard (1983, 1988), very few detailed sets of measurements have been made in the entrainment layer, owing to the severe practical difficulties involved. Some measurements combining balloon-cable sensors and Doppler-radar scanning have indicated the complicated and highly intermittent nature of the layer (Rayment and Readings 1974), with upcurrent-induced hummocks, entrained warm downcurrents, augmented wind shear and small-scale Kelvin-Helmholtz breakdown at the tops of the hummocks, and in some conditions (Carson and Smith 1974) overturning and engulfment of the hummocks themselves. A survey is given in the review by Caughey (1982).

Detailed measurements of this complicated three-dimensional structure must be far more difficult than those within the CBL itself, where the major convective elements have at least the simplification of being elongated in the vertical.

It is worth noting that the practical advance mentioned in (ii) has been achieved by the introduction of some physical insight, in the recognition of the difference in transfer relationships for upward and downward transfer. We may expect that successively more elaborate physical understanding will be essential for achieving the future goals of formulating the intra-CBL processes and (with still more difficulty) the complicated entrainment processes.

Convective Surface Layer Structure

Let us consider some observational evidence on the physical structure of the CBL, starting with the convective surface layer up to height $|L|$. The magnitude of the Obukhov length L is typically around 15-30 m in moderate winds over short grass.

The convective elements in the surface layer are the buoyant plumes evidenced by the familiar ramp (sawtooth) pulses seen in temperature trace recordings (Fig. 2). Detailed inspection of such traces recorded in the laboratory (Chen and Blackwelder 1978) and in the atmosphere shows that these plumes have their origins at quite low heights, $-z/L$ well below 0.03, where they are generated as part of the wind-shear turbulence (Webb 1984). Undoubtedly the plumes correspond to the "ejections" or "bursts" of the coherent structures which are a major feature of the turbulent shear boundary layer (e.g., Beljaars et al. 1981).

An interesting sidelight is that for the turbulent shear boundary layer the simple Prandtl similarity relationships work extremely well, despite their elementary derivation on a different basis from that now known involving the large-scale coherent structures. This simplification then carries over more generally to the Monin-Obukhov scaling for the buoyancy-stratified surface layer.

The significance of a value of the Richardson number Ri, or z/L, about -0.03 was first noted by Priestley (1955). He found from measurement data that as -Ri increases through 0.03 the heat flux-gradient relationship changes from the forced-convection to the free-convection form. These forms have potential temperature gradient proportional, respectively, to z^{-1} (the Prandtl relationship) and to $z^{-4/3}$ (free convection relationship due, independently, to Prandtl, Obukhov, and Priestley). See also Priestley (1959).

Measured potential temperature profiles do indeed show the transition from z^{-1} to $z^{-4/3}$ form of the gradient, and also show at greater heights (-z/L above about 0.3) an intermittent vanishing of the gradient over time intervals of up to several minutes (Webb 1958, 1982, 1984). This vanishing corresponds to the intermittent encountering of CBL air with its near-zero, usually mildly stable, stratification (see later). Wyngaard (1988) mentions that some temperature profile analyses show a negative height index greater than 4/3; it seems certain that this is the result of averaging when the actual profiles follow the $z^{-4/3}$ form with intermittent vanishing at the upper levels.

The onset of the $z^{-4/3}$ form at a value of -z/L as small as 0.03, i.e., in conditions far short of free convection, is believed to be explained by a process of composite convection in the $z^{-4/3}$ range (Webb 1962, 1984). Briefly, there are two opposing effects of wind-shear turbulence - initial generation of convective elements and then diffusive erosion of these rising elements - which remain in approximate counterbalance as they both diminish at increasing heights.

It is important to recognize the limitation of heights to which the buoyant plumes extend: they generally reach only up to about height -L, or a little greater. Thus, the plumes transport heat into the convective surface layer but not beyond.

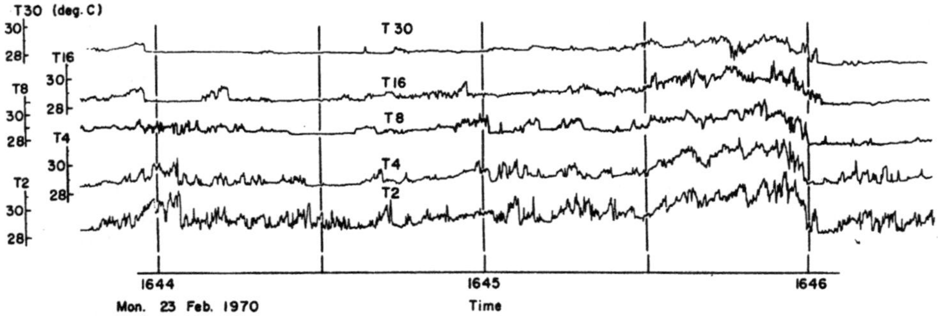

Fig. 2. Example of recorded fluctuations of temperature T at heights from 2 to 30 m as indicated. Measurements over flat dry grassland at Hay, Australia, February 1970. Conditions, for 30-min run 1634-1704: $U_2 = 4.5$ m s^{-1}, $Ri_2 = -0.17$, $L = -18.5$ m.

This termination of heat flux is strikingly apparent in the temperature-quiescent "flat bases" (Fig. 2). Their duration increases with increasing -z/L, and at a measurement height of 30 m it is not uncommon for the temperature to remain undisturbed over an interval of 2 or even 3 min. Over this interval there is clearly no convective heat flux to greater heights, despite the continuing heat input from the underlying surface.

In Fig. 2 the major temperature pulse feature between times 1645 and 1646 represents a different type of event from the plumes just discussed. It represents the passage of a convective line thermal or "wall" extending to heights far above the surface layer, as discussed below.

Role of the Surface Layer in the CBL

A poster presentation by the author in this Symposium[1] showed some results from an experimental investigation conducted over flat grassland at Hay, N.S.W., Australia, in February 1970. The objective was to gain better understanding of the convection process, in particular to elucidate how the surface layer (lowest 20 m or so) gives rise to the extended boundary-layer convection with thermal columns up to heights of 1 or 2 km. A brief outline has been given earlier (Webb 1984). After long postponement, the final stages of analysis are now proceeding. The temperature traces shown in Fig. 2 above are from this source.

As well as turbulent fluctuations and mean wind and temperature up to 32 m being recorded, thermals passing across the site were tracked by means of circling gliders at a height of around 800 m. In addition, the pattern of surface air flow was recorded by an array of seven wind speed-direction sensors (height 1.5 m) set in line across a 600 m span transverse to the wind direction; the derived horizontal convergence or divergence indicates the pattern of major ground-based upcurrents or downcurrents.

In moderate winds (up to about 5 m s^{-1} at height 2 m) the surface flow patterns indicate convective upcurrents in the form of walls about 50–100 m thick, arranged in an irregular net of polygonal cells about 3 km across. An example is shown in Fig. 3. The walls carry warm surface-layer air up through the CBL, as indicated by associated temperature pulses such as those seen terminating at time 1646 in Fig. 2.

The convection walls evidently extend up to heights of around 300 m, where, probably at the junctions of walls, they channel the warm air into the bases of the columnar thermals which extend up to the top of the CBL.

In the central areas surrounded by the cell walls, the subsiding cooler air is entirely devoid of temperature fluctuations, and is the source of the "flat base" temperature behaviour often observed down to heights of 8 m and even lower. This air is not itself affected by the heat input from the surface below, until it sinks into the surface layer where it encounters the population of plumes and turbulent fluctuations of temperature.

The whole multi-polygonal pattern travels in the general wind direction, probably at a speed which is nearly that of the wind in the CBL.

Other observations have shown features consistent with the polygonal cellular pattern - spaced temperature sensors at height 4 m showing temperature pulse line features oriented roughly transverse to the wind direction (Wilczak and Tillman 1980); dual-Doppler radar chaff observations showing cellular patterns

[1]EK Webb (1987) Modes of convection in the lower atmosphere. Symposium poster paper

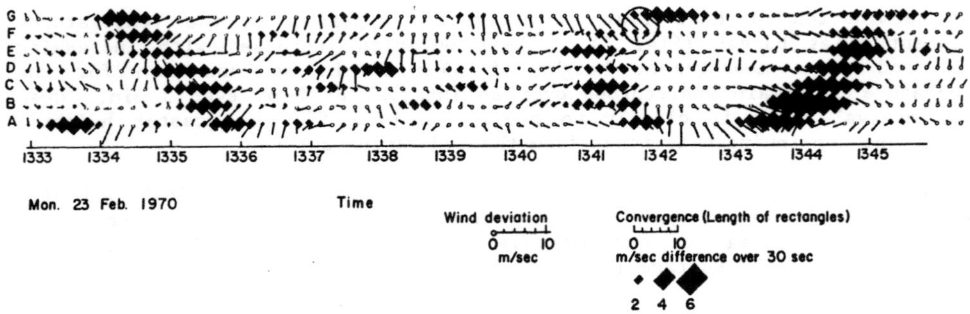

Fig. 3. Surface flow pattern during a run with mean wind speed $U_2 \simeq 4$ m s^{-1}. Time scale represents equivalent translated length, i.e. this is like a plan view of the strip of air passing across the 600 m span of the seven-instrument array. Vector deviations from mean wind are shown. Superimposed are black rectangles whose length indicates calculated convergence in directions at 45^0 to the mean wind direction. Circle shows position of glider indicating thermal passing above. (Webb 1984)

(Frisch et al. 1976; Kropfli and Hildebrand 1980); direct radar observations of insects carried up in the cell walls (Hardy and Ottersten 1969; Schaefer 1976); and visual observations of water vapour condensation patterns in cold air flowing over warm water (Lyons and Pease 1972).

In the case of laboratory convection in fluid layers between horizontal plates (no shear flow), the polygonal cellular form is observed near the upper and lower boundaries for Rayleigh numbers in the range 10^5–10^6 (e.g., Nielsen and Sabersky 1973). While the convection itself is apparently similar to that in the atmosphere, the surface layer which provides its buoyancy is governed by molecular transfer, whereas in the atmospheric case it is the Monin-Obukhov layer of turbulent transfer including the wind-shear generation.

In a laboratory model of Willis and Deardorff (1979) (again with no shear flow), the fluid (water) was heated from below and had a mixed region with a stable region above. They observed a convection pattern similar to that in the atmosphere, with open cellular, roughly polygonal, form near the lower surface, and thermal upcurrents rising mainly from the junctions.

From the observed structure in the atmosphere, we can conclude that the two modes of convection, surface layer and extended CBL, are completely decoupled as far as individual convective elements are concerned. The small-scale turbulence and plumes transfer heat from the ground into the surface layer, while this heated surface layer as a whole gives rise to the boundary-layer convection walls. The warm air flowing into the base of each wall carries with it its content of plumes, as can be seen in the data of Wilczak and Tillman (1980); the counterpart of this has also been noted by Willis and Deardorff (1979) in their laboratory model.

It seems likely that the width of the walls would be scaled to L, while the height at which they give way to columnar thermals would depend on their width and also their upward speed and their length between wall junctions, and hence on H_0 and z_i as well as L. But the height of the total CBL system of walls and columns is of course scaled to z_i, which leads to the simplicity of the usual CBL free-convection scaling.

This picture of decoupled atmospheric convection seems to shed light on a question posed by Wyngaard (1983, 1988). He has noted that for quite a range of $-z_i/L$ sufficiently large for free convection to be well established in the CBL, the turbulent kinetic energy generation by wind shear through the whole CBL depth nevertheless exceeds that by buoyancy. This seems to be resolved if we consider that nearly all the wind-shear production is in the lower part of the surface layer, where it is involved in the transfer of heat into the surface layer. The small-scale turbulence fluctuations generated there do not contribute to driving the CBL convection, but are simply carried up in the convection walls. It is then a case of free convection in the CBL, driven solely by buoyancy from the warm surface layer as a whole.

Convection Within the CBL

From the observational evidence such as that reviewed by Caughey (1982) and Webb (1984), there can be no doubt that in a well-developed CBL the main convective elements are columnar thermals. The width of these is likely to be scaled to the surface-layer Obukhov length L, since they arise from the polygon walls. However, as already mentioned, the height of the system of walls and columns is scaled to z_i.

The CBL environmental air through which the thermals rise usually has a mildly stable stratification. It is readily seen that this can arise from the entrainment of warmer air from above into the CBL.

However, Telford (1970) and Wyngaard (1987) have given an interesting argument, with theoretical formulations, for the steadily warming CBL even without entrainment. As the surface-layer temperature is increasing with time, the air entering a thermal upcurrent arising from it will be progressively warmer as time proceeds, and thus the air in the upcurrent will have potential temperature decreasing with height. Similarly the sinking environmental air will have potential temperature decreasing with decreasing height. Since the proportion of area occupied by thermals is only about 0.2 or so, updraught speed is greater than sinking speed, so the unstable potential temperature gradient in the updraughts will be smaller than the stable gradient in their environment. For this reason, as well as the small proportion of horizontal area occupied by the thermals, the horizontally averaged air is stably stratified.

A theory of thermal columns in the CBL, including mixing between the thermals and their environment, has been formulated by Telford (1972).

DISCUSSION: TURBULENCE IN THE CBL

Much of the discussion centred on the nature and role of turbulence in transport processes in the CBL. John Wyngaard (personal communication 1987) clarified what he meant in his lecture by the term "turbulence engineering" - the development of tools for predicting the *effects* of turbulence in real situations. He added that turbulence engineering has given us simple, effective models of turbulence (e.g., second-order closures) that can often be used successfully in design. These models are physically plausible parameterizations for the effects of turbulence, rather than statements about the *physics* of turbulence.

John Finnigan (personal communication 1987) asked about the role of approaches to turbulence engineering more sophisticated than large-eddy simulation (LES) - two-point closure, for example. He also pointed out that there has been a good deal of interest recently, at least among mathematicians, in deterministic chaos and the relevance of strange attractors to turbulence, and asked Wyngaard if this is what he meant when he asked for an investigation of the stochastic nature of turbulence.

On the first point, Wyngaard replied that the principal advantage of LES is its minimal reliance on parameterizations in calculating turbulence dynamics in the energy-containing range. By contrast, single-point (i.e. second-order) closures rely heavily on parameterizations, which can fail, as he discussed for scalar transport. Two-point closures are intermediate in this regard. From discussions with his colleague J Herring, Wyngaard further noted that because two-point closures are very cumbersome to apply to inhomogeneous problems, they are unlikely to see much direct use in turbulence engineering; they are more likely to be used in shaping and guiding the construction of single-point closures from databases assembled with LES and full turbulence simulation.

On the stochastic nature of turbulence, Wyngaard said that until now we have dealt primarily with ensemble-mean properties of turbulence; in traditional engineering applications this is all we needed. In atmospheric problems, however, we increasingly need information on the turbulence within individual realizations in order to predict turbulent dispersion episodes or short-term weather, or to understand the results of a given experiment. These needs go beyond the usual ensemble-mean statistics. As H Tennekes said in discussing the short-range forecasting problem, we need to begin forecasting forecast skill.

Peter Manins (personal communication 1987) reiterated that in the CBL the concept of an eddy diffusivity is inappropriate, because the turbulent motions are of comparable size to the layer. He noted that this reinforces statements made earlier in the Symposium regarding indiscriminate use of eddy diffusivities in soil solute transport and transport in plant canopies. Wyngaard added that while the failure of eddy diffusivity in the particular case he discussed (Wyngaard 1988) is due, he believes, to the unusual physics of convective turbulence, the message is clear: Transport in turbulence cannot be reliably assumed, in general, to be an eddy diffusion process.

Finally, on the point of the observed regularity in convection cells, John Finnigan reported that in a recent experiment designed to produce a climatology of gravity waves, he and his coworkers found that convective structures had a periodic signature that made them indistinguishable from waves generated in the lower boundary layer during stable times. They are very coherent events.

REFERENCES

Beljaars ACM, Prasad KK, de Vries DA (1981) A structural model for turbulent exchange in boundary layers. J Fluid Mech 112: 33-70

Carson DJ, Smith FB (1974) Thermodynamic model for the development of a convectively unstable boundary layer. Adv Geophys 18A: 111-124

Caughey SJ (1982) Observed characteristics of the atmospheric boundary layer. In: Nieuwstadt FTM, van Dop H (eds) Atmospheric turbulence and air pollution modelling. Reidel, Dordrecht Boston London, pp 107-158

Chen C-HP, Blackwelder RF (1978) Large-scale motion in a turbulent boundary layer: a study using temperature contamination. J Fluid Mech 89: 1-31

Frisch AS, Chadwick RB, Moninger WR, Young JM (1976) Observations of boundary-layer convection cells measured by dual-Doppler radar and echosonde, and by microbarograph array. Boundary-Layer Meteorol 10: 55-68

Hardy KR, Ottersten H (1969) Radar investigations of convective patterns in the clear atmosphere. J Atmos Sci 26: 666-672

Kropfli RA, Hildebrand PH (1980) Three-dimensional wind measurements in the optically clear planetary boundary layer with dual-Doppler radar. Radio Sci 15: 283-296

Lyons WA, Pease SR (1972) Steam devils over Lake Michigan during a January Arctic outbreak. Mon Weath Rev 100: 235-237
Moeng C-H, Wyngaard JC (1984) Statistics of conservative scalars in the convective boundary layer. J Atmos Sci 41: 3161-3169
Nielsen RC, Sabersky RH (1973) Transient heat transfer in Bénard convection. Int J Heat Mass Transfer 16: 2407-2420
Priestley CHB (1955) Free and forced convection in the atmosphere near the ground. Q J R Meteorol Soc 81: 139-143
Priestley CHB (1959) Turbulent transfer in the lower atmosphere. University of Chicago Press, Chicago
Rayment R, Readings CJ (1974) A case study of the structure and energetics of an inversion. Q J R Meteorol Soc 100: 221-233
Schaefer GW (1976) Radar observations of insect flight. In: Rainey RC (ed) Insect flight. Symp R Entomol Soc Lond, No 7, pp 157-197
Telford JW (1970) Convective plumes in a convective field. J Atmos Sci 27: 347-358
Telford JW (1972) A plume theory for the convective field in clear air. J Atmos Sci 29: 128-134
Webb EK (1958) Vanishing potential temperature gradients in strong convection. Q J R Meteorol Soc 84: 118-125
Webb EK (1962) Thermal convection with wind shear. Nature 193: 840-842
Webb EK (1982) Profile relationships in the superadiabatic surface layer. Q J R Meteorol Soc 108: 661-688
Webb EK (1984) Temperature and humidity structure in the lower atmosphere. In: Brunner FK (ed) Geodetic refraction: Effects of electromagnetic wave propagation through the atmosphere. Springer, Berlin Heidelberg New York Tokyo, pp 85-141
Wilczak JM, Tillman JE (1980) The three-dimensional structure of convection in the atmospheric surface layer. J Atmos Sci 37: 2424-2443
Willis GE, Deardorff JW (1979) Laboratory observations of turbulent penetrative-convection planforms. J Geophys Res 84: 295-302
Wyngaard JC (1983) Lectures on the planetary boundary layer. In: Lilly DK, Gal-Chen T (eds) Mesoscale meteorology - theories, observations, and models. Reidel, Dordrecht, pp 603-650
Wyngaard JC (1985) Structure of the planetary boundary layer and implications for its modeling. J Climate Appl Meteorol 24: 1131-1142
Wyngaard JC (1987) A physical mechanism for the asymmetry in top-down and bottom-up diffusion. J Atmos Sci 44: 1083-1087
Wyngaard JC (1988) Convective processes in the lower atmosphere. This volume, pp
Wyngaard JC, Brost RA (1984) Top-down and bottom-up diffusion of a scalar in the convective boundary layer. J Atmos Sci 41: 102-112

TURBULENT WAKES IN THE ATMOSPHERIC BOUNDARY LAYER

Peter A. Taylor

Atmospheric Environment Service, 4905 Dufferin St., Downsview, Ontario, M3H 5T4, Canada

Summary. The wakes behind moving bodies and the associated 'fluid resistance' were studied intensively by fluid dynamicists from about 1920 onwards. Many of their studies were stimulated by the rapidly evolving field of aeronautics. Detailed studies of flow past cylinders and other bodies led to some understanding of flow separation and the characteristics of the 'near wake', while boundary-layer approximations, conservation laws and linearisation of the perturbation momentum equations led to 'far wake' solutions for laminar and, with appropriate closure hypotheses, for turbulent wakes.

There are additional complications when we study the flow around and behind surface-mounted obstacles, including hills, imbedded within a pre-existing boundary layer, and much of the theoretical work on these flows is relatively recent. Wind engineers have a special interest in the detailed flow around buildings, especially at ground level where high winds can cause considerable discomfort or even danger to pedestrians. Wind-tunnel studies have been the usual technique for investigating these problems. Wind engineers and agricultural meteorologists have both been interested in windbreaks or shelterbelts and much of what we know about wakes in the atmospheric boundary layer stems from their studies. In the last ten years there has been renewed interest in wind energy and several studies of the wakes behind wind turbines.

This review will attempt to summarise results on wakes and drag for surface-mounted obstacles and hills., As for free wakes, theoretical results are somewhat easier to obtain, and rather more general, for the far wake region, well downstream of the obstacle itself. Momentum budget considerations lead to some interesting conjectures. Close to the obstacle, the detailed shape becomes important and quantitative results become rather 'obstacle-specific', and generally complex.

INTRODUCTION AND 'FREE WAKES'

The literature on wakes in turbulent boundary layers is fairly limited in total 'volume' but is spread over many years and is quite diverse. The subject matter has traditionally been of significant practical importance to agriculturalists, architects and wind engineers but such wakes are also of basic interest to fluid dynamicists as an example of a perturbed turbulent boundary layer. There is a close link with the study of boundary layer flow over hills, especially with respect to the diffusion and dispersion of pollutants and in the determination of the overall surface drag on the atmosphere in complex terrain. The diversity of the literature and the fact that some of it is rather 'grey' will undoubtedly mean that I have overlooked or simply not been aware of some important contributions. Let me apologise in advance to the authors of any such papers but hope that this brief review will serve to draw attention to a few aspects of the topic which, as far as I have been able to determine, are not fully resolved and could benefit from further research. Earlier review articles and reports which I have found particularly useful in preparing this paper include those by Jensen (1954), van Eimern et al. (1964), Lemberg (1973) and Hosker (1984).

Many of our ideas concerning wakes in the atmosphere boundary layer stem from the existing theory of wakes behind obstacles in a nonturbulent free stream. There are two basic differences to be considered in the boundary layer case. First, there is the effect of upstream turbulence and velocity shear which will influence growth rates etc., and second is the presence of the wall, which will need to be taken into consideration both as a barrier and as a momentum sink. In this respect there are some close parallels with the turbulent wall jet, which Launder and Rodi (1983) claim 'arguably provides more puzzles for those seeking an ordered set of rules to describe turbulence than any other class of shear layer'. To set the scene let us first review some of the fundamental ideas from 'free wake' theory.

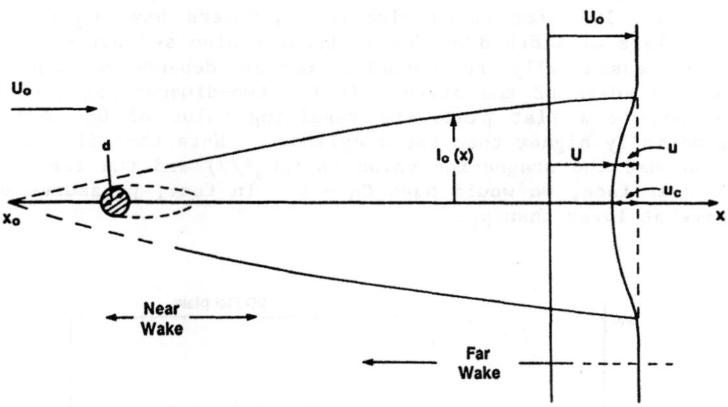

Fig. 1. Definition sketch for a 'Free Wake'

Batchelor (1967, p 348-353), or Schlichting (1968) derive the basic integral constraint for the momentum budget that, for flow parallel to the x-axis, as in Fig. 1,

$$\iint_{-\infty}^{\infty} (U_0-U)\,dy\,dz = Q \quad , \tag{1}$$

where Q is the wake 'strength'. The drag on the obstacle, D, in a uniform stream with velocity, U_0, is then given by

$$D = \rho U_0 Q \quad , \tag{2}$$

where ρ is the air density.

Determination of the magnitude of this drag was one of the prime objectives of early work, especially as it applied to cylinders and spheres in a uniform stream. Goldstein (1938) reviews many of these studies. Results are traditionally presented in terms of a drag coefficient, C_D, defined by

$C_D = D/(0.5\rho U_0{}^2 d)$ for a cylinder of diameter d, or

$C_D = D/(0.5\rho U_0{}^2 \pi a^2)$ for a sphere of radius a.

The drag D is per unit length in the case of the cylinder. For both spheres and circular cylinders there is a distinct pattern of behaviour with increasing Reynolds number, $Re = U_0 d/\nu$, where ν is the molecular kinematic viscosity; for air, $\approx 10^{-5}\ m^2s^{-1}$. The variations are caused in part by changes in the location on the body at which the flow separates. Drag coefficients are shown in Fig. 2, adapted from Batchelor (1967). Note that these are for smooth-surfaced bodies and that the variation with Reynolds number for rough spheres and cylinders will be modified, see, for example, Sachs (1978, p 79). The situations of interest in the atmosphere are usually for very high values of Re, i.e., $Re > 10^6$, for which circular cylinders have $C_D \approx 0.7$. For an infinite flat plate of width d or for a circular disc set perpendicular to the flow there is essentially no Reynolds number dependence since the flow separates at the edges of the plate. In the two-dimensional case of flow at normal incidence to a flat plate the resulting value of C_D is ≈ 2.0 (Sachs 1978), substantially higher than for a cylinder. Note that if the pressure on the front face had the stagnation value $(p_0+\rho U_0{}^2/2)$ and the free stream value (p_0) on the rear face, we would have $C_D = 1$. In fact, pressures on the rear face are somewhat lower than p_0.

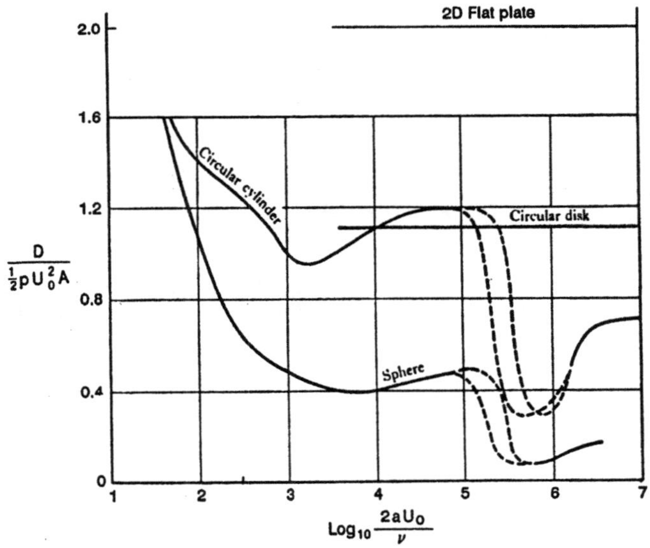

Fig. 2. Drag coefficients normalised by cross-sectional area for circular cylinders, spheres and discs. (Batchelor 1967). The value for a flat plate at normal incidence has been added.

For comparison with the wakes behind porous windbreaks, to be discussed later, we note that Castro (1971) and Graham (1976) discuss the wakes behind porous plates in a free stream. In this context we should formally define porosity, φ, as the ratio of open area to total area in an essentially plane fence or windbreak. For other windbreaks, e.g. hedges or rows of trees, this may be harder to define and a resistance coefficient may be more practical. Hoerner (1965, Ch. 3) defines a resistance coefficient, k_r, in terms of the pressure drop across a screen or windbreak if it were placed completely across a wind tunnel and empirically relates this to porosity. Hoerner also discusses the drag on finite sized circular porous disks in a free stream, where, of course, the flow can go both through and around the disk. Castro's drag coefficients for two dimensional porous flat plates show a steady increase in C_D as φ decreases to a maximum of about 1.85 at $\varphi = 0$. He finds two distinct flow regimes with a transition near $\varphi = 0.2$. At low porosity, vortex shedding dominates the wake while it is absent at higher values of φ.

The concept of a 'far wake' region, where the direct effects of the presence and detailed shape of the obstacle have effectively disappeared, is described in several standard texts, along with arguments for the use of boundary-layer approximations in the far wake. Townsend (1976), in particular, makes use of similarity and self-preserving flow ideas to obtain results for the rate of spread of the wake and the decay rate for the velocity perturbation. If x_0 is the effective origin for the far wake, then the wake 'thickness', ℓ_0, is found to increase at a rate proportional to $(x-x_0)^a$ while the centreline velocity deficit, u_c, decays as $(x-x_0)^b$. Values for a and b for two- and three-dimensional laminar and turbulent wake flows are given in Table 1. Note that Eq. (1) requires that $a + b = 0$ for the two-dimensional cases and $2a + b = 0$ for three dimensions. Table 1 also contains values for c, a power law coefficient for the rate of decay of the shear stress distribution with x. To obtain this we assume

$$\tau = (x-x_0)^c f(y/\ell_0, z/\ell_0) \quad . \tag{3}$$

For laminar flow with, in Cartesian coordinates, $\tau_{xz} = \nu \partial U/\partial z$, this leads to $c = b - a$. For turbulent, self-preserving flows we assume that all components of the perturbation Reynolds stress tensor show a similar behaviour. Self-preserving requirements for free wakes then lead to $c = 2b$.

Table 1. Power law coefficients for wake thickness, velocity deficit and shear stress in laminar and turbulent two- and three-dimensional free wakes in a uniform stream. Values of a and b are from Schlichting (1968) and Townsend (1976).

	Laminar Theory			Turbulent Flow		
	a	b	c	a	b	c
Two-dimensional	1/2	-1/2	-1	1/2	-1/2	-1
Three-dimensional	1/2	-1	-3/2	1/3	-2/3	-4/3

The simple results above will form a basis for illustrating some of the differences between wakes as free turbulent shear flows and as perturbations within pre-existing, neutrally-stratified turbulent boundary layers. We will find a much more complex situation in this case without even touching on the additional difficulties associated with the effects of thermal stratification which, as we all know, are often present in the atmospheric boundary layer. In their 1974 paper, Counihan, Hunt and Jackson stated that 'at present little is known about the wake behind surface (mounted) obstacles in turbulent boundary layers'. This was perhaps a slight exaggeration. Thirteen years later some progress has certainly been made but the flow situation remains a complex one and our understanding is still far from complete.

LOCAL FORCES ON SURFACE-MOUNTED OBSTACLES

By analogy with the free wake situation, some of the first questions to be asked about the flow past surface-mounted obstacles concern the drag. How large is it and at what point can it be considered to act? To answer the second part of the question requires a knowledge of the couple exerted by the flow on the obstacle, usually thought of in terms of the moment about the point of contact with the surface. For simple obstacles, trees, fences, buildings, etc., these are fairly straightforward questions but for a hill, which can for some purposes be viewed as a surface-mounted obstacle, it may be difficult to define the horizontal extent of the obstacle and to separate the drag on the obstacle from the surface shear stress on the surrounding area.

Windbreaks

For simple two-dimensional solid or porous fences or windbreaks, of height h, Plate (1971) reviewed some of the information on drag coefficients available at that time. The choice of scaling velocity for the drag coefficient is not unique in a sheared flow, $U_0(z)$, and two alternatives are commonly used. They are

$$C_h = D/(0.5\rho U_0{}^2(h)h) \tag{4}$$

or

$$C_* = D/(0.5\rho u_*{}^2h) \quad , \tag{5}$$

where D is the drag/unit length on the obstacle. For neutral thermal stability and geometrically similar windbreaks, Plate argues that C_* should be a function of (h/z_0) only, where z_0 is the surface roughness length for the upstream flow. Plate appears to favour the use of C_* over C_h but they are simply related via

$$C_h = C_*\kappa^2/(\ell n(h/z_0))^2 \quad , \tag{6}$$

where κ (=0.4) is the von Karman constant, so that, as Seginer and Sagi (1972) note, C_h should likewise be a function of (h/z_0). In fact, the dependence of C_h on h/z_0 appears to be fairly weak (see Seginer and Sagi and also Wilson's (1985) numerical model calculations) and it is perhaps the more practical of the two coefficients. For porous windbreaks C_h is, of course, a function of porosity. In this context we should note that Perera (1981) found that the form of construction was not important so that for uniform windbreaks the single parameter, φ, will presumably suffice.

For $\varphi = 0$ Sachs (1978, Table 3.2) gives $C_D = 1.2$ for a flat plate with one edge on the ground but it is not clear whether this is within a fully developed boundary layer. Good and Joubert (1968) conducted an extensive study of smooth-wall turbulent boundary-layer flow past wall-mounted flat plates. They found that the drag results were most consistent in terms of C_* but indicate values of 1.0 to 1.1 for C_h. Raine and Stevenson (1977), however, found a value of 1.3 to 1.4 in wind-tunnel tests with a smooth wall. Ranga Raju et al.

(1976) reanalysed Good and Joubert's data to account for blockage effects and also conducted extensive experiments of their own which appear to show that C_* can, for many purposes, be considered a function of h/z_0 only. However, Castro and Fackrell (1978) questioned this and were unable to obtain the same result in their own experiment. Converting from Ranga Raju et al.'s C_* values to C_h gives values varying between 1.18 at $h/z_0 = 10^2$ to 0.95 at $h/z_0 = 10^5$. Typical windbreaks would have h/z_0 between 10^2 and 10^3 for which $C_h \approx 1.1$ (with $\varphi = 0$).

In a paper on stability effects on windbreak shelter and drag, Seginer (1975) reports C_h values for a 50% porosity windbreak with $h/z_0 \approx 600$. For neutral stability he gives a value of 0.77. There is possibly a slight stability dependence, but it is weak and rather less than the scatter in the observations. Seginer also presents wind-tunnel results obtained by Tani and by de Bray for a range of porosities. These are shown in Fig. 3. Tani's results were originally given in terms of a drag coefficient based on the average wind speed over the height of the windbreak. Here they have been adjusted to C_h assuming a 1/7 power law profile and they then match de Bray's results reasonably well. Wilson (1985) has recently made some numerical model calculations of flow through windbreaks and presents results on the variation of drag coefficient with resistance coefficient and with h/z_0. His definitions of both coefficients are nonstandard. We can, however, note that the variation with h/z_0, over the range studied (100-1200), is quite small but shows C_h increasing with h/z_0. We have adjusted Wilson's C_F and k_r definitions and related k_r to porosity using Hoerner's equation, slightly stretching its intended range of applicability. The results for $h/z_0 = 600$ have been added to Fig. 3. We have also added a short line based on Ranga Raju et al.'s (1987)

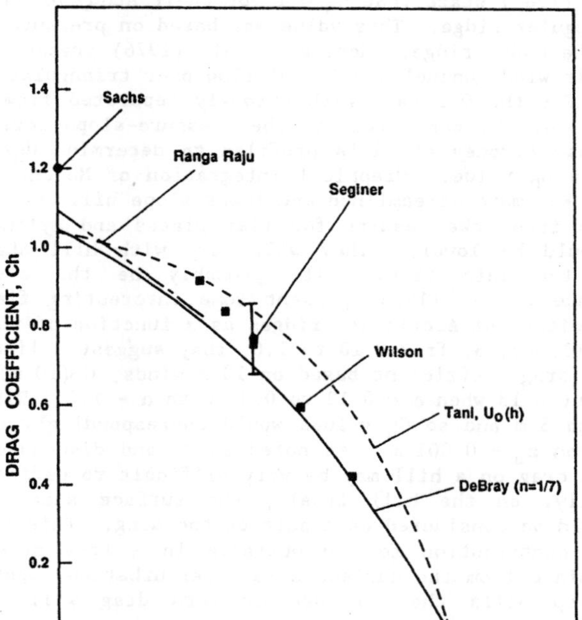

Fig. 3. Results for the drag coefficient of a porous windbreak. Adapted from Seginer (1975) with adjustments to Tani's results and additional data from Wilson (1985) and Ranga Raju et al. (1987).

recent work where they find a dependence of drag coefficient on porosity of the form $C_D(\varphi) = C_D(\varphi=0)(1+\varphi)^{-1}$ for porosities up to about 0.4. The result would clearly be inappropriate for higher porosities since we would expect $C_D = 0$ for $\varphi = 1$. In general, there appears to be a reasonable level of agreement between wind-tunnel results, Wilson's numerical model predictions and Seginer's field measurement. There is perhaps a need for additional field verification at different porosities but the situation seems reasonably satisfactory. The simple approximation,

$$C_h = 1.2 - 0.4\varphi - 0.8\varphi^2 \quad , \tag{7}$$

provides a reasonable fit to most of the results shown in Fig. 3.

For a free wake the 'strength', Q, was simply related to the drag through Eq. (2). In the surface-mounted case, Counihan et al. (1974) suggest that the wake strength is more closely related to the couple exerted by the fluid on the body. Seginer's (1975) study also measured the couple acting on the windbreak. These results showed that the effective height of application of the drag force for a uniform windbreak of 50% porosity was at 0.55h. Similar results for other porosities might be useful.

Two-Dimensional Hills and Ridges

Information on drag coefficients for neutrally-stratified flow over hills or ridges appears to be very limited, despite several excellent wind-tunnel studies of the flow over idealised two- and three-dimensional model hills. Arya and Shipman (1981) state that $C_h \approx 1$ for their study of flow over a steep (slope = 2) triangular ridge. This value was based on pressure measurements on the surface of the model ridge. Meroney et al. (1976) present static pressure results from their wind-tunnnel studies of flow over triangular ridges of slope 0.167 and 0.5. For the 0.5 case with strongly separated flow the C_p results are shown in Fig. 4. We can integrate the pressure-slope correlation to give the form drag, use Meroney et al.'s profiles to determine U_h/U_0 (= 0.73) and hence calculate a C_h value. Graphical integration of Meroney et al.'s curve gives $C_h = 1.1$. For more streamlined and lower slope hills we would expect, by analogy with the free wake results for flat plates and cylinders, that drag coefficients should be lower. They will vary with hill shape and surface roughness, but the main factor will probably be the steepness of the topography. Banke et al. (1976) present some interesting field data on the form drag coefficients of Arctic ice ridges as a function of their slope. For a range of mean slopes, α, from 0.18 to 1.0, they suggest a linear relationship between the form drag coefficient based on 10 m winds, $C_F(10\text{ m})$, and α with C_F varying from about 0.13 when $\alpha = 0.18$ to 0.55 with $\alpha = 1.0$. Ridge heights vary from about 1 m to 3 m and so C_h values would correspondingly range from $1.8C_F$ to $1.3C_F$, based on $z_0 = 0.001$ m. As noted above and discussed further below, the total or net drag on a hill may be very difficult to define and determine. Even very locally, on the hill itself, the surface shear stress will be modified and could be considered as a part of the drag. This is similar to the 'skin friction' contribution for an obstacle in a free stream, but is now difficult to isolate from the surface stress perturbations upstream and in the wake. For steep hills the pressure or form drag will be the dominant contribution.

Taylor and Gent (1974) and Taylor (1977) present numerical-model results of form drag for nonseparating flow over various two-dimensional gentle topographies. Table 2 shows drag coefficients, C_h, computed from their results for 'Conformal' and 'Gaussian' hills with characteristic horizontal length scale, $\beta = 5000z_0$. The flow is 'just attached' for the steepest cases shown. Since these have $C_h \sim 0.1$ compared to $C_h \sim 1.0$ for steep triangular ridges,

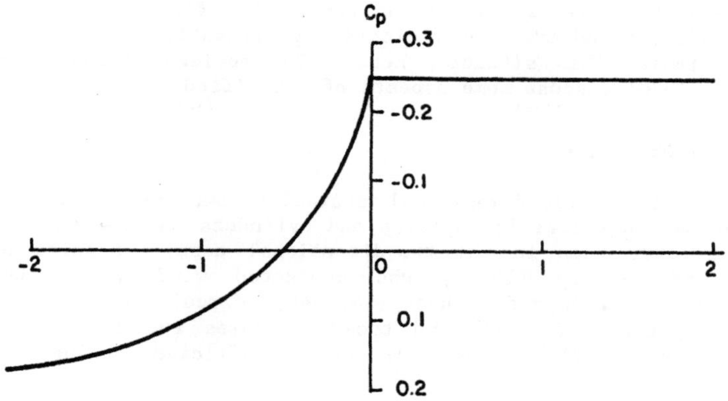

Fig. 4. Static pressures over the surface of a triangular 1:2 ridge. (Meroney et al. 1976).

there does seem to be a need for additional study of the situation. This is especially true for determining drag coefficients or effective roughness lengths to represent hilly and mountainous terrain in Numerical Weather Prediction and Global Circulation models. Mason (1986) and Carson (1986) discuss this aspect and suggest that for topographies with representative slopes < 0.2, drag coefficients can be estimated from linear theories (which will give results similar to those of Table 2), while for slopes > 0.2 they propose, 'from an order of magnitude point of view',

$$C_D \approx 0.3 \quad \text{for} \quad 0.2 < \text{slope} < 1.0$$
$$\approx 0.7 \quad \text{for} \quad \text{slope} > 1.0$$

In this context they define C_D in terms of the wind speed at height h/2, and appear to suggest that the estimates will be valid for both two- and three-dimensional topographies. For low and intermediate slopes their suggestions are based in part on Newley's (1985) model computations.

Table 2. Form drag coefficients, C_h, for flow over idealised two-dimensional gentle hills. Based on results from Taylor (1977).

Hill Shape	Maximum Slope	C_h(Form Drag Only)
'Conformal Hill'	0.07	0.02
	0.14	0.04
	0.32	0.09
'Gaussian Hill'	0.09	0.03
	0.17	0.06
	0.35	0.12

If the upstream flow is stably stratified, the flow situation can be substantially modified and drag coefficients significantly increased. We will not attempt to review this situation here. The reviews by Hunt (1980) and Taylor and Gent (1980) discuss some aspects of stratified flow over hills.

Three-Dimensional Obstacles

Drag coefficients for three-dimensional obstacles can, by analogy with the differences between C_D values for spheres and cylinders in free wake flows, be expected to be lower than in the two-dimensional case. Results available include those of Lemberg (1973), who conducted wind-tunnel studies on boundary-layer wakes behind model cubes (S4, S4D), circular cylinders (C4, C6) and rectangular prisms (S6, S6D) and those on domes and tents reported by Johnson and Surry (1987). Lemberg's drag coefficient values, D^*, are defined by

$$D^* = \mathrm{Drag}/(\rho U_0 U_h A) \quad , \tag{8}$$

where A is the frontal area of the obstacle and U_0 is the free stream velocity in the tunnel; they are given in Table 3. Note that S4D and S6D are results for the cube and prism mounted diagonally to the flow.

Table 3. Drag and moment coefficients from Lemberg's (1973) study of wakes behind bluff bodies in turbulent boundary layers. C_h coefficients have been added. Re = 10^5, $h_{cp}/h = M^*/D^*$ = normalised height of centre of pressure.

Model	S4	S4D	C4	S6	S6D	C6
M^*	0.26	0.24	0.15	0.28	0.25	0.16
D^*	0.34	0.29	0.18	0.37	0.32	0.20
h_{cp}/h	0.76	0.82	0.83	0.76	0.78	0.80
C_h	0.89	0.76	0.47	0.91	0.79	0.49

We have also computed C_h values in accord with the definition given in Eq. (4) using Lemberg's velocity profile and boundary-layer depth information. The Reynolds number for these studies, based on the frontal width of the obstacles, was 10^5. Comparison with Fig. 2 shows that the corresponding drag coefficient for a smooth two-dimensional circular cylinder in a free stream at this Reynolds number is 1.2, and that the flow is in a 'subcritical' range of near-constant drag coefficient. The flow Reynolds number in Lemberg's and other circular cylinder studies must be considered in applications to atmospheric situations, as pointed out recently by Sabransky and Melbourne (1987), but the sharp-edged model values should be directly applicable. Similar results for a range of flow orientations relative to a cube and for a range of h/δ values (where δ is the boundary-layer depth) are reported by Sakomoto and Arie (1982), using slightly different definitions. Lemberg's results are for $\delta/h = 3$ or 4, approximately 1/7 power law velocity profiles and equivalent $h/z_0 = 2000$ or 3000. These conditions should be approximately correct for many full scale applications. Table 3 also gives values for the height of the centre of pressure, h_{cp}, which is $\approx 0.8h$ in all cases. This is substantially higher than that noted earlier for a 50% porous windbreak (0.55h).

Johnson and Surry (1987), in a study of wind loads on tents, present some information on drag coefficients for surface-mounted hemispheres, semi-cylinders and domes in addition to specific tent designs. Table 4 contains information extracted from their report. They ran wind-tunnel tests at Re ~ 10^5 for all models and at Re ~ 10^6 for the H = R/2 dome. The hemispheric dome result, C_h ~ 0.8, is clearly high and strongly affected by the Reynolds number. The somewhat square CF TEMS tent has a C_h value compatible with Lemberg's cubes (0.89), the semi-cylindrical arch-roof and relatively low CF 10 tent give lower values and the H = R/2 dome (at high Re) offers the least air resistance of the models studied.

A systematic wind-tunnel investigation of the drag at high Re (say > 10^6) on surface-mounted, rough-surfaced domes, cones and pyramids would be a useful aid to parameterising the drag on small, but maybe steep, hills and other topographic features. Nothing very dramatic may come from such an investigation but it would be nice to know the numbers with a little more certainty than we appear to at present.

Table 4. Drag coefficients for tents, arch-roof and domed structures. Data from Johnson and Surry (1987). C_D' is based on plan area, A_p, and roof height wind speed, U_h. A_* is the cross-sectional area normal to the wind.

Structure	C_D'	A_*/A_p	C_h
CF TEMS tent	0.66	~0.71	0.94
CF 10 tent	0.22	~0.46	0.48
Semi-Cylindrical Arch-Roof, L=D	~0.2	0.5	0.4
Hemispherical Dome, H=R, Re≈10^5	~0.4	0.5	0.8
Dome, H=R/2, Re=4×10^4	~0.18	0.22	~0.8
Dome, H=R/2, Re=10^6	~0.06	0.22	~0.27

NET FORCES

My own interest in boundary-layer wakes started when I read a paper by Seginer and Sagi (1972) in which they attempted to calculate the drag on a windbreak from the momentum budget for a control volume surrounding it. They argued that the drag on the windbreak would be balanced by surface shear stress reductions upstream and in the wake and that they could thus calculate the drag on the windbreak if they could determine the surface stress distribution from a knowledge of near-surface winds. A corollary of their method is that a single windbreak or other surface-mounted obstacle would have no net effect on the areally-averaged horizontal force between the atmosphere and the earth's surface. I found this unlikely but it took me some while to resolve the issue. Let us look at it in a little more detail.

Figure 5 illustrates the situation for a two-dimensional obstacle, which could be solid or porous, mounted on the surface beneath a constant-stress turbulent boundary layer above an otherwise uniform, flat surface. If the control volume is taken to be sufficiently large that the flow at the outflow boundary has readjusted to the same logarithmic, contant-stress layer that existed upstream, and the upper boundary is sufficiently high that the shear stress there is not affected, then Seginer's hypothesis, and the corollary, would appear to hold. Counihan et al. (1974) presented essentially this argument and obtain the result

$$D + \int_{-\infty}^{\infty} (\tau-\tau_0)_s dx = 0 \quad . \tag{9}$$

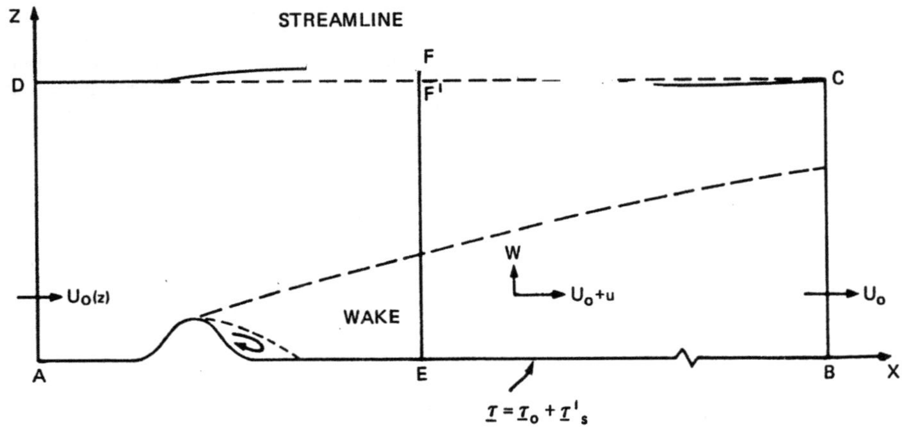

Fig. 5. Definition sketch for control volume with surface-mounted obstacle.

Are they correct or is there a flaw in the argument? Perhaps we should say that they may be completely correct in the limit as the ratio of boundary-layer depth to obstacle height approaches infinity but in most practical atmospheric situations the finite depth of that portion of the boundary layer which can be considered as having constant stress requires us to reassess the assumptions. Taylor (1983) used various models for eddy viscosity variation in the wake to show that the momentum flux perturbation decayed extremely slowly with distance downstream, as Seginer and Sagi had suspected. This was confirmed by Seginer's (1972) more detailed calculations and is also pointed out by Townsend (1965a). As a result the outflow boundary of the control volume described above should often be tens or even hundreds of kilometers downstream of the obstacle if equation (9) is to hold. On these scales Coriolis forces and transverse flow components need to be taken into consideration and Eq. (9) has to be replaced by

$$N_x - D_x + \int_A^B (\tau_x')_s dx = \iint_{ABCD} fv dz dx$$

$$N_y - D_y + \int_A^B (\tau_y')_s dx = - \iint_{ABCD} fu dz dx \quad , \tag{10}$$

which include terms involving the Coriolis parameter, f. Here D_x and D_y are the components of the local drag force on the obstacle (D_y will be 0 in the simple two-dimensional case) and N_x, N_y is the net change in the total horizontal force on the surface due to the presence of the obstacle. Note that N_y may be non-zero even in the case of a two-dimensional obstacle. The surface stress perturbations, $(\tau_x')_s$ and $(\tau_y')_s$, are perturbations relative to those applying for an equilibrium planetary boundary layer (PBL), i.e.

$$(\tau_x)_s = f \int_0^\infty (V_0-V_g)dz \quad ; \quad (\tau_y)_s = -f \int_0^\infty (U_0-U_g)dz \quad , \tag{11}$$

where $U_0, V_0(z)$ are the equilibrium velocity components and U_g, V_g are the components of the geostrophic wind.

The paper by Taylor (1983) includes computations with a linearised two-dimensional PBL wake model and estimates of the fraction of the drag on a surface-mounted obstacle which is balanced by shear stress reduction in the wake. The results are dependent on the form assumed for the initial wake profile just downstream of the obstacle, and on the turbulence closure model used, but should give at least some indication of the overall stress increase to be expected. This depends on a number of factors, but for a typical roughness Rossby number, $Ro = U_g/fz_0 = 10^7$, and obstacles about 10 m high, 57% of the drag was balanced by surface stress reduction. Table 4 shows similar results for a range of Ro values. The initial wake velocity deficit profile is assumed to be of the form

$$u = -\alpha U_0 e^{-kz} - \delta dU_0/dz \quad ; \quad v = -\beta V_0 e^{-kz} - \delta dV_0/dz \quad , \tag{12}$$

where U_0 and $V_0(z)$ are the upstream, undisturbed velocity components, α and β are scaling constants and $\delta = \alpha(1-e^{-kz})/k$ is the streamline displacement. The initial depth of the wake is characterised by k^{-1}. The results given in Taylor (1983) are all for relatively deep initial wakes ($k^{-1} > 10$ m) for direct application to small windbreaks and fences but the model could, in principle, be run for shallower wakes to extend the range for which results are available.

Table 4: Percentages of the drag on surface-mounted obstacles which are balanced by surface shear stress reductions in the wake. Based on results from Taylor (1983). z_0 and k^{-1} values are for $U_g = 10$ m s^{-1} and $f = 10^{-4}$s^{-1}. The geostrophic wind is parallel to the x-axis.

		k^{-1}, m			
z_0, cm		100	50	25	10
		k Ro			
	Ro	10^3	2×10^3	4×10^3	10^4
10	10^6	0.44	0.55	0.63	
1	10^7	0.35	0.44	0.51	0.57
0.1	10^8	0.27	0.36	0.42	0.46

FAR-WAKE BEHAVIOUR

In the far wake region, which can generally be thought of as $x > 10h$ for two-dimensional obstacles or perhaps $x > 5h$ in some three-dimensional cases, where h is the obstacle height, we are interested in the asymptotic, self-preserving forms for the velocity deficit and turbulence perturbations, if indeed they are self-preserving, and in the growth rate for the wake, usually expressed in the form of power laws, as indicated for free wakes above. Two-dimensional and three-dimensional wakes will again have different behaviours; we will consider two-dimensional wakes first. To focus our ideas we can also consider only wakes behind small or moderate sized obstacles, which may of course be porous, and assume $z_0 \ll h \ll \delta$, where z_0 is the surface roughness and δ is now the PBL depth. If we regard 1 km as a typical value for δ and 1 cm as a typical roughness length, then $1 \text{ m} < h < 10 \text{ m}$ will meet these criteria nicely. In the present context we are not so concerned with the overall momentum budget and can afford to ignore Coriolis effects and consider the idealised case of wake perturbations within a constant flux surface layer. The range of downstream distance that we are considering here will typically be 5h < z < 100h, although at x = 5h we may not yet be into the far-wake region, especially in the case of a solid two-dimensional obstacle.

Let us assume then that our undisturbed upstream flow has a constant stress ρu_*^2, and a surface roughness z_0. The next statement would normally be that the corresponding velocity profile is logarithmic and of the form

$$U_0 = (u_*/\kappa)\ell n((z+z_0)/z_0) \quad . \tag{13}$$

In some cases, however, it has been mathematically expedient to assume a power-law profile of the form

$$U_0 = u_1(z/z_1)^n \tag{14}$$

with a corresponding eddy viscosity

$$K_0 = k_1(z/z_1)^m \tag{15}$$

in order to obtain simple analytic solutions. For a constant stress layer the 'conjugate power law' requires m + n = 1.

Analyses using power law profiles have been carried out by Lemberg (1973), Counihan et al. (1974) and Taylor and Richards (1979). Both Lemberg and Counihan et al. introduce the notion that the 'couple'

$$M = \int_0^\infty zuU_0(z)dz \tag{16}$$

is a constant for the flow (i.e., independent of x), equal to the base moment acting to overturn the obstacle (see also Hunt 1971). This is true if m = 0, which Counihan et al. assume within their 'mixing' region, but is not satisfied exactly in general. With m = 0, Counihan et al. are unable to satisfy the conjugate power law. Lemberg (1973) also selects m values which do not necessarily meet this criterion and further argues that m = n (in the present notation) on the basis of comparing his three-dimensional wake data to the model predictions. We would argue that m + n = 1 is a condition which should be retained for full-scale atmospheric boundary-layer wakes and that the analysis given by Taylor and Richards (1979) (see also Taylor 1983) is the most appropriate. In place of Eq. (10) above, they argue that

$$\int_0^\infty (z^{n+1-m}\, u - z^{n-m}\, m\, \Psi)dz = \text{constant} \quad , \tag{17}$$

where Ψ is a nondimensional stream function, should be used as the integral constraint. They show how similarity-variable solutions of the form $\Psi = x^d f(\eta)$, where $\eta = z/x^a$, exist provided $a = 1/(2+n-m)$ and $d = -a(1+n-m)$. The corresponding velocity perturbation will be of the form

$$u = x^{d-a}\, f'(\eta) = x^{-1}\, f'(\eta) \quad , \tag{18}$$

while the surface shear stress perturbation will have c = am-1-a. For the traditional values of n = 1/7 and m = 6/7 these results give power law coefficients a = 7/9, b = -1 and c = -10/9, substantially different from the free wake case. Note also that we no longer have a + b = 0 and, in fact, the momentum flux deficit decays slowly with x.

Turning to other theoretical estimates of a and b for boundary-layer wakes, Taylor (1983) notes that Townsend's (1965a,b) analyses, which are based on the assumption of self-preserving flow perturbations, lead, for large x and after some additional assumptions, to a length scale, ℓ_0, which satisfies

$$\ell_0 \; \ell n(\ell_0/z_0) = 2\kappa^2 x \tag{19}$$

(cf. Jackson and Hunt's 1975 inner-layer depth equation for flow over hills). In addition, the 'wake component' velocity perturbation is shown, in the simplest case, to behave asymptotically as

$$u = x^{-1}(\ell n(\ell_0/z_0))^{-1} \; f(z/\ell_0) \quad . \tag{20}$$

We can loosely interpret this as corresponding to a = 1-ε, b = -1-ε, where ε is small. Counihan et al. (1974) also find b = -1 in their 'mixing regime' but in contrast they obtain a = 1/(n+2). For n = 1/7 this gives a = 7/15, distinctly different from the other two theories discussed above. The difference is primarily due to Counihan et al.'s assumption of an effectively constant eddy viscosity, i.e., m = 0, within their mixing region.

Table 5. Power-law coefficients for wake thickness, velocity deficit and stress perturbations in turbulent two-dimensional boundary-layer wakes.

Theories	a	b	c
Taylor and Richards (n = 1/7, m = 6/7)	7/9	-1	-10/9
Townsend	1-ε	-1-ε	
Counihan et al. (n = 1/7, m = 0)	7/15	-1	-22/15
Lemberg (m = n = 1/7)	1/2	-15/14	-20/14
Wilson (numerical model results)	1/2	-1	
Field Observations			
Bradley and Mulhearn (1983)	1/2	-1	-5/2
Jacobs (1984)	0.4*	-0.91	
Wind Tunnel			
Castro (1979)	0.45	-1	~-1
Perera (1981)	1/(n+2)	-1	-1
Arya and Shipman (1981)	1/2	-1	-1
Counihan et al. (1974) (n = 1/8)	8/17	-1	~-3/2

Values of a, b and c from these and other theories are listed in Table 5, together with estimates from field observation and wind-tunnel studies. There is general agreement that the value of b (for the decay of the velocity perturbation) is ≈ -1; that there are two possibilities (≈ 0.5 or ≈ 1) for a (wake depth); and that values for c range from -1 to -5/2. The -5/2 estimate was introduced by Bradley and Mulhearn (1983) based on Fig. 7 of their paper. Alternative interpretation of that figure could lead to an estimate closer to -1.5, in better agreement with other results. The determination of the 'correct' values of a, b and c and the resolution of possible differences between wind-tunnel and full scale values should be a fairly simple matter but it seems to be causing some difficulty.

For three-dimensional wakes there is a distinct 'signal to noise' problem with observations, especially in the field, since the perturbations decay more rapidly than in the two-dimensional case. In principle, we could assume nonisotropic wakes with self-preserving forms which were scaled with separate y and z lengths, ℓ_{0y} and ℓ_{0z}, with corresponding power law coefficients a_y and a_z. Lemberg's (1973) study of three-dimensional wakes, however, gave $a_y = a_z = 0.5$. He also found b = -1.6, in good agreement with Hunt's (1971) value of -1.5, but a relatively slow decay rate for turbulence intensity perturbations with a power law coefficient of -0.3 along the y = 0 centreline. However, Arya and Gadiyaram (1986), in their detailed wind-tunnel study of the wakes behind cones, found decay rates for both mean velocity and turbulence perturbations with power-law coefficients of ~-1.6. Their wake depths, represented by the height of perturbation maxima, lead to a_z values of 0.35 and 0.49 for the steep (slope = 0.5) and shallow (slope = 0.3) cones respectively.

NEAR-WAKE BEHAVIOUR

Although the far-wake may be more amenable to analysis, for most practical applications we are often more keenly interested in the region fairly close to the obstacle. For solid, surface-mounted obstacles or hills the occurence of flow separation is often of particular concern, while for windbreaks and shelterbelts the protection afforded is greatest for x/h < 20. The problem in dealing with this region is that results tend to become highly 'obstacle specific'.

There have been quite a number of wind-tunnel studies of neutrally stratified boundary-layer flow over two-dimensional barriers and idealised triangular and other hills. Good examples are provided by the work of Pearse et al. (1981), Arya and Shipman (1981) and Arya et al. (1987). These and other studies provide information on the mean flow and turbulence distributions in the wake of triangular and other simple hill shapes. For steep triangular hills (slopes > 0.25), Pearse et al. observed a strong wake region characterised by low mean velocities and high turbulence intensities while their triangular ridge studies with slopes of 0.05 and 0.15 did not exhibit this behaviour. In these cases we presume that the flow does not separate at the summit, or if it does, then the separation region is extremely short and shallow. Arya et al. studied the flow over three model two-dimensional symmetric hills with maximum slopes of 0.49, 0.29 and 0.18 (see Fig. 6a). For the steepest hill the flow separated vigorously and induced a separated 'cavity' region with depth ~1h and length ~5.5h (Fig. 6b). The flow did not separate over the low slope hill while for the intermediate one, with maximum slope = 0.29, there was intermittent separation with a very shallow cavity. Wake characterisitics are thus critically dependent on hill slope, as, of course, we should expect from analogies with the streamlining of obstacles in free streams. Nonseparating flow behind two-dimensional hills will still exhibit velocity deficits and perturbations to the turbulence fields but in general these will be considerably smaller than in the vigorously separating cases. Arya et al. provide details of the mean flow and shear stress profiles in both the near and far wake regions, and also report on the dispersion from point sources located in various positions near the hills.

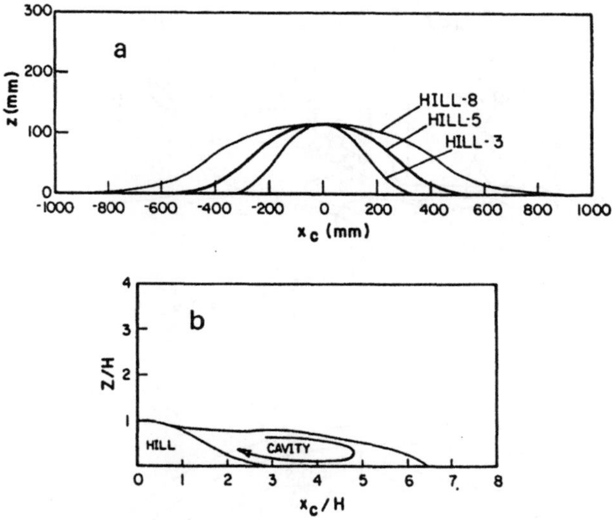

Fig. 6. (a) Hill profiles studied by Arya et al. (1987). (b) The recirculating cavity in the lee of hill-3.

Numerical-model studies by Taylor and Gent (1974), Taylor (1977) and others are really limited to nonseparating flows but they can give an indication of the slopes or curvatures at which separation first occurs. As a rough guide we should expect separation to occur for essentially two-dimensional hills or ridges when the maximum slopes are of order 0.3. There will be a dependence on surface roughness and on the detailed shape of the hill. As a general rule increased roughness will tend to encourage separation slightly as will sharp curvature at the hilltop. The model computations of Mason and Sykes (1979) include solutions for separated, neutrally-stratified Ekman layer wake flows. They find steady attached flow in the lee of two-dimensional 'cosine squared' hills or ridges of maximum slope 0.05, steady separated flow when the maximum slope is 0.1 and unsteady separated flow over hills of maximum slope 0.2 and higher. However, we should remark that they, in effect, assume a constant eddy-viscosity and direct application to turbulent boundary-layer flows is questionable, especially in respect of the slope at which flow separation may occur. Taylor (1981) studied flow over the same topography with his PBL model and found no flow separation in a case with maximum slope 0.23, as well as other differences between the predictions of the two models.

Wakes behind three-dimensional obstacles are particularly complex, since flow separation and reattachment can take many different forms. Where in two-dimensions we could talk of separation points on the surface, we now have separation lines. In steady two-dimensional flow, separation went hand-in-hand with a region of flow reversal. This is not necessary in three dimensions and separation can sometimes be very hard to detect or identify unambiguously. Hunt et al. (1978) give some examples of the range and variety of three-dimensional flow separation. Figure 7a shows a sketch from their paper of the flow around a surface-mounted cube, while Fig. 7b is a sketch of mean streamlines in the centre plane of flow over an axisymmetric hump. Mason and Morton (1987) have recently presented a detailed study of the vortices in the wakes of surface-mounted obstacles, further emphasising the range and complexity of possible flow patterns.

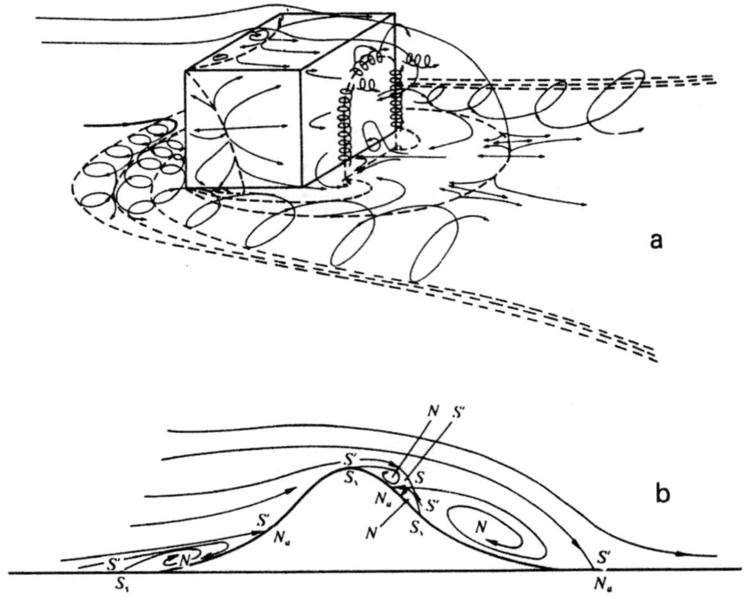

Fig. 7. (a) Sketch of the flow pattern around a surface-mounted cube in a turbulent boundary layer flow. (b) Mean streamlines and singular points on the centre-line plane of the flow over an axisymmetric hump in a turbulent boundary layer. (Hunt et al. 1978).

Wind tunnel studies of flow over conical hills by Pearse (1982), Castro and Snyder (1982), Arya and Gadiyaram (1986) and others give us some information on the near-wake region in these flows. For instance, Arya and Gadiyaram report a well defined recirculation zone behind a steep cone (slope = 0.5) but none for a cone with a slope of 0.3. Castro and Snyder studied a similar cone and present some information on the near-wake turbulence profiles.

It must be stressed that the results discussed are for neutrally stratified flows and that stratification, especially stable stratification, can substantially modify the near wake flow. Taylor and Gent (1980) discuss the conditions under which stratification becomes important while, for example, Brighton (1978) considers some of the details of the flow.

APPLICATIONS

Windbreaks

The widespread use of windbreaks in agriculture and architecture is so commonplace that we do perhaps tend to overlook their importance. The thesis by Jensen (1954) and the WMO report by van Eimern et al. (1964) document many studies of the shelter produced by windbreaks and their impact on the local thermal energy and moisture budgets, on soil erosion and on crop yield. In terms of wind reduction they report on the extent of zones with 25%, 50% and 75% reduction in speed from different field measurements and wind-tunnel studies, but it is clearly difficult to provide a simple theory in view of the different factors involved, especially for natural shelterbelts of hedgerows or trees. What is clear, is that potentially beneficial reductions in wind speed can occur for short distances (2-4h) upstream of the windbreak and over much longer distances in the lee (20-40h). Recent demonstrations of this are provided by Jacobs' (1984) field observations of the flow around a closed (zero porosity) fence and Wilson's (1987) studies on 50% porous windbreaks. Figure 8 shows Jacobs' results for near-neutral flow perpendicular to his fence. As a general rule wind speed reductions immediately downwind of the windbreak are greatest with a solid or closed fence but the greatest overall shelter is obtained with windbreaks of approximately 50% porosity.

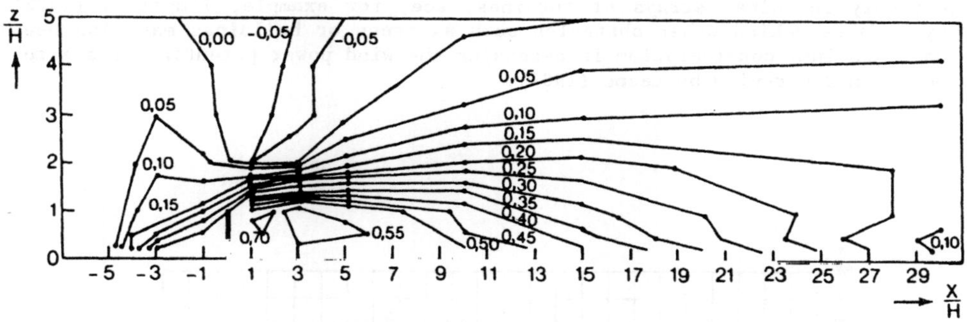

Fig. 8. Wind speed reduction caused by a closed fence. (Jacobs 1984).

The 'shelter' provided by a windbreak can be defined in a variety of ways. Locally it is usual to define s as the percentage reduction in wind speed at a given height relative to the upstream value at the same height above ground, $u_0(z)$. Curves of s versus x/h for selected z/h values are then the simplest way to describe the effect of the windbreak. Figure 9, from Jensen's work, illustrates this. It also shows that the shelter is relatively insensitive to the detailed structure for a given porosity and that significant sheltering can exist even at x/h = 50.

In order to obtain an overall value of shelter, one could, in principle, integrate s, as a fraction rather than a percentage, over the x,z plane ($z > 0$) for a long or two-dimensional windbreak or even over the half-space $z > 0$ for the finite length case. Jensen integrates s with respect to x/h for fixed z/h (= 0.4) to obtain a 'total shelter', S(total), for model windbreaks of different porosities. Values obtained with a rough tunnel wall, and for $x > 0$ only, range from S = 12.8 for $\varphi = 0$ to 14.3 for $\varphi = 0.36$ and then decline with increasing porosity. With a smooth tunnel wall the maximum was 21.2 (cf. Fig. 9). Upstream contributions were also computed in some cases and found to be approximately 10% of the total.

Windbreaks are widely used in architecture and are often designed to provide specific wind microclimates. Gandemer (1979) and Gandemer and Guyot (1981) provide interesting overviews of this application and give illustrations of a wide range of windbreak patterns. In many situations the windbreaks will be of limited length and end effects have to be taken into account. Similarly the direction of the wind relative to the windbreak is often a factor in their effectiveness, see Mulhearn and Bradley (1977).

Wind Turbine Wakes and Windfarm Applications

Wind turbines are a rather special type of surface-mounted obstacle whose wakes have attracted considerable interest over the past decade or so. In some ways a wind turbine can be considered as rather like a porous disc with a size of the swept area of the turbine, and indeed this has formed the basis for several wind-tunnel studies. Much of the interest in wind turbine wakes stems from a need to estimate power loss due to wake effects in wind turbine arrays, since one approach to the cost-efficient harvesting of wind energy has been to group turbines together in centralised, logistically efficient windfarms. The problem then arises that some turbines will be operating in the wakes of others and will thus experience lower wind speeds and higher turbulence. Similar considerations have been taken into account in estimating the wind energy resource of areas which could, in principle, be covered by very extensive, effectively infinite, arrays of turbines, see, for example, Templin's (1974) study. Wakes behind other obstacles such as trees or buildings may also need to be taken into consideration in assessing the wind power potential of a site, as noted in the review by Leene (1987).

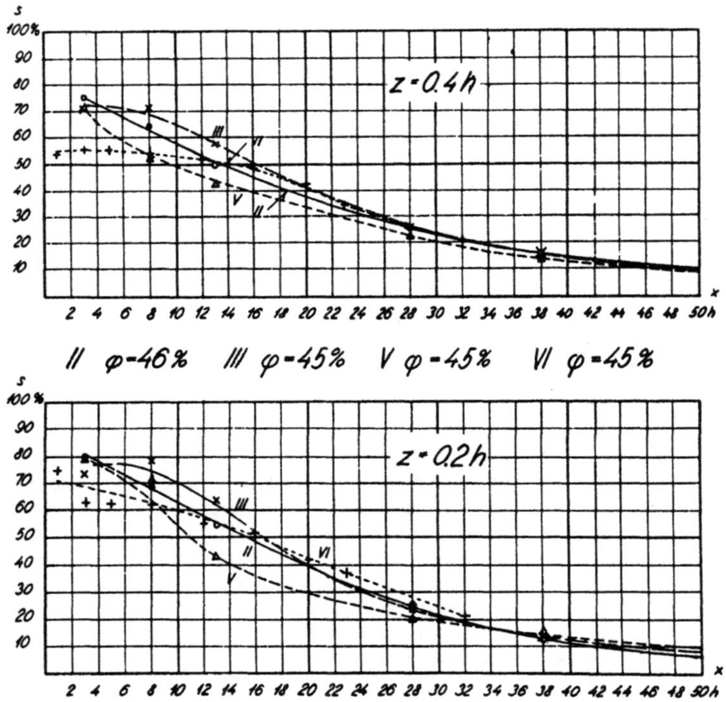

Fig. 9. Shelter effect of different screens of porosity v = 40%, smooth tunnel wall. From Jensen (1954). III - vertical palings, IV - screen of horizontal and vertical battens, V - screen with circular openings.

For finite arrays, Lissaman (see, for example, Builtjes and Smit 1978 or Lissaman 1979) has developed a type of 'Gaussian wake' model, analogous to the Gaussian plume models used for pollutant dispersal studies. This has been used quite widely. De Vries (1979) presents results obtained by Crafoord with an extended version of Templin's theory while Taylor (1980) describes applications of his linear model for boundary-layer wakes to row spacing in windfarms. Wind-tunnel modelling by Builtjes (1978), Builtjes and Milborrow (1980) and others have shown how the power per turbine is reduced as the turbine density is increased. For large, dense arrays the power output per turbine (proportional to U^3) can be substantially reduced relative to that for a single isolated machine. Builtjes (1978) gives an example where a regular array of 97 turbines with a minimum separation of 4.25d, where d is the turbine diameter, has an overall efficiency of only about 50%. Arrays with separations of order 10d perform rather better with average efficiencies of order 90%.

Field studies of wind turbine wakes have been carried out by several groups, including a collaborative study under the auspices of the International Energy Agency programme of R & D on Wind Energy Conversion Systems (IES R&D WECS) at Nibe in Denmark, see Taylor et al. (1985). The presence of two similar turbines at Nibe with a 5d separation has allowed measurements of power loss due to wake interaction effects in addition to detailed full scale measurements of the near-wake structure.

CONCLUSIONS

The wake behind a surface-mounted obstacle in a boundary-layer can be a highly complex flow and our quantitative understanding of the region is still far from complete. There appear to be gaps in our knowledge of simple bulk parameters such as the drag coefficient for some surface mounted obstacles and some uncertainty in the growth rates for the far-wake region. We note that the local drag on a surface-mounted obstacle is not necessarily equal to the net increase in drag on the surface since it can be partially offset by reductions in surface shear stress in the wake. This complicates the problem of parameterising the drag on small scale hills and other topography for large scale atmospheric flow models, even in the simple case of near-neutral thermal stratifiction. Flow separation is an important aspect of the flow over streamlined obstacles such as low hills and it appears, as we might expect, that the drag increases substantially once the flow separates. Practical issues involving wakes include the shelter provided by windbreaks and power losses due to wake interaction in wind turbine arrays.

ACKNOWLEDGEMENTS

I am grateful to many individuals for their help in sending me preprints and reports of their recent work, and for helping me track down hard-to-find reference material. In particular, Hans Teunissen and Dave Surry provided valuable comments and John Wilson made some additional calculations with his windbreak model to provide input to Table 5.

REFERENCES

Arya SPS, Gadiyaram PS (1986) An experimental study of flow and dispersion in the wakes of three-dimensional low hills. Atmos Environ 20: 729-740

Arya SPS, Shipman MS (1981) An experimental investigation of flow and diffusion in the disturbed boundary-layer over a ridge - I. Mean flow and turbulence structure. Atmos Environ 15: 1173-1184

Arya SPS, Capuano ME, Fagen LC (1987) Some fluid modelling studies of flow and dispersion over two-dimensional low hills. Atmos Environ 21: 753-764

Banke EG, Smith SD, Anderson RJ (1976) Recent measurements of wind stress on Arctic sea ice. J Fish Res Board Can 33: 2307-2317

Batchelor GK (1967) An introduction to fluid dynamics. Cambridge University Press, 615 pp

Bradley EF, Mulhearn PJ (1983) Development of velocity and shear stress distributions in the wake of a porous shelter fence. J Wind Eng Indust Aerodyn 15: 145-156

Brighton PWM (1978) Strongly stratified flow past three-dimensional obstacles. Q J R Meteorol Soc 104: 289-307

Builtjes PJH (1978) The interaction of wind turbine wakes. Proc 2nd Int Symp on Wind Energy Systems, Amsterdam, BHRA Fluid Engineering, Cranfield

Builtjes PJH, Milborrow DJ (1980) Modelling of wind turbine arrays. Proc 3rd Int Symp on Wind Energy Systems, Copenhagen, BHRA Fluid Engineering, Cranfield, pp 417-430

Builtjes PJH, Smit J (1978) Calculation of wake effects in wind turbine parks. Wind Eng 2: 135-145

Carson DJ (1986) Issues concerning the evaluation of effective surface roughness of heterogeneous terrain. Dynamical Climatology Technical Note No 35, UK Meteorological Office, Bracknell

Castro IP (1971) Wake characteristics of two-dimensional perforated plates normal to an air stream. J Fluid Mech 46: 599-609

Castro IP (1979) Relaxing wakes behind surface-mounted obstacles in rough-wall boundary layers. J Fluid Mech 93: 631-659

Castro IP, Fackrell JE (1978) A note on two-dimensional fence flows with emphasis on wall constraint. J Ind Aerodyn 3: 1-20

Castro IP and Snyder WH (1982) A wind-tunnel study of dispersion from sources downwind of three-dimensional hills. Atmos Environ 16: 1869-1887

Counihan J, Hunt JCR, Jackson PS (1974) Wakes behind two-dimensional surface obstacles in turbulent boundary-layers. J Fluid Mech 64: 529-563

de Vries O (1979) Fluid dynamic aspects of wind energy conversion. Agardograph No 243, AGARD, NATO, Neuilly sur Seine

Gandemer J (1979) Wind shelters. J Ind Aerodyn 4: 371-389

Gandemer J, Guyot A (1981) La protection contre le vent. CSTB publication, Paris, 132 pp

Goldstein S (ed) (1938) Modern developments in fluid mechanics. Dover edition (1965) Dover Publications, New York, 702 pp in 2 volumes

Good MC, Joubert PN (1968) The form drag of two-dimensional bluff plates in turbulent boundary-layers. J Fluid Mech 31: 547-582

Graham JMR (1976) Turbulent flow past a porous plate. J Fluid Mech 73: 565-591

Hoerner SF (1965) Fluid-dynamic drag. Hoerner Fluid Dynamics, Brick Town, NJ

Hosker RP Jr (1984) Flow and diffusion near obstacles. In: Randerson D (ed) Atmospheric science and power production, US Department of Energy publication 241-326

Hunt JCR (1971) The effect of single buildings and structures. Phil Trans R Soc London, Ser A 269: 457-467

Hunt JCR (1980) Wind over hills. In: Wyngaard JC (ed) Workshop on the planetary boundary-layer, Am Meteorol Soc, Boston, MA, pp 107-144

Hunt JCR, Abell CJ, Peterka JA, Wu H (1978) Kinematic studies of the flow around free or surface-mounted obstacles; applying topology to flow visualisation. J Fluid Mech 86: 179-200

Jackson PS, Hunt JCR (1975) Turbulent wind flow over a low hill. Q J R Meteorol Soc 101: 929-955

Jacobs A (1984) The flow around a thin closed fence. Boundary-Layer Meteorol 28: 317-328

Jensen M (1954) Shelter effect. The Danish Technical Press, Copenhagen, 264 pp

Johnson GL, Surry D (1987) Unsteady wind loads on tents. Boundary-Layer Wind Tunnel Laboratory, Report BLWT-SS5-1985, Univ of Western Ontario

Launder BE, Rodi W (1983) The turbulent wall jet - measurements and modelling. Ann Rev Fluid Mech 15: 429-459
Leene JA (1987) Review of previous studies on wake effects of obstacles in relation to wind turbine siting at MT-TNO. Report No 87-145 TNO, Apeldorn
Lemberg R (1973) On the wakes behind bluff bodies in a turbulent boundary-layer. Research report BLWT-3-73, Faculty of Engineering Science, Univ of Western Ontario, 162 pp
Lissaman PBS (1979) Energy effectiveness of arbitrary arrays of wind turbines. 17th AIAA Aerospace Meeting, New Orleans
Mason PJ (1986) On the parameterisation of orographic drag. Boundary-Layer Research Turbulence and Diffusion Note No 177, UK Meteorological Office, Bracknell
Mason PJ, Morton BR (1987) Trailing vortices in the wake of surface-mounted obstacles. J Fluid Mech 175: 247-293
Mason PJ, Sykes RI (1979) Separation effects in Ekman layer flow over ridges. Q J R Meteorol Soc 105: 129-146
Meroney RN, Sandborn VA, Bouwmeester R, Rider M (1976) Wind-tunnel simulation of the influence of two-dimensional ridges on wind speed and turbulence. Proc Int Symp on Wind Energy Systems, Cambridge, BHRA Fluid Engineering, Cranfield, A6: 89-104
Mulhearn PJ, Bradley EF (1977) Secondary flows in the lee of porous shelter-belts. Boundary-Layer Meteorol 12: 75-92
Newley TMJ (1985) Turbulent air flow over hills. Ph.D. Thesis, University of Cambridge
Pearse JR (1982) Wind flow over conical hills in a simulated atmospheric boundary-layer. J Wind Eng Indust Aerodyn 10: 303-313
Pearse JR, Lindley D, Stevenson DC (1981) Wind flow over ridges in simulated atmospheric boundary-layers. Boundary-Layer Meteorol 21: 77-92
Perera MDAES (1981) Shelter behind two-dimensional solid and porous fences. J Wind Eng Indust Aerodyn 8: 93-104
Plate EJ (1971) The aerodynamics of shelterbelts. Agric Meteorol 8: 203-222 See also Aerodynamic characteristics of atmospheric boundary-layers. US Atomic Energy Commission Critical Review Series Publication
Raine JK, Stevenson DC (1976) Wind protection by model fences in a simulated atmospheric boundary-layer. J Ind Aerodyn 2: 159-180
Ranga Raju KG, Loeser J, Plate EJ (1976) Velocity profiles and fence drag for a turbulent boundary-layer along smooth and rough flat plates. J Fluid Mech 76: 383-399
Ranga Raju KG, Garde RJ, Singh SK, Singh N (1987) Experimental study on characteristics of flow past porous fences. Proc Aachen Wind Engineering Conf, pp 229-238
Sabransky JJ, Melbourne WH (1987) Design pressure distribution on circular silos with conical roofs. J Wind Eng Indust Aerodyn 26: 65-84
Sakamoto H, Arie M (1982) Flow around a cubic body immersed in a turbulent boundary-layer. J Wind Eng Indust Aerodyn 9: 275-293
Sachs P (1978) Wind forces in engineering 2nd ed. Pergamon Press, Oxford, 400 pp
Schlichting H (1968) Boundary-layer theory 6th ed. McGraw-Hill, New York, 747 pp
Seginer I (1972) Windbreak drag calculated from the horizontal velocity field. Boundary-Layer Meteorol 3: 87-97
Seginer I (1975) Atmospheric stability effect on windbreak shelter and drag. Boundary-Layer Meteorol 8: 383-400
Seginer I, Sagi R (1972) Drag on a windbreak in two-dimensional flow. Agric Meteorol 9: 323-333
Taylor GJ, Milborrow DJ, McIntosh DN, Swift-Hook DT (1985) Wake measurements on the Nibe windmills. In: Wind energy conversion 1985, Proc 7th BWEA Conf, Mechanical Engineering Publications Ltd, London, pp 67-74
Taylor PA (1977) Some numerical studies of surface boundary-layer flow above gentle topography. Boundary-Layer Meteorol 11: 439-465

Taylor PA (1980) On wake decay and row spacing for WECS farms. Proc 3rd Int Symp on Wind Energy Systems, BHRA Fluid Engineering, Cranfield

Taylor PA (1981) Model predictions of neutrally stratified planetary boundary-layer flow over ridges. Q J R Meteorol Soc 107: 111-120

Taylor PA (1983) On wakes and the net forces produced by surface-mounted obstacles in neutrally stratified atmospheric boundary-layers. Boundary-Layer Meteorol 27: 393-412

Taylor PA, Gent PR (1974) A model of atmospheric boundary-layer flow above an isolated two-dimensional hill, an example of flow above gentle topography. Boundary-Layer Meteorol 7: 349-362

Taylor PA, Gent PR (1980) Modification of the boundary-layer by orography. In: Hide R, White PW (eds) Orographic effects in planetary flows, WMO GARP publication No 23, pp 143-165

Taylor PA, Richards KJ (1979) On wakes and drag in the neutrally-stratified atmospheric boundary-layer. Report ARQL 7/77 AES, Downsview

Templin RJ (1974) An estimate of the interaction of windmills in wide-spread arrays. NRC,NAE Report LTR-LA-171, Ottawa

Townsend AA (1965a) Self-preserving flow inside a turbulent boundary-layer. J Fluid Mech 22: 773-797

Townsend AA (1965b) The response of a turbulent boundary-layer to abrupt changes in surface conditions. J Fluid Mech 22: 799-822

Townsend AA (1976) The structure of turbulent shear flow. Cambridge University Press, 429 pp

van Eimern J, Karschon R, Razumova LS, Robertson GW (1964) Windbreaks and shelterbelts. Technical Note No 59, WMO No 147, WMO, Geneva

Wilson JD (1985) Numerical studies of flow through a windbreak. J Wind Eng Indust Aerodyn 21: 119-154

Wilson JD (1987) On the choice of a windbreak porosity profile. Boundary-Layer Meteorol 38: 37-49

TURBULENT WAKES IN THE ATMOSPHERIC BOUNDARY LAYER: COMMENTARY

W.H. Melbourne

Department of Mechanical Engineering, Monash University, Clayton, Vic., 3168, Australia

Summary. This commentary discusses some effects of incident turbulence on wind flows. It introduces some very recent work on wake flows from another discipline, and highlights the importance of incident turbulence in situations where there are separating shear layers. It also notes the need to measure peak gust wind speed information in studies of flow in the natural environment.

INTRODUCTION

Dr Taylor's review (Taylor 1988) and two poster papers highlighting applications set the scene for a discussion of turbulence. The complementary ingredient, largely missing from these papers, is the effect of incident turbulence on flow over bluff bodies and various topographical features, and critical statistical information concerning the peak wind speeds in the various turbulent flows. This commentary highlights some of these effects of turbulence and poses some related questions.

EFFECTS OF TURBULENCE ON BLUFF AERODYNAMICS

The main effects of turbulence and the scale of turbulence on the flow around structures (including trees) and the loading on them can be put into two categories.

(i) Where separation occurs, the wake development and distance to reattachment (if it occurs), and by implication, the mean radius of curvature of the separated (and wildly fluctuating) shear layer, are critically dependent on the incident turbulence. This dependence is observed in the way the wake develops generally; in pressures on the streamwise surface, which affect cross-wind loading; and in the peak and mean wind speeds which occur in the wake and, more importantly, in the zones in which the shear layer is fluctuating.

(ii) The along-wind loading and response are dependent both on the turbulent energy available directly in the incident flow to excite resonant and background response and on the effects of turbulence on the development of the wake and hence the wake and base pressures which contribute to the along-wind loading.

In terms of environmental wind flows and natural phenomena, the effect of the incident turbulence on the behaviour of the separated shear layer is perhaps most important, and will be the subject of most of the comments made in this paper. The direct effect of turbulence on the response of structures is well known and is described by many of the papers dealing with the Gust Factor

Approach to wind loading, in particular, Davenport (1967) and Vickery (1971). This, incidentally, is the main mechanism of wind loading on tall trees. The loss of a number of eucalyptus trees at the southeast corner of the Melbourne Botanical Gardens some 15 years ago was thought to be due to the increased turbulence generated by a new tall building adjacent to the Gardens.

EFFECTS OF TURBULENCE ON THE STANDARD SHEAR LAYER

Early descriptions of these effects are given by Gartshore (1973) and Melbourne (1975).

(i) Gartshore's conclusions:

> "Small scale freestream turbulence can influence the drag, fluctuating lift, base pressure coefficient and galloping behaviour of bluff bodies by increasing the rate of entrainment of the shear layers shed from them. The freestream turbulence, after severe distortion by the mean flow field, produces the increased mixing and entrainment by its presence in the shed shear layers, probably in the form of local concentrations of stream-wise vorticity. Only turbulence approaching a bluff body along its front stagnation line is required to produce the major effects of freestream turbulence on the flow around the body, so that a narrow wake effectively replaces grid-produced turbulence for experimental purposes. This observation also emphasises the importance of nearby upstream structures when considering the flow around tall buildings in the natural wind.
>
> Turbulence can both decrease and increase drag and produce unstable galloping oscillations in shallow rectangular bodies ($d/b < 0.5$) which are stable in the absence of turbulence. It can also decrease the drag and stabilize bodies of deeper section ($d/b > 0.6$) which are unstable without turbulence. Here b is the upstream face width and d is the streamline face depth.
>
> The effect of turbulence scale is small provided this scale is significantly smaller than the body dimension."

Gartshore's graphical description of these effects is reproduced in Fig. 1.

(ii) Melbourne's conclusions:

> "The shear layers and pressure underneath fluctuate. For certain conditions of freestream turbulence, angle of attack, (and radius of leading edge curvature) the shear layer commences to occasionally reattach onto the wall. As this occurs the cavity region under the shear layer is no longer vented and any decrease in pressure causes the initial radius of curvature of the shear layer to decrease further which in turn increases the local freestream velocity outside the shear layer and further decreases pressure. This is an unstable process which proceeds until the shear layer breaks up into a complete separation again and the cavity becomes vented. During this unstable process, very low, intermittent pressures can occur on the surface under the shear layer near the leading edge."

The development of this hypothesis, shown diagrammatically in Fig. 2, was influenced by the earlier work of Gartshore (1973), which led to the observation that if the freestream turbulence is increased, and in particular if the small scale turbulence energy is increased, the increased rate of entrainment into the shear layer would cause earlier reattachment; the smaller

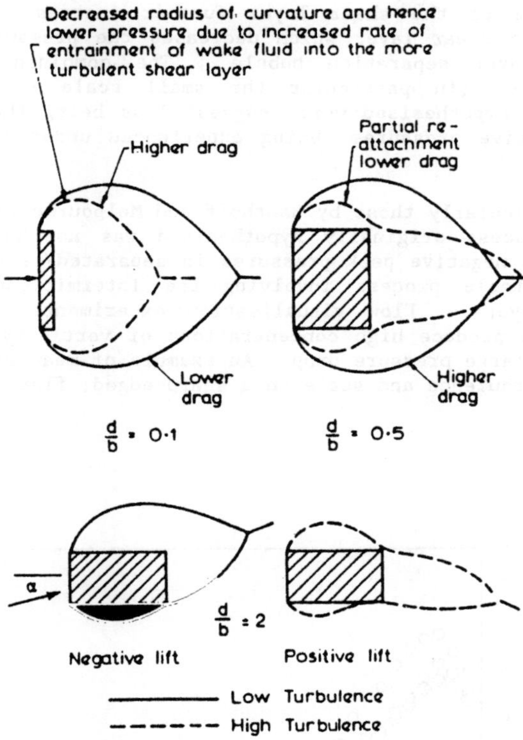

Fig. 1. Sketches of expected effect of turbulence on dividing streamlines as it affects drag and lift of rectangular sections. (Gartshore 1973).

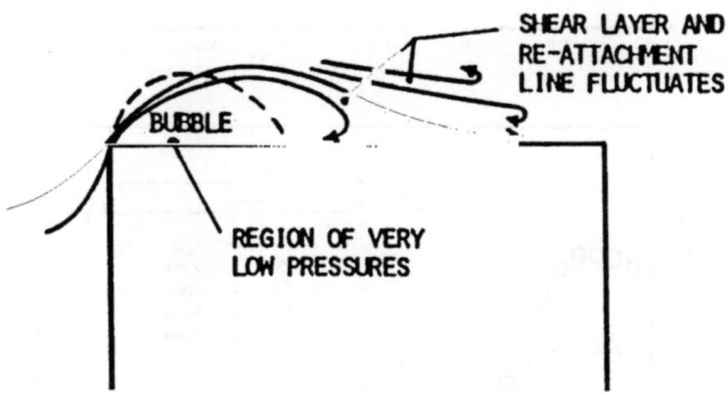

Fig. 2. Diagrammatric representation of an hypothesis on the cause of high negative pressures near the leading edge under a separating and reattaching shear layer. 1. Shear layer reattachment effectively seals off bubble from direct venting to freestream. 2. Freestream flow over the front of the shear layer accelerates; the increased velocity is accompanied by a decrease in pressure. 3. The combined effects of entrainment into the shear layer from inside the bubble and the decrease of pressure at the boundary causes the shear layer to move to reduce bubble volume and internal pressure (dotted line). 4. This process continues in an unstable way to reduce surface pressures under the bubble until it is vented by bursting or by artificial means. (Melbourne 1979).

radius of curvature of the shear layer further increases local freestream velocity outside the shear layer which decreases the pressure on the surface under the shear layer separation bubble. The combined effects of high freestream turbulence (in particular the small scale turbulence) and the instability process hypothesised were suggested as being the major causes of the very high negative pressures being experienced under shear layers which are reattaching.

Later studies, particularly those by Saathoff and Melbourne (1988), showed that the instability process originally hypothesised was not true, and that the generation of large negative peak pressures in separated/reattaching flows was related to an unstable process involving the intermittent roll-up of the separated shear layers. Flow visualisation experiments revealed that the roll-up process can produce high concentrations of vorticity near the surface and the subsequent large pressure drop. An example of mean and peak pressures, as a function of turbulence and scale on a sharp-edged, flat plate, is shown in Fig. 3.

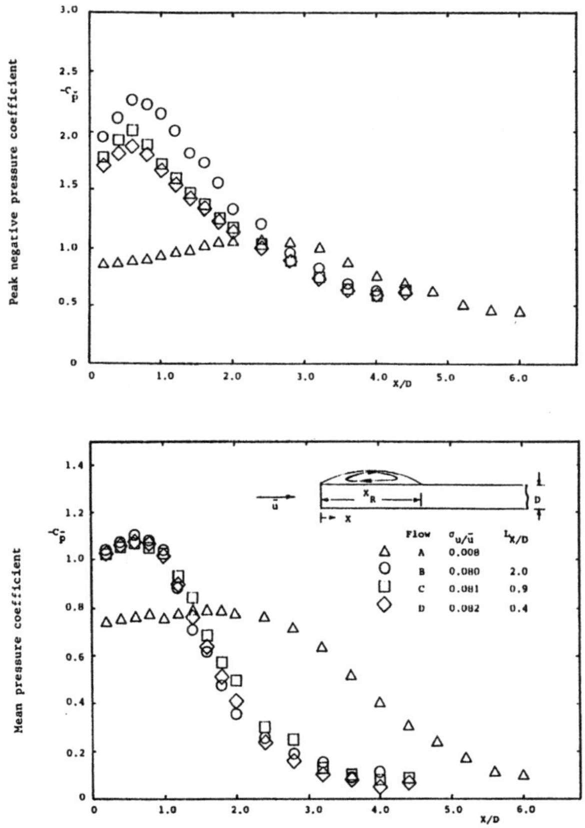

Fig. 3. Distributions of mean and peak pressures in turbulent and smooth flows. (Saathoff and Melbourne 1988).

FLOW OVER AN ESCARPMENT

One of the best examples of the effect of turbulence on an environmental wind flow is that of flow over a relatively sharp escarpment, where the flow is separating from the top. In his review paper, Taylor (1988) covered the rounder hills where turbulence effects are not so large. For a sharp escarpment, the phenomena discussed in the previous section become significant, not so much with respect to surface pressures, but with respect to peak wind speeds in which structures or crops may be exposed. Measurements by Vickery (1976) and unpublished measurements by D Sharp and WH Melbourne have shown that the effect of turbulence on the reattachment of a boundary layer flow over a step (or escarpment) is very significant, and, as shown in samples of this data in Fig. 4, can cause increases to design wind speed which effectively double the design load for relatively low rise structures placed in critical areas near an escarpment. The high and low turbulence levels referred to in Fig. 4 are those relating to flow over suburban or wooded terrain and over open country (airfield) terrain, respectively.

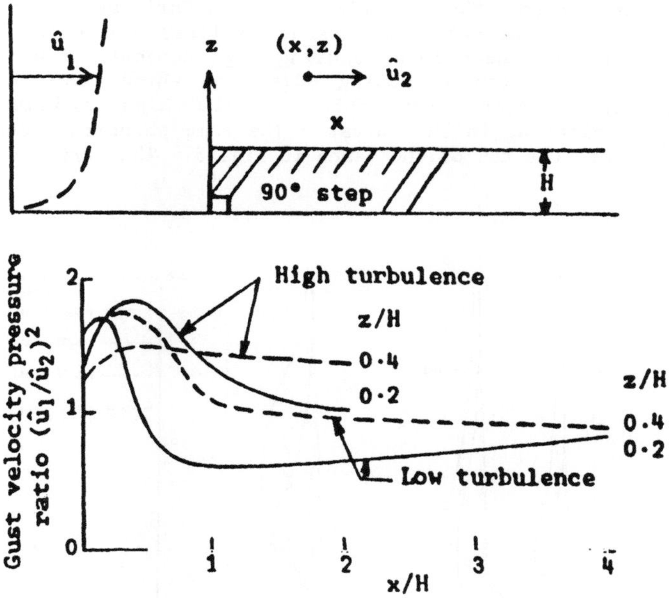

Fig. 4. Gust velocity ratios downstream of a vertical step. High turbulence lines from Sharp and Melbourne, unpublished data; low turbulence lines from Vickery (1976).

PERFORMANCE OF WINDBREAKS

The poster paper on turbulence statistics within and above a multiple windbreak array[1] gave information about mean and standard deviation of windspeed and the skewness and kurtosis of the distributions measured in full scale, but often the most critical piece of information required with respect to the

[1]KJ McAneney, MJ Judd (1987) Turbulence statistics within and above a multiple windbreak array. Symposium poster paper

effectiveness of a windbreak is the reduction effected in the peak gust wind speed. In this case the objective was to provide protection for kiwifruit, which are small, and as such the criterion for windbreak effectiveness would clearly be the reduction in peak gust wind speed. Unfortunately, this information was not available. An example of the effect of fence porosity on the ratio of peak gust wind speeds above and below a single windbreak are given for approach flow over open country terrain in Fig. 5.

The poster paper on the turbulent wake effects around a clearing in a vegetation canopy[2] gave a very interesting example in which a number of phenomena discussed earlier in this paper could well be involved. A similarly slow response of turbulence characteristics to the initial change of roughness from rough to smooth terrain has been observed by the author and is the subject of equations which will be made available for general use in the new Australian wind loading code.

However, of more interest in the present context is the observation from these studies that there is increased turbulence downstream of the clearing, that increased tree damage is observed a few canopy heights downstream of leading edges, and that wind problems are a likely cause of the nesting problems experienced by the seabird, the Abbott's Booby, on Christmas Island. It seems highly likely that the mechanism discussed earlier, where the intermittent roll-up of a separating shear layer causing high concentrations of vorticity some step heights back from a leading edge, and which cause high negative surface pressures, is also responsible for the high turbulence observed downstream of the clearing in the forest. The same phenomenon can be seen to occur downstream of the top of the fence in Fig. 5. The strength of the high

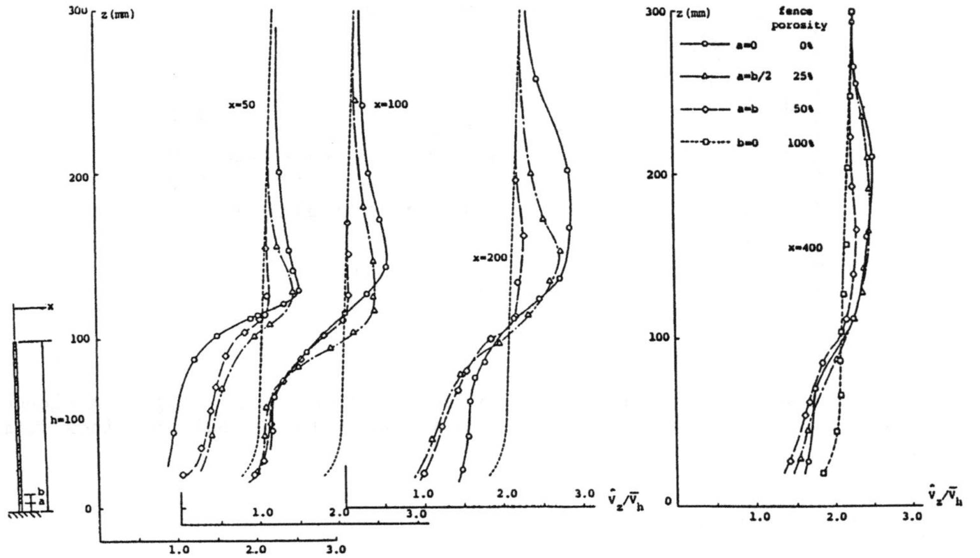

Fig. 5. Ratio of peak velocity at height z to upstream reference mean velocity at height h traversed at different distances x behind fences of different porosities. (Cheung and Melbourne 1988).

[2]MR Raupach, H Ghadiri, EF Bradley (1987) Turbulent wake effects around a clearing in a vegetation canopy. Symposium poster paper

concentration of vorticity is, from the work of Saathoff and Melbourne (1988), closely related to the incident turbulence level. This would be very high over the approach tree canopy, and not reduced significantly for the open clearing; hence the strength of the vorticity in the separating (and reattaching) shear layers in this case would be expected to be high. It would be very interesting to obtain some peak wind speed measurements on the wind-tunnel model of the forest clearing.

REFERENCES

Cheung JCK, Melbourne WH (1988) Solution to wind environment problems at the base of tall buildings. Proc 4th Int Conf on Tall Buildings, Hong Kong Inst of Eng, in press

Davenport AG (1967) Gust loading factors. J Struc Div Proc ASCE Vol 93, No ST3, pp 11-34

Gartshore IS (1973) The effects of freestream turbulence on the drag of rectangular two-dimensional prisms. BLWT Report 4/73, University of Western Ontario, Canada

Melbourne WH (1975) Cross-wind response of structures to wind action. 4th Int Conf on Wind Effects on Buildings & Structures, Heathrow, pp 343-358

Melbourne WH (1979) Turbulence effects on maximum surface pressures - A mechanism and possibility of reduction. 5th Int Conf on Wind Eng, Vol I, Session V-5-1, Fort Collins, CO, pp 541-552

Saathoff PJ, Melbourne WH (1988) The generation of peak pressures in separated/reattaching flow. To be published

Taylor PA (1988) Turbulent wakes in the atmospheric boundary layer. This volume, pp

Vickery BJ (1971) On the reliability of gust loading factors. C E Trans I E Aust, Vol CE13, No 1, pp 1-9

Vickery BJ (1976) Wind loads on low rise buildings. Research report commissioned by the Australian Department of Construction

PARTICULATE TRANSPORT IN THE ATMOSPHERE: WIND EROSION

B.B. Willetts

Department of Engineering, University of Aberdeen, Kings College, Aberdeen AB9 2UE, UK

Summary. A general description is given of the transport process mechanics in the saltation layer and below, including the effect of the sand on the wind in time-averaged terms. Recent analyses of saltation and of intersaltation collision are reviewed, as is work on surface grain re-sorting. Attention is drawn to the lack of adequate knowledge of momentum transfer mechanisms in the saltation layer and the resulting instantaneous velocity field.

The evolution of sand transport models from that of Bagnold is described, including a requisite model not yet implemented for establishing the stable grain activity level in a given wind. The research needs of the subject are related to the problems of implementing this model.

Soil transport is discussed as a more complicated extension of sand transport. The research requirements of semi-arid agriculture are outlined and expressed in terms of models of clean air flow over complex terrain and of the responsive particle-laden boundary layer.

INTRODUCTION

Wind erosion is important in many contexts. The three in which it presents large scale problems are in coastal zone management, in semi-arid zone agriculture and in deserts. The last two are strongly linked, since imprudent grazing or cultivation in semi-arid zones leads to desertification. What follows is written with an eye mainly to this link, which has had devastating consequences for rural populations in many parts of the world, not least in this century. It may appear paradoxical that we shall begin with an account of some work done on the behaviour of sand, rather than soil. Later in the paper we will make an attempt to develop arguments relating what is known about sand transport to the more complicated behaviour of soil. Of particular interest is the scope for using the methods developed to study sand behaviour in studies of soil blowing.

While people have been fascinated for a very long time by the scale and the aesthetic features of aeolian sand processes (Cornish 1914), Bagnold's (1941) magisterial account of his desert and wind-tunnel observations is a prominent milestone in our progress towards a proper understanding. His most important contributions were to show the importance of basing study on the physical principles of grain and fluid mechanics. Whilst acknowledging the contribution of earlier students of the subject, it is with this account we shall start. After pursuing some of the strands of subsequent work, we sketch the present state of knowledge. We then examine the application of this knowledge to soil erosion problems and conclude with a shopping list of outstanding requirements.

GENERAL DESCRIPTION OF SAND TRANSPORT

All sand-transporting winds are turbulent. The turbulence structure depends on upwind terrain, including roughness, and the air and ground (or sea) temperature fields. In this account we assume throughout that the profile of time-averaged velocity of air which is free of sand can be described by

$$u = Au_* \log \frac{z}{z_0} \quad , \qquad (1)$$

where u is the mean streamwise velocity at height z above the ground; A is a coefficient; z_0 is a roughness length; and u_* is the friction velocity. (The significance of the time variation of u, the gustiness, will be discussed later).

The key feature linking grain and air motions, once sand begins to be transported, is saltation. Grains leave the surface and execute discrete paths under the influence of gravity, drag, and perhaps Magnus effect, descending to strike the bed obliquely at considerable speed. This has a disturbing effect on the sand of the bed, as we shall see later, and also modifies the near-bed air flow considerably. Bagnold described the resulting time-averaged air velocity profile thus:

$$u = Au_*' \log \frac{z}{z_0'} + u_t \quad , \qquad (2)$$

where z_0' is a vertical coordinate somewhat analogous to roughness height (Fig. 1) associated with the presence of a cloud of saltating grains; u_t is the velocity at height z_0' at threshold of motion; and u_*' is the velocity gradient with the logarithm of height in the presence of saltation.

The difference between the distribution of Eq. (1) and of Eq. (2) is illustrated in Fig. 1, which shows that, above a "threshold" value of u_*, velocity profiles remain logarithmic but appear to focus approximately at some distance above the original common focus for sand-free air. This results from the transfer of momentum from the air to saltating grains, which exhibits itself in the reduction of near-bed velocity, which can be observed in Fig. 1. In consequence the air velocity in the immediate vicinity of the bed is reduced considerably from the value it would have were there no saltation. Indeed, Fig. 1 suggests that zero air velocity exists very near bed level at high values of u_*'. This would imply that no grains in this regime are dislodged by wind-generated forces. If this were true, then the maintenance of a population of moving grains must depend on the exchange of momentum at the collision between saltating grains and the surface grains of the bed. Owen (1964), who showed that a point focus was an oversimplified view of velocity distribution, contended that wind-generated forces remain influential in dislodgement even when transport is well established.

Saltation is one of three transport modes of sand grains. The others are suspension and creep. Suspension is the transport mode of some fine material which is entrained by turbulent motions of the air and undertakes airborne excursions which are extended in height, distance and duration. Creep is the forward movement of grains which remain in contact with the bed. They are usually the coarser grains of the bed material. Figure 2 indicates the grain size susceptible to each transport mode at a range of wind speeds.

Very soon after sand begins to move, ripples develop and the grain size distribution is modified at the surface. Both these features influence subsequent mobility and the boundary conditions for wind flow over the surface. In many circumstances dunes are formed which may be large in size (many metres in height) and of several different types (Cooke and Warren 1973). Most dunes develop in such a way that flow separates at or near the crest. This gives rise to a discontinuity in the grain transport process which interacts with the dune shape, and a wake flow which produces disturbed patterns of grain transport downwind.

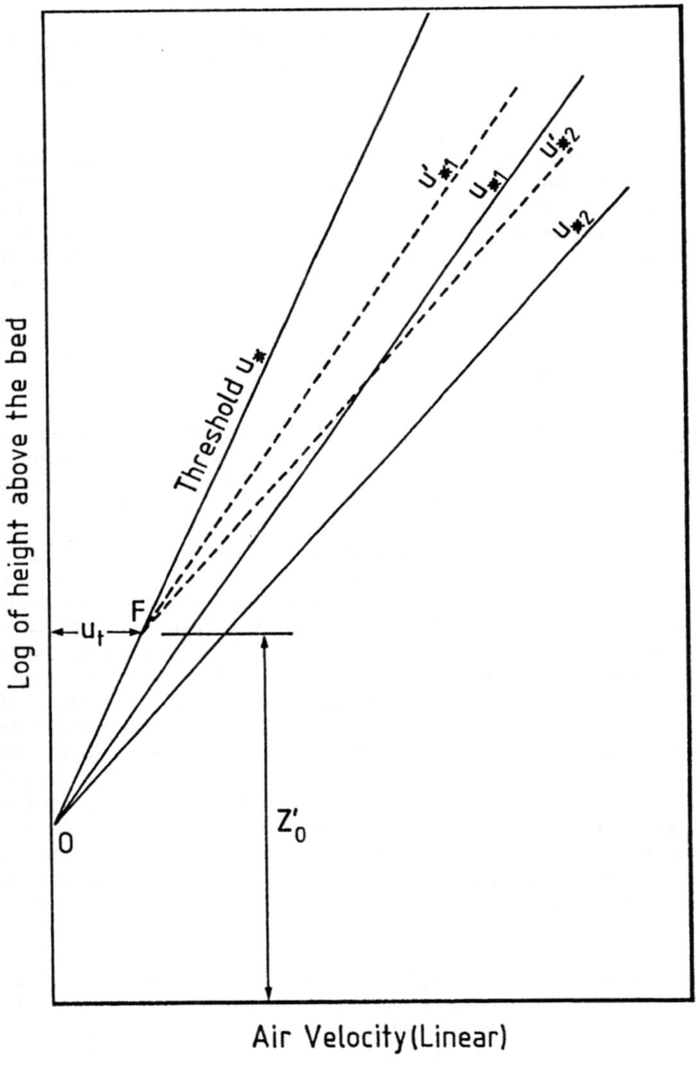

Fig. 1. The change in velocity profile when sand is driven by the wind. The dashed profiles (which are sand-laden) appear to converge at F rather than at O.

Both ripples and dunes involve a combination of wind and gravity stimulated grain motion. The migration of ripples and of transverse and barchan dunes progressively exposes the whole volume of the bed feature to wind action. Changes of shape of longitudinal and star dunes do likewise in a less systematic way. Since wind action winnows out fine material into suspension, the result of bed-form movement is to re-sort the material with loss of fines down to the base level of the bed forms, whether this is a couple of centimetres, in the case of ripples, or several metres, as in many dune systems. We see the outcome of this process in sand deserts. The segregation is of great practical importance whether expressed as loss of fertility or as unwanted dust deposition.

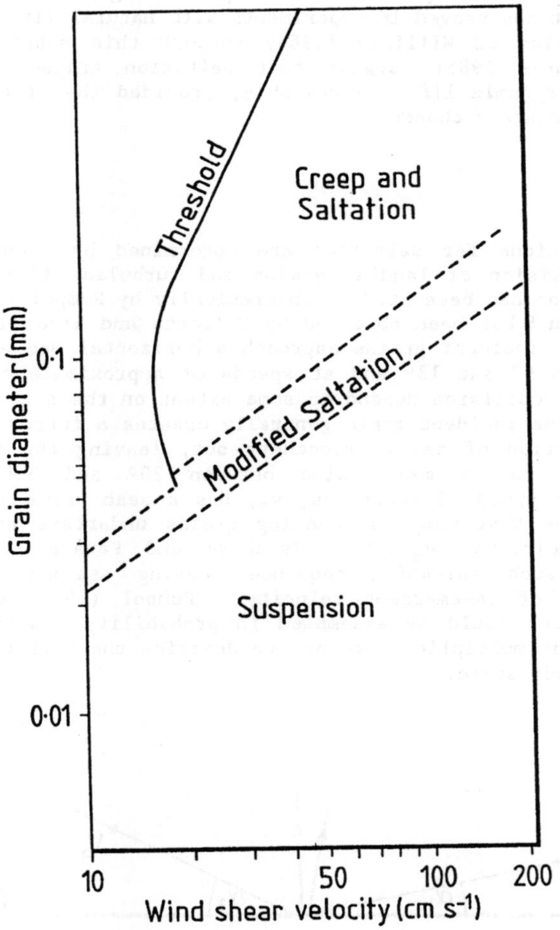

Fig. 2. Transport modes of quartz spheres (Bagnold 1941).

RECENT WORK ON GRAIN TRANSPORT PHYSICS

Saltation Trajectories

Saltation trajectories have been calculated in several studies (e.g. Owen 1964; White and Schultz 1977; Jensen and Sorensen 1986; Anderson and Hallet 1986). In most of these, the force terms in the equations of motion were grain weight and the drag associated with the air velocity relative to the grain. However, White and Schulz (1977) included a term for the fluid dynamic lift induced by particle spin (Magnus effect) and have investigated its significance. This investigation has been extended to Martian and Venusian circumstances (White 1979, 1981) in which spin is respectively less and more influential than it is on earth. While White and Schulz obtained better fit with observed trajectories when the lift term was incorporated in calculations, subsequent workers have not generally found it necessary to include such a term. It seems probable that grain shape influences the significance of spin and that its effect was greater in the experiments with spherical grains conducted by White and Schultz than it has proved in experiments with natural (irregularly shaped) grains. The results of Williams (1964) support this supposition. Recent comparisons (Nalpanis 1985) suggest that saltation trajectory calculations which exclude aerodynamic lift are reliable, provided that the initial launch velocity is appropriately chosen.

Collision

The initial conditions for saltation are determined by events at the bed. These include collision of landing grains and turbulent fluctuations of air velocity. Collision has been studied theoretically by Rumpel (1985) and Werner and Haff (1986), and has been observed by Willetts and Rice (1985a, 1986) and by Werner (1987). Incident grains approach a horizontal bed at an angle, α_1, which lies between 9° and 13° and at speeds of approximately $10u_*$ (Fig. 3). The details of the collision depend to some extent on the size of the incident grain. However, the incident grain generally creates a furrow in the bed and, in a large proportion of cases, ricochets out, leaving the bed again at an angle, α_3, which has a mean value between 20° and 35°, but a broad distribution. The speed of emergence, v_3, has a mean value in the region of $6u_*$. This implies that many ricocheting grains undertake another saltation with initial conditions (v_3,α_3). Tsuchiya and Kawata (1972) have made calculations of such saltation sequences, using assumptions rather than experimental data of re-emergent velocity. Rumpel (1985) demonstrated how collision statistics could be assembled in probability distribution matrices which could then be multiplied together to describe the Markhovian progress to statistically steady state.

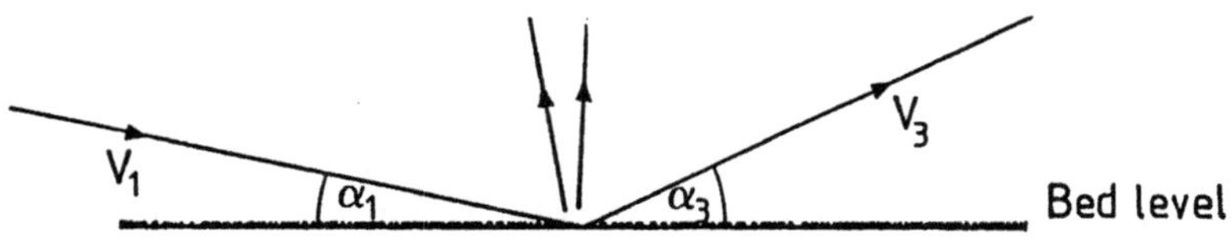

Fig. 3. The geometry of collision in a vertical plane parallel to the air flow.

Other grains are disturbed by the passage of the ricocheting grain. In many collisions some of these are ejected clear of the bed. However, the mean speed of ejection is of the order u_* and therefore, while many of such grains undertake small saltations, the probability is lower than for ricocheting grains that the emerging grain will undertake a sequence of saltations and collisions. Anderson (1987) has used the term "reptation" for feeble movements clear of the bed, in contrast with the vigorous movements of saltations. He has developed an explanation of the spontaneous growth of ripples on the postulation of a steady, uniform saltation cloud driving the reptation and creep populations downwind. The transport rate of the latter exhibits instability, giving rise to the spatial fluctuations of transport rate associated with ripple patterns.

It is very difficult, and probably quite artificial, to be specific in defining the distinction between saltation, reptation and creep. In reality there is a continuous spectrum rather than three distinct classes of movement. However, the current, rather vague, definitions are probably worth preserving while they are helpful in conceptualisation and model building.

The Mobile Population

Analyses of saltation mechanisms and the statistics of impact lead readily to information about the composition of the saltation cloud, but not its density. Yet proper understanding of the interaction between grain movement and wind structure, and of the rate of entrainment of fine material in suspension, depends critically on knowledge of the saltation process rate. This may be expressed in terms of the number and character of launches per unit time and bed area; or as the vertical profile of saltation transport rate together with the velocity vector of each transported grain.

Williams (1964) measured transport rate at different heights in the saltation cloud (for grain populations of different characteristic shapes). His data for quartz sand were analysed by Jensen and Sorensen (1986) with the object of describing the saltation cloud by means of their trajectory model and his data. In view of the height restriction in Williams' wind tunnel, they undertook additional measurements of transport rate at various heights above the bed, both in a larger wind tunnel and in the field (Sorensen 1985). As yet there is no useful data on the velocity of grains entering the compartments of the trap used to make the measurements. However, a momentum transducer, very recently introduced by Stockton and Gillette, affords a means of obtaining such data.

Tracer experiments by Barndorff-Nielsen et al. (1985) and Willetts and Rice (1985b) provide information respectively on the duration of rest periods of a single tagged grain and on the rate at which a population of coloured surface grains is depleted. The latter experiment has recently been repeated (Willetts and Rice 1987) and now gives a crude estimate of dislodgement rate of grains in two size ranges as a function of u_* (Fig. 4). More work is needed on the manner of dislodgement of the lost grains and their subsequent mode of transport.

The Saltation Laden Boundary Layer

The effect on the wind of the saltating grains is understood only in the general terms established by Bagnold (1941). Work is in progress, in Denmark (KR Rasmussen, personal communication 1987), to monitor in more detail the velocity structure deep in the saltation layer. The problems of doing this are considerable because the environment is destructive of hot wire or film probes and the opacity of the layer makes the use of laser techniques difficult. It may be that the latter may become usable in conjunction with optical fibre conductors. Meanwhile Gerety (1985) has studied the problems of

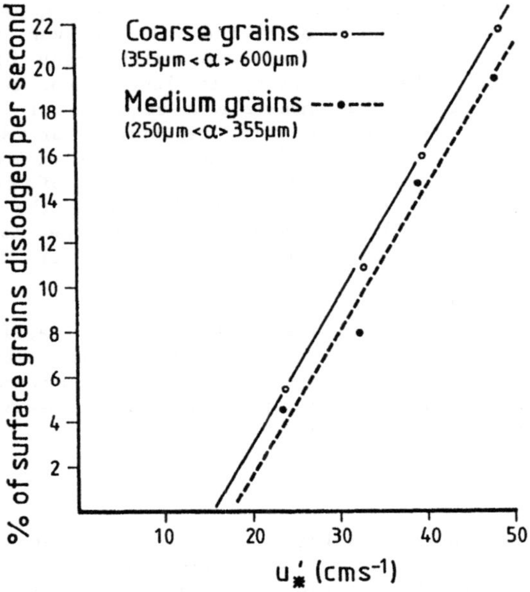

Fig. 4. Rate of dislodgement of grains from a flat bed.

evaluating u_*' from velocity measurements in the saltation layer and Chamberlain (1983) has explored the equivalent roughness length of a surface with saltating sand (and snow).

Better information on the turbulence structure in the saltation layer is badly needed in order to improve our understanding of the establishment of equilibrium in the process rates of grain transport and also of the entrainment process of fine material. It seems plain, intuitively, that the near-bed region of the boundary layer must be shot through by the wakes of saltating grains, and therefore must have a different turbulence structure from that of clear air flow. We do not know the extent to which the features of clear air turbulence survive in this region. It has been suggested, plausibly, that fine material is carried upwards through it in the wakes of saltation grains, and released into the free turbulence above the saltation layer, but evidence is absent.

Surface Re-sorting

Surface re-sorting has been studied in the field by Barndorff-Nielsen et al. (1982). Barndorff-Nielsen and Christiansen (1986) have described distributions in terms of the parameters of a hyperbolic curve fitted to sieve (or settling tube) data plotted on logarithmic axes, grain size and proportion by weight occurring. They have shown that if two such invariant parameters, expressing respectively skewness and kurtosis, are plotted on Cartesian axes, all possible distributions are found in a triangular domain. Furthermore, processes of erosion and aggradation produce sorting changes which are represented by different and characteristic shifts in this domain.

Shelter

Shelter may be provided by topographical features or by porous screens (natural or artifical) which create zones in which the friction velocity is lower than the generally prevailing value. The result may be to produce net deposition in a generally erosive or equilibrium environment, or to produce subthreshold conditions in which no movement occurs. The effect of porous screens is quite well understood in practical terms (Phillips and Willetts 1979; Iverson 1981). Topographical shelter involves the study of micrometeorology in the lower atmosphere, which is reviewed elsewhere in this volume.

The ground surface may also be sheltered by a pattern of superficial roughness such as that provided by growing plants. Wasson and Naninga (1986) have proposed a modification of the sand transport rate formula of Bagnold (1941) to account for vegetative cover. Lyles and Allison (1976) produced, from wind-tunnel studies, an equation for predicting the degree of protection provided to soil by standing crop residues and immobile soil aggregates. Hagen and Armbrust (1985) have reported the effectiveness of field ridges in reducing soil transport.

A particularly fascinating and important form of shelter is that afforded by topographical features (dunes) formed in the sand itself. The pattern of shelter associated with the dune field promotes a related pattern of sand movement, the nonuniformities of which induce further topographical developments. Thus, the wind and the landscape interact in a complicated, pleasing and sometimes impressive way which is of great interest to geomorphologists but which will have to be neglected in this review.

THE EVOLUTION OF MODELS OF GRAIN TRANSPORT

Transport rate prediction is usually made at present with the aid of formulae which are based on dimensional and simple physical arguments concerning the saltation process (Bagnold 1941; Owen 1964). Creep transport is calculated pro-rata. The form of such formulae is

$$q_s = K\, u_*'^{\,n} \quad , \tag{3}$$

or

$$q_s = L\, u_*'^{\,n_1}(u_*' - u_{*c}')^{n_2} \quad . \tag{4}$$

q_s is the sand transport rate; K and L are coefficients embodying the influence of bed geometry, texture, moisture content and grain shape; u_*' is the prevailing shear velocity and u_{*c}' its value at the threshold of motion. n, n_1 and n_2 are exponents which are assigned integer values (3, 1 and 2 respectively) by physical reasoning. When empirical values are fitted, they are found to vary significantly with sorting and grain shape.

Useful versions of (3) and (4) depend on experimental values of the coefficients, if not the exponents (in some circumstances it is possible to use the original calibration of the author of the formula). All such experiments measure transport rate averaged over a substantial period (usually greater than 100 s) and many of them have used time-averaged measurements of velocity from which to evaluate u_*'. Because transporting flows are always turbulent, velocities are not constant, but vary with time at any location. Furthermore, the bed condition changes as the bed is worked by the wind. Equation (3) might thus be written

$$\bar{q}_s = \frac{1}{T}\int_0^T K_t\, u_*(t)^3\, dt \quad , \tag{5}$$

where the bar indicates averaging over time T, and the subscripts t indicate the variation with time of both the local wind and the bed condition. While the variation of K_t can be thought of as a gradual drift, that of u_*' is irregular and complex, depending on the turbulence structures present at a particular location. It follows from the form of Eq. (5) that mean sand transport rate will, in general, be different for two flows which have the same time-averaged value of u_*', but different turbulence characteristics. It is therefore prudent to use a calibration made in circumstances similar to those of the application and, in particular, to be sceptical about wind-tunnel calibration if the application is to be made in the field.

The recent improvement in knowledge of saltation and collision reported above now makes possible some improvement in the representation of grain activity. Using a probability distribution for the immediate outcome of each collision and calculated trajectories for the intervening saltations, and doing this fraction by fraction for different grain sizes, it is feasible to construct a model of the saltation cloud in constant (or slowly changing) wind conditions. This model replicates the activity in the saltation layer proportionally (i.e., without the population density information). Its weakness is that the air velocity field in which the saltations are computed is invariant. It is not retarded by the momentum exchange with the grains, which is known to be an important feature of the physics of the process. Therefore, the vigour of the saltation process is not moderated by the wind change and the model does not stabilise at an equlibrium process rate. The necessary feed-back loop is missing in the calculation. Owen (1964) postulated a feed-back mechanism as did Bagnold (1941), but no successful demonstration of its operation has been achieved.

The loss of forward momentum of grains at collision with the bed is clearly linked with creep transport, and simple calculations can be made of the creep which might be expected to be driven by a postulated collision density. However, this line of argument, with interesting implications for ripple development, is also impeded by the gap in our knowledge of the saltation process rate and therefore the frequency of collision.

The next stage of model development should be aimed at representing the interaction between grains and wind. This will involve calculations not only of the forward momentum transfer from wind to grains but also of the vertical transfer of fluid momentum in the presence of the saltation cloud. The calculation of the latter is dependent on physics which is imperfectly understood at present. Research is therefore required into the fluid mechanics of the saltation layer, a difficult region to explore because the environment is too hostile for most measurement probes. However, the rewards of attaining such a model would be great, because it would be capable of predicting the equilibrium process rate in a range of bed and wind conditions.

SOIL PROCESSES

Soils of even the most simple structure are more complex than sands. They contain a greater proportion of fine material (< 50 μm) and a significant proportion of organic material, which may not be granular at all. Consequently, their behaviour is much more sensitive than that of sands to the presence of water. This is true even of very sandy soils.

However, many soils in transport can be seen to exhibit common features with sand. In particular, saltation remains an important mechanism, ripples form in the same way, and re-sorting (or size segregation) occurs at the surface. It seems probable, therefore, that theory developed for pure sand will apply, in modified form, to some soils at least. Unfortunately, the difficulties of observing particle behaviour, or even of measuring transport rate, are made even greater than they are in sand experiments by the fine material present in soil. Given that the theoretical treatment of sand transport is still impeded by observational difficulties, it would be unduly optimistic to expect a reliable modified theory for soils in the near future.

For the present it seems that empirically based methods, such as those developed at the US Department of Agriculture Wind Erosion Research Laboratory (Fryrear and Lyles 1977), will be the basis of prediction of soil behaviour in many circumstances. The hazards of applying them in new circumstances have already been noted. Meanwhile, improvements in the theory of sand transport should be extended progressively to broadly sorted sands, to bimodal sands, and to sandy soils as fast as this can be done dependably. Each such broadening of the theoretical base will lend confidence in the interpretation of empirical formulae for applications which have no direct precedent.

RESEARCH NEEDS RELATED TO SEMI-ARID ZONE AGRICULTURE

Prudent agricultural practice on semi-arid terrain requires the control of grazing intensity and the selection of appropriate crops and crop sequences. The soil is most vulnerable when fallow between crops. All management systems involve a degree of risk. In order to take rational decisions about the acceptable risk level, it is necessary to be able to predict the following.

1. Wind climate throughout the agricultural year: in particular, terrain-related wind strength exceedance curves, season by season.
2. The threshold wind condition for each soil type encountered in each of several surface conditions.
3. The rate at which grain transport, loss of fines and segregation, and surface texture changes proceed when the threshold is exceeded.

For the poorest and most fragile soils, the context of prediction is very similar to that for sands (often they have been segregated in the past, and have a very high sand content). Thus, the most acute practical problems concern soils for which the difficulties of prediction have more attainable solutions than those for more complex soils. The case for undertaking research is therefore a strong one, particularly as wind action on these fragile sandy soils leads quickly to desertification.

The research needs can be built explicitly around the three prediction requirements listed above. Clearly, there is distinct but limited value in site-specific and soil-specific solutions. Because of the widespread distribution of the problems, a general method capable of versatile application is greatly to be prized. This demands a sound theoretical base. Such a base would be filled by an adequate knowledge of flow over complex terrain (Finnigan 1988) together with an interactive model of the ground surface/saltation layer, discussed earlier in this paper, applicable to sandy soils.

REFERENCES

Anderson RS (1986) Sediment transport by wind: saltation, suspension, erosion and ripples. PhD dissertation, Univ of Washington

Anderson RS, Hallett B (1986) Sediment transport by wind: toward a general model. Geol Soc Am Bull 97: 523-535
Bagnold RA (1941) The physics of blown sand and desert dunes. Methuen & Co., London (Republished by Chapman and Hall Ltd), p 265
Barndorff-Nielsen OE, Dalsgaard K, Halgreen C, Kuhlman H, Moller JJ, Schou G (1982) Variation in particle size distribution over a small dune. Sedimentology 29: 53-65
Barndorff-Nielsen OE, Jensen JL, Nielsen HL, Rasmussen KR, Sorensen M (1985) Wind tunnel tracer studies of grain progress. In: Barndorff-Nielsen OE et al. (eds) Proc Int Wkshop on the Physics of Blown Sand, Memoirs 8, Dept Theoretical Statistics, Aarhus Univ, Denmark, p 243
Barndorff-Nielsen OE, Christiansen C (1986) Erosion deposition and size distributions of sand. Research Rep 149, Dept Theoretical Statistics, Aarhus Univ, Denmark
Chamberlain AC (1983) Roughness length of sea, sand and snow. Boundary-Layer Meteorol 25: 405-409
Cooke RM, Warren A (1973) Geomorphology in deserts. Batsford, London
Cornish V (1914) Waves of sand and snow. T Fisher Unwin, London
Finnigan JJ (1988) Air flow over complex terrain. This volume, pp
Fryrear DW, Lyles L (1977) Wind erosion research accomplishments and needs. Trans ASAE 20: 916-918
Gerety KM (1985) Problems with determination of u_* from wind-velocity profiles measured in experiments with saltation. In: Barndorff-Nielsen OE et al. (eds) Proc Int Wkshop on the Physics of Blown Sand, Memoirs 8, Dept Theoretical Statistics, Aarhus Univ, Denmark, p 271
Hagen LJ, Armbrust DV (1985) Effects of field ridges on soil transport by sand. In: Barndorff-Nielsen OE et al. (eds) Proc Int Wkshop on the Physics of Blown Sand, Memoirs 8, Dept Theoretical Statistics, Aarhus Univ, Denmark, p 563
Iverson JD (1981) Comparison of wind-tunnel model and full-scale fence drifts. J Wind Eng Indust Aerodyn 8: 231-249
Jensen JL, Sorensen M (1986) Estimation of some aeolian saltation transport parameters: a re-analysis of Willaims' data. Sedimentology 33: 547-558
Lyles L, Allison BE (1976) Wind erosion: the protective role of simulated standing stubble. Trans ASAE 19: 61-64
Nalpanis P (1985) Saltating and suspended particles over flat and sloping surfaces II. Experiments and numerical simulations. In: Barndorff-Nielsen OE et al. (eds) Proc Int Wkshop on the Physics of Blown Sand, Memoirs 8, Dept Theoretical Statistics, Aarhus Univ, Denmark, p 37
Owen PR (1964) Saltation of uniform grains in air. Fluid Mech 20: 225-242
Phillips CJ, Willetts BB (1979) Predicting sand deposition at porous fences. J Waterway, Port, Coastal and Ocean Div ASCE 105, No WW1: 15-31
Rumpel DA (1985) Successive aeolian saltation: studies of idealised collisions. Sedimentology 32: 267-280
Sorenson M (1985) Estimation of some aeolian saltation transport parameters from transport rate profiles. In: Barndorff-Nielsen OE et al. (eds) Proc Int Wkshop on The Physics of Blown Sand, Memoirs 8, Dept Theoretical Statistics, Aarhus Univ, Denmark, p 141
Tsuchiya Y, Kawata Y (1972) Characteristics of saltation of sand grains by wind. Proc 13th Int Coastal Eng, pp 1617-1625
Wasson RJ, Naninga PM (1986) Estimating wind transport of sand on vegetated surfaces. Earth Surf Proc and Landforms 11: 505-514
Werner BT (1987) A physical model of wind-blown sand transport. PhD thesis, California Inst of Tech
Werner BT, Haff PK (1986) The impact process in aeolian saltation two dimensional studies. Report B3-50, Div Physics Maths and Astronomy, California Inst of Tech, 43 pp
White BR (1979) Soil transport by winds on Mars. J Geophys Res 84: 4643-4651
White BR (1981) Venusian saltation. Icarus 46: 226-232
White BR, Schulz JC (1977) Magnus effect in saltation. J Fluid Mech 81: 497-512

Willetts BB, Rice MA (1985a) Inter-saltation collisions. In: Barndorff-Nielsen OE et al. (eds) Proc Int Wkshop on The Physics of Blown Sand, Memoirs 8, Dept Theoretical Statistics, Aarhus Univ, Denmark, p 83
Willetts BB, Rice MA (1985b) Wind-tunnel tracer experiments using dyed sand. In: Barndorff-Nielsen OE et al. (eds) Proc Int Wkshop on the Physics of Blown Sand, Memoirs 8, Dept Theoretical Statistics, Aarhus Univ, Denmark, p 225
Willetts BB, Rice MA (1986) Collision in aeolian transport: the saltation/creep link. In: Nickling W (ed) Aeolian Geomorphology, Allen and Unwin, p 1
Willetts BB, Rice MA (1987) Particle dislodgement from a flat sand bed by wind. Earth Surf Proc and Landforms, in press
Williams G (1984) Some aspects of the aeolian saltation load. Sedimentology 3: 257-287

PROCESSES OF WATER EROSION

C.W. Rose and P.B. Hairsine

School of Australian Environmental Studies, Griffith University, Brisbane, Qld., 4111, Australia

Summary. This paper considers some Australian research on the various processes of erosion and deposition involved in net loss of soil in water erosion. Detachment of soil by rainfall is non-size-selective, and in the presence of the highly size-selective process of deposition, a sediment rich in fine sediment is initially produced by these processes. The nature of the changing character of the soil surface sediment is discussed. It is suggested that the development of a deposited layer coarser than the original soil may be the reason for the commonly-observed general decline in sediment concentration toward a constant value under steady rainfall. A distinction is made between entrainment and the re-entrainment of sediment deposited during the erosion event, though both processes are due to the same shear stresses between overland flow and the land surface. New ways of mathematically representing these processes are developed, and the results compared with some other previously-reported methods. Approximate equilibrium theory is outlined for the simultaneous operation of entrainment, re-entrainment and deposition processes. This approximate theory is extended to allow interpretation of erosion data measured on a ridge-furrow system, and applied to the experiments of Meyer and Harmon (1985).

INTRODUCTION

The largest intensive experimental program on water erosion research remains that which led to the summary of the extensive data base collected in the USA, known as the Universal Soil Loss Equation, USLE (Wischmeier and Smith 1978). This was, and remains, a major guide in achieving soil conservation objectives. This work is similar in modelling methodology to that commonly used in agronomy. Experiments are carried out on the variety of soil types of concern and the major management factors, such as tillage type and stubble cover, are varied. What is measured (rainfall rate rather than runoff rate in the USLE) is affected by what, at the time, was thought to be the major driving variable. The model used is largely the outcome of the structure of statistical analysis applied to the collected data. A limitation to this research methodology, which was very successful in the USA, is that the data summarized for USA soils could not be transferred. Thus a similar investment of research was required in any other country wishing to use this approach. This was attempted in New South Wales in Australia (K Edwards, personal communication 1987). Because of the somewhat arid climate, some 30 years of data collection proved necessary to provide the needs of the USLE methodology, which is restricted to the long-term average annual soil loss. The analysis of these data is still being completed. Because of the long time scale needed by the USLE methodology (in semi-arid agricultural regions), or the large scale of research effort involved, interest has risen in alternative approaches.

The main alternative approach being explored is to develop and test models of the processes at work in erosion by water. This approach allows analysis of data collected on a single erosion event. This paper considers some research in this active area going on in Australia, and does not attempt to be comprehensive in its coverage. A data set from the USA is analysed in terms of the theory outlined in the paper.

RAINFALL DETACHMENT AND SEDIMENT DEPOSITION

Sediment Deposition

Whatever the processes which add to sediment concentration, sediment of size class i with settling velocity v_i and concentration c_i' will, by definition of the terms, experience a rate of deposition d_i (kg $m^{-2}s^{-1}$) given by

$$d_i = v_i c_i' \ . \tag{1}$$

In (1), c_i' is the sediment concentration adjacent to the depositing surface, which in deeper flow may differ from the average sediment concentration c_i by a factor α_i. Thus, (1) may also be written

$$d_i = \alpha_i v_i c_i \ . \tag{2}$$

The parameter α_i may be taken as unity in shallow flows, but if rills occur of depth greater than about 1-2 cm, α_i may be significantly greater than unity (Croley 1982).

The most appropriate method for determining the settling velocity distribution for sediment is a bottom withdrawal tube (Anonymous 1943). Lovell and Rose (1986) have described modification of the equipment and technique described in Anonymous (1943) which removes significant errors. Lovell and Rose (1988a,b) have developed theory to analyse such data, eliminating further error introduced by approximate corrections used in earlier analysis.

Rainfall Detachment

The rate of rainfall detachment for sediment of size class i (c_i) can be given (Rose et al. 1983a) by

$$e_i = a_1 C_e P/I \ , \tag{3}$$

where a_1 is the detachability of soil to rainfall of rate P assuming that the possible exponent on P is unity, C_e is the fraction of the soil surface exposed to rainfall, and I is the arbitrary number of size classes of equal mass into which the sediment can be conceptually divided. It is known that a_1 depends on soil type and condition, depth of water layer above the soil surface, and C_e.

Simultaneous Detachment and Deposition

Consider sediment transport on a plane land element where volumetric water flux per unit width of flow is q ($m^3m^{-1}s^{-1}$), and flow depth is D (m). Let x represent distance down the plane and t represent time. Then mass conservation requires that

$$\frac{\partial}{\partial x}(qc_i) + \frac{\partial}{\partial t}(Dc_i) = e_i - d_i \ . \tag{4}$$

Consider application of (4) to the simple situation where the soil is level and initially covered by a shallow layer of clear water, so that $c_i = 0$, $t \leq 0$, for all x. At $t = 0$, this water-covered soil is subject to simulated rainfall of steady rate P, as in experiments with a horizontal flume. Since the solution of (4) obtained for this situation will be shown to quickly reach an asymptotic value, it will also be assumed that the water depth D changes little before sediment concentration reaches its asymptotic value. We denote the average value of D over this limited time period by $\bar{D}$, which is also independent of x since $q = 0$ for a horizontal soil bed.

With these assumptions

$$\frac{\partial}{\partial x}(qc_i) = 0 \tag{5}$$

and (4) with (2) and (3) give

$$\frac{dc_i}{dt} + c_i\left[\frac{Q + \alpha_i v_i}{\bar{D}}\right] = \frac{a_1 C_e P}{I\bar{D}} \ . \tag{6}$$

The solution to (6) is

$$c_i = \frac{a_1 C_e P}{I(Q + \alpha_i v_i)}\left\{1 - \exp\left[-\left[\frac{Q + \alpha_i v_i}{\bar{D}}\right]t\right]\right\} , \tag{7}$$

where the constant of integration is evaluated using the initial condition that $c_i = 0$ at $t = 0$. Equation (7) shows c_i as rising to the asymptotic value of $a_1 C_e P/I(Q + \alpha_i v_i)$. The rate of rise is faster for particles with larger v_i, but is typically rapid for soils used in agriculture, the asymptotic value being achieved essentially in a period of minutes at most.

Notice that the asymptotic value of c_i is higher the lower the value of v_i. Thus, as experimentally found at times soon after the commencement of rainfall, the sediment is richer in finer fractions than is the original soil. As this fine-rich sediment is transported by overland flow (assumed to move sufficiently slowly not to add sediment itself by such flow), then a deposited layer accumulates which is coarser than the sediment transported above it. As explained by Rose (1985), this phenomenon must take place when any non-size selective mechanism (such as rainfall detachment) removes sediment, in the inevitable presence of the strongly size-selective process of deposition (Eq. (1)). Thus, Eq. (3) requires modification, and so Eq. (4), and hence (7) is only adequate to describe the outcome of rainfall detachment and deposition in the early stages following the commencement of deposition. To describe changes thereafter, (3) must be restricted to the fraction of the soil surface not covered by the coarser deposited layer, and another expression added to describe rainfall detachment from the coarser deposited layer, which continues to grow in fractional area coverage (under constant rainfall) until a dynamic equilibrium is achieved. The time required to achieve this equilibrium depends on soil and rainfall factors, but can be of order 10-30 min. At this dynamic equilibrium just a sufficient fraction of the original soil surface remains exposed to provide the deposited layer with sediment which deposits at a rate equal to its rate of loss by rainfall detachment. Thus, the fractional area coverage by the deposted layer at equilibrium must be less than unity. The rainfall detachability of the original soil surface and of the coarser deposited layer will in general not be the same. Thus, prior to the achievement of the dynamic equilibrium, sediment concentration will vary with time. Following the rather rapid initial increase in concentration if the initial conditions giving rise to (7) are satisfied, if rainfall rate is constant, concentration is generally found to decline toward a constant value corresponding to dynamic equilibrium.

ENTRAINMENT AND RE-ENTRAINMENT

Qualitative Description

Entrainment is the term adopted to describe the initial removal of soil material from its surrounding soil environment due to the mutual shear stresses which exist between overland flow and the surface across which it is flowing. Flow causing entrainment at a substantial rate is normally turbulent, so that the shear stresses are due to turbulent momentum transport toward the soil-water interface. Since eddy shear stresses will continually and randomly vary in intensity at any particular position, there will be the opportunity for sediment to settle, even at the position where, at a previous time, soil was entrained. This implies some limit to the frequency of shear stress of such intensity as to cause entrainment, since sediment deposition is observed to occur, though the settling velocity of finer fractions of sediment may be too slow to achieve deposition between the re-occurrence of high local eddy shear stresses. This may be the basis for the common distinction between "bed-load" and "suspended-load" in streams, or between the saltating layer and the suspended load above it in wind erosion.

Deposition of entrained sediment occurs to form, however temporarily at a given position, a thin layer of deposited sediment. The same eddy shear stresses responsible for entrainment of the original soil surface material can also remove the sediment deposited in the current erosion event. Such removal of sediment previously deposited in the current erosion event is termed re-entrainment. Thus, as for detachment and deposition under constant rainfall, a dynamic equilibrium will be achieved when the rate of re-entrainment of the deposited layer balances the deposition rate, so that the mass per unit area of the deposited layer is constant, as is the fractional coverage area it provides.

Because of the complexity of the processes involved, there has been little attempt to separate the processes of entrainment and re-entrainment. Indeed, most mathematical representations have effectively attempted to describe the net outcome of the three processes of entrainment, re-entrainment, and deposition. Recognising the aggregate nature of the description thus attempted, one type of approach has been to employ aggregate stream properties such as stream power (Bagnold 1977; Rose et al. 1983a) or unit stream power (Yang 1972; Moore and Burch 1986). Whilst such approaches appear to have some advantages over the more commonly employed formulations depending on effective discharge or shear stress, it remains a desirable objective to provide a theoretical framework within which to describe the processes. What follows in this section is an attempt to provide simple but physically-based descriptions of the rate of each of the two sediment removal processes: entrainment and re-entrainment. If such rate descriptions are then combined with those of other processes affecting sediment concentration in the context of mass conservation of sediment and water, then, as illustrated by Rose et al. (1983a), solution of the resulting differential equation allows experimental testing of the adopted rate process descriptions.

Entrainment Rate (r_e)

It follows from the definition of entrainment given earlier that we assume that no contribution to sediment concentration is made by processes such as bank collapse or landslides. Consider overland flow on a plane land surface.

Let r_e equal the mass rate of soil entrained (kg $m^{-2}s^{-1}$). In entrainment we assume that mechanical work is done dominantly against those soil strength-related forces which resist the initial removal of soil from its neighbouring soil material. Let J (J kg^{-1}) equal the amount of work expended per unit mass of soil in the entrainment process. Then the rate at which work is expended per unit area to sustain an entrainment rate r_e is r_eJ (J $m^{-2}s^{-1}$).

The rate at which energy is available (per unit area) from the flow to either perform mechanical work or be dissipated as heat or noise is the stream power, Ω (W m^{-2}), the rate of working of the shear stress (Bagnold 1966). Let Ω_0 (W m^{-2}) be the threshold value of Ω beyond which entrainment commences. Suppose that some fraction F of the excess stream power $(\Omega - \Omega_0)$ is available for the performance of "useful" work-entrainment in this case. Then it follows that

$$r_eJ = F(\Omega - \Omega_0) \ . \tag{8}$$

Suppose also that a fraction (H) of the soil surface is covered by the coarser deposited layer described above, exposing fraction (1-H) of the original soil surface. Assume further that entrainment rate is the same for each of the I size classes of equal mass into which the soil can be conceptually divided, i.e. that the eddy shear stresses causing entrainment are not discriminating in respect to sediment size. Then, denoting the rate of entrainment of size class i by r_{ei}, it follows that

$$r_{ei} = (1 - H) \frac{F}{IJ} (\Omega - \Omega_0) \ , \tag{9}$$

where $\Omega = \rho g S q$, (10)

if q, the volumetric flow rate per unit width, is spatially uniform, with ρ the fluid density, g gravitational acceleration, and S slope (the sine of the angle of inclination of the land surface).

In (9) there are three physically defined parameters (H, F and J). Parameters H and J are potentially measurable, though no standard method exists for their determination. It will be shown below that by using dynamic equilibrium assumptions H can be eliminated, but F and J remain unknown. Their determination will be illustrated in a later section.

For a short time following commencement of entrainment, especially if the entrained sediment is not coarse, then H will be small compared to unity, and may be neglected in (9). If the time-varying term $\partial(Dc_i)/\partial t$ in the expression of sediment mass conservation is also neglected, then the solution of the resulting differential equation

$$\frac{d}{dx} (qc_i) = r_{ei} - d_i \ , \tag{11}$$

where $c_i = 0$ at the distance downslope on a plane (x_0) at which the threshold for entrainment is exceeded, is

$$c = \frac{F\rho g S}{IJ} \left[x \sum_{i=1}^{I} \frac{1}{\gamma_i + 1} - x_0 \sum \gamma_i \left[\frac{\gamma_i + 2}{\gamma_i + 1}\right]\right] \tag{12}$$

where $\gamma_i = 1 + \alpha_i v_i / Q$. (13)

Equation (12) indicates that c_i is approximately proportional to $1/(\gamma_i+1)$, which implies that the resulting sediment is very much finer than the soil from which it originates. An approximate linear dependence of c on x is also indicated by (12) for the restrictive assumptions made in deriving it. These restrictions are such that the purpose of presenting (12) is to demonstrate the consequences of entrainment and deposition alone. In practice, re-entrainment will commence virtually immediately, which will modify the form of Eq. (12).

Re-entrainment Rate (r_r)

Let r_{ri} be the mass of soil of size class i re-entrained from the deposited layer per square metre per second. At dynamic equilibrium the rate of re-entrainment must equal the deposition rate of any size class i, so that from (2)

$$r_{ri} = \alpha_i v_i c_i \ . \tag{14}$$

The strength of the deposited layer is expected to be very low. It will therefore be assumed that in re-entrainment, mechanical work is expended dominantly in raising the sediment against its immersed weight, that involved in sediment removal being negligibly small in comparison.

The mass per unit area of sediment in size class i in overland flow on a plane land surface is Dc_i. Thus, the immersed weight per unit area of this sediment is $Dc_i\ g(\sigma-\rho)/\sigma$ where σ is the mean sediment density. The rate $\partial W_i/\partial t$ at which work is done in continuously raising this immersed weight of size class i sediment at velocity v_i is therefore

$$\frac{\partial W_i}{\partial t} = \alpha_i c_i \left[\frac{D}{\alpha_i}\right] g \left[\frac{\sigma - \rho}{\sigma}\right] v_i$$

and from (14):

$$= r_{ri}\left[\frac{D}{\alpha_i}\right] g \left[\frac{\sigma - \rho}{\sigma}\right] \ . \tag{15}$$

The fraction F of excess stream power $(\Omega - \Omega_0)$ available for mechanical work will operate in fractional area H covered by the deposited layer. However, this deposited layer has a different size distribution to the original soil, so that we cannot assume, as we did for entrainment, that fraction 1/I of the effective excess stream power is available for each size fraction. Rather, we will assume that the energy absorbed by sediment of size fraction i is in direct proportion to the mass fraction (M_{di}/M_{dt}) of that fraction, where M_{di} is the mass per unit area of sediment of size class i in the deposited layer, and M_{dt} is the total mass per unit area of this layer $(= \Sigma M_{di})$. From these assumptions

$$\frac{\partial W_i}{\partial t} = HF(\Omega - \Omega_0)\ \frac{M_{di}}{M_{dt}} \ . \tag{16}$$

From (15) and (16)

$$r_{ri} = \frac{HF\alpha_i}{g}\ \left[\frac{\sigma}{\sigma - \rho}\right]\left[\frac{\Omega - \Omega_0}{D}\right]\ \frac{M_{di}}{M_{dt}} \ . \tag{17}$$

At longer times from the commencement of runoff, and at larger distances downslope, $\Omega >> \Omega_0$, and H in (17) will be constant and can be close to unity, especially if soil is in a cultivated and well-aggregated condition. With these approximations, summing (17) over all size ranges gives

$$r_r = \frac{F\overline{\alpha}_i}{g}\ \left[\frac{\sigma}{\sigma - \rho}\right]\ \frac{\Omega}{D} \tag{18}$$

since $\sum_{i=1}^{I} (M_{di}/M_{dt}) = 1$ by definition, and $\overline{\alpha}_i$ is the mean value of α_i.

From (10) it follows that the term Ω/D in (18) is equal to ρgSV, since $q = DV$,

V being the velocity of flow. Since SV is unit stream power (Yang 1972), then (18) shows r_r as being proportional to unit stream power, in contrast to r_e, which is proportional to stream power (9). At the longer times and distances assumed in obtaining (18), for which H is close to unity, r_e will be small compared to r_r because (1-H) is small. Under these assumed conditions, a useful approximate expression for c should be obtained from

$$\frac{d}{dx}(qc_i) = r_{ri} - d_i = 0 \ , \quad \text{so that } r_{ri} = d_i \ ,$$

which summing over all size classes i, neglecting Ω_0, and taking H = 1, yields

$$c = \frac{F\rho}{(\Sigma v_i/I)} \left[\frac{\sigma}{\sigma - \rho}\right] SV \ . \tag{19}$$

The assumptions made in reaching this approximate solution imply that it should be a good approximation at longer distances downslope for a cultivated, well-aggregated soil, especially if runoff rate is not too variable with time. Equation (19) can be compared with the simplified version of the equations of Yang (1972) employed by Burch et al. (1986), namely

$$c = [5.435 - 0.286 \log(wd/\nu) - 0.457 \log(U*/w](VS)^{\beta} \tag{20}$$

$$\text{where } \beta = 1.799 - 0.409 \log(wd/\nu) - 0.314 \log(U*/w) \ . \tag{21}$$

Here c is in ppm by weight, w is the (single value) settling velocity in water, d is the median particle size diameter of the bed material, ν the kinematic viscosity of water, and $U* = (gDS)^{0.5}$ is the average shear velcoity.

Term β in (20) and (21) can be close to unity, and thus broad similarities between (19) and (20) with (21) are apparent, despite differences.

Equation (19) can also be compared with the equation used by Rose et al. (1983a) to describe the net concentration resulting from entrainment and re-entrainment, namely

$$c = 0.276 \ \rho g S \eta (1 - \Omega_0/\Omega) \ . \tag{22}$$

Parameter η (where $0 \leqslant \eta \leqslant 1$) is a measure of the net efficiency of the erosion processes. This equation results from the assumption (which has some experimental support) that sediment flux is proportional to excess stream power $(\Omega - \Omega_0)$. As Ω increases, c in (22) reaches an asymptotic limit, not increasing indefinitely with V as in (19) or (20). There are field data supporting both these conclusions, but less data supporting an asymptotic limit to c.

Approximate Theory of Combined Entrainment and Re-entrainment

Overland flow due to excess rainfall on a plane land element will continue to be assumed, with the upstream boundary condition

$$q = 0 \text{ at } x = 0, \ t \geqslant 0 \ , \tag{23}$$

with the volumetric flux per unit width of plane q given by the approximate solution for overland flow on a plane (Rose et al. 1983b)

$$q = Qx \ , \tag{24}$$

where Q is the runoff rate per unit area of plane. We will also neglect the time varying term $\partial(Dc_i)/\partial t$ in the sediment mass conservation equation, a justifiable assumption in application to the data analysed in the next section.

Thus, we seek a solution of

$$\frac{d}{dx}(qc_i) = r_{ei} + r_{ri} - d_i \tag{25}$$

where q, r_{ei}, r_{ri} and d_i are given by equations (24), (9), (17) and (2) respectively. Neglect of the time-varying term limits the solution to the special but important case where a dynamic equilibrium is achieved, which involves no time variation in the mass per unit area or size distribution of the deposited layer, of fractional area H. This equilibrium assumption allows H to be expressed analytically, and thus H disappears as an explicit term in the solution. Equation (25) then takes the form, when summed over all c_i:

$$\frac{dc}{dx} + \frac{c}{x}\left[1 + \frac{gD\Sigma v_i/I}{QJ\ \sigma/(\sigma - \rho)}\right] = \frac{F}{J}\left[\frac{\rho g S q - \Omega_0}{q}\right] \tag{26}$$

Solution of (26) is particularly simple if the power m = 2 is adopted in

$$q = KD^m , \tag{27}$$

where $K = S^{\frac{1}{2}}/n$ with n the Manning's roughness coefficient. Taking m = 2 is a choice intermediate between the value of 5/3 appropriate for turbulent flow and the value 3 appropriate for laminar flow.

Taking m = 2, and neglecting Ω_0, the solution to (26) is

$$c = \varphi\left[x^{\frac{1}{2}} - \frac{3}{2\beta}\left[1 - \frac{1}{\beta x^{\frac{1}{2}}} + \frac{1}{2\beta^2 x}\right]\right] \tag{28}$$

$$\text{where } \varphi = \frac{F\rho S(KQ)^{\frac{1}{2}}\sigma/(\sigma - \rho)}{\Sigma v_i/I} , \tag{29}$$

$$\text{and } \beta = \frac{F\rho g S}{J\varphi} . \tag{30}$$

From (28) and (30) it follows that c decreases as soil strength (to which J is related) increases. Also sediment concentration is dependent on the overall settling velocity characteristics of the soil. The term $\Sigma v_i/I$ is dominated by v_i from the coarser fraction of the soil. Although equation (28) indicates that as x becomes indefinitely large c will tend toward $\varphi x^{\frac{1}{2}}$, in practice the values of x for this to be so are so large (of order hundreds of metres) that this value will not be approached.

Equation (28) provides an expression for sediment concentration where that concentration is controlled by the soil strength, which determines J in β (Eq. (30)). From (28) and (30) it follows that as J increases, the terms subtracted from $x^{\frac{1}{2}}$ in (28) also increase, resulting in a decrease in c. This indicates how an increase in soil strength affects sediment concentration. Less obviously, an increase in soil strength also reduces the dependence of sediment concentration on slope S. From (28), $c \propto \Phi \propto SK^{\frac{1}{2}} \propto S^{1.25}$ (Eq. (29)). However, since $\beta \propto S$ (Eq. (30)), the presence of terms in $(1/\beta)$ in terms subtracted from $x^{\frac{1}{2}}$ in (28) can substantially reduce dependence of c on S, especially if soil strength is significant.

Equation (28) may therefore be described as the appropriate equation if the entrainment or detachment of soil is the process limiting the concentration of sediment. In contrast, (19) contains no term in J and is therefore not influenced directly by soil strength. Thus, (19) describes a situation, however temporary it may be, where the sediment concentration is determined by the capacity of flow at velocity V on slope S to transport sediment whose settling velocity characteristic determines $\Sigma v_i/I$. Sediment concentration is therefore partly determined by sediment characteristics not directly related to strength, also depending on slope and flow characteristics. In the circumstances for which (19) is applicable, sediment concentration may be described as "transport limiting" for brevity.

Perhaps more commonly the strength of the soil matrix will affect sediment concentration by restraining the entrainment rate, in which case (28) is relevant. In these circumstances sediment concentration may be described as "entrainment limiting". Foster (1982) has used the term "detachment limiting" in a very similar way. To avoid confusion we suggest use of the term "detachment limiting" in a generic manner, with limitation by entrainment or by rainfall detachment as two specific examples.

It is possible, and perhaps common, for conditions to fluctuate during an erosion event between the upper and lower limits to sediment concentration set by the transport limit and entrainment limit respectively. If so, then sediment concentration would behave somewhat as a limit cycle, though perhaps commonly biassed in time towards the lower detachment limit. These limits may therefore be understood as defining the envelope for sediment concentration for any given soil in any particular condition and hydrologic context. For sediment concentration to be sustained at the transport limit requires a continuous sediment supply unlimited by any strength which the soil matrix supplying the sediment may have. This may apply approximately with sand or river-bed sediment for an indefinite period, as suggested by experimental support for the equation of Yang (1972) (see (20) and comment on its similarity with (19)). However, in soil erosion by overland flow the strength of the soil would normally be expected to play a role, though any accelerated dislodgement could well increase sediment concentration to the transport limit. Bank collapse of incised rills has been found to rapidly increase sediment concentration to such an upper limit, followed by decline towards the lower, and perhaps more common, entrainment or detachment limit.

On this interpretation sediment concentration is a variable which in any situation may cycle between physically defined upper and lower limits. If so, then the mean concentration measured over a time period which is long compared to any individual cycle will depend on the magnitude of these limits, and on the detailed time history of excursions in concentration between or within these limits. The determination of true values of parameters F and J introduced into this analysis requires a knowledge of sediment concentration at these two limits. If, as is commonly the case, only mean sediment concentration for the duration of an erosion event is known, then the values of F and J obtained by fitting the theory to such data may not be true values, but effective mean values which summarise the result of detailed variation in sediment concentration during the event.

ANALYSIS OF EROSION DATA

The experiments of Meyer and Harmon (1985)

Meyer and Harmon (1985) reported the results of erosion experiments on bedded rows established in the field on a Loring silt loam soil (Typic Fragiudalfs). The ridge-furrow system had side slopes of 23%, a row length of 9.3 m, and ridge-to-ridge width of 0.97 m. Five furrow gradients from 0.5 to 6.5% were formed. Plots in a bare, tilled condition were subject to two initial rainstorms of rainfall rate 70 mm h^{-1} and durations of 60 min and 30 min. Finally, the experimental plots were subject to simulated rainfall at consistent rates of 26, 70 and 108 mm h^{-1} for 15 min. It is results from these final experiments which are considered here. Near the end of each experiment the constant rates of loss of sediment and water were measured for duplicate plots. Sediment delivery rate from sideslopes to furrow was measured on adjacent short (0.91 m long) row segments. Here the furrow was a metal collection trough which delivered sediment eroded from the sideslopes to measurement containers. The results from these short sideslope plots were assumed to indicate the rate of sediment delivery from sideslopes to furrows in the longer plots.

The geometry of the ridge-furrow system employed in these experiments was approximated by a shallow vee or triangular shape. Equations (19) and (28) for uniform flow were therefore appropriately modified to apply to this particular geometry. Theory for particular geometrics can be applied to interpret local features of eroding surfaces, such as rills, and is therefore referred to as "local theory". Also, the results of the local theory presented are for $m = 5/3$ (appropriate if flow is turbulent). Lateral input of sediment from sideslope erosion adds a major concentration term which was included in the analysis.

Since the later input of sediment from the row sideslope into the flow in the furrow is a measured quantity, no assumptions concerning the erosion mechanisms involved in this flow need be made. It will be assumed, however, that the depth of water in furrow flow is sufficient that rainfall detachment can be neglected, the only erosion processes in such flow being entrainment and re-entrainment.

Table 1 provides a summary of data from the experiments of Meyer and Harmon (1985) considered here. For a furrow slope of 0.5% Table 1 shows that more sediment reached the furrow from the sideslopes than was delivered at exit from the furrow. Hence, there would be a continuous accumulation of a deposited layer within the furrow. In the theory for entrainment/re-entrainment given in the previous section the simplifying assumption was made that the deposited layer had achieved a dynamic equilibrium, so that $\partial M_{di}/\partial t = 0$. Data in Table 1 indicates that this assumption will not be satisfied at all when furrow slope is 0.5%, since clearly $\partial M_{di}/\partial t > 0$.

Since only mean values of sediment concentration are available from the data of Meyer and Harmon (1985), then as discussed in the theory section for entrainment and re-entrainment, the values of F and J which satisfy the data will not necessarily be true values. In particular, the value obtained from F from mean concentration data will tend to be low if this is obtained from data where the transport limit for sediment concentration exceeds the entrainment limit. To minimise this source of error, the value of F in this data was evaluated as follows. When sediment concentration in the furrow flow is equal to the measured concentration for sideslope erosion, there must be no sediment entrained from the furrow; then the value of J is irrelevant and this situation will describe a point on the transport limiting relationship.

For a rainfall rate of 26 mm h^{-1} Fig. 1 shows sediment concentration measured at exit from the furrow for both replicate experiments at each of the five furrow slopes investigated. If a curve is fitted to the mean sediment concentration of the replicates, it intersects the mean concentration from the slideslope ($\bar{c}_{sideslope} = 11.0$ kg m^{-3}) at the point shown in Fig. 1. The value of F was then determined by fitting the transport limiting curve to this point. This is achieved by solving for F using the local theory equivalent of (19), and the values of c, S and V appropriate to the intersection point described above. This gave $F = 0.115 \pm 0.01$ approximately. It was then assumed that this value of F was the same for the other two experiments with higher rainfall rate, an assumption which will be tested in the following figures. The resulting transport limiting curve is shown to pass well above the mean experimental furrow concentrations at higher slopes. That the detachment limiting curve would pass above the furrow concentration for 0.5%

Table 1. The net difference in sediment loss (furrow - row side slope) in t $ha^{-1}h^{-1}$ (adapted from Meyer and Harmon 1985).

Furrow Slope %	Replicate	Rainfall Rate		
		26 mm h^{-1}	70 mm h^{-1}	108 mm h^{-1}
0.5	1	-1.3	-7.5	-23.1
	2	-0.4	-2.6	-13.3
2.0	1	-0.3	1.2	0.1
	2	0.8	5.6	6.2
3.5	1	0.0	2.1	0.5
	2	0.9	4.9	7.8
5.0	1	0.5	3.3	3.6
	2	1.1	4.1	7.8
6.5	1	0.5	8.1	11.5
	2	2.2	13.0	16.4

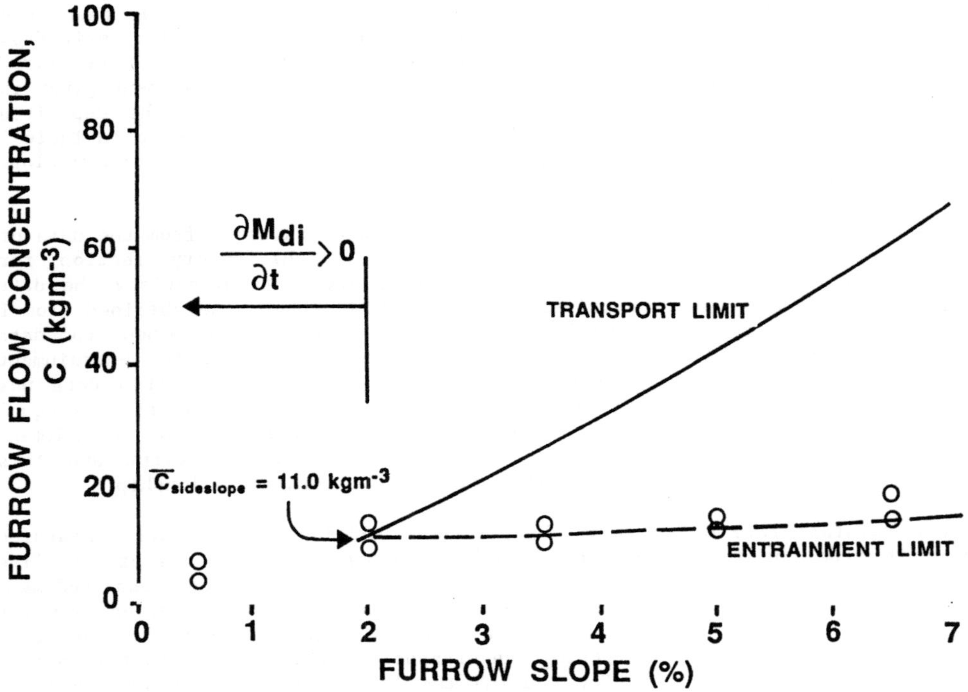

Fig. 1. Sediment concentration c measured at exit from furrows at a range of slopes under rainfall of rate 26 mm h^{-1} in the experiments of Meyer and Harmon (1985). Mean concentration of sediment from furrow sideslopes is shown. The transport and entrainment limits calculated as described in the text are superimposed on the measurement data.

slope is an indication that, as mentioned above, the assumption of $\partial M_{di}/\partial t = 0$ in the theory is not satisfied.

Having thus obtained a value for F, the unknown strength-related parameter J was chosen so that the theoretical entrainment limiting curve provided a lower boundary to the experimental data, as it should. The value of J thus obtained was 84 ± 10 J kg^{-1}, and as for F, this will also be assumed to apply for the experiments at other rainfall rates. The entrainment limiting curve was obtained using the local theory equivalent of (28), but for $m = 5/3$ rather than the value of 2 used in deriving (28). Figure 1 shows furrow concentrations remaining very close to the fitted entrainment limiting curve as slope increases.

Figure 2 (for a rainfall rate of 70 mm h^{-1}) and Fig. 3 (for 108 mm h^{-1}) show the corresponding set of results for exit furrow concentration measured by Meyer and Harmon (1985). Also shown on these figures is the set of transport and entrainment limiting curves assuming the same values of F and J as was obtained using the data in Fig. 1. It can be seen that these concentration limiting curves continue to provide an envelope which encompasses the experimental data, though the slope below which $\partial M_{di}/\partial t > 0$ is shown to rise as rainfall rate increases. This rise in slope value is not supported by the data in Table 1. This apparent discrepancy may be understood in terms of change in flow cross-section during the experiments.

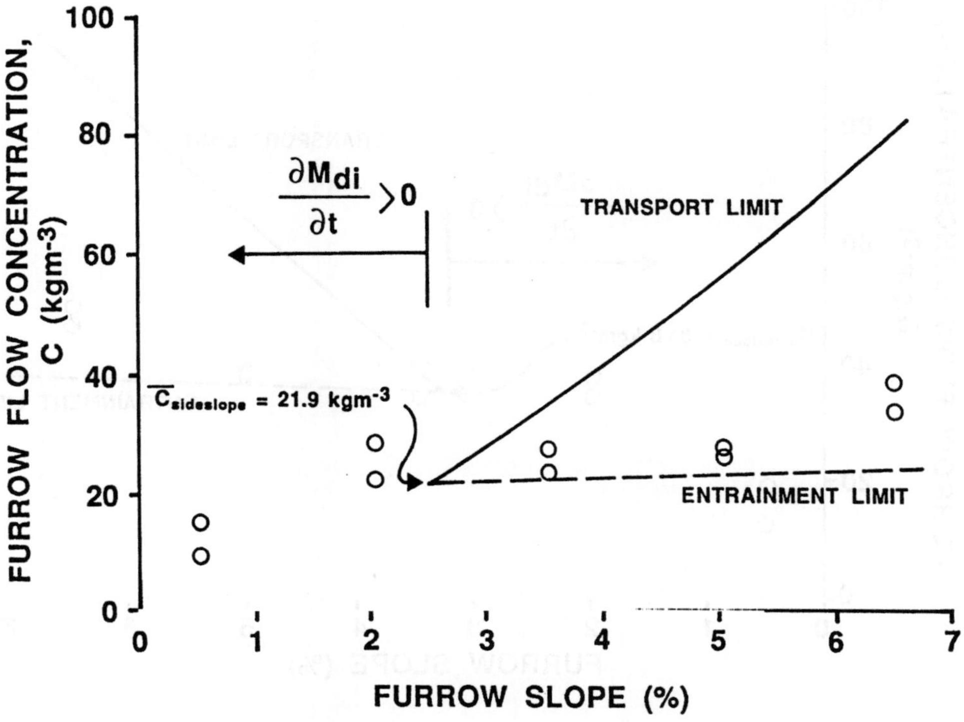

Fig. 2. As for Fig. 1 but for rainfall rate of 70 mm h^{-1}.

Especially for a furrow slope of 6.5%, Figs. 2 and 3 show furrow concentration rising significantly above the entrainment limit, though these mean concentrations are still well below the transport limit. This result is not inconsistent with the possibility that short period excursions in concentration to the transport limit may have occurred.

DISCUSSION AND CONCLUSIONS

Even when rainfall rate and ponded depth of water are constant, it is a common observation that the sediment concentration resulting from the simultaneous operation of rainfall detachment rate and deposition rate varies with time, though a constant value is eventually achieved. Equation (7) describes the rate of rise in sediment concentration following the initiation of rainfall. Except for fine sediment (7) predicts this rise to be rapid, which is no doubt why it is not always observed. However, of more practical importance is the subsequent decline in sediment concentration with time towards an equilibrium value if all factors are constant. Further analysis describing the buildup with time of a deposited layer of sediment, coarser than the original soil, promises to provide an explanation of this common observation. Not only the deposited layer, but the sediment in the water layer, also changes its size or settling velocity characteristics with time.

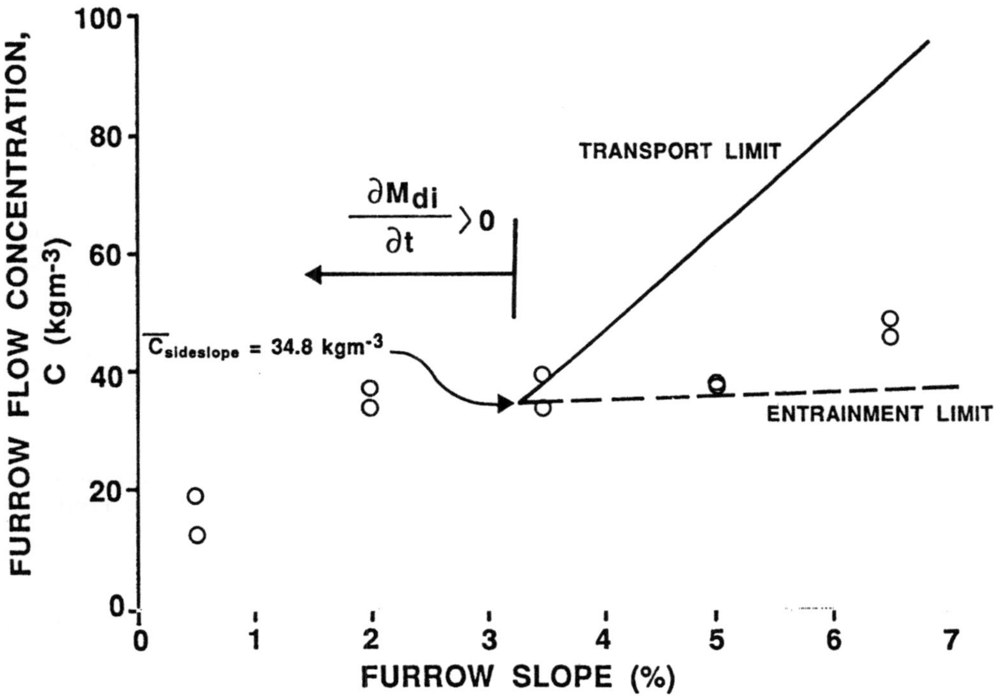

Fig. 3. As for Fig. 1 but for rainfall rate of 108 mm h^{-1}.

Whilst it is of the nature of research to continually improve our understanding of processes, it appears that our ability to adequately describe entrainment and re-entrainment processes is in a state of flux. Indeed, the distinction in this paper between entrainment and re-entrainment has not previously been drawn in such explicit and quantitative terms. Since entrainment immediately leads to sediment concentration in overland flow, deposition also commences immediately following the commencement of entrainment. Further exploration of the representation of these processes is needed, and the results compared with good data. Comparison with the data set of Meyer and Harmon (1985), whilst useful, is clearly not definitive. How the processes of entrainment and re-entrainment interchange through time requires further study.

More investigation is also needed of the relative contribution to sediment loss of the variety of erosion mechanisms. Present knowledge indicates that the relative contributions of rainfall detachment and runoff entrainment/re-entrainment mechanisms can vary greatly with soil type. Complexity of surface geometry in the field, and changing contributions through time under natural rainfall, leave a place here for controlled studies.

It is encouraging and yet challenging that we appear to be on the threshold of better insight into change, not only in sediment concentration, but also in the size or settling velocity characteristics of the resultant eroded or deposited sediment. As illustrated by Rose and Dalal (1987), this is essential if we are to interpret the consequences of erosion for the loss of soil nutrients.

The more fundamental progress in understanding the various erosion processes which is outlined in this paper also has implications for methods of modifying soil erosion loss which are not considered here. In such consideration due attention must be given to the spatial and temporal complexity in land surfaces in the field. This results in variability in factors such as flow depth due to natural topographic variation, cultivation or vegetation effects, and preferred channel organisation of flow. Broader three-dimensional features of landscapes also have significant consequences for flow concentration and so erosion hazard, as has been well developed by Moore and Burch (1986) and Burch et al. (1986), for example.

ACKNOWLEDGEMENTS

The contribution through their experimental work and discussions by the following post-graduate students in the School of Australian Environmental Studies, Griffith University, is gratefully acknowledged: C.J. Lovell, A.P.B. Proffitt, G. Okwach and R. Palis. Some of their work was supported by funds from the National Soil Conservation Project of the Commonwealth Government (Grant No. 1984/1986 No. 4), Rural Credits Development Fund Grant (GU/8607), the Australian Centre for International Agricultural Research and the Australian Development and Assistance Bureau. Discussions with Dr G.R. Foster on the data of Meyer and Harmon (1985) was most helpful.

REFERENCES

Anonymous (1943) A study of new methods of size analysis of suspended sediment samples. Report No. 7: Office of Indian Affairs, Bureau of Reclamation T.V.A., Corps of Engineers Geological Survey, Dept of Agric and Iowa Institute of Hydraulic Res, University of Iowa

Bagnold RA (1966) An approach to the sediment transport problem from general physics. U.S. Geol Survey Professional Paper 442-I 5

Bagnold RA (1977) Bed load transport by natural rivers. Water Resour Res 13: 303-311

Burch GJ, Moor ID, Barnes CJ, Aveyard JM, Barker PJ (1986) Modelling erosion hazard: A total catchment approach. In: Hydrology and water resources symposium 1986 "River Basin Management". The Institution of Engineers, Australia, Canberra, ACT, p 345

Croley II TE (1982) Unsteady overland sedimentation. J Hydrol 56: 325-346

Foster GR (1982) Modelling and erosion process. In: CT Haan (ed) Hydrologic modelling of small watersheds. Am Soc Agric Eng Monogr No 5. St Joseph, MI, p 297

Lovell CJ, Rose CW (1986) Measurement of the settling velocities of soil aggregates using a Modified Bottom Withdrawal Tube. AES Working Paper 4/86. Griffith University, Brisbane, Australia

Lovell CJ, Rose CW (1988a) Measurement of soil aggregate settling velocities I. A Modified Bottom Withdrawal Tube. Aust J Soil Res, in press

Lovell CJ, Rose CW (1988b) Measurement of soil aggregate settling velocities II. Sensitivity to sample moisture content and implications for studies of structural stability. Aust J Soil Res, in press

Meyer LD, Harmon WC (1985) Sediment loss from cropland furrows of different gradients. Trans ASAE 28: 448-453

Moore ID, Burch GJ (1986) Sediment transport capacity of sheet and rill flow: Application of unit stream power theory. Water Resour Res 22: 1350-1360

Rose CW (1985) Developments in soil erosion and deposition models. In: Advances in soil science, Vol. 2. Springer-Verlag, New York, pp 1-63

Rose CW, Dalal RC (1987) Erosion and runoff of nitrogen. In: Wilson JR (ed) Advances in nitrogen cycling in agricultural ecosystems. Commonwealth Agricultural Bureaux (CAB International), in press

Rose CW, Williams JR, Sander GC, Barry DA (1983a) A mathematical model of soil erosion processes. I. Theory for a plane land element. Soil Sci Soc Am J 47: 991-995

Rose CW, Parlange J-Y, Sander GC, Campbell SY, Barry DA (1983b) A kinematic flow approximation to runoff on a plane: an approximate analytic solution. J Hydrol 62: 363-369

Yang CT (1972) Unit stream power and sediment transport. J Hydraul Div Am Soc Civ Eng 98 (HY 10): 1805-1826

Wischmeier WH, Smith DD (1978) Predicting rainfall erosion losses - a guide to conservation planning. Agric Handb No 537. USDA Govt Printing Office, Washington, DC

ONE-DIMENSIONAL SOLID-LIQUID SEPARATION

D.E. Smiles and J.M. Kirby

CSIRO Division of Soils, GPO Box 639, Canberra, ACT 2601, Australia

Summary. We describe an approach to separation of finely dispersed solid from liquid which has general application for many practically important suspensions. We propose that the separation of the liquid from the solid in particulate suspensions can usefully be described in terms of the liquid flow relative to the solid. The combination of this proposition with continuity equations for the liquid and the solid then gives rise to a flow equation that may be expressed in either spatial or material coordinates. For both filtration and sedimentation, the latter formulation is more convenient since it gives rise to a nonlinear Fokker-Planck equation for which solutions are known for many sets of initial and boundary conditions important in solid-liquid separation processes.

The theory is macroscopic in the sense that it deals with material properties that represent averages taken over a material volume (or cross-section) large enough to contain many particles of the solid. It applies to systems in which the energetics and hydrodynamics of the liquid phase are fully characterised by readily measurable functions that relate the chemical potential of the liquid and the permeability of the system to the liquid content. The theory is restricted to one-dimensional processes, although this is not seen as a serious restriction on its application because many separation processes approximate the one-dimensional condition. Finally, we describe a range of tests of the theory, present experimentally-derived material properties which might be used in illustrative calculation for a number of practically important situations, and identify initial and boundary conditions for many important separtion procedures.

INTRODUCTION

Theories of filtration and sedimentation have their origins in chemical engineering (e.g., Shirato et al. 1986), in civil engineering (e.g., Gibson et al. 1981), fluid mechanics (Wakeman 1986) and in soil science (Philip and Smiles 1982). There are, however, difficulties which restrict application in each discipline and, in fact, conventional wisdom in one may await revelation in another.

These problems are based, in part, in semantics and the use of different functional variables in different disciplines. An example is the use of water potential in soil physics to describe what in chemical engineering is the negative of the solid compressive pressure, or in civil engineering the negative of the effective stress. Another problem arises when empircial relations are used prematurely in the derivation of the flow equations appropriate to the processes under consideration. For example, the early substitution of the Kozeny-Carman relation, which purports to relate the permeability (or hydraulic conductivity) to the void ratio (or volume fraction of water) of the system, in the derivation of a flow equation often greatly complicates the form of that equation and tends to obscure the principles on which it is based.

A second important difficulty arises because the demands of engineering practice often preclude the opportunity for generation or application of innovative theory. The continued use of Terzaghi's approximate theory of consolidation in soil mechanics, and its recent introduction into chemical engineering when "more correct" formulae are now available, is an example of this problem.

This paper addresses three issues:

i) A filtration/sedimentation equation is derived from continuity equations of the liquid and the solid, and Darcy's Law. In the process, the physical rationale for material coordinates (which much simplify the solution of flow problems) is set out, together with the reason for using volume fraction in physical space, and void ratio in material space.

ii) The material properties which are necessary and sufficient to specify the material are then examined. These are required to predict various important aspects of filtration and sedimentation.

iii) Finally, the paper discusses the initial and boundary conditions for which the flow equation must be solved for a range of important separation procedures.

THEORY

The theory developed here is for essentially one-dimensional flow in a system of two phases (liquid and solid). The development was helped by discussions with PAC Raats, who presents elsewhere (Raats 1984) a more general dissertation on rational coordinates. The theory is also macroscopic (Philip 1972) and deals with measurable physical quantities averaged over a volume great enough to contain many particles.

Without loss of generality, we take the liquid to be water. We thus deal with flow in a system in which θ_s and θ_w are the volume fractions of solid and water, so that

$$\theta_s + \theta_w = 1 \ . \tag{1}$$

Material Balance Equations

In this system, during one-dimensional, unsteady flow, equations of continuity may be written for the water (2) and for the solid (3), viz.

$$\left[\frac{\partial \theta_w}{\partial t}\right]_z = -\left[\frac{\partial F_w}{\partial z}\right]_t \ , \tag{2}$$

$$\left[\frac{\partial \theta_s}{\partial t}\right]_z = -\left[\frac{\partial F_s}{\partial z}\right]_t \ . \tag{3}$$

In these equations F_w and F_s are the volume fluxes of water and solid, respectively, relative to an external observer, and z and t are distance and time, respectively.

During flow involving water content change, both the water and the solid are in motion, so the flux of water relative to an observer has a component, u, relative to the solid particles, and a "convective" component associated with the moving particles. Thus, F_w may be written

$$F_w = u + \theta_w F_s/\theta_s = u + \vartheta F_s \ . \tag{4}$$

In (4), F_s/θ_s is the average velocity of the solid and $\theta_w/\theta_s = \vartheta$ is the moisture ratio (volume of water per unit volume of solid). In a saturated system, ϑ is identical to the void ratio e: the distinction is maintained to anticipate situations where air may enter the system and $\vartheta < e$ (Philip and Smiles 1969).

If we now substitute for F_w from (4) in (2), and also substitute $\vartheta\theta_s$ for θ_w, we obtain

$$\left[\frac{\partial(\vartheta\theta_s)}{\partial t}\right]_z = - \left[\frac{\partial u}{\partial z}\right]_t - \left[\frac{\partial(\vartheta F_s)}{\partial z}\right]_t \ . \tag{5}$$

Differentiation by parts, followed by the elimination of two terms using (3), and division by θ_s then yield the equation

$$\left[\frac{\partial\vartheta}{\partial t}\right]_z = -(1/\theta_s) \left[\frac{\partial u}{\partial z}\right]_t - (F_s/\theta_s) \left[\frac{\partial\vartheta}{\partial z}\right]_t \ . \tag{6}$$

Material Coordinates

Equation (6) provides a basis for an Eulerian analysis of unsteady flow problems expressed in terms of the space coordinate z (Philip 1968; Wakeman 1978). Alternatively, a Lagrangian analysis may be used if we recognize that the left-hand side of (6), together with the second term on the right, represent the differential of ϑ following the motion of the solid (e.g., Bird et al. 1963; McNabb 1960; Smiles and Rosenthal 1968), that is

$$\left[\frac{\partial\vartheta}{\partial t}\right]_z + (F_s/\theta_s) \left[\frac{\partial\vartheta}{\partial t}\right]_t = \left[\frac{\partial\vartheta}{\partial t}\right]_m \ , \tag{7}$$

with m(z,t) a material coordinate defined by the equations

$$\frac{\partial m}{\partial z} = \theta_s \quad \text{and} \quad \frac{\partial m}{\partial t} = -F_s \ , \tag{8}$$

so that

$$dm = \theta_s \, dz - F_s \, dt \ . \tag{9}$$

For both filtration and sedimentation, the m-coordinate is determined by integrating (9), taking advantage of the fact that at the filter membrane or the base of the vessel containing the sedimenting system (z = 0), the flux of solid, F_s, is zero. Thus

$$m(z,t) = \int_0^z \theta_s(\tilde{z},t) \, d\tilde{z} - \int_0^t F_s(0,\tilde{t}) \, d\tilde{t}$$

$$= \int_0^z (1 + \vartheta)^{-1} \, dz \ , \tag{10}$$

and m is therefore the cumulative volume of the solid, per unit area of cross-section, measured into the system from $z = 0$.

It follows from (10) that a system of initial length, L, and water content ϑ_n will have a fixed material length $M = L/(1 + \vartheta_n)$.

Note that the equations (8) satisfy the continuity equation (3) for the solid.

Substitution of (7) in (6) and the use of the first of (8) yield the continuity equation for the water in material space, viz.

$$\left[\frac{\partial \vartheta}{\partial t}\right]_m = -\left[\frac{\partial u}{\partial m}\right]_t . \tag{11}$$

Further development of the theory concentrates on (11), and in particular on the laws of flow necessary to define u.

Darcy's Law for Colloidal Systems

In the system we describe, Darcy's Law describes the volume flux of the water *relative to the particles* in response to a *space gradient* of piezometric head (Zaslavsky 1964). Here we use the total potential Φ of the water rather than the piezometric head, although the two are simply related (as we shall show). Darcy's Law then becomes

$$u = -k(\vartheta)\ \upsilon^{-1}\ \partial\Phi/\partial z . \tag{12}$$

In (12) u has units ($m\ s^{-1}$), υ is the kinematic viscosity of water ($m^2\ s^{-1}$), and $k(\vartheta)$ is the water content dependent permeability. If Φ is expressed as energy per unit mass of water, with SI units $J\ kg^{-1}$, then the permeability takes units of m^2. The form of $k(\vartheta)$ in systems that change their volume with ϑ is well known and Carman (1937), in particular, explored methods for its prediction. Here we assume that $k(\vartheta)$ is well-defined.

Total Potential of Water

In a two phase swelling material, the total potential of the water is given by

$$\Phi = \Psi(\vartheta) + \Omega + gz . \tag{13}$$

In (13) g is the acceleration due to gravity so gz is the gravitational potential of water at z relative to a convenient datum; Ω is the overburden potential; and $\Psi(\vartheta)$ is the water content dependent potential that arises as a result of interaction of the water with the solid surfaces and their geometry. $\Psi(\vartheta)$ is readily measured (Black 1965).

The overburden potential is defined (Philip 1969a) by

$$\Omega = g \int_z^{z_1} \gamma_w\, dz + P , \tag{14}$$

where γ_w is the wet specific gravity of the system, $z = z_1$ is its upper surface, and P is any normal surface load.

Combination of (13) and (14) yields

$$\Phi = \Psi(\vartheta) + g \int_{z}^{z_1} \gamma_w \, dz + P + gz$$

$$= p_w(z) + gz \ . \tag{15}$$

In (15) $p_w(z)$ is the water pressure measured with a manometer fitted with a membrane that permits passage of water but not solid. According to (15), $\Psi(\vartheta)$ is the negative of the "effective stress" of civil engineering theory (Aitchison 1961) or the interparticle or solid compressive pressure of filtration theory (Shirato et al. 1970).

Substitution of (13) in (12) and the inclusion of (14) then yield Darcy's Law in the form

$$u = -k(\vartheta)\upsilon^{-1} \left[\frac{\partial\Psi}{\partial z} + g(1 - \gamma_w)\right]$$

$$= -\frac{k(\vartheta)}{\upsilon(1 + \vartheta)} \left[\frac{\partial\Psi}{\partial m} + g(1 - \gamma)\right] \ . \tag{16}$$

Here γ is the specific gravity of the solid component of the system, and we have used (8), noting that $\theta_s = (1 + \vartheta)^{-1}$, to derive the second equality. The group $k(\vartheta)\upsilon^{-1}(1 + \vartheta)^{-1}$, units (s), plays the same role in material space as the hydraulic conductivity $K(\vartheta)(= k(\vartheta)\upsilon^{-1})$, units (s), plays in physical space.

Equation of Unsteady Vertical Flow in a Two Component System

A general equation of flow now follows if we substitute for u from (16) in (11):

$$\frac{\partial\vartheta}{\partial t} = \frac{\partial}{\partial m} \left[\frac{k(\vartheta)}{\upsilon(1 + \vartheta)} \frac{\partial\Psi}{\partial m}\right] - g(\gamma - 1) \frac{\partial}{\partial m} \left[\frac{k(\vartheta)}{\upsilon(1 + \vartheta)}\right] , \tag{17}$$

which may be written as the nonlinear Fokker-Planck equation (Philip 1969b)

$$\frac{\partial\vartheta}{\partial t} = \frac{\partial}{\partial m} \left[D(\vartheta) \frac{\partial\vartheta}{\partial m}\right] - E(\vartheta) \frac{\partial\vartheta}{\partial m} , \tag{18}$$

in which the moisture diffusivity D is given by

$$D = \frac{k(\vartheta)}{\upsilon(1 + \vartheta)} \frac{d\Psi}{d\vartheta} \tag{19}$$

and the coefficient E by

$$E = g(\gamma - 1) \frac{d}{d\vartheta} \left[\frac{k(\vartheta)}{\upsilon(1 + \vartheta)}\right] . \tag{20}$$

The moisture diffusivity D ($m^2 s^{-1}$) is a coefficient of consolidation in civil engineering terms, and an expression coefficient in filtration theory. The coefficient E ($m s^{-1}$), which embodies the effects of gravity both directly and through the overburden component of potential, does not generally appear in filtration theory.

The fact that both D and E vary with ϑ complicates the solution of (20). There is, nevertheless, a substantial literature, in soil physics and hydrology in particular, devoted to analytical, quasi-analytical and numerical methods of solving this equation subject to relevant conditions.

In practice, however, this full solution may be unnecessary for many situations and we show later that for a material with $\gamma = 2.6$ and $\vartheta = 25$, the "effect of gravity" is substantially reduced. The flow process, in consequence, approximates more closely that of a "gravity free" system for which (18) becomes the nonlinear diffusion equation

$$\frac{\partial\vartheta}{\partial t} = \frac{\partial}{\partial m}\left[D(\vartheta)\,\frac{\partial\vartheta}{\partial m}\right] . \tag{21}$$

The importance of gravity also diminishes as the imposed pressure P_i is increases [cf. (15)], and the major features of constant pressure filtration, for example, are well predicted by solutions to (21).

MATERIAL PROPERTIES AND THEIR MEASUREMENT

It is evident in the derivation of (17) and (18) that the conductivity and permeability functions $K(\vartheta)$ and $k(\vartheta)$, the water potential $\Psi(\vartheta)$ and the specific gravity of the solid must be specified if the theory is to be used. These macroscopic functions, or combinations such as the diffusivity $D(\vartheta)$, may be measured, but are not generally predictable.

In this section we present examples of these properties measured for suspensions of bentonite in water with initial water contents $\vartheta_n \cong 40$ and $\vartheta_n \cong 400$. The different initial water contents present problems which force us to use different methods of measurement. We thus have an opportunity to test the basis of theory in these sets of data.

Initial Water Content, $\vartheta_n = 40$

<u>Water Potential - Water Content Relation $\Psi(\vartheta)$</u>: The $\Psi(\vartheta)$ relation may be determined using well-established methods. The data here were derived from pressure cell experiments in which a thin sample of a clay-water mixture is subjected to a constant gas pressure P in excess of atmospheric pressure. The water but not the clay may escape through a filter membrane to a pool of free water at atmospheric pressure in the plane of the pressure cell membrane. Equilibrium is assumed to exist when no more water leaves the cell, and when in a thin sample (according to (15)), $\Psi = -P_i$. The sample is then recovered and the water content determined by drying at 105°C.

Figure 1 shows the results of a series of such experiments on suspensions of bentonite in equilibrium with three different NaCl solutions. For each solution concentration, three different temperature regimes were used. Table 1 sets out chemical properties of the three samples of clay used.

Each $\Psi(\vartheta)$ has been fitted to the equation

$$\ell n(-\Psi) = \alpha + \beta\vartheta \tag{22}$$

with coefficients given in Table 1. It is clear from the table and the figure that while the equilibrium salt concentration significantly affects the $\Psi(\vartheta)$ relation, there is no signficant temperature effect. Diffuse double layer concepts are qualitatively consistent with these data, but quantitative prediction of the three $\Psi(\vartheta)$ relations is not possible for structural reasons set out by Bolt (1982). The data shown in Fig. 1, however, provide an unambiguous and sufficient characterization of the clay-water interaction for macroscopic purposes, and an easily determined macroscopic relationship for testing integration of microscopic theory.

It should also be noted that while the data in Fig. 1 extend from $\vartheta = 25$

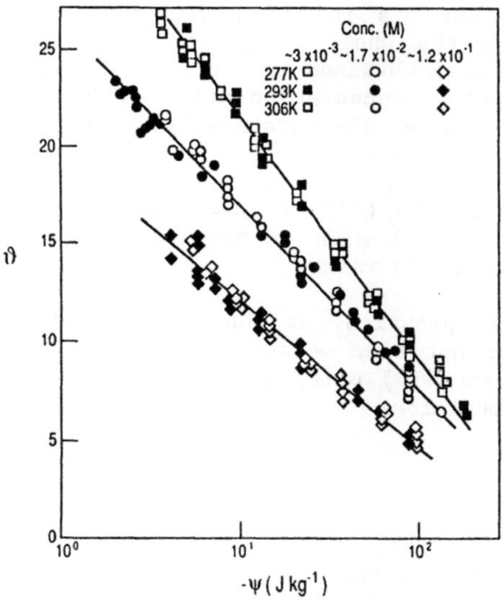

Fig. 1. Experimental data relating the water content ϑ to the water potential Ψ (J kg⁻¹) for samples of bentonite in equilibrium with three solution salt (NaCl) concentrations at three temperatures. The straight lines represent the regression equation $\ell n(-\Psi) = \alpha + \beta\vartheta$. (Smiles 1986).

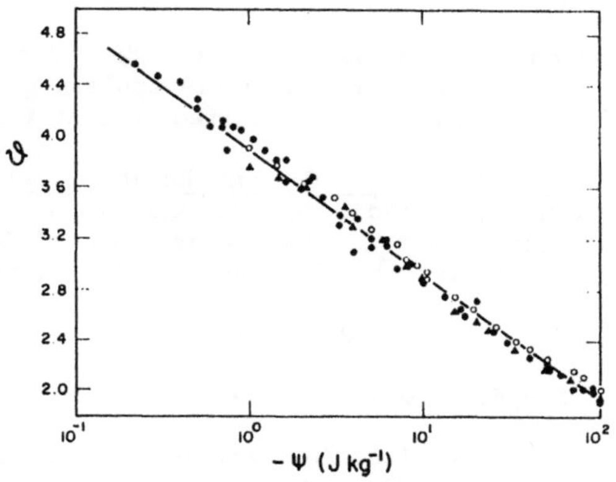

Fig. 2. Experimental data relating the water content ϑ to the water potential Ψ (J kg⁻¹) for "red mud". (Smiles 1986).

Table 1. Coefficients of the regression equation $\ell n(-\Psi) = \alpha + \beta\vartheta$ used to fit the experimental $\Psi(\vartheta)$ data shown in Fig. 1 (Smiles et al. 1985).

NaCl Concentration Molarity	Temperature, °K 277	293	306	Overall Average
	Value of α			
4×10^{-3}	6.321	6.307	6.428	6.353
1.6×10^{-2}	6.419	6.718	6.159	6.514
1.2×10^{-1}	6.107	6.199	5.891	6.060
	Value of $-\beta$			
4×10^{-3}	0.182	0.179	0.184	0.182
1.6×10^{-2}	0.233	0.255	0.219	0.243
1.2×10^{-1}	0.310	0.311	0.293	0.305
	Value of regression coefficient r^2			
4×10^{-3}	0.994	0.992	0.996	0.994
1.6×10^{-2}	0.994	0.983	0.997	0.985
1.2×10^{-1}	0.964	0.989	0.983	0.972

($\theta_w = 0.96$) to $\vartheta = 5$ ($\theta_w = 0.83$), the technique was initially developed for nonswelling porous media and applies through the whole range of suspension concentrations. Figure 2, for example, shows data obtained in a manner similar to that for Fig. 1. This data is for "red mud" produced as an effluent during the Bayer extraction of alumina from bauxite. Again, the data may be fitted to (22).

It must be emphasized that the fit of these data to (22) is irrelevant to the major argument; the analysis applies generally, and independently of the form of $\Psi(\vartheta)$: all that is required is that $\Psi(\vartheta)$ be single valued.

Diffusivity - Water Content Relation $D(\vartheta)$: In practice, at the initial water content discussed here, the $k(\vartheta)$ relation has not been measured directly, but has been inferred from measurements of $D(\vartheta)$ and $\Psi(\vartheta)$ using (19). Measurement of $D(\vartheta)$ relies on solutions to (21) for the conditions

$$\vartheta = \vartheta_n \qquad m > 0 \qquad t = 0 \tag{23}$$

$$\vartheta = \vartheta_0 \qquad m = 0 \qquad t > 0 \ , \tag{24}$$

which are realized in experiments in which a slurry, initially at uniform water content, is filtered at constant pressure. The diffusivity $D(\vartheta)$ is then calculated from the water content profile, using the equation

$$D(\vartheta)\ d\vartheta/d\lambda = -\int_{\vartheta_n}^{\vartheta} \lambda/2\ d\vartheta \tag{25}$$

in which $\lambda = mt^{-\frac{1}{2}}$. Figure 3 shows typical water content profiles for the material in equilibrium with $\sim 3 \times 10^{-3}$ M NaCl and $\Psi(\vartheta)$ shown in Fig. 1.

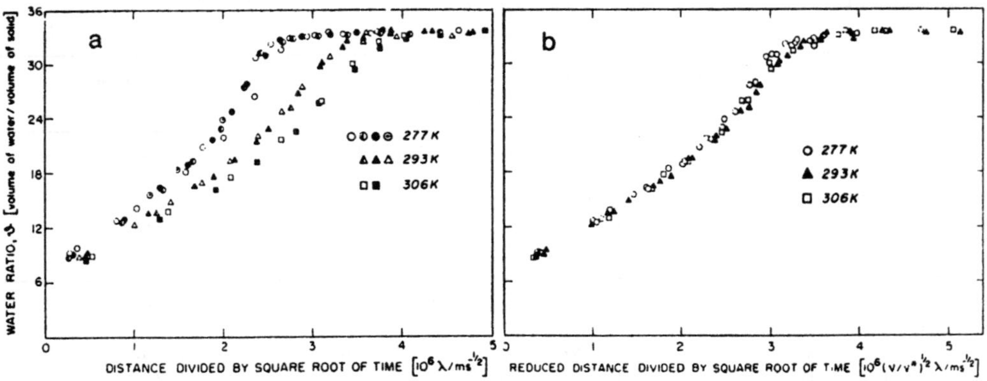

Fig. 3. Water content profiles observed during constant pressure desorption of bentonite in 1.2×10^{-1} M NaCl. Profiles at each of three temperatures were determined in a number of experiments terminated at different times as described in the text. The variable $\lambda = mt^{-\frac{1}{2}}$ reduces profiles obtained at the same temperature to a single curve, while $(\upsilon/\upsilon_*)^{\frac{1}{2}}$ scales the profiles to 293°K according to the temperature dependence of the kinematic viscosity of water. (Smiles et al. 1985).

These data were generated in a series of constant pressure filtration experiments all at the same imposed pressure, but terminated at different times to test the similarity behaviour implied in the use of the λ-coordinate, and at three different temperatures (277, 293, 306°K). As required by the analysis, the plot of the data against the similarity variable λ lies on the single curve $\vartheta(\lambda)$ for a given temperature.

Furthermore, for all temperatures the plot of ϑ against $(\upsilon/\upsilon_*)^{\frac{1}{2}}\,\lambda$, shown in Fig. 3, is unique. In this figure υ_* is the kinematic viscosity of water at datum temperature 293°K, so scaling with $\upsilon^{\frac{1}{2}}$ implies that regarding the process as one of viscous flow wholly accounts for the observed temperature effect.

Figure 4 shows the $D(\vartheta)$ relations computed from the data of Fig. 3 by means of (25). Also shown are $D(\vartheta)$ for clay in equilibrium with 1.6×10^{-2} M and 1.2×10^{-1} M NaCl. Although the calculation of $D(\vartheta)$ involves differentiation of the experimental $\vartheta(\lambda)$ data, and is therefore liable to amplification of errors, the resulting curves are well-defined and vary systematically with salt concentration.

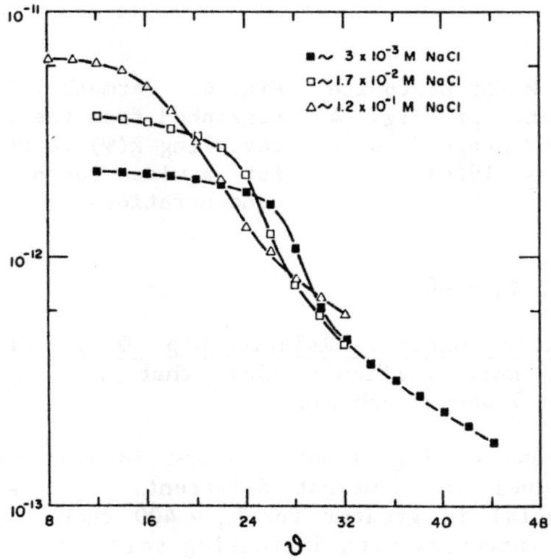

Fig. 4. Moisture diffusivity D of bentonite at 293°K and 1.2×10^{-1} M NaCl calculated from the data of Fig. 3 and graphed as a function of ϑ. The graph also shows $D(\vartheta)$ for the other two salt concentrations, derived in the same manner. (Smiles 1986).

Permeability - Water Content Relation $k(\vartheta)$: Figure 5 shows the $k(\vartheta)$ relations calculated from the data of Figs. 1 and 4 through (19). Their accuracy depends on that of the $\Psi(\vartheta)$ and $D(\vartheta)$ relations from which they are derived. Since λ scales as $\upsilon^{\frac{1}{2}}$, k is independent of υ (and of temperature), as it should be. The dependence of $k(\vartheta)$ on salt concentration is qualitatively consistent with the predictions of double layer theory on configuration of clay particles; but no reliable quantitative microscopic analysis is available to estimate $k(\vartheta)$. We note here that $k(\Psi)$ calculated using the data of Fig. 5 and Fig. 1 is apparently unique, and independent of solution salt concentration. We graph this curve as Fig. 6 and return to the point later.

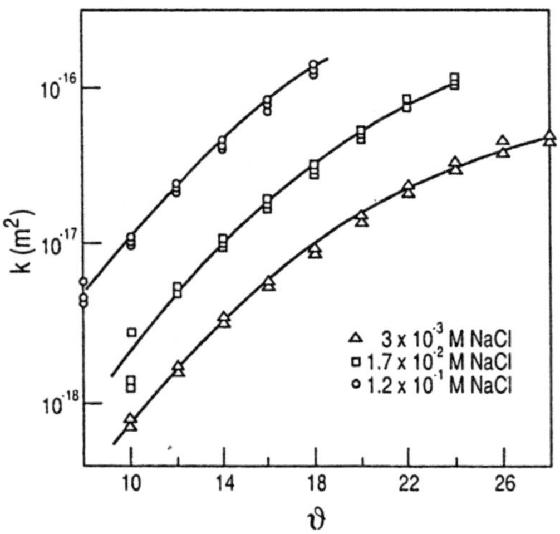

Fig. 5. Permeability k for bentonite calculated from the data of Figs. 4 and 1 using (19) and graphed as a function of ϑ. (Smiles 1986).

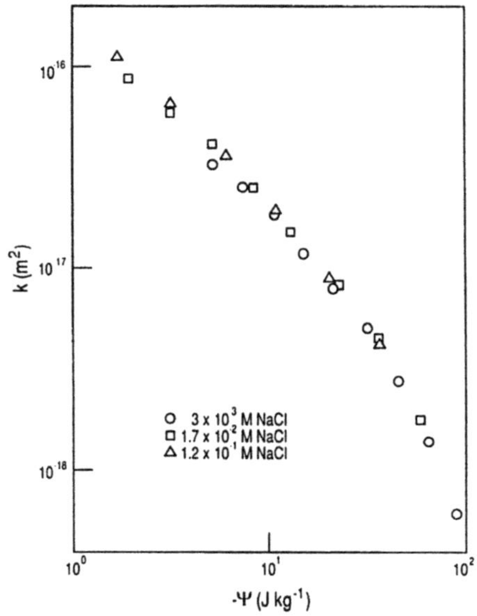

Fig. 6. Permeability k for bentonite regraphed from the data of Figs. 5 and 1 revealing $k(\Psi)$ to be effectively unique for a wide range of solution salt concentrations.

Initial Water Content, $\vartheta_n = 400$

Water Potential - Water Content Relationship $\Psi(\vartheta)$: Measurement of $\Psi(\vartheta)$ for $\vartheta_n = 400$ is no more difficult than that for $\vartheta_n = 40$ for P_i such that $\vartheta_0 \lesssim 30$. Figure 7 shows such data.

Trends similar to those of Fig. 1 are evident in relation to solution salt concentration but curves are somewhat different. In particular, the water content at any potential is greater for $\vartheta_n = 400$ than for $\vartheta_n = 40$. We also note that the effect increases with increasing salt concentration, and implies the existence of structure in the 'matrix' of clay platelets which becomes more dense as the solution salt concentration is increased (e.g., Rieke and Chilingarian 1974; Bolt 1982).

Permeability - and Diffusivity - Water Content Relations, $k(\vartheta)$ and $D(\vartheta)$: Calculation of $D(\vartheta)$ using (15) requires measurement of $\vartheta(\lambda)$, and by implication a material coherent enough to sample. This is not the case when $\vartheta_n = 400$. Two alternative methods are discussed here: the first measures $D(\vartheta)$; the second measures $k(\vartheta)$ directly.

In the first of these methods, conditions (23) and (24) were imposed in a series of filtration experiments in which ϑ_0 was varied between experiments. In each experiment the outflow, $-i$, was measured as a function of t. Linearity of $-i$ vs $t^{\frac{1}{2}}$ is required by the solution of (21) subject to (23) and (24). Furthermore, the slope S of that line is related to an integral involving $D(\vartheta)$ in the range $\vartheta_0 \leqslant \vartheta \leqslant \vartheta_n$, so that $\partial S/\partial \vartheta_0$ can be used to derive $D(\vartheta)_{\vartheta_0}$.

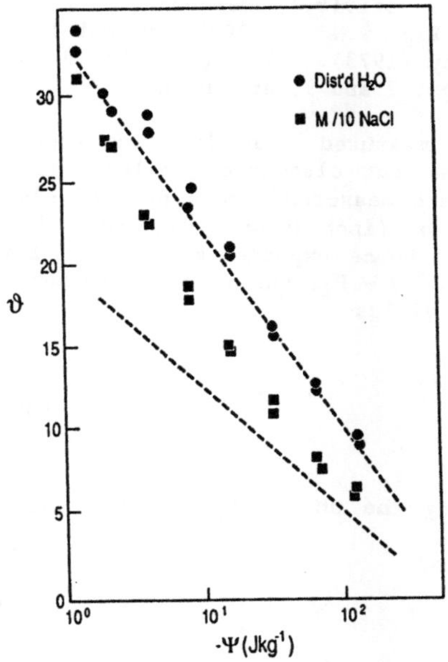

Fig. 7. Experimental $\Psi(\vartheta)$ data similar to that shown in Fig. 1, but for clay with initial water content $\vartheta_n = 400$. The dashed lines are the corresponding regressions from Fig. 1.

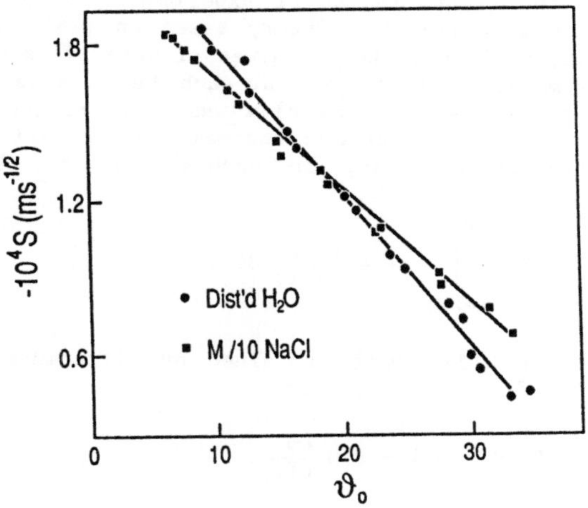

Fig. 8. Sorptivity S of bentonite with an initial water content $\vartheta_n = 400$, graphed as a function of the final water content ϑ_0.

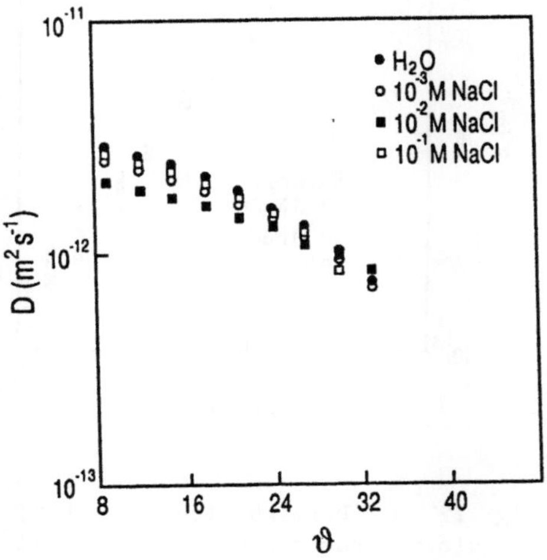

Fig. 9. Moisture diffusivity $D(\vartheta)$ derived from outflow experiments, as described in text, on clay with with $\vartheta_n = 400$.

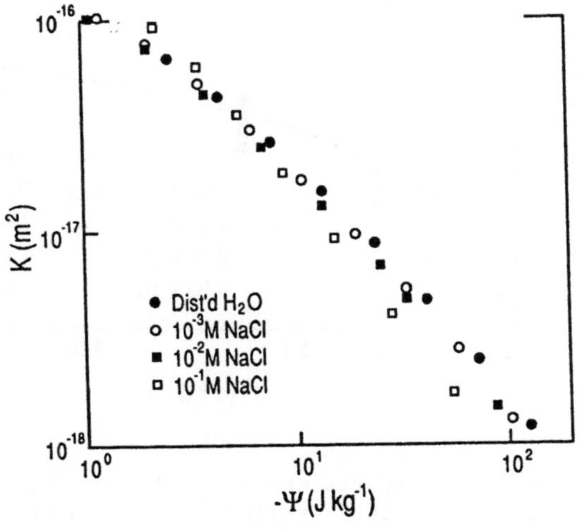

Fig. 10. Permeability $k(\Psi)$ calculated from data shown in Figs. 7 and 9 for $\vartheta_n = 400$. Note the close correspondence with the data of Fig. 6.

Figure 8 shows S (which Philip (1957) calls sorptivity), measured in these experiments, graphed as a function of ϑ_0 and Fig. 9 shows $D(\vartheta)$ derived from these data using the method of Smiles and Harvey (1973). The $k(\Psi)$ relation of Fig. 10 was then calculated from the data of Figs. 7 and 9, and using (19).

In the second series of experiments, $k(\Psi)$ was measured directly using steady flow experiments. Theory based on (16) is set out elsewhere (Smiles 1969), but, in brief, steady values of water flux u were measured across pads of clay of material thickness M at each one of a series of (increasing) pressures. The upper surface of the clay was unrestrained and hence experienced $\Psi = 0$; the lower surface, on a filter membrane, experienced $-\Psi = P_i$, the imposed pressure. Integration of (16), for each steady flow, then yields

$$uM = -\int_{\Psi=-P_i}^{\Psi=0} \upsilon \frac{k(\vartheta)}{(1+\vartheta)} d\Psi , \tag{26}$$

whence $k(\vartheta)$ may be obtained by differentiating the uM vs P_i relation using Fig. 7 and the equation

$$k(\vartheta) = \upsilon(1+\vartheta)M \frac{\partial u}{\partial P_i} . \tag{27}$$

Figure 11 shows experimental uM vs P_i data while Fig. 12 shows $k(\Psi)$ calculated from them using (27) and Fig. 7.

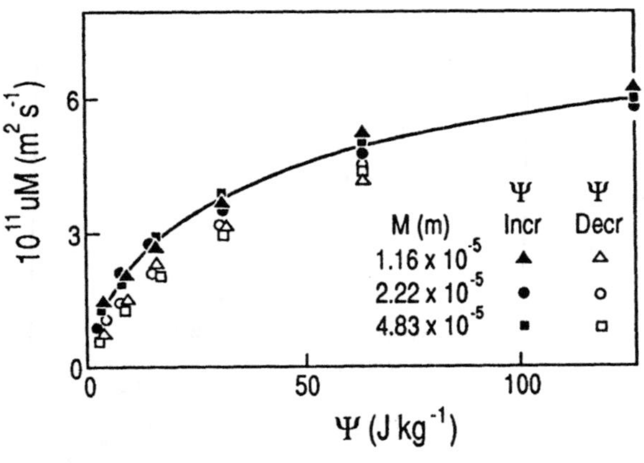

Fig. 11. Experimental data showing the relation during steady flow of water flux u and potential drop Ψ across pads of clay of material length M.

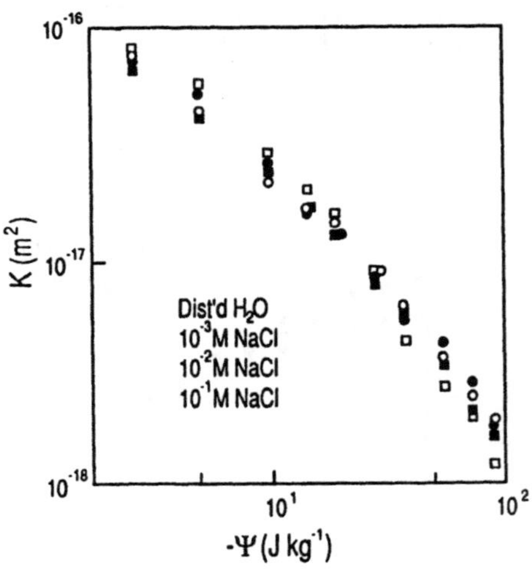

Fig. 12. Permeability $k(\Psi)$ calculated from the data of Fig. 11 using (7). Note the correspondence with Figs. 6 and 10.

Again we observe that $k(\Psi)$ is apparently unique for Na-bentonite in equilibrium with NaCl with concentrations c in the range $0 < c/M < 10^{-1}$. It is then important to compare Figs. 6, 10 and 12. Figures 6 and 10 were derived using different transient flow methods and with differential initial water contents of $\vartheta = 40$ and 400, respectively. Figure 12 was derived from steady state experiments on very thin samples of clay (< 2 mm) with initial water content $\vartheta = 400$. Furthermore, the $\Psi(\vartheta)$ data corresponding to Figs. 10 and 12 differ significantly from those for Fig. 6. The three sets of $k(\Psi)$ data are practically coincident, however, and they therefore represent strong support for the theoretical approach presented in this paper.

It is not our intention to explore specific issues in relation to these data since this is done elsewhere. It is, however, important to note tht the data set presented in this paper for bentonite suspensions is probably the most complete in existence.

Finally, we reiterate that the $k(\Psi)$ relations are not predictable but are readily measured, and the fact that $k(\Psi)$ appears to be unique over three orders of magnitude of electrolyte concentration has great practical advantage, since it implies that only the dependence on electrolyte of $\vartheta(\Psi)$ need be specified to fully characterize the system. Clearly, the issue merits further examination.

APPLICATION OF THEORY

The advantage of this macroscopic analysis lies in its predictive value and, with $\Psi(\vartheta)$ and $k(\vartheta)$ functions known, the solution of (18) or (21) for appropriate initial and boundary conditions permits prediction of the important features of solid-liquid separation. A number of these processes have been analyzed in addition to that outlined above of constant pressure filtration through a nonimpeding filter membrane described by (23) and (24).

Constant-Pressure Filtration Through an Impeding Membrane

Here, an initially uniform system is subject to a constant pressure P_i and the water escapes through a membrane that significantly impedes the flow. The initial and boundary conditions governing (21) are

$$\begin{aligned} &t = 0 \quad m \geqslant 0 \quad \vartheta = \vartheta_n \\ &t > 0 \quad m = 0 \quad D(\vartheta)\frac{\partial\vartheta}{\partial m} = c[\Psi_0(t) - \Psi_i] \ , \end{aligned} \tag{28}$$

in which c ($m^{-1}s$) is the constant transfer coefficient of the membrane, $\Psi_i = P_i$ and $\Psi_0(t)$ is the time-dependent potential at the clay-membrane interface. This boundary condition is analogous to the radiation boundary condition in heat conduction. The analysis (Smiles et al. 1982) permits prediction of the evolution of the water content profiles and the time course of both Ψ_0 and the surface water flux. The analysis can embrace particular forms of time-dependence of c, if required.

Constant-Rate Filtration

In this case, an initially uniform clay paste is compressed at a constant rate. The initial and boundary conditions governing (21) here are

$$\begin{aligned} &t = 0 \quad m \geqslant 0 \quad \vartheta = \vartheta_0 \\ &t > 0 \quad m = 0 \quad D(\vartheta)\frac{\partial\vartheta}{\partial m} = -V_0 \ . \end{aligned} \tag{29}$$

The analysis (Smiles 1978) permits calculation of both the evolution of the water content profile for any given rate of outflow of water $-V_0$ and also the rate of increase of pressure necessary to sustain it.

Self-Weight Filtration With a Given Rate of Accretion of Slurry

In this analysis (Smiles et al. 1979), a column of clay filtering as a result of self-weight has added to it further material (or an increasing load) at a constant rate. The initial and boundary conditions governing (18) are

$$t = 0 \quad m \geqslant 0 \quad \vartheta = \vartheta_n$$

$$t > 0 \quad m = 0 \quad \vartheta = \vartheta_*(t) \;, \tag{30}$$

where ϑ_* is the ϑ-value corresponding to $\Psi_*(t) = g\int_0^t(\gamma_w R - V_0)\,dt$. Here, γ_w is the wet specific gravity of the material arriving at the surface of the slurry at the rate R ($m\ s^{-1}$), and $V_0(t)$ the rate of escape of liquid through the membrane.

Sedimentation

The process may be described by solutions of (18). The equation is subject to the following initial and boundary conditions:

$$t = 0 \quad 0 \leqslant m \leqslant M \quad \vartheta = \vartheta_n$$

$$t > 0 \quad m = 0 \text{ and } M \quad \frac{\partial\vartheta}{\partial m} = g(\gamma_w - 1)\frac{\partial\vartheta}{d\Psi} \;. \tag{31}$$

Note from (16) that the second of (31) represents no-flux conditions at the upper and lower boudaries, m = 0 and M. Integral behaviour and the evolution of profiles of liquid content and particle concentration are predicted fairly well by an approximate solution of (18) subject to conditions (31). Furthermore, sedimentation in a centrifugal field of strength G ($m\ s^{-2}$) is described by solutions of (18) subject to conditions (31) with G substituted for g in both (14) for E and conditions (31).

In addition, the final equilibrium profiles of water and solid content for sedimentation may be calculated by integrating (15) to find $z(\vartheta)$ with the integration constant found from the material conservation condition (Smiles 1976). The calculation of centrifugal equilibrium is similar with G in place of g.

CONCLUDING REMARKS

We have shown how a macroscopic approach developed originally in the context of soil water flow may be generalized and extended to apply to filtration processes. The central concepts in the analysis relate to the energetics and the flow of the liquid phase, and lead to a nonlinear Fokker-Planck equation which often may be simplified to a diffusion equation.

The approach applies to systems that are characterized by two macroscopic functions, $\Psi(\vartheta)$ and $k(\vartheta)$ (or $D(\vartheta)$) which may be measured in simple experiments. If these functions are known, then the analysis permits prediction of the important features of filtration subject to a range of initial and boundary conditions, and to sedimentation.

Difficulties arise both because of the often strong dependence of Ψ and k on ϑ, and then because of the sometimes complicated initial and boundary conditions experienced in filtration processes. Here, we concern ourselves with the development and test of the basic mechanisms of flow. We pursue problems for which at least quasi-analytical methods of solution are available.

We illustrate these issues with experimental data derived for bentonite in equilibrium with a wide range of NaCl solution concentrations, a range of temperatures, a range of methods and widely differing initial water contents. These data collectively represent the most complete experimental set available, and they provide strong support for the theory.

The generality and adaptability of the approach is emphasized, it being necessary and sufficient only that $\Psi(\vartheta)$ and $k(\vartheta)$ (or $D(\vartheta)$) exist and be measurable for the material to be filtered.

Finally, we reiterate that the approach is basically macroscopic, and free from abstractions and simplifications required for development from a microscopic scale. All the information required is found in $\Psi(\vartheta)$ and $k(\vartheta)$, and these may be reliably measured in relatively simple experiments.

REFERENCES

Aitchison GD (1961) Relationships of moisture stress and effective stress functions in unsaturated soils. In: Conf on Pore Pressure and Suction in Soils. Butterworths, London, pp 47-52

Bird RB, Stewart WE, Lightfoot EN (1963) Transport phenomena. Wiley, New York

Black CA (ed) (1965) Methods of soil analysis Part 1 Physical and mineralogical properties, including statistics of measurement and sampling. Am Soc Agron Inc, Madison, Wisconsin

Bolt GH (1982) Soil chemistry. B. Physico-chemical models. Elsevier, Amsterdam, p 527

Carman PC (1937) Fluid flow through granular beds. Trans I Chem E 15: 150-166

Gibson RE, Schiffman RL, Cargill KW (1981) The theory of one-dimensional consolidation of saturated clays II. Finite non-linear consolidation of thick, homogeneous layers. Can Geotech J 18: 280-293

McNabb A (1960) A mathematical treatment of one-dimensional soil consolidation. Q Appl Math 17: 337-347

Philip JR (1957) The theory of infiltration: 4 sorptivity and algebraic infiltration equations. Soil Sci 84: 257-264

Philip JR (1968) Kinetics of sorption and volume change in clay-colloid pastes. Aust J Soil Res 6: 249-267

Philip JR (1969a) Moisture equilibrium in the vertical in swelling soils. I. Basic theory. Aust J Soil Res 7: 99-120

Philip JR (1969b) Theory of infiltration. Adv Hydrosci 5: 215-296

Philip JR (1972) Hydrostatics and hydrodynamics in swelling media. In: Fundamentals of transport phenomena in porous media. IAHR, Elsevier, Amsterdam, pp 341-350

Philip JR, Smiles DE (1969) Kinetics of sorption and volume change in three component systems. Aust J Soil Res 7: 1-19

Philip JR, Smiles DE (1982) Macroscopic analysis of the behaviour of colloidal suspensions. Adv Colloid Interface Sci 17: 83-103

Raats PAC (1984) Appendix 5D. In: Truesdell C (ed) Rational thermodynamics, 2nd edn. Springer-Verlag, New York

Rieke HH, Chilingarian GV (1974) Compaction of argillaceous sediments. Elsevier, Amsterdam, p 424

Shirato M, Kato H, Kobayashi K, Sakazaki H (1970) Analysis of settling of thick slurries due to consolidation. J Chem Eng Jap 3: 98-104

Shirato M, Murase T, Iwata M, Kurita T (1986) Principles of expression and design of membrane compression-type filter press operation. In: Cheremisinoff NP (ed) Encyclopedia of fluid mechanics, vol 5, Slurry flow technology. Gulf Pub Co, Houston, pp 905-964

Smiles DE (1969) Steady flow experiments in saturated clays. Aust J Soil Res 7: 91-98
Smiles DE (1976) Sedimentation and filtration equilibria. Separ Sci 11: 1-16
Smiles DE (1978) Constant rate filtration of bentonite. Chem Eng Sci 33: 1355-1361
Smiles DE (1986) Principles of constant pressure filtration. In: Cheremisinoff NP (ed) Encyclopedia of fluid mechanics, Vol 5, Slurry flow technology. Gulf Pub Co, Houston, pp 905-964
Smiles DE, Harvey AG (1973) Measurement of moisture diffusivity of wet swelling systems. Soil Sci 116: 391-399
Smiles DE, Rosenthal MJ (1968) The movement of water in swelling materials. Aust J Soil Res 6: 237-248
Smiles DE, Barnes CJ, Gardner WR (1985) Water relations of saturated bentonite: Some effects of temperature and solution salt concentration. Soil Sci Soc Am J 49: 66-69
Smiles DE, Knight JH, Nguyen-Hoan TXT (1979) Gravity filtration with accretion of slurry at a constant rate. Separ Sci & Technol 14: 175-192
Smiles DE, Raats PAC, Knight JH (1982) Constant pressure filtration: the effect of a filter membrane. Chem Eng Sci 37: 707-714
Wakeman RJ (1978) A numerical integration of the differential equations describing the formation of and flow in compressible filter cakes. Trans I Chem E 56: 258-265
Wakeman RJ (1986) Theoretical approaches to thickening and filtration. In: Cheremisinoff NP (ed) Encyclopedia of fluid mechanics, vol 5, Slurry flow technology. Gulf Pub Co, Houston, pp 649-684
Zaslavsky D (1964) Saturated and unsaturated flow equation in an unstable porous medium. Soil Sci 98: 317-321

ONE-DIMENSIONAL SOLID-LIQUID SEPARATION: COMMENTARY

C.J.D. Fell and R.M. McDonogh

School of Chemical Engineering and Industrial Chemistry, University of New South Wales, PO Box 1, Kensington, N.S.W. 2033, Australia

Summary. A brief review of the current industrial methods for quantifying filtration, expression and gravity thickening is given. The new treatment of Smiles and Kirby (1988) is an attempt to develop a unified mathematical approach to the three processes based on experimental measurement of water content - compressive stress and permeability - compressive stress relationships. Whilst the limited experimental data provide confirmation that the theory holds in the subatmospheric region, its extrapolation to pressure filtration and expression situations remains unproven. Moreover, the theory suffers from the constraint of being one-dimensional and does not account for the nonplug-flow phenomena (channelling and funnelling) occurring in full scale plant. It is relatively easy to relate the parameters of the Smiles and Kirby treatment to those used in industrial practice. Use of the Smiles and Kirby approach ensures that effects such as varying moisture content throughout the cake and the effect of solids movement are accounted for.

INTRODUCTION

Filtration and sedimentation remain the principal means used in the processing industries for separation of solids from liquids. Both are relatively inexpensive operations, relying on physical size difference or differences in specific gravity to effect the separation. Thus, filtration uses a filter medium or septum to exclude particles beyond a certain size, with the passage of liquid being induced by downstream vacuum or upstream pressure. In sedimentation, the solid particles in the feed slurry are allowed to settle under gravity or under a centrifugally induced field.

In both processes the key parameter of interest is the water content of the solids-rich underflow. For conventional sedimentation tanks this varies from 10-30 weight percent solids, depending on the particle size in the slurry, nature of the solid (discrete particles or flocs) and allowed settling time. In filtration, the forces exerted by the applied pressure assist in the dewatering process and can lead to higher degrees of dewatering. However, this is at the cost of considerably greater energy input.

Attempts to augment the extent of dewatering in sedimentation and thickening include the use of deep cone thickeners, where a large burden of concentrated solids is established to assist in dewatering by the imposition of mechanical stress, and the development of raking devices designed to compact the concentrated solid stream in the base of the thickener.

In filtration there has been a move towards dewatering filters, particularly those of the belt type, where the slurry, after initial dewatering, is expressed between moving belts, pressed together as they pass over sets of rollers. Other filtration devices relying on continuous expression are in a state of development.

Some idea of the size of the equipment in use in industry can be gained from the filtration and thickening requirements of Australian resource processing industries. In alumina plant, for example, up to 10^6 tonnes of alumina trihydrate must be filtered and several times this quantity of red mud must be settled annually. Similar quantities are handled annually in each of Australia's coal washeries, with coarse and fine coal undergoing filtration and coal tailings or slimes containing difficult-to-process clays being settled and thickened. Other industries making significant use of filtration and thickening are the sugar, paper and fertiliser industries. Here, however, the size of units is smaller. The dewatering of sewage sludge remains a highly significant problem worldwide. Sewage sludges from municipal treatment works in large cities are generated in high volumes. They are difficult to dewater and are the subject of much research in Japan in particular, where Shirato (1986) has estimated the volume as 2-3 x 10^6 m^3y^{-1}.

The layers of concentrated solids formed in filtration and sedimentation devices vary both in depth and permeability. In filtration equipment the cake layer is seldom more than 100 mm thick. In sedimentation units the concentrated solids layer may be several metres thick. Cake permeabilities experienced in filtration range from 10^{-9} to 10^{-11} m^2 (high permeability) through to 10^{-11} to 10^{-12} m^2 (intermediate permeability) and 10^{-12} to 10^{-16} m^2 (low permeability). Low permeability filtration cakes are generally associated with fine solids, particularly those forming flocs that tend to compress.

PRESENT INDUSTRIAL DESIGN BASES FOR FILTRATION AND SEDIMENTATION EQUIPMENT

Filtration Equipment

The classical analysis of filter performance is based on the equation of Ruth (1946), which relates the rate of filtrate collection to the applied pressure (vacuum) and the hydraulic resistance of the cake and filter medium:

$$\frac{1}{A}\frac{dV}{d\theta} = \frac{p}{\mu(\alpha(W/A) + R_m)} \quad , \tag{1}$$

where symbols are as defined in Notation. Here α is the *average* specific cake resistance, which is often assumed to vary with pressure as

$$\alpha = \alpha_0\ p^{s'} \tag{2}$$

with s' varying from 0.1 to 0.8, suggesting a degree of cake compression with most solids.

To design a full scale filter, a batch filtration test is carried on the slurry in question with the aim of determining α and R_m. This is done by plotting filtration rates in terms of θ/V versus V, which yields a straight line, from the slope and intercept of which α and R_m can be recovered.

It has been recognised for some time that the specific cake resistance α varies from the surface of the filter medium to the surface of the cake (Tiller and Shirato 1964). Also, it is necessary to correct for the effect of solids movement with respect to the filtrate. Shirato (1986) has shown how both effects can be allowed for by introducing a correction term J_s such that

$$\alpha = J_s\ \alpha_R \tag{3}$$

with

$$J_s = \int_0^1 1 - \frac{(\epsilon_x - \epsilon_{avx})(m' - 1)}{(1 - \epsilon_x)\epsilon_{av}(1 - m's)} \, s\left[\frac{x}{L}\right] d\left[\frac{w_x}{w}\right] \quad . \tag{4}$$

The term J_s depends on filtration pressure and slurry concentration. The effect of slurry movement can be neglected when the ratio of the fraction of solids in the slurry to the fraction in the surface layer is less than 0.7 (Shirato 1986). For a moderately compressible material (ignition plug slurry) the value of J_s was reported as 0.835.

Shirato's method of correction allows for the effect of the scale to be taken into account and can compensate for different cake thicknesses and different feed concentrations.

Expression

The application of increased filtration pressure alone may not give a significant improvement in cake dewatering because of the variation in moisture content throughout the filter cake. Hence it is necessary to use a mechanical means (e.g., belt filtration) to express the remaining moisture.

The analytical technique used for quantifying equipment needs in expression is derived from the literature of soil mechanics. For constant pressure expression of semi-solid materials, Shirato et al. (1974) have shown that the Terzaghi-Voigt model can be used. This model is based on the consolidation equation:

$$\frac{\partial p_s}{\partial \theta_c} = C_e \frac{\partial^2 p_s}{\partial \omega^2} \tag{5}$$

and uses three parameters C_e, B and η to characterise the behaviour of the slurry. The first, C_e, the average consolidation coefficient, relates to the primary compression phase, while B and η characterise the creep behaviour of the solid which subsequently occurs with time under a constant pressure p_s.

With the Terzaghi-Voigt combined model, the average consolidation ratio U_c can be expressed as

$$U_c = \frac{L_1 - L}{L_1 - L_\infty} = B(1 - \exp(-\eta\theta_c))$$

$$+(1-B)\left\{1- \sum_{N-1}^{\infty} \frac{8}{(2N-1)^2\pi^2}\exp\left[-\frac{(2N-1)^2\pi^2 i^2 C_e\theta_c}{4\omega_0^2}\right]\right\} \quad . \tag{6}$$

Suggesting that U_c is a unique function of

$$T_c = i^2 \, C_e\theta_c/\omega_0^2 \quad , \tag{7}$$

C_e, B and η can be determined by tests using the slurry of interest in a pressure cell.

Such tests determine the void ratio e and specific resistance α as a function of compressive pressure p_s. Using the following empirical relationships for e and α,

$$e = E_0 - C_c \, \ell n[p_s] \tag{8}$$

and

$$\alpha = \alpha_0 + \alpha_1 \, p_s^{n'} \quad , \tag{9}$$

the average consolidation coefficient C_e may be written as

$$C_e = \frac{1}{\mu \rho_s \alpha_{av} (-\Delta e/\Delta p_s)} \quad . \tag{10}$$

Equations (8)-(10) provide a link between filtration and expression theory. If the value of n in (9) is high, the slurry will be a good candidate for dewatering by expression.

The design of a full scale expression unit thus involves a series of tests on a Terzaghi compression-permeability cell and the use of the data thus obtained to predict the dewatering characteristics of the slurry on the full scale unit.

Thickening

Whilst adequate theories have been developed for the particulate settling and zone settling regimes of clarifying thickeners (see, for example, Perry et al. (1984)), analysis of the compression zone is still largely empirical and the subject of considerable disagreement between authors.

The empirical approaches of Coe and Clevenger (1916) and Kynch (1952) are still used to determine the cross- sectional area of the unit. Both involve batch scale settling experiments.

Where the underflow concentration is governed by the height of the compression zone, it is necessary to carry out deep cylinder tests. Such tests involve the determination of the settling speed of the interface.

Whilst laboratory experiments can yield interesting data on rates of thickening, scale-up of thickeners is notoriously difficult because they are sensitive to the effects of impurities, bypassing and flow profiles within the unit and the effectiveness or otherwise of rakes.

Dixon (1979) has presented theory for gravity thickening and clarification. The relevant equations, which relate the local concentration to the applied compressive stress, have been expressed in both Eulerian (equipment fixed) and Lagrangian (particle fixed) reference frames. Dixon concludes that the bulk of the weight of sedimenting particles is supported by upflow of the displaced liquid, except in very deep thickeners (e.g., deep cone), where the weight of the particle burden becomes important. To characterise his sludges, Dixon has used two parameters, permeability K_q and interparticle compressive stress τ. K_q may be related to the 'permeability' K_x used in soil mechanics as $K_x = K_q/c$. Using both K_q and τ determined by a series of batch experiments, it is possible to determine the settling rate of the slurry u for design purposes.

In his analysis of thickening, Dixon (1979, 1980) has particularly drawn attention to the role of funnelling in commercial units. This, he claims, explains why laboratory data frequently do not match up with full scale equipment performance and makes scale-up hazardous.

PLACING SMILES AND KIRBY (1988) IN PERSPECTIVE

The above summaries of the current thinking on filtration, expression and gravity thickening show that to date each has been treated as a separate phenomenon. However, clearly the fundamental mechanisms in each are similar.

Particles subject to a compressive stress move to an interparticle proximity that is determined by their individual characteristics. The permeability of such a particle bed is dependent on interparticle spacing. Hence it should be possible to derive a generalised theory in which the two primary relationships are those of compressive stress - interparticle spacing and interparticle spacing - permeability. This proposition has been put forward by Wakeman (1981) and Dixon (1979) as well as by Smiles (1976).

The particular approach used by Smiles and Kirby (1988) is to cast the problem in a reference frame in which material coordinates are used. The governing equation then becomes

$$\frac{\partial e}{\partial t} = \frac{\partial}{\partial m}\left[D(e)\frac{\partial e}{\partial m}\right] - E(e)\frac{\partial e}{\partial m} \quad , \tag{11}$$

where e is the void ratio for a saturated system. The moisture diffusivity D(e) (pressure dependent) is given by

$$D(e) = \frac{k(e)}{\nu(1 + e)}\frac{d\psi}{de} \tag{12}$$

and the coefficient E (gravity-dependent) is given by

$$E = g_c(\nu - 1)\frac{d}{de}\left[\frac{k(e)}{\nu(1 + e)}\right] \quad . \tag{13}$$

The solution of (12) and (13) depends on the boundary conditions appropriate to the operation.

By determining $\psi(e)$ and D(e) (or k(e)) curves for the slurry in question in pilot scale tests, it is therefore possible to predict the performance of a full-scale unit, irrespective of the type of operation (filtration, expression or thickening). All that is required is that the $\psi(e)$ and D(e) curves be continuous over the range under investigation.

Smiles and Kirby have used filtration, flow and compression experiments to measure the $\psi(e)$ and D(e) characteristics of bentonite and red mud slurries of varying concentrations. Both $\psi(e)$ and D(e) show a dependence on the ionic content of the continuous phase. This is to be expected from double layer theory. However, when permeability k (recovered from D(e)) is plotted against compressive stress ψ, a unique curve is obtained. Smiles and Kirby do not explain why this is so. One possible explanation is that electo-osmotic effects are significant in the slurries they have examined. These will be greatest at low ionic strengths when interparticle spacing is highest, and will act to retard the flow of the continuous phase, hence bringing permeabilities at different ionic strengths more into line.

It is instructive to examine the range of experimental variables used by Smiles and Kirby in the light of the parameter values discussed earlier. Water content e values of 4, 40 and 400 correspond to solid mass fractions of 0.2, 0.084 and 0.009 respectively. The last two of these represent very dilute slurries which would not normally be encountered in conventional filtration, expression or thickening operations, except where the solid was highly flocculated.

Applied compressive stresses ψ range from 0.1 to 10 J kg^{-1}. This corresponds to applied pressures of 1 kPa to 100 kPa (0.01 - 1.0 atm).

Using the relationships

$$\alpha = \frac{k(1 - \epsilon)^2 S_0{}^2}{\epsilon^3} \qquad (14)$$

and

$$k = \frac{g_c}{\mu\alpha} , \qquad (15)$$

permeabilities and specific resistances may be calculated for Smiles and Kirby's Fig. 12 to give the values reported in Table 1.

Table 1. Calculated permeabilities and specific resistances.

Compressive Stress ψ J kg^{-1}	Pressure p atm	Permeability k m s^{-1}x10^{-10}	Specific Resistance α m kg^{-1}x10^{-13}
1	0.01	11.0	1.0
2	0.02	8.0	1.4
3	0.03	5.0	2.2
5	0.05	3.0	3.6
8	0.08	1.8	6.1
10	0.1	1.5	7.3
20	0.2	0.8	14.0
30	0.3	0.5	22.0
50	0.5	0.3	37.0
80	0.8	0.2	55.0
100	1.0	0.15	73.0

The calculated values of specific resistance suggest that the bentonite slurries studied by Smiles and Kirby are of low permeability. When values of specific resistance α are plotted on logarithmic coordinates against the applied pressure (Fig. 1), the relationship is substantially linear, with a slope of 0.6 (s' in Eq. (2)), suggesting a high level of compressibility.

The experimental values presented by Smiles and Kirby span only a modest range of the applied pressures and are not applicable to pressure filtration or expression. Without experimental data at higher pressures it is not possible to verify that the approach used can be extended to these operations.

LIMITATIONS OF SMILES AND KIRBY THEORETICAL TREATMENT

Issues raised in the discussion period following presentation of the Smiles and Kirby paper included comment: that the treatment may be limited because it was one-dimensional; that the formulation of the one-dimensional case had already appeared in the literature; and that more could be gained from a better understanding of the water-solid interactions than 'retreating into a model'. Later comment suggested that the extension of the Smiles and Kirby approach to

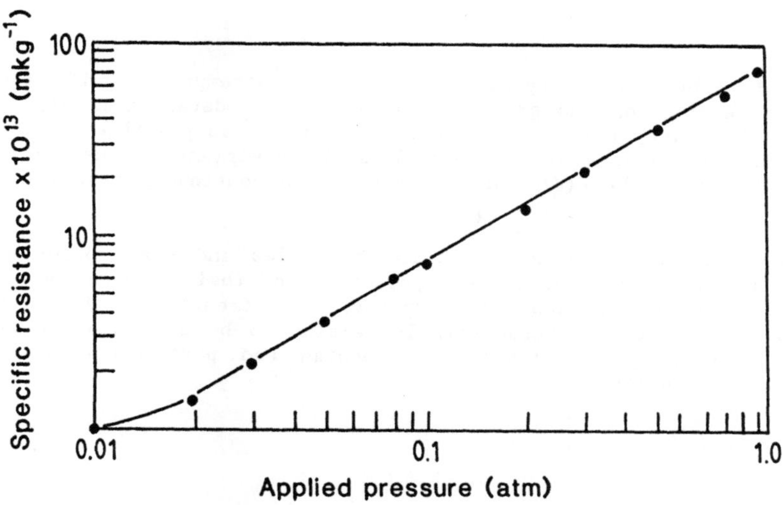

Fig. 1. Specific resistance versus applied pressure for bentonite slurries.

two and three dimensions was feasible, but that the subsequent equations would prove difficult to solve.

Certainly, it is fair to say that previous authors have presented models based on material coordinate systems (Dixon 1979) and have recognised the need to correct for varying permeability and relative movement between solids and liquid. They have used different parameters to characterise the slurry than those used by Smiles and Kirby. However, in principle, these different parameters can be related back to those (e, k, ψ) used by Smiles and Kirby.

The treatment presented by Smiles and Kirby has as its principal advantage its applicability to three different processing operations. But it has yet to be established that parameters measured in one operation can be satisfactorily used to predict the performance in another operation if the applied pressure is higher than atmospheric.

What is perhaps most of concern in the Smiles and Kirby approach is its current limitation to one-dimensional situations. Dixon (1980) has pointed out that the funnelling that occurs in commercial thickeners greatly affects the underflow composition. Hence it is necessary to treat such systems by two- or three-dimensional models. It is likely that a similar approach would be needed for a commercial expression unit, for the flow of expressed liquid will be multidimensional. There is thus a very real danger that an undisciplined use of the Smiles and Kirby treatment could yield inaccurate predictions of full-scale performance.

CONCLUSION

Smiles and Kirby (1988) represents an elegant attempt to unify theory for filtration, expression and gravity thickening. By determining $e(\psi)$ and $k(\psi)$ plots for the slurry (cake) under investigation, it is possible, in theory, to predict the performance of commercial scale equipment. The only crucial assumption is that the $e(\psi)$ and $k(\psi)$ curves are continuous over the range of interest.

Whilst many of the concepts raised in the Smiles and Kirby paper have been traversed by previous authors, and application of their approaches will yield similar predictions for scale-up purposes, the strength of the Smiles-Kirby treatment lies in its universality. It remains to be seen whether it can be effectively used by industry in the design and performance appraisal of commercial scale units.

NOTATION

Roman

A	area of filter (m^2)
B	slurry creep parameter in expression
c	volume fraction particle concentration
C_e	average consolidation coefficient (m^2s^{-1})
C_c	parameter in (8)
$D(e)$	moisture diffusivity (m^2s^{-1})
e	local void ratio
$E(e)$	parameter in (11) ($m\ s^{-1}$)
E_0	local void ratio parameter in (8)
g_c	gravitational constant ($m\ s^{-2}$)
i	number of drainage surfaces in expression
J_s	correction factor to account for cake thickness
k	permeability ($m\ s^{-1}$)
$k(e)$	water content dependent permeability (m^2)
K_q	particle coordinate permeability ($kg^{-1}m^3s$)
K_x	position coordinate permeability ($kg^{-1}m^3s$)
L	thickness of cake (m)
L_1	thickness of cake at $\theta = 0$ (m)
L_∞	thickness of cake at $\theta = \alpha$ (m)
m	material coordinate of Smiles and Kirby (1988)
m'	ratio of mass of wet to dry cake
n'	exponent in (9)
p_s	applied pressure in (5) (Pa)
p	applied pressure (Pa)
R_m	filter medium resistance (m^{-1})
s'	exponent on applied pressure in (2)
s	mass fraction of solids in slurry
S_0	specific particle surface area per unit volume (m^{-1})
t	time in (11)
T_c	dimensionless time constant in (7)
u	settling rate of slurry ($m\ s^{-1}$)
U_c	average consolidation ratio
V	volume of filtrate (m^3)
w	mass of cake solids per unit area ($kg\ m^{-2}$)
w_x	mass of cake solids per unit area in distance x from medium ($kg\ m^{-2}$)
W	weight of filter cake (kg)
x	distance from filter medium (m)

Greek

α	specific cake resistance (m kg^{-1})
α_0	specific cake resistance at zero applied pressure (m kg^{-1})
α_1	coefficient in empirical specific resistance equation (9)
α_{av}	average specific cake resistance (m kg^{-1})
α_R	specific cake resistance defined by Ruth (m kg^{-1})
ϵ	porosity of cake
ϵ_x	local value of porosity at distance x from medium
ϵ_{av}	average porosity for entire cake
ϵ_{avx}	average proosity for portion of cake between medium and distance x
η	Terzaghi-Voigt slurry creep parameter in (6)
θ	time (s)
θ_c	consolidation time in (5) (s)
μ	viscosity (Pa.s)
ν	kinematic viscosity (m^2s^{-1})
ρ_s	density of solid (kg m^{-3})
τ	interparticle compressive stress (Pa)
ψ	effective stress (J kg^{-1})
ω	solid volume per unit section area (m)
ω_0	total solid volume per unit section area in (6) (m)

REFERENCES

Coe HS, Clevenger GH (1916) Method for determining the capacity of slime settling tanks. Trans Am Inst Min Eng 55: 356-381

Dixon DC (1979) Theory of gravity thickening. In: Wakeman RJ (ed) Filtration and separation I, Elsevier, Amsterdam

Dixon DC (1980) Effect of sludge funneling in gravity thickeners. AIChE J 26(3): 471-477

Kynch GJ (1952) A theory of sedimentation. Trans Faraday Soc 43: 166-176

Perry RH, Green D (1984) Chemical engineers handbook, McGraw Hill NY p 19-53 - 19-65

Ruth BF (1946) Correlating filtration theory with industrial practice. Ind Eng Chem 38: 564

Shirato M (1986) Course notes for Workshop on Filtration and Expression, CSIRO Division of Energy Technology, Clayton

Shirato M, Murase T, Tokuraga A, Yamada, O (1974) Calculations of consolidation period in expression operations. J Chem Eng Jap 7: 229

Smiles DE (1976) Sedimentation: Integral behaviour. Chem Eng Sci 31: 243-245

Smiles DE, Kirby JM (1988) One-dimensional solid-liquid separation. This volume, pp

Tiller FM, Shirato M (1964) The role of porosity in filtration: VI. New definition of filtration resistance. AIChE J 10: 61

Wakeman RJ (1981) Thickening and filtration: Review and evaluation of recent research. Trans I Chem E 59: 147-160

ATMOSPHERIC DISPERSION - OBSERVATIONS AND MODELS

J.C. Weil*

National Center for Atmospheric Research**, Boulder, CO 80307, U.S.A.

Summary. Over the past two decades, substantial progress has occurred in our understanding of atmospheric turbulence and dispersion, both for the convective boundary layer (CBL) and for the stable boundary layer (SBL). This paper highlights the recent advances in dispersion following a brief review of early work.

For the CBL, major insights into dispersion came from the laboratory experiments of Willis and Deardorff (1976, 1978) and the numerical simulations of Lamb (1978, 1982). They showed the rapid increase in dispersion with downwind distance or travel time and its dependence on the turbulence velocity and length scales - the convective velocity scale w_* and the CBL height h. In addition, they demonstrated the important differences in dispersion from surface and elevated sources as well as the non-Gaussian nature of the vertical dispersion. These results have been used to develop new models for dispersion applications, which have given better predictions of the downwind concentration field than have older models based on empirical parameterizations.

For the SBL, the emphasis has been on modelling the vertical dispersion σ_z, which has been explored theoretically and experimentally. The small turbulence length scale in the SBL has led to the application of the diffusion equation to both surface and elevated sources. For the elevated source and a sufficiently large time t, this approach yielded $\sigma_z \propto t^{\frac{1}{2}}$ (Venkatram et al. 1984). However, the random force theory of Pearson et al. (1983) predicted that σ_z could approach a constant at large t. Both types of behaviour have been observed; further work is necessary to determine when each should occur. Due to the uncertainty in σ_z and in estimating the turbulence structure, these results have seen limited use in applications.

Improved physics in simple models has led to better model performance, but the deviation between predicted and observed concentrations is still large. This has motivated the study of concentration fluctuations.

INTRODUCTION

Dispersion in the atmospheric boundary layer (ABL) is controlled by turbulence which varies strongly with the stratification. In the unstable or convective boundary layer (CBL), turbulence is characterized by large coherent eddies

* Current affiliation: Cooperative Institute for Research in Environmental Sciences, University of Colorado, Boulder, CO 80309, U.S.A. Former affiliation: Martin Marietta Laboratories, Baltimore, MD 21227, U.S.A.

** The National Center for Atmospheric Research is sponsored by the National Science Foundation.

which scale in size with the CBL depth, i.e., typically 1 to 2 km. In the stable boundary layer (SBL), turbulence is much weaker with typical eddy sizes of the order of tens of metres or less. These differences lead to widely different dispersion rates. For example, in the CBL, a tall stack plume can be brought to the surface in high concentration within a few kilometres from the source (Venkatram 1980a; Weil and Brower 1984). However, in the SBL, elevated plumes often remain aloft with negligible surface impact for tens of kilometres (Hunt 1982).

This paper reviews atmospheric dispersion research over the past 30 years with attention focussed on dispersion from a point source in level, homogeneous terrain; interest is restricted to distances of 20 km or less and to short-term (~1 h) average concentrations. Prior to about 1970 most investigations were limited to sources in the lowest 100 m of the ABL and to downwind distances of a few kilometres (e.g., see Pasquill and Smith 1983). This limitation was caused by (1) the difficulty and expense of measuring turbulence and concentrations aloft, and (2) the lack of information about the overall turbulence structure. A brief review of work before about 1970 is given in the following section.

Major progress in understanding dispersion through the full depth of the ABL began in the 1970s. A significant contribution to this was new knowledge of the ABL turbulence structure which became available through numerical modelling (e.g., Deardorff 1972; Brost and Wyngaard 1978), field observations (Kaimal et al. 1976; Caughey et al. 1979), and laboratory simulations (Willis and Deardorff 1974). This knowledge advanced more rapidly for the CBL (Wyngaard 1983). Other major contributions toward our understanding of dispersion came from the laboratory simulations by Willis and Deardorff (1976, 1978) and the numerical simulations by Lamb (1978, 1979). They showed that the dispersion of passive tracers in the CBL was much different than predicted by the Gaussian plume model, a popular modelling tool. In later sections, we discuss the major theoretical and experimental advances since 1970 and show how this new knowledge has been used to improve simple models for dispersion applications.

Model developers have found that the incorporation of improved physics in simple models has led to better overall agreement between predicted and observed concentrations than achieved with older models (e.g., Weil et al. 1986). However, the deviation between the prediction, which is an estimate of the ensemble mean concentration for given conditions, and an individual realization - the observation - is still large, i.e., of the order of the prediction. A major portion of this deviation is believed to be caused by stochastic variability in the atmosphere. This has given a major impetus to the study of concentration fluctuations (Sykes 1988).

EARLY STUDIES OF DISPERSION IN THE LOWEST 100 METRES

Theoretical Developments and Data Analyses

Traditionally, dispersion in the lowest part of the ABL was pursued along three lines - the diffusion equation, Lagrangian similarity theory, and statistical theory. The first two approaches were useful in describing vertical dispersion from a surface release; the third was applied to vertical dispersion from an elevated source and to lateral dispersion at any height.

Diffusion equation: The eddy diffusion concept is limited to problems in which the local plume width is of the order of or large compared to the characteristic eddy size. Short-range dispersion from a surface release satisfies this limitation. The two-dimensional diffusion equation is appropriate for modelling the crosswind-integrated concentration (CWIC) from a point source. However, this equation cannot be solved analytically for a realistic logarithmic or log-linear wind profile. Thus, Roberts (see Calder 1949) solved it for power law forms of wind speed u(z) and eddy diffusivity K(z): $u(z) = u_r(z/z_r)^m$ and $K(z) = K_r(z/z_r)^n$, where z is height above the surface and u_r and K_r are reference values at the height z_r. His solution for the CWIC, C^y, in the form adopted by van Ulden (1978), is

$$C^y(x,z) = \frac{AQ}{\bar{u}\bar{Z}} \exp\left[-\left[\frac{Bz}{\bar{Z}}\right]^s\right] , \tag{1}$$

where Q is the emission rate, s (= m − n + 2 > 0) is the shape factor, A and B are constants dependent on s, $\bar{Z}$ is the mean plume height, and $\bar{u}$ is the mean wind speed at $\bar{Z}$.

Calder (1949) was the first to adapt this solution to dispersion from a surface release in the neutral surface layer. From his power law fit to the logarithmic wind profile, he obtained s = 1.37 and $C^y \propto x^{-0.86}$. The s was in good agreement with later values (~1.4 − 1.5) found by Elliot (1961) and Nieuwstadt and van Ulden (1978) in extensive analyses of concentration data from the Prairie Grass experiments. However, the exponent on x (−0.86) disagreed with the value (−1) representative of experimental data because Calder chose $K \propto z^{1-m}$ instead of $K \propto z$ (see Pasquill and Smith 1983). Early attempts to adapt Roberts' solution to non-neutral conditions were not successful (Sutton 1953), but more recent ones were (van Ulden 1978).

Lagrangian similarity theory: This theory applies to vertical dispersion of passive particles from a surface release (Monin 1959; Gifford 1962). The main hypothesis is that the Lagrangian velocity statistics of particles depend on the same parameters that govern the Eulerian velocity statistics − the friction velocity u_*; the surface heat flux $\overline{w\theta}_0$; and the buoyancy parameter g/Θ_0. w and θ are the deviations of vertical velocity and temperature about their means, Θ_0 is the mean potential temperature, and g is the gravitational acceleration. On dimensional grounds, the ensemble average vertical $\bar{Z}$ and horizontal $\bar{X}$ displacements of particles are assumed to depend on time t according to

$$\frac{d\bar{Z}}{dt} = bu_*\Phi(\bar{Z}/L), \quad \frac{d\bar{X}}{dt} = u(c\bar{Z}) . \tag{2}$$

b and c are assumed to be universal constants and Φ is a universal function; L is the Monin-Obukhov (M-O) length, $-u_*^3\Theta_0/kg\overline{w\theta}_0$, and k is the von Kármán constant. The relationship between $\bar{X}$ and $\bar{Z}$ is obtained by eliminating t and integrating the equation $d\bar{X}/d\bar{Z} = u(c\bar{Z})/bu_*\Phi$.

In testing the theory, Pasquill (1966) obtained a good correlation between the observed vertical dispersion σ_z and the Richardson number at 2 m. However, the theoretical predictions of σ_z based on Monin's Φ disagreed with the observed behaviour, mainly by underestimating the departures from neutral conditions. This deficiency was corrected by Chaudhry and Meroney (1973), who identified Φ with φ_h based on a diffusion equation analysis; φ_h is the dimensionless temperature gradient in the M-O theory and was obtained from the 1968 Kansas experiment. Their reanalysis of Pasquill's data resulted in impressive agreement between the predicted variation of $\bar{Z}$ with L and the observed. In another analysis, Klug (1968) used the Lagrangian theory to predict the axial concentration (not the CWIC) downwind of a surface source and found that a supposed universal function (not Φ) was inversely proportional to σ_v/u. This suggested that σ_v was not given by the M-O theory, at least for all cases; in particular, we now know that σ_v for the strongly unstable cases should be determined by w_*, the convective velocity scale, as discussed later.

The popularity of Lagrangian similarity theory in the 1960s was stimulated, I believe, by the success of M-O theory in describing flow structure in the surface layer. However, on balance it did not offer any real advantage over the diffusion equation. Both gave acceptable predictions of $\overline{Z}$ and C^y in the surface layer as shown by later work (Horst 1979; Nieuwstadt and van Ulden 1978; van Ulden 1978). Furthermore, the Lagrangian theory was dependent on diffusion theory for the determination of Φ, b, and c. The values of b (= k = 0.4) and c(= 0.56) were found by Ellison (see Pasquill and Smith 1983) and Chatwin (1968), respectively, using the diffusion equation.

Statistical theory: Taylor's (1921) statistical theory was well known by the 1960s, but it was in this decade and the previous one that atmospheric data became available for testing it. The theory applies to the dispersion of passive particles from a point source in homogeneous, stationary turbulence. Here, the main results of interest are predictions of the short- and long-time limits of dispersion, which are given below for σ_y, the lateral dispersion parameter:

$$\sigma_y = \sigma_v t \text{ if } t \ll T_L^y, \quad \sigma_y = (2\sigma_v{}^2 T_L^y t)^{\frac{1}{2}} \text{ if } t \gg T_L^y . \tag{3}$$

T_L^y is the Lagrangian integral time scale for the lateral turbulence.

For σ_y, Cramer (1957) demonstrated that the angular plume width at short range (x < 800 m) from a surface release was well correlated with the standard deviation σ_A of wind direction, $\sigma_A = \sigma_v/u$. This is consistent with the short-time limit of statistical theory upon replacing t by the plume travel time x/u. The larger data summary compiled by Slade (1968) showed good correlation between σ_y/σ_A and x for distances as large as 10 km.

For vertical dispersion from elevated sources, the few available data sets showed general agreement with statistical theory. For example, Hay and Pasquill's (1957) measurements from a 150-m high source supported the short-range limit out to 500 m. Hogstrom's (1964) observations of smoke puffs released at heights of 50 and 87 m verified both limits of the theory out to 5 km; the long-range limit was attained in stable conditions. In the most stable case, there was a region between the short- and long-time limits where σ_z appeared to level off, i.e. vary little with distance. This "levelling off" of σ_z in strongly stable conditions was more evident in some of the experiments conducted by Hilst and Simpson (1958). These results were consistent with Csanady's (1964) theory, which was probably not much appreciated at the time but is now; this theory is discussed later.

In summary, it was well known in the 1960s that statistical theory could describe the lateral dispersion of point source plumes from arbitrary release heights and vertical dispersion from elevated releases. However, predictions required the direct measurement of turbulence at the source height, which was difficult and expensive to do. Simple methods to estimate turbulence throughout the ABL from routine measurements of wind speed, insolation, etc. had to await the developments of the 1970s.

Practical Schemes for Dispersion Applications

In the 1960s, several simple approaches were introduced to estimate dispersion parameters (σ_y,σ_z) for point source emissions (e.g., Pasquill 1961; Singer and Smith 1966). All were used in combination with the Gaussian plume model to predict the ground-level concentrations (GLCs), C, downwind of a source:

$$C(x,y,0) = \frac{Q}{\pi U\sigma_y\sigma_z} \exp\left[-\frac{z_s{}^2}{2\sigma_z{}^2} - \frac{y^2}{2\sigma_y{}^2}\right] , \qquad (4)$$

where z_s is the source height. By far, the most widely used approach was Pasquill's. His original method gave estimates of the angular plume spread in the lateral and vertical directions as a function of distance for six "stability categories", ranging from the most unstable (A) to the most stable (F). Gifford's (1961) conversion of the angular spreads to σ_y and σ_z resulted in the so-called "Pasquill-Gifford" (PG) curves. The σ_z curve for neutral stability was based on Calder's (1949) eddy-diffusion model; those for non-neutral conditions were determined from the Prairie Grass data (Cramer 1957), which were limited to $x < 800$ m. The σ_y was calculated from statistical theory using 3-min average wind statistics from a height of 16 m. In summary, the curves applied to surface releases, short-range dispersion ($x < 1$ km), and small roughness heights (~1 cm); Pasquill's extension of them to distances exceeding 1 km was based on limited data. As pointed out earlier (Hanna et al. 1977), the PG curves have been applied to problems far beyond their intended purpose, e.g., tall stacks.

An attractive feature of Pasquill's scheme was the limited meteorological input required to use it - the wind speed at 10 m, insolation, and cloudiness. The wind speed was a measure of turbulence production due to wind shear; the insolation and cloudiness were indicators of the surface heat flux and the turbulence production or destruction due to buoyancy forces. However, the correspondence between the stability categories and turbulence measurements was not unique, as shown by Luna and Church (1972). Another weakness was the subjective choice of insolation. In an effort to resolve this, Turner (1964) gave an objective interpretation of the categories in terms of routinely available weather data from airports. The combined approach, known as the "Pasquill-Gifford-Turner" or PGT method, is the most widely used today despite the subsequent advances in our understanding.

DISPERSION IN THE CONVECTIVE BOUNDARY LAYER

The rapid dispersion of plumes in the CBL results from the large-scale updraughts and downdraughts that are driven by the surface heating. These convection elements extend from the surface to the top of the CBL, which is capped by an inversion. They have dimensions that scale with h, the CBL depth, and velocities that are proportional to w_*, the convective velocity scale:

$$w_* = \left[\frac{g\overline{w\theta_0}h}{\Theta_0}\right]^{1/3}. \qquad (5)$$

The success of w_* and h in ordering turbulence data was first shown by Deardorff's (1972) large-eddy simulations and later by laboratory experiments (Willis and Deardorff 1974) and field observations (Kaimal et al. 1976; Caughey 1982). Typical field values of h and w_* are 1 to 2 km and 2 m s^{-1} respectively.

Mechanical or shear-generated turbulence scales with u_* and is confined mostly to heights, $-z < L$. Since convection extends to h, Deardorff (1972) suggested $-h/L$ as a stability parameter characterizing the relative importance of convective and mechanical turbulence throughout the CBL; an alternative is w_*/u_*. When $-h/L > 10$, the bulk of the CBL is dominated by convection.

The CBL often can be idealized as horizontally homogeneous but vertically inhomogeneous. The largest gradients in u, Θ, and σ_w exist in the surface layer ($z < 0.1h$), where u follows the M-O similarity profile (Garratt et al. 1982). In the surface layer, the vertical and horizontal turbulence components can be estimated from the following expressions (Panofsky et al. 1977):

$$\frac{\sigma_w}{u_*} = 1.3[1 + 3(-z/L)]^{1/3} , \tag{6a}$$

$$\frac{\sigma_u}{u_*} = \frac{\sigma_v}{u_*} = [12 + 0.5(-h/L)]^{1/3} . \tag{6b}$$

The σ_w expression interpolates between the neutral limit, $\sigma_w \propto u_*$, and the free-convection limit, $\sigma_w = 1.34(g\overline{w\theta_0}z/\Theta_0)^{1/3}$ (Wyngaard et al. 1971). Likewise, the expression for σ_u, σ_v interpolates between the neutral limit, $\sigma_v \propto u_*$, and the strongly convective limit, $\sigma_v \propto w_*$. Thus, as Eq. (6b) shows, the horizontal components do not follow M-O similarity in the strong convective limit, i.e., they are not a function of z/L, but are determined by the velocity scale w_* of the large eddies; this is discussed further in Wyngaard (1988). Above the surface layer, σ_w exhibits a weaker variation with z while u, Θ, and σ_v are nearly constant; in applications, the turbulence is often assumed to be uniform with height with σ_u, σ_v, $\sigma_w \sim 0.6w_*$.

For dispersion, an important characteristic of the CBL is the nonGuassian nature of vertical velocity statistics. Numerical simulations (Lamb 1982) and field data (Caughey et al. 1983) show that the probability density function (p.d.f.) of w is positively skewed; this is consistent with the observation that downdraughts occupy a greater fraction (0.6) of the horizontal area than do updraughts.

Dispersion Characteristics

Passive plumes - overall patterns: The salient features of dispersion in the CBL have been demonstrated through laboratory experiments (Willis and Deardorff 1976, 1978, 1981), numerical simulations (Lamb 1978, 1979, 1982), and field observations (Moninger et al. 1983; Nieuwstadt 1980). They show that dispersion depends on source height z_s and can be grouped broadly into surface-layer sources, $z_s < 0.1h$, and elevated sources, $z_s > 0.1h$.

Figure 1 shows isopleths of the CWIC in an x-z plane from the Willis and Deardorff experiments. The CWIC, made dimensionless using Q/Uh, where U is the mean wind speed in the CBL, is shown as a function of the nondimensional height, $Z = z/h$, and distance, $X = w_*x/Uh$. X can be interpreted as a nondimensional time; it is the ratio of travel time x/U to the convective time scale h/w_*, which is typically 5 to 10 min.

For the near surface source (Fig. 1a), the average plume centreline, as defined by the locus of maximum concentration, ascends after a short distance, $X \sim 0.5$. In contrast, the centreline from the elevated source descends until it reaches the ground. These behaviours can be understood in terms of the release of material into either updraughts or downdraughts. For a surface source, material emitted into the base of an updraught begins rising almost immediately, whereas that released into downdraughts remains near the ground and moves horizontally. Since downdraughts occupy about 60% of the horizontal area, more material remains near the ground initially as does the maximum concentration. Once a significant amount of material is swept out of downdraughts and into neighbouring updraughts, the centreline begins to rise (Lamb 1982). For the elevated source (Fig. 1b), the centreline descent is explained by the greater areal coverage of downdraughts and, thus, the higher probability of material being released into them. In addition, the downdraughts are long-lived so that material emitted into them persists in its travel to the surface. Results for a source at $z_s/h = 0.5$ are quite similar to those in Fig. 1b.

Aside from minor differences, the overall features of dispersion calculated numerically by Lamb (1982) were in good agreement with the laboratory data; these calculations were based on a Monte Carlo model using Deardorff's numerical data as input. The field observations of Moninger et al. (1983) for $z_s/h = 0.5$ were quite similar to the numerical and laboratory simulations for the same height. More recently, the above dispersion patterns have been reproduced using the Langevin equation (Sawford and Guest 1987), which promises to be a useful simulation approach. The consistency of the results from these widely differing investigations reinforces our understanding of dispersion in the CBL.

In contrast to Fig. 1, the Gaussian plume model predicts that an elevated plume centreline remains elevated until a sufficient number of particle reflections occurs at the surface; the centreline then moves toward the surface (Lamb 1982). The difference between the "Gaussian" and observed behaviour is the positive skewness in the "real" p.d.f. of w.

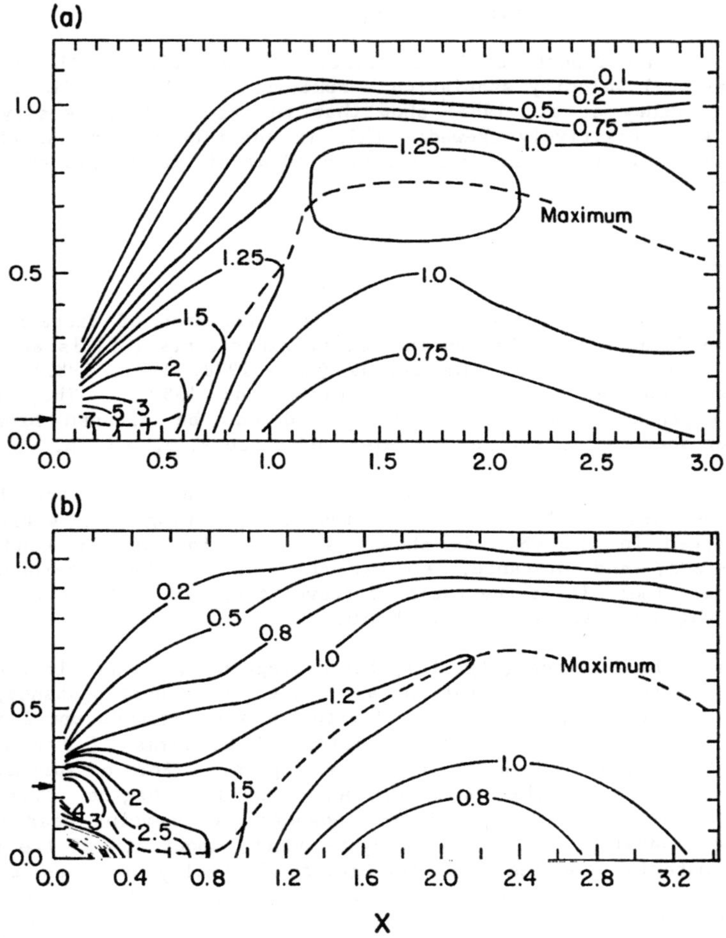

Fig. 1. Laboratory convection tank results showing contours of nondimensional crosswind-integrated concentration as a function of dimensionless height Z and downwind distance X; horizontal arrows denote the release height. (a) $z_s/h = 0.067$ (Willis and Deardorff 1976); (b) $z_s/h = 0.24$ (Willis and Deardorff 1978).

<u>Passive plumes - dispersion parameters</u>: Numerical simulations and experiments have provided much needed information on the behaviour of the dispersion parameters σ_y and σ_z. Figure 2 shows Lamb's results for vertical dispersion. The σ_z/h data for the elevated sources collapse to a nearly universal curve, $\sigma_z/h = 0.5X$, at short range, $X < 0.7$. This is consistent with statistical theory, assuming $\sigma_w = 0.5w_*$, and is attributed to the quasi-homogeneity of turbulence in the upper 90% of the CBL. The constant (0.5) is close to that (0.6) estimated from σ_w observations, thus establishing a link between numerical predcitions, simple theory, and field data. For $X > 1$, the σ_z/h curves approach constants due to the trapping of material below $z = h$, and to the homogenization of concentrations in the vertical. The latter behaviour is exhibited by the approach of C^yUh/Q to 1 at $X \sim 2\text{–}3$ (Fig. 1). The numerical predictions in Fig. 2 are in good agreement with laboratory data as shown by Lamb (1982).

For the lowest source in Fig. 2, σ_z/h increases at a greater-than-linear rate with X for small X, in qualitative agreement with Yaglom's (1972) similarity theory. This theory predicts that $\sigma_z/h \propto X^{3/2}$ and applies to a true surface release, $z_s = 0$. An elevated plume in the surface layer should initially grow linearly with t or X according to statistical theory because of the nonzero σ_w at the source height, i.e., $\sigma_w = 1.34(g\overline{w\theta_0}z_s/\Theta_0)^{1/3}$. At greater times, the increase of σ_w with z should accelerate the vertical spread such that $\sigma_z \propto X^{3/2}$. Thus, an approximate power law fit to the σ_z/h vs X curve over both of these regimes would be expected to yield an exponent on X between 1 and 3/2 as found by Lamb (1979).

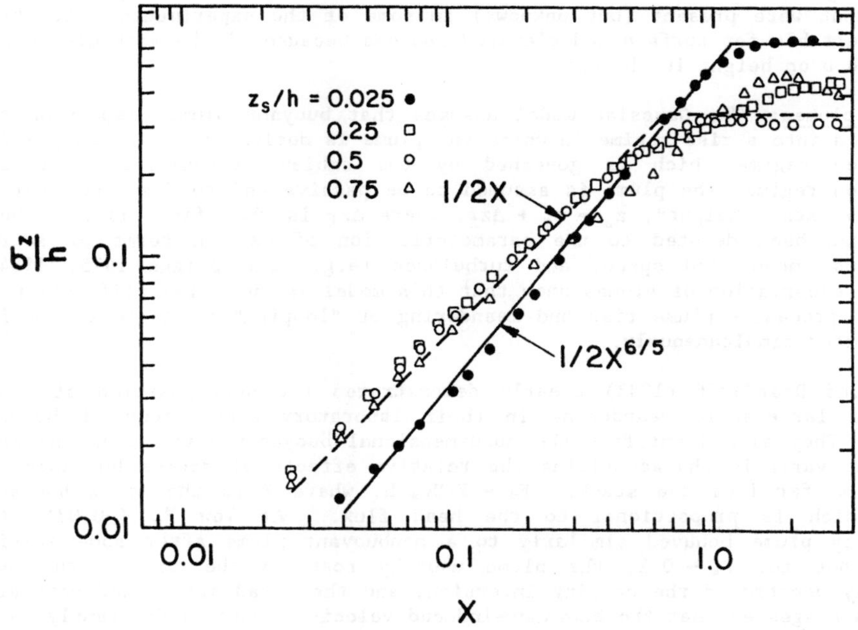

Fig. 2. Numerically calculated dispersion parameter as a function of dimensionless distance for a point source in a convective boundary layer. (Lamb 1982).

Recent field observations support the general behaviour in Fig. 2. From elevated releases of oil fog and radar chaff, Briggs (1986) found that $\sigma_z/h = 0.6X$ at short range, in agreement with Lamb's result except for the replacement of his 0.5 by 0.6. For $X > 1$, σ_z/h tended toward a constant, ~ 0.3, as in Fig. 2. Earlier, Carras and Williams (1983) obtained measurements of vertical spread in weakly buoyant plumes from release heights $0.06 < z_s/h < 0.15$; they showed that vertical depth ($= 4.3\sigma_z$) was bounded by Lamb's surface and elevated source curves, as adjusted to yield the plume depth. For surface sources, Nieuwstadt (1980) provided indirect support for Yaglom's (1972) result from an analysis of the surface CWIC measurements from the Prairie Grass experiments. Assuming a Gaussian vertical distribution, which is a useful approximation for $X < 0.5$ (Deardorff and Willis 1975), he argued that $C^yUh/Q \sim (\sigma_z/h)^{-1}$ if Yaglom's theory was correct. His plot of the dimensionless CWIC indeed showed an approximate $X^{-3/2}$ dependence for $0.03 < X < 0.3$. Similar results were obtained by Briggs (1986) for surface releases.

Field, laboratory, and numerical data show that the crosswind concentration distribution is approximately Gaussian (e.g., Lamb 1982). For σ_y, a useful expression that matches the short and long time limits of statistical theory is (e.g., Weil 1985)

$$\frac{\sigma_y}{h} = \frac{0.6X}{(1 + \alpha X)^{1/2}}, \qquad (7)$$

where $\alpha = 0.5\, h/T_L^y w_*$ but has a somewhat uncertain value. Briggs (1985) estimated $\alpha = 2$ as an upper bound based on his analysis of passive tracer data. However, his data and Weil's (1983) suggest that 0.7 is more typical value; this α may include the effects of wind direction shear and large mesoscale eddies that were present (but unknown) in some of the experiments. Equation (7) is suitable for surface and elevated sources because of the weak dependence of σ_v and u on height in the CBL.

Buoyant plumes: The Gaussian model assumes that buoyant plume dispersion can be divided into a rise regime in which the plume is dominated by buoyancy and a dispersion regime which is governed by the ambient turbulence. In the dispersion regime, the plume is assumed to be passive and to disperse from an "effective stack height", $z_e = z_s + \Delta z_f$, where Δz_f is the "final rise". Much effort has been devoted to the parameterization of Δz_f in terms of source conditions, mean wind speed, and turbulence (e.g., see Briggs 1975, 1984). However, observation of plumes show that this model is an oversimplification of the real process - plume rise and meandering or "looping" by large convective eddies occur simultaneously.

Willis and Deardorff (1983) clearly demonstrated the superposition of plume rise and large-scale meandering in their laboratory simulations of buoyant plumes. They also identified the nondimensional buoyancy flux F_* as the most important variable characterizing the relative effects of source buoyancy and convection far from the stack. $F_* = F/Uw_*^2h$, where F is the stack buoyancy flux, which is proportional to the heat flux. At low F_* (~ 0.03), the laboratory plume behaved similarly to a nonbuoyant plume after some initial rise. But for $F_* \sim 0.1$, the plume usually rose to the top of the CBL, partially penetrated the capping inversion, and then gradually mixed downward. The data suggested that the buoyancy-induced velocities were sufficiently large to overcome the downdraught velocities over a considerable range of distances.

From photographs of plume rise, Willis and Deardorff inferred that the vertical concentration distribution was skewed and that with increasing F_*, smaller amounts of effluent reached the surface at a given X. The experiments by Willis and Hukari (1984) quantitatively confirmed the skewness and the reduction in the surface CWIC with F_*, but they were limited to $F_* < 0.12$. Willis and Deardorff (1987) extended the experiments to $F_* = 0.54$ and found more significant reductions in the CWIC at a given X; e.g., at $X = 5$, the plume with $F_* = 0.54$ did not even approach a uniform vertical distribution whereas it did with $F_* = 0.11$. They also showed that plume material was temporarily "trapped" above $z = 0.8h$ and that this varied dramatically with F_* over the range $0.1 < F_* < 0.6$.

Applied Dispersion Models - Passive Plumes

We first consider vertical dispersion from elevated sources and then from surface-layer sources. The vertical dispersion determines the CWIC field from which one can find the GLC assuming a Gaussian crosswind distribution. The GLC along the plume centreline is $C = C^y(X,0)/(2\pi)^{\frac{1}{2}}\sigma_y$, where σ_y is obtained from Eq. (7).

Elevated sources: The simulations discussed earlier motivated the development of a new dispersion model - the p.d.f. approach - based on the p.d.f. of w, p_w, which is not Gaussian (Weil and Furth 1981; Misra 1982; Venkatram 1983). One assumes that passive particles from a source are emitted into a train of convective elements - updraughts and downdraughts - that are travelling with the mean wind speed U. The w in each element is assumed to be a random variable. A key assumption is that the Lagrangian time scale is effectively infinite so that a particle remains within an element until it reaches the surface or the capping inversion, where it reverses direction. Assuming that w is uniform with height in a given element, a particle follows a straightline trajectory: $z_c = z_s + wx/U$, where z_c is the particle height. For a distribution of w governed by p_w, a particle can follow an infinite number of trajectories.

The C^y is found by equating the mass flux of particles through an elemental distance Δz in a vertical (y-z) plane, $UC^y(x,z)\Delta z$, with $Qp_z(z_c)\Delta z$, where p_z is the p.d.f. of particle height. This balance results in $C^y = Qp_z/U$. Given the relationship between z_c and w, p_z can be found from p_w through $p_z = p_w[w(z_c)]\left|dw/dz_c\right|$, where $dw/dz_c = U/x$. Thus,

$$C^y(x,z) = Q\frac{p_w[U(z_c-z_s)/x]}{x} \quad . \tag{8}$$

An expression for C^y can be derived even if w depends on z (Misra 1982).

For simplicity, p_w is assumed to be a superposition of two Gaussian distributions, which has been shown to provide a good fit to observations and simulations of p_w in the CBL (Misra 1982; Baerentsen and Berkowicz 1984; Weil 1988). Assuming perfect reflection at $z = 0$, the dimensionless CWIC at the surface is

$$\frac{C^y U h}{Q} = \frac{\Gamma_1}{X}\exp\left[-\frac{R_1}{2}\left[\frac{Z_s}{a_1 X} - 1\right]^2\right] + \frac{\Gamma_2}{X}\exp\left[-\frac{R_2}{2}\left[\frac{Z_s}{a_2 X} + 1\right]^2\right] \quad , \tag{9}$$

where $\Gamma_i = 2P_i/(2\pi)^{\frac{1}{2}}b_i$, $a_i = \overline{w}_i/w_*$, $b_i = \sigma_i/w_*$, $R_i = (a_i/b_i)^2$, $Z_s = z_s/h$, and $i = 1,2$. $P_i, \overline{w}_i$, and σ_i are the total probability, mean vertical velocity, and standard deviation of distribution i. Equation (9) applies to the C^y from particles coming directly from the source. Additional image source terms are necessary to account for particle reflections at $z = h$.

Figure 3 compares the p.d.f. model prediction of the surface C^y with numerical and laboratory data for $Z_s = 0.5$. The model agrees well with the simulations; similar results are found for $Z_s = 0.24$ (Fig. 4). For reference, we show the

Gaussian plume model result with $\sigma_z/h = 0.5X$, which is based on Lamb's (1979) data. As can be seen, the model peaks at the correct distance, but it underestimates the maximum C^y due to the absence of skewness.

Surface-layer sources: The major problem in modelling dispersion in the surface layer is the vertical inhomogeneity in σ_w. One simple approach is to use the Gaussian plume model but with a suitably chosen σ_z. Deardorff and Willis (1975) found that a Gaussian vertical distribution was an approximate fit to their laboratory measurements of C^y for $X < 0.5$ and suggested the following interpolation formula:

$$\frac{\sigma_z}{h} = \left[1.8Z_s^{2/3}X^2 + 0.25X^3 + (\lambda u_* X/w_*)^2\right]^{\frac{1}{2}} \quad , \tag{10}$$

where λ is a coefficient dependent on u_*x/Uz_s . In appropriate ranges of X, Z_s, etc., Eq. (10) satisfies the short-time limit of statistical theory, Yaglom's (1972) similarity model, and the Lagrangian similarity theory for the neutral surface layer. Weil (1988) shows that this σ_z yields reasonable estimates of the surface C^y in the laboratory experiments for $X < 1$, but overestimates C^y by as much as a factor of 2 for $1 < X < 2$.

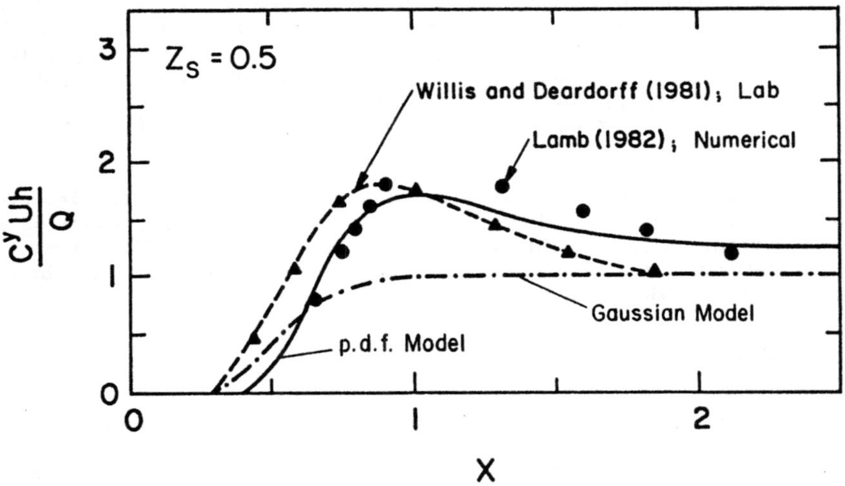

Fig. 3. Nondimensional crosswind-integrated concentration at the surface versus dimensionless downwind distance for a passive release in the convective boundary layer. (Weil 1988).

The eddy-diffusion model is another way to estimate the short-range vertical dispersion from a surface source (Nieuwstadt and van Ulden 1978; van Ulden 1978). Using Roberts' solution (Eq. 2), van Ulden obtained an analytical expression for C^y; he related $\bar{u}$ and $\bar{Z}$ to u_*, L, z_0, and x by an appropriate matching of the power law forms of u(z) and K(z) to the M-O similarity profiles. Van Ulden demonstrated the applicability of this approach using the Prairie Grass data. More recently, Gryning et al. (1987) successfully applied the model to vertical dispersion from surface-layer sources and gave operational methods for implementing it. This approach is limited to downwind distances such that $\bar{Z}$ is still within, or mostly within, the surface layer.

Applied Dispersion Models - Buoyant Plumes

The following models incorporate some of our current understanding of dispersion in the CBL.

Gaussian model: Weil and Brower (1984) and Berkowicz et al. (1986) improved the Gaussian model by including dispersion parameters and stability estimation methods based on existing knowledge of CBL turbulence. Weil and Brower adopted Briggs empirical dispersion curves, which were intended for elevated releases and were consistent with the short-range limit of statistical theory. They defined the applicability of the very unstable (A) to neutral (D) curves in terms of U/w_* using the short-range limits of the curves and interpolation expressions for σ_v and σ_w. The expressions were of the form $\sigma_w = (\sigma_{wn}^2 + \sigma_{wc}^2)^{1/2}$, where $\sigma_{wn}(\propto U)$ and $\sigma_{wc}(\propto w_*)$ are the σ_w in the neutral and strongly convective limits. Their model only applied to elevated releases, which were defined as those with $z_e > 0.1h$.

Berkowicz et al. developed expressions for σ_y and σ_z explicitly in terms of w_*, u_*, h, z_s, and x. They used statistical theory for elevated sources and an adaptation of it for surface-layer sources. Their model, as well as Weil and Brower's, included Briggs' (1975, 1984) formulations of final rise in convective conditions and plume penetration of elevated stable layers.

Weil and Brower evaluated their model with SO_2 GLCs downwind of Maryland power plants, and showed that it performed much better than the CRSTER model (EPA, 1977), which is based on the PGT method. The data were obtained during daytime convective periods. The better performance was attributed primarily to use of the Briggs A σ_z curve, when applicable, and to the more realistic distribution of stability - 62%, 30%, 8%, and 0% for the A to D categories, respectively. In contrast, the CRSTER model predicted neutral stability in more than 50% of these daytime convective cases. Recently, the Weil/Brower model was modified to better predict the dispersion of highly buoyant plumes ($F_* > 0.1$) that "loft" near the elevated inversion and resist downward mixing. The performance of the Berkowicz et al. model was quite similar to Weil and Brower's.

P.d.f. model: The p.d.f. model for buoyant plumes (Weil et al. 1986) differs for "low" and "high" buoyancy fluxes, which are defined by $F_* < 0.1$ and $F_* > 0.1$, respectively. This difference is motivated by the laboratory investigations discussed earlier.

For low buoyancy fluxes, Weil et al. superpose the vertical displacements due to source buoyancy and convection to find the "instantaneous" plume centreline height as

$$z_c = z_s + 1.6\frac{F^{1/3}}{U} x^{2/3} + \frac{wx}{U} \quad . \tag{11}$$

The second term on the right-hand side is the rise due to buoyancy and is given by the familier "two-thirds law" (Briggs 1975). The CWIC, C_b^y, in the instantaneous plume is assumed to have a Gaussian distribution about z_c. The ensemble-averaged C^y is then found by integrating over all possible plume heights

$$C^y = \int C_b^y(x,z,z_c)p_z(z_c)dz_c \quad , \tag{12}$$

where the p_z is found from p_w as in the case of passive plumes, but with z_c given by Eq. (11).

The nondimensional CWIC at the surface due to plume segments arriving directly from the stack is given by

$$\frac{C^y Uh}{Q} = \frac{1.2}{(2\pi)^{\frac{1}{2}}\sigma^*_{z1}} \exp\left[\frac{-h_1^{*2}}{2\sigma^{*2}_{z1}}\right] + \frac{0.8}{(2\pi)^{\frac{1}{2}}\sigma^*_{z2}} \exp\left[\frac{-h_2^{*2}}{2\sigma^{*2}_{z2}}\right] \quad , \tag{13}$$

where

$$\begin{aligned} \sigma^*_{zi} &= [1.3\beta'^2 F_*^{2/3} X^{4/3} + b_i^2 X^2]^{1/2} \\ h_i^* &= Z_s + 1.6 F_*^{1/3} X^{2/3} + a_i X \end{aligned} \quad . \tag{14}$$

The a_i and b_i are the same coefficients used in Eq. (9), $i = 1,2$ and $\beta'(= 0.4)$ is the ratio of the buoyant plume radius to the rise. The terms involving buoyancy appear separately in the above expression so that as $F_* \to 0$, the solution for a nonbuoyant source is recovered. Additional terms are included to satisfy the no-flux condition at $z = h$.

At high F_*, the plume is assumed to rise to the top of the CBL, where it "lofts" or lingers until individual cross-sections are mixed to the surface by downdraughts. There are two key assumptions. First, a plume cross-section remains within the same downdraught until it is uniformly mixed in the vertical i.e., $C^y = Q/Uh$. Second, plume segments become uniformly mixed only when a downdraught has sufficient kinetic energy $\rho w^2/2$ to overcome the potential energy difference between the plume and the environment $\Delta\rho gh$; $\Delta\rho$ is the density difference between the plume and the environment. The specific criterion for uniform mixing is $\Delta\rho gh = \alpha_m \rho w^2/2$, and this occurs at a distance x_m; α_m is empirically found to be 3.1.

The ensemble-averaged CWIC at the surface is Q/Uh times the probability P_m that uniform mixing has occurred at x, i.e., the probability that $x_m < x$. Thus, $C^y = QP_m/Uh$. P_m is given by an integral over the p.d.f. of x_m, which is found from p_w, the above energy criterion for uniform mixing, and a relationship between $\Delta\rho$ and x. This relationship is obtained by assuming that the buoyancy flux in the elevated plume is equal to the flux at the source F and from assumptions about the growth of the elevated plume prior to "mixing-down".

The above assumptions together with an assumed Gaussian p_w at the top of the CBL, as suggested by observations, result in the following expression for the CWIC at the surface:

$$\frac{C^y Uh}{Q} = 1 - \mathrm{erf}\left[\frac{2A_m}{\sqrt{2}(X/F_*)^{1/3}}\right] \quad . \tag{15}$$

The finding that the CWIC is a single function of X/F_* is in accord with the analyses of field and laboratory data by Briggs (1985) and Willis and Deardorff (1987), respectively. However, Eq. (15) tends to underestimate these data for $X/F_* > 10$; this requires further examination.

Evaluation of the p.d.f. model with laboratory measurements near the surface shows good agreement for both nonbuoyant and buoyant plumes (Fig. 4). Model evaluation with SF_6 GLCs downwind of the Kincaid power plant (Fig. 5) shows fair to good agreement on average. The geometric mean (GM) and geometric standard deviation (GSD) of C_{pred}/C_{obs} are 1.1 and 2.1, respectively, where C_{pred} and C_{obs} are the predicted and observed GLCs. In addition, 68% of the predictions are within a factor of 2 of the observations. However, further work is necessary to understand the relatively low variance explained by the model - $r^2 = 0.34$, where r is the correlation coefficient between $\ell n\, C_{obs}$ and $\ell n\, C_{pred}$. For this problem, stochastic variability and concentration fluctuations are quite important. For reference, the CRSTER model is compared with the Kincaid data (Fig. 5). As can be seen, the results are poorer, as

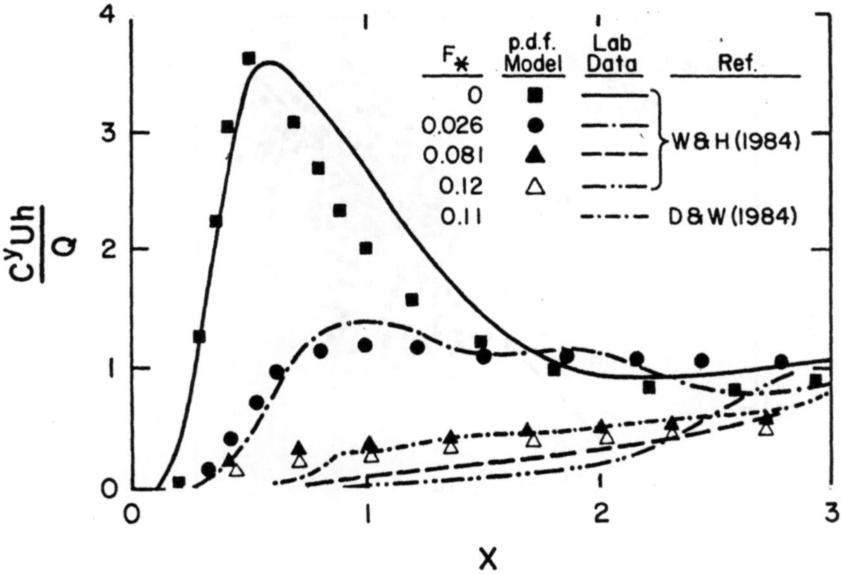

Fig. 4. Nondimensional crosswind-integrated concentration at the surface versus dimensionless downwind distance for passive and buoyant plumes; $z_s/h = 0.24$ and 0.14 for $F_* = 0$ and $F_* > 0$, respectively. W & H (1984) is Willis and Hukari (1984); D & W (1984) is Deardorff and Willis (1984). (Weil et al. 1986).

demonstrated by greater scatter (GSD = 4.6), fewer predictions within a factor of 2 of C_{obs} (33%), and an r^2 of 0.02.

Best et al. (1986) described a model originated by Spillane that is very similar to the p.d.f. approach given here. In addition, they demonstrated good performance of the model using the Willis and Deardorff laboratory data and GLC measurements around several Australian power plants.

Other models: Venkatram's (1980a) impingement model is similar to the p.d.f. approach in that plume rise is considered relative to updraughts and downdraughts, but the focus is only plume segments caught in downdraughts. The GLC is found from an assumed p.d.f. of the impingement distance, which is where plume segments in downdraughts first touch the ground. The mean impingement distance is obtained from Briggs' (1975) touchdown model. Evaluation of Venkatram's model around tall stacks showed good performance; it is probably most applicable for moderately buoyant sources, $F_* < 0.1$.

For highly buoyant plumes ($F_* > 0.1$), Briggs' (1985) semi-empirical expression is another method for estimating the CWIC and GLC. This expression is based on an intuitive argument that X/F_* is the relevant nondimensional distance for such plumes and an analysis of GLC data around two power plants.

DISPERSION IN THE STABLE BOUNDARY LAYER

Turbulence in the SBL is generated by wind shear and destroyed by buoyancy forces and viscous dissipation. It is characterized by small velocity fluctuations (~0.1 m s⁻¹) and length scales as well as strong vertical

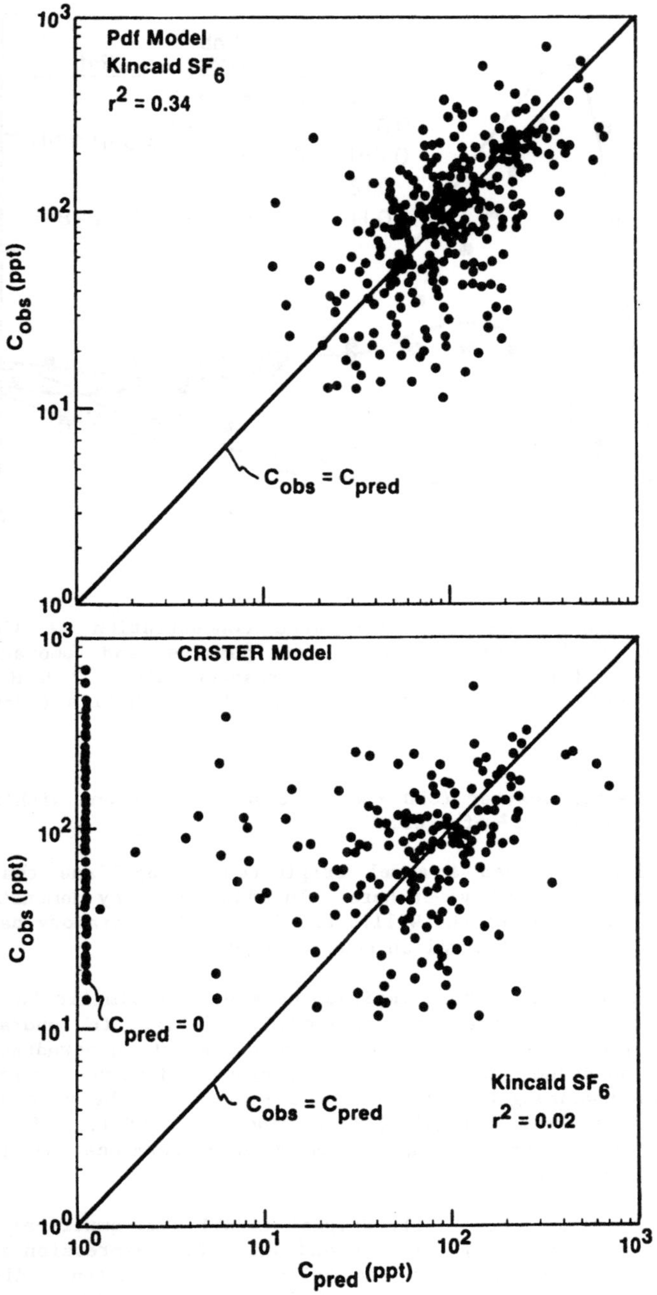

Fig. 5. Observed versus predicted ground-level SF_6 concentrations for the p.d.f. and CRSTER models at the Kincaid power plant. (Weil et al. 1986).

inhomogeneities. The SBL ranges from approximately 10 m to a few hundred metres in depth. In general, we know less about turbulence in the SBL than inthe CBL due to the greater measurement problems and to the complications caused by gravity waves, drainage flows, and intermittency (Wyngaard 1983).

Current knowledge is most complete for a continuously turbulent SBL in which waves and drainage effects are ignored. This is based largely on observations (Caughey et al. 1979; Nieuwstadt 1984a) and calculations, e.g., Brost and Wyngaard's (1978) second-order closure model and Nieuwstadt's (1984a) local scaling hypothesis. These show that near the surface, the turbulence length and velocity scales are z and u_*, and the mean wind and temperature follow the M-O similarity profiles. Far from the surface, the length scale attains a limiting value $\ell_B \sim \sigma_w/N$, due to the strong stable stratification, where N is the Brunt-Väisälä frequency $[(g/\Theta)(\partial\Theta/\partial z)]^{\frac{1}{2}}$; ℓ_B is typically 10 m. Using the local scaling hypothesis and assuming a stationary flow, Nieuwstadt solved the Brost-Wyngaard closure model for turbulence profiles. For σ_w, he obtained

$$\sigma_w/u_* = 1.4(1 - z/h)^{3/4} , \tag{16}$$

where $h = 0.4(u_*L/f)^{\frac{1}{2}}$, as found earlier by Zilitinkevich (1972); f is the Coriolis parameter. Nieuwstadt showed that, with the exception of temperature fluctuations, the predicted profiles were in reasonable agreement with measurements in the The Netherlands. However, they were restricted to cases without strong gravity wave effects.

The above expression for h applies to stationary conditions, but the SBL usually evolves slowly with time (Brost and Wyngaard 1978). Thus, in general, h should be described by a rate equation model such as that of Nieuwstadt and Tennekes (1981); however, this requires an initial value of h, which is not readily predictable. Fortunately, it has been shown that rough estimates can be obtained from the steady state model: $h = 0.4(u_*L/f)^{\frac{1}{2}}$ (Venkatram 1980b; Garratt 1982; Nieuwstadt 1984b), which is also a simple formula for applications. Further work is necessary to fully understand the effects of nonstationarity, waves, and intermittency and to deal with them appropriately in models.

Dispersion Models and Observations

Our knowledge of passive plume dispersion in the SBL has been gained mainly by theoretical and analytical modelling and a few experiments. The investigations have been limited mostly to vertical dispersion.

<u>Vertial dispersion - surface sources</u>: The eddy-diffusion model is a suitable representation of dispersion in the surface layer and should be more applicable in stable than unstable conditions because of the small length scale ℓ_B in the upper part of the SBL. Nieuwstadt and van Ulden (1978) numerically solved the diffusion equation using the M-O similarity profiles for u(z) and K(z) obtained by Businger et al. (1971). They predicted $\overline{Z}$ versus L in good agreement with experimental data and showed that the shape factor s was ~ 2 in very stable conditions. Gryning et al. (1983) found that s increased with x in stable conditions to values greater than 2.5; similar results were reported by Ogawa et al. (1985) based on wind tunnel data. The s behaviour is caused by the increase of wind shear with z and by the K profile shape: K initially increases with z and then decreases (e.g., see Brost and Wyngaard 1978). Nieuwstadt's (1984b) numerical calculations using a representative K(z) for the entire SBL confirmed the evolutionary nature of the vertical concentration profile. They showed that the profile had a near-Gaussian shape at small time and was uniform at large time.

Approximate expressions for the vertical dispersion and the surface CWIC were found by van Ulden (1978) and Venkatram (1982, 1988). Using K theory, van Ulden obtained analytical expressions for $u(\bar{Z})$ and $x(\bar{Z})$ for use in Eq. (2) and found good agreement between predictions and observations of C^y. Venkatram combined statistical theory with similarity arguments to obtain asymptotic expressions for σ_z and the CWIC. Starting from an alternative form of Taylor's theory, $d\sigma_z^2/dt = 2K(t)$, he assumed that $K = K_h$ with K_h given by the Businger et al. (1971) expression. By evaluating K_h at $z = \bar{Z}$ and assuming $\sigma_z \propto \bar{Z}$, he obtained $d\sigma_z^2/dt \sim K_h(\sigma_z/L)$. In the limit of small σ_z/L, $\sigma_z \sim u_*t$ and $C^y \sim Q/u_*x$, consistent with the neutral limits of similarity and eddy-diffusion theories. For $\sigma_z/L >> 1$, $\sigma_z \sim L^{2/3}x^{1/3}$ and $C^y \sim Q/u_*L^{1/3}x^{2/3}$. Venkatram showed good correlation between these C^y expressions and observations.

Vertical dispersion - elevated sources: The observations of Hogstrom (1964) and Hilst and Simpson (1958) motivated a new theory to explain the constant σ_z at intermediate and long times in the SBL.

Csanady (1964) and Pearson et al. (1983) developed model equations for the vertical velocity w_L and density fluctuation ρ' following a fluid particle; the fluctuation is about the local mean density ρ_0. The Pearson et al. equations are

$$\frac{dw_L}{dt} = -2\beta N w_L - \frac{g\rho'}{\rho_0} + H(t) \ , \tag{17a}$$

$$\frac{d\rho'}{dt} = -\frac{\partial\rho_0}{\partial z}w_L - \gamma N\rho' \ , \tag{17b}$$

where $-2\beta N w_L$ is a wave-damping term due to the motion of the particle, and $H(t)$ represents the random forces due to the motion of all the other fluid particles; β is of order 1. Together, $-2\beta N w_L$ and $H(t)$ represent the pressure gradient and viscous forces acting on a fluid particle. The term $-\gamma N\rho'$ models the diffusion of heat from a fluid particle where γN^{-1} is a time scale for a fluid element to change its density via molecular mixing. In the absence of molecular mixing ($\gamma = 0$), a fluid particle retains its initial density and moves through a vertical distance of $\sim \sigma_w/N$ before the buoyancy force returns it to its equilibrium position. However, for $\gamma > 0$, a particle can become warmed or cooled if moving up or down; the buoyancy force is then weakened, and the particle can be displaced a distance greater than σ_w/N.

For short times, $t << N^{-1}$, the Pearson et al. σ_z prediction is $\sigma_w t$ as in Taylor's theory. For $t >> N^{-1}$, the prediction is

$$\sigma_z = (\sigma_w/N)(\zeta_z^2 + 2N^2tT_L^z)^{\frac{1}{2}} \ , \tag{18}$$

where ζ_z is a dimensionless parameter (~1) dependent on the turbulence spectrum and $T_L^z \sim \gamma^2/N$. If $\gamma \sim 0.1$, σ_z approaches a constant over a considerable range of time, but if $\gamma > 0.3$, σ_z exhibits a parabolic growth as in statistical theory. Figure 6 compares these predictions to the data of Hilst and Simpson and shows the two limiting cases of σ_z growth.

In the above comparison, γ was not measured, but was chosen to illustrate the limiting behaviour. Hunt (1982, 1985) discussed methods for estimating γ from heat flux measurements. However, I believe that the meaning of γ, including its dependence on small-scale turbulence as well as molecular diffusion, must be clarified for this theory to be well-understood and appreciated.

Venkatram et al. (1984) assumed that statistical theory described the vertical dispersion of an elevated plume in the SBL. They adopted an interpolation formula for σ_z that matched the short- and long-time limits of the theory:

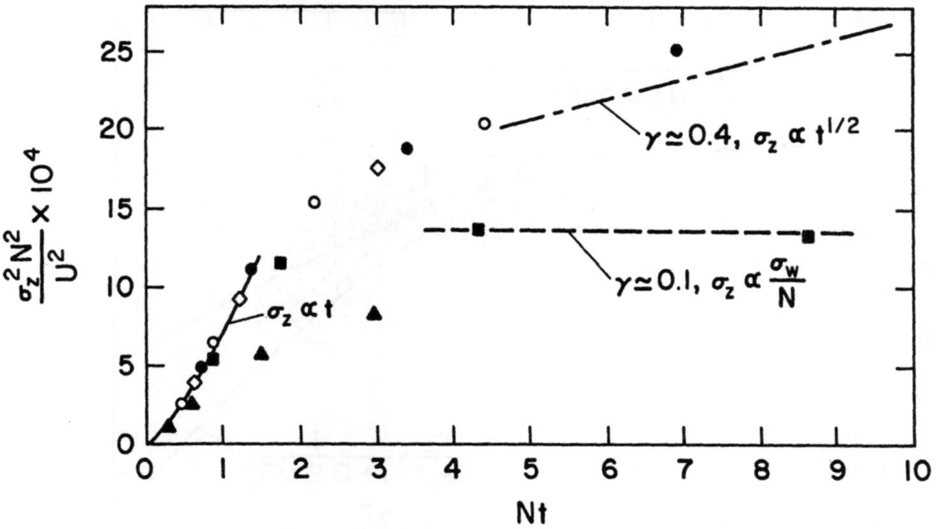

Fig. 6. Vertical dispersion parameter nondimensionalized by U/N as a function of the dimensionless time Nt in a stable boundary layer. Pearson et al. (1983) theoretical predictions (lines) compared with field data of Hilst and Simpson (1958). (Pearson et al. 1983).

$$\sigma_z = \frac{\sigma_w t}{(1 + t/2T_L^z)^{\frac{1}{2}}}, \qquad (19)$$

where σ_w is evaluated at the source height. The T_L^z was evaluated from $T_L^z = \ell/\sigma_w$, where ℓ is the turbulence length scale given by the interpolation expression $\ell^{-1} = \ell_n^{-1} + \ell_s^{-1}$; this is similar to that of Brost and Wyngaard (1978). ℓ_n and ℓ_s denote the scales in the neutral and strongly stable limits; $\ell_n = \mu z_s$ and $\ell_s = \gamma^2\sigma_w/N$, where γ is assumed to be independent of molecular diffusion. Venkatram et al. found μ (0.36) and γ (0.52) by matching the assumed $K = \sigma_w\ell = K_h$ with the small and large z/L limits of the Businger et al. K_h expression.

Figure 7 compares a dimensionless form of Eq. (19) with field observations; z_s was typically 50 m and the observations extended to about to 1 km downwind. The model gives a reasonable description of the decrease of $\sigma_z/\sigma_w t$ with t/T_L^z, but overestimates σ_z at large times. The dashed line is an interpolation expression, $\sigma_z = \sigma_w t/(1 + Nt/\zeta_z)$ based on Eq. (18) with $\zeta_z = 1.3$, N replaced by γ^2/T_L^z, and $2N^2tT_L^z$ omitted. This expression provides a lower limit to the data at large t/T_L^z. Further measurements are necessary to resolve the situations under which each approximate expression should apply.

Lateral dispersion: Lateral dispersion due to turbulence can be expressed by Eq. (19) but with σ_w and T_L^z replaced by σ_v and T_L^y, respectively. However, there are no general scaling relationships for σ_v or T_L^y. σ_v can have a significant component from the large horizontal, two-dimensional eddies that are generated by terrain inhomogeneities and stable stratification; models such as Brost and Wyngaard's do not describe this component. Observations suggest that σ_v, especially in light winds, is highly site specific with typical values ranging from 0.3 to 1 m s^{-1} (Hanna 1983; Venkatram 1988; Weil 1985). In addition to turbulence, wind direction shear can signficantly increase σ_y. The total σ_y is often expressed as $\sigma_y = (\sigma_{yt}^2 + \sigma_{ys}^2)^{\frac{1}{2}}$, where σ_{yt}

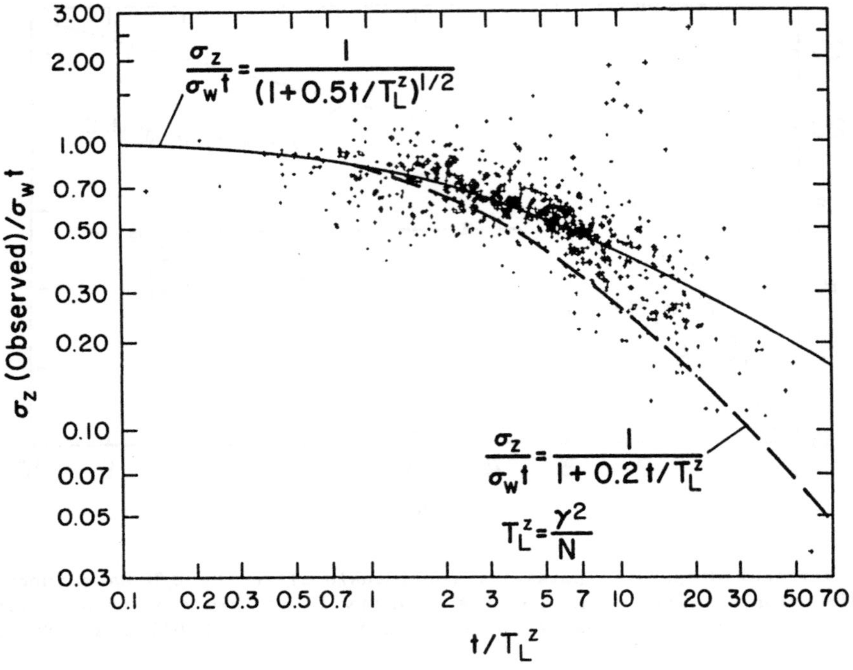

Fig. 7. Vertical dispersion parameter nondimensionalized by $\sigma_w t$ as a function of the dimensionless time t/T_L^z in a stable boundary layer. Dashed line shown for $\gamma = 0.52$ and $\zeta_z = 1.3$. Adapted from Venkatram et al. (1984).

and σ_{ys} are the components due to turbulence and directional shear, respectively. This expression as well as those for estimating the shear contribution are discussed by Venkatram (1988).

Dispersion Applications

For passive plumes, the vertical dispersion and the CWIC can be estimated from the models described above assuming a Gaussian distribution for the vertical concentration profile for an elevated plume for downwind distances such that $\sigma_z/h << 1$. Likewise, a Gaussian distribution can be assumed for the crosswind profile. To apply these models, one requires u_*, $\overline{w\theta}_0$, h, $\partial\theta/\partial z$, u, σ_w, σ_v, and the wind direction shear. Due to the variability of the SBL, these quantities should be measured. However, if this is not possible, rough estimates of some variables can be obtained using wind and temperature data from a 10-m tower and cloud cover observations. For example, u_* and $\overline{w\theta}_0$ can be estimated from a surface energy balance coupled with the M-O similarity expressions for u(z) and θ(z) (van Ulden and Holtlsag 1985; Venkatram 1980b). Van Ulden and Holtslag (1985) showed that u(z) follows the M-O expression over most of the SBL, but with φ_m somewhat different than that found by Businger et al. (1971). h can be determined from $0.4(u_*L/f)^{\frac{1}{2}}$ and σ_w from Eq. (16) (see van Ulden and Holtslag 1985 and Venkatram 1988 for further discussion).

For elevated buoyant plumes, high GLCs can occur in a weakly stable environment coupled with high winds. Venkatram and Paine (1985) addressed this problem with a model akin to the Gaussian plume, but they took into account the vertical inhomogeneity of σ_w. This was done by computing σ_z separately above and below the plume centreline assuming $d\sigma_z/dt = \sigma_w(z')$; z' corresponded to the lower and upper plume "edge" for dispersion below and above the centreline. Their model evaluation with GLC data around tall stacks demonstrated good performance and much better than that attained with the CRSTER model.

CONCENTRATION FLUCTUATIONS

Stochastic variability is one of the main causes of the difference between C_{obs} and $\overline{C}_{pred}$ in Fig. 5. It leads to a concentration uncertainty $\sigma_c = \langle(\overline{C} - \langle\overline{C}\rangle)^2\rangle^{\frac{1}{2}}$, where the overbar denotes a time average and the brackets an ensemble average (Weil 1985). σ_c arises because the details of the flow are not the same in individual realizations even though the gross conditions defining the ensemble are. Such conditions include source characteristics, meteorology, integral time scale, etc. In controlled laboratory experiments, an ensemble definition is straightforward, but in the atmosphere it is not because conditions change from hour to hour. Thus, some investigators define the ensemble in terms of the model inputs (Venkatram 1982). When $\sigma_c/\langle C\rangle$ is ~1, estimates of σ_c are necessary to predict the potential *range* of concentrations experienced at some downwind location. Our understanding of σ_c and its application to real problems is in a developmental stage and is only briefly covered here; see Sykes (1988) for a comprehensive review.

In a stationary flow, σ_c can be determined in principle from the autocorrelation function of C and the ensemble variance $\sigma_{ci}{}^2$ defined in terms of the instantaneous concentration (e.g., see Sykes 1984). For large averaging times T_{avg} such that $T_{avg}/T_c \gg 1$, this yields $\sigma_c^2 \simeq 2\sigma_{ci}{}^2T_c/T_{avg}$, where T_c is the integral time scale of the concentration. This formalism identifies the important quantities - the concentration fluctuation σ_{ci} and T_c. However, models or estimates of σ_{ci} and T_c are required.

For an elevated plume, σ_{ci} arises from concentration inhomogeneities within the plume, i.e., over distance scales less than the instantaneous width σ_r and from the meander σ_m of the entire cross-section by the large ambient eddies. Meandering is important as long as σ_r is less than the turbulence length scale. In the CBL, vertical meandering dominates near the source because of the large eddy size and the long time scale (~10 min). One of the first and most useful models for the ensemble variance was Gifford's (1959) fluctuating plume. It ignored small-scale fluctuations and treated only those due to the meander. On the plume centreline, it yielded $\sigma_{ci}/\langle C\rangle = \sigma_m{}^2/\sigma_r(2\sigma_m{}^2 + \sigma_r{}^2)^{\frac{1}{2}}$. The model was supported by Fackrell and Robins (1982) wind tunnel data. A useful extension to treat time-averaged concentrations was made by Sykes (1984). Simple models such as these as well as more computationally demanding ones based on Lagrangian, two-particle simulations and ensemble closure are reviewed by Sykes (1988).

Laboratory simulations of concentration fluctuations have contributed significantly toward our understanding of them. The Fackrell and Robins experiments, conducted in a neutral wind tunnel flow, clearly showed that the fluctuation intensity $\sigma_c/\langle C\rangle$ at the ground was much greater for an elevated than a surface source due to the greater plume meander. Large intensities were also revealed by Deardorff and Willis' (1988) laboratory simulations of passive and buoyant plumes in the CBL. Figure 8 shows that the intensity is indeed large at ground level and close to the source (X < 3). This supports the contention that a major fraction of the variance in Fig. 5 is stochastic in nature. Averaging times much longer than 1 h are required to substantially reduce the variance.

One of the real problems in constructing and verifying σ_c models with field

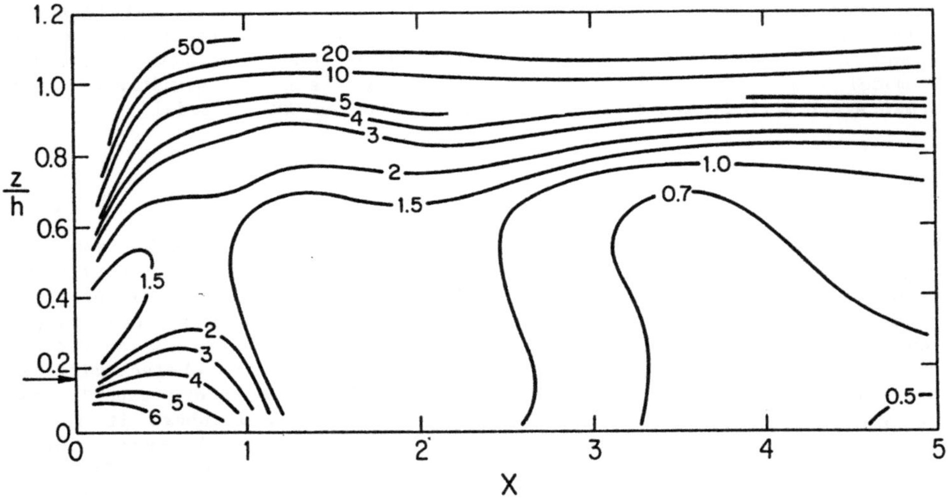

Fig. 8. Laboratory convection tank results showing contours of $\sigma_c/\langle C \rangle$ as a function of dimensionless height and downwind distance X for a buoyant plume, $F_* = 0.03$; horizontal arrow denotes release height. (Deardorff and Willis 1988).

observations is that the required data base is enormous and perhaps unfeasible. A second problem is the nonstationarity of the atmosphere. Thus, I believe that the fluctuation problem must be pursued first using numerical and laboratory simulations; this should be feasible for the CBL. Simple models for σ_c can be developed from such simulations as they were for $\langle C \rangle$ for the CBL. With these models and understanding, one would then be in a position to assess quantitatively what fraction of the variance in Fig. 5 is stochastic and how much is due to incorrect model physics.

ACKNOWLEDGEMENTS

I wish to thank Hope Hamilton for her skillful production of the original manuscript. I am grateful to the U.S. Army Research Office and the Electric Power Research Institute for their support during the prepartion of the manuscript.

REFERENCES

Baerentsen JH, Berkowicz R (1984) Monte Carlo simulation of plume dispersion in the convective boundary layer. Atmos Environ 18: 701-712

Berkowicz R, Olesen HR, Torp U (1986) The Danish Gaussian air pollution model (OML): Description, test and sensitivity analysis in view of regulatory applications. In: De Wispelaere C, Schiermeier FA, Gillani NV (eds) Air pollution modeling and its application V, Plenum, New York, pp 453-481

Best PR, Kanowski M, Stumer L, Green D (1986) Convective dispersion modelling utilising acoustic sounder information. Atmos Res 20: 173-197

Briggs GA (1975) Plume rise predictions. In: Haugen DA (ed) Lectures on air pollution and environmental impact analyses, Am Meteorol Soc, Boston, pp 59-111

Briggs GA (1984) Plume rise and buoyancy effects. In: Randerson D (ed), Atmospheric science and power production, U.S. Dept. of Energy DOE/TIC-27601, pp 327-366
Briggs GA (1985) Analytical parameterizations of diffusion: The convective boundary layer. J Climate Appl Meteorol 24: 1167-1186
Briggs GA (1986) Convective diffusion field measurements compared with laboratory and numerical experiments. 5th Joint Conf on Applications of Air Pollution Meteorology, Am Meteorol Soc, Boston, pp 340-343
Brost RA, Wyngaard JC (1978) A model study of the stably stratified planetary boundary layer. J Atmos Sci 35: 1427-1440
Businger JA, Wyngaard JC, Izumi Y, Bradley EF (1971) Flux-profile relationships in the atmospheric surface layer. J Atmos Sci 28: 181-189
Calder KL (1949) Eddy diffusion and evaporation in flow over aerodynamically smooth and rough surfaces: A treatment based on laboratory laws of turbulent flow with special reference to conditions in the lower atmosphere. Q J Mech App Math II: 153-176
Carras JN, Williams DJ (1983) Observations of vertical plume dispersion in the convective boundary layer. 6th Symp on Turbulence and Diffusion, Am Meteorol Soc, Boston, pp 249-252
Caughey SJ (1982) Observed characteristics of the atmospheric boundary layer. In: Nieuwstadt FTM, van Dop H (eds) Atmospheric turbulence and air pollution modelling, Reidel, pp 107-158
Caughey SJ, Kitchen M, Leighton JR (1983) Turbulence structure in convective boundary layers and implications for diffusion. Boundary-Layer Meteorol 25: 345-352
Caughey SJ, Wyngaard JC, Kaimal JC (1979) Turbulence in the evolving stable boundary layer. J Atmos Sci 36: 1041-1052
Chatwin PC (1968) The dispersion of a puff of passive contaminant in the constant stress region. Q J R Meteorol Soc 94: 350-360
Chaudhry FH, Meroney RN (1973) Similarity theory of diffusion and the observed vertical spread in the diabatic surface layer. Boundary-Layer Meteorol 3: 405-415
Cramer HE (1957) A practical method of estimating the dispersal of atmospheric contaminants. Conference on Applied Meteorology, Am Meteorol Soc, Boston
Csanady GT (1964) Turbulent diffusion in a stratified fluid. J. Atmos. Sci. 21: 439-447
Deardorff JW (1972) Numerical investigation of neutral and unstable planetary boundary layers. J Atmos Sci 29: 91-115
Deardorff JW, Willis GE (1975) A parameterization of diffusion into the mixed layer. J Appl Meteorol 14: 1451-1458
Deardorff JW, Willis GE (1984) Groundlevel concentration fluctuations from a buoyant and a non-buoyant source within a laboratory convectively mixed layer. Atmos Environ 18: 1297-1309
Deardorff JW, Willis GE (1988) Concentration fluctuations within a laboratory convectively mixed layer. In: Venkatram A, Wyngaard JC (eds) Lectures on air pollution modeling, Am Meteorol Soc, Boston, in press
Elliot WP (1961) The vertical diffusion of gas from a continuous source. Int J Air Water Pollut 4: 33-46
EPA, 1977: User's manual for single source (CRSTER) model. U.S. Environmental Protection Agency, Research Triangle Park, NC, Ref. No. EPA-450/2-77-013, 285 pp
Fackrell JE, Robins AG (1982) Concentration fluctuations and fluxes in plumes from point sources in a turbulent boundary layer. J Fluid Mech 17: 1-26
Garratt JR (1982) Surface fluxes and the nocturnal boundary-layer height. J Appl Meteorol 21: 725-729
Garratt JR, Wyngaard JC, Francey RJ (1982) Winds in the atmospheric boundary layer - prediction and observation J Atmos Sci 39: 1307-1316
Gifford FA (1959) Statistical properties of a fluctuating plume dispersion model. Adv Geophys 6: 117-138
Gifford FA (1961) Uses of routine meteorological observations for estimating atmospheric dispersion. Nucl Safety 2: 47-51

Gifford FA (1962) Diffusion in the diabatic surface layer. J Geophys Res 67: 3207-3212

Gryning SE, Holtslag AAM, Irwin JS, Sivertsen B (1987) Applied dispersion modelling based on meteorological scaling parameters. Atmos Environ 21: 79-89

Gryning SE, van Ulden AP, Larsen S (1983) Dispersion from a continuous ground-level source investigated by a K model. Q J R Meteorol Soc 109: 355-364

Hanna SR (1983) Lateral turbulence intensity and plume meandering during stable conditions. J Appl Meteorol 22: 1424-1430

Hanna SR, Briggs GA, Deardorff J, Egan BA, Gifford FA, Pasquill F (1977) AMS workshop on stability classification schemes and sigma curves - Summary of recommendations. Bull Am Meteorol Soc 58: 1305-1309

Hay JS, Pasquill F (1957) Diffusion from a fixed source at a height of a few hundred feet in the atmosphere. J Fluid Mech 2: 299-310

Hilst GR, Simpson CL (1958) Observations of vertical diffusion rates in stable atmospheres. J Meteorol 15: 125-126

Hogstrom U (1964) An experimental study on atmospheric diffusion. Tellus 16: 205-251

Horst TW (1979) Lagrangian similarity modeling of vertical diffusion from a ground-level source. J Appl Meteorol 18: 733-740

Hunt JCR (1982) Diffusion in the stable boundary layer. In: Nieuwstadt FTM, van Dop H (eds) Atmospheric turbulence and air pollution modelling, Reidel, pp 231-274

Hunt JCR (1985) Diffusion in the stably stratified atmospheric boundary layer. J Climate Appl Meteorol 24: 1187-1195

Kaimal JC, Wyngaard JC, Haugen DA, Coté OR, Izumi Y, Caughey SJ, Readings CJ (1976) Turbulence structure in the convective boundary layer. J Atmos Sci. 33: 2152-2169

Klug W (1968) Diffusion in the atmospheric surface layer: comparison of similarity theory with observation. Q J R Meteorol Soc 94: 555-562

Lamb RG (1978) A numerical simulation of dispersion from an elevated point source in the convective planetary boundary layer. Atmos Environ 12: 1297-1304

Lamb RG (1979) The effects of release height on material dispersion in the convective boundary layer. 4th Symp on Turbulence, Diffusion and Air Pollution, Am Meteorol Soc, Boston, pp 27-33

Lamb RG (1982) Diffusion in the convective boundary layer. In: Nieuwstadt FTM, va Dop H (eds) Atmospheric turbulence and air pollution modelling, Reidel, pp 159-229

Luna RE, Church HW (1972) A comparison of turbulence intensity and stability ratio measurements to Pasquill stability classes. J Appl Meteorol 11: 663-669

Misra PK (1982) Dispersion of nonbuoyant particles inside a convective boundary layer. Atmos Environ 16: 239-243

Monin AS (1959) Smoke propagation in the surface layer of the atmosphere. Adv Geophys 6: 331-341

Moninger WR, Eberhard WL, Briggs GA, Kropfli RA, Kaimal JC (1983) Simultaneous radar and lidar observations of plumes from continuous point sources. 21st Conf on Radar Meteorology, Am Meteorol Soc, Boston, pp 246-250

Nieuwstadt FTM (1980) Application of mixed-layer similarity to the observed dispersion from a ground-level source. J. Appl. Meteor. 19: 157-162

Nieuwstadt FTM (1984a) The turbulent structure of the stable nocturnal boundary layer. J Atmos Sci 41: 2202-2216

Nieuwstadt FTM (1984b) Some aspects of the turbulent stable boundary layer. Boundary-Layer Meteorol 30: 31-54

Nieuwstadt FTM, Tennekes H (1981) A rate equation for the nocturnal boundary-layer height. J Atmos Sci 38: 1418-1428

Nieuwstadt FTM, van Ulden AP (1978) A numerical study of vertical dispersion of passive contaminants from a continuous source in the atmospheric surface layer. Atmos Environ 12: 2119-2124

Ogawa Y, Diosey PG, Uehara K, Ueda H (1985) Wind tunnel observation of flow and diffusion under stable stratification. Atmos Environ 19: 65-74

Panofsky HA, Tennekes H, Lenschow DH, Wyngaard JC (1977) The characteristics of turbulent velocity components in the surface layer under convective conditions. Boundary-Layer Meteorol 11: 355-361

Pasquill FA (1961) The estimation of the dispersion of windborne material. Meteorol Mag 90: 33-49

Pasquill FA (1966) Lagrangian similarity and the vertical diffusion from a source at ground level. Q J R Meteorol Soc 92: 185-195

Pasquill FA, Smith FB (1983) Atmospheric diffusion (3rd ed.) Wiley, New York, 437 pp

Pearson HJ, Puttock JS, Hunt JCR (1983) A statistical model of fluid-element motions and vertical diffusion in a homogeneous stratified turbulent flow. J Fluid Mech 129: 219-249

Sawford BL, Guest FM (1987) Lagrangian stochastic analysis of flux-gradient relationships in the convective boundary layer. J Atmos Sci 44: 1152-1165

Singer IA, Smith ME (1966) Atmospheric diffusion at Brookhaven National Laboratory, Int J Air Water Pollut 10: 125-135

Slade DH (ed) (1968) Meteorology and atomic energy. U.S. Atomic Energy Commission, Div. Tech. Inf.

Sutton OG (1953) Micrometeorology, McGraw-Hill, New York, 333 pp

Sykes RI (1984) The variance in time-averaged samples form an intermittent plume. Atmos Environ 18: 121-123

Sykes, RI (1988) Concentration fluctuations in dispersing plumes. In: Venkatram A, Wyngaard JC (eds) Lectures on air pollution modeling, Am Meteorol Soc, Boston, in press

Taylor GI (1921) Diffusion by continuous movements. Proc London Math Soc Ser 2 20: 196-212

Turner DB (1964) A diffusion model for an urban area. J Appl Meteor 3: 83-91

van Ulden AP (1978) Simple estimates for vertical dispersion from sources near the ground. Atmos Environ 12: 2125-2129

van Ulden AP, Holtslag AAM (1985) Estimation of atmospheric boundary layer parameters for diffusion applications. J Climate Appl Meteorol 24: 1196-1207

Venkatram A (1980a) Dispersion from an elevated source in the convective boundary layer. Atmos Environ 14: 1-10

Venkatram A (1980b) Estimating the Monin-Obukhov length in the stable boundary layer for dispersion calculations. Boundary-Layer Meteorol 19: 481-485

Venkatram A (1982) A framework for evaluating air quality models. Boundary-Layer Meteorol 24: 371-385

Venkatram A (1983) On dispersion on the convective boundary layer. Atmos. Environ 17: 529-533

Venkatram A (1988) Dispersion in the stable boundary layer. In: Venkatram A, Wyngaard JC (eds) Lectures on air pollution modeling, Am Meteorol Soc Boston, in press

Venkatram A, Strimaitis D, Dicristofaro D (1984) A semiempircal model to estimate vertical dispersion of elevated releases in the stable boundary layer. Atmos Environ 18: 923-928

Venkatram A, Paine R (1985) A model to estimate dispersion of elevated releases into a shear-dominated boundary layer. Atmos Environ 19: 1797-1805

Weil JC (1983) Application of advances in planetary boundary layer understanding to diffusion modeling. 6th Symp on Turbulence and Diffusion, Am Meteorol Soc, Boston, pp 42-46

Weil JC (1985) Updating applied diffusion models. J Climate Appl Meteorol 24: 1111-1130

Weil JC (1988) Dispersion in the convective boundary layer. In: Venkatram A, Wyngaard JC (eds) Lectures on air pollution modeling, Am Meteorol Soc, Boston, in press

Weil JC, Brower RP (1984) An updated Gaussian plume model for tall stacks. J. Air Pollut Control Assoc 34: 818-827

Weil JC, Corio LA, Brower RP (1986) Dispersion of buoyant plumes in the convective boundary layer. 5th Joint Conf of Applications of Air Pollution Meteorology, Am Meteorol Soc, Boston, pp 335-338

Weil JC, Furth WF (1981) A simplified numerical model of dispersion from elevated sources in the convective boundary layer. 5th Symposium on Turbulence, Diffusion and Air Pollution, Am Meteorol Soc, Boston, pp 76-77

Willis GE, Deardorff JW (1974) A laboratory model of the unstable planetary boundary layer. J Atmos Sci 31: 1297-1307

Willis GE, Deardorff JW (1976) A laboratory model of diffusion into the convective planetary boundary layer. Q J R Meteorol Soc 102: 427-445

Willis GE, Deardorff JW (1978) A laboratory study of dispersion from an elevated source within a modeled convective planetary boundary layer. Atmos Environ 12: 1305-1312

Willis GE, Deardorff JW (1981) A laboratory study of dispersion from a source in the middle of the convectively mixed layer. Atmos Environ 15: 109-117

Willis GE, Deardorff JW (1983) On plume rise within the convective boundary layer. Atmos Environ 17: 2435-2447

Willis GE, Deardorff JW (1987) Buoyant plume dispersion and inversion entrapment in and above a laboratory mixed layer. Atmos Environ 21: 1725-1735

Willis GE, Hukari N (1984) Laboratory modeling of buoyant stack emissions in the convective boundary layer. 4th Joint Conf on Applications of Air Pollution Meteorology, Am Meteorol Soc, Boston, pp 24-26

Wyngaard JC (1983) Lectures on the planetary boundary layer. In: Lilly DK, Gal-Chen T (eds) Mesoscale Meteorology-Theories, Observations and Models, NATO ASI Series, Reidel, pp 603-650

Wyngaard JC, Coté OR, Izumi Y (1971) Local free convection, similarity, and the budgets of shear stress and heat flux. J Atmos Sci 28: 1171-1182

Wyngaard JC (1988) Convective processes in the lower atmosphere. This volume, pp

Yaglom AM (1972) Turbulent diffusion in the surface layer of the atmosphere. Izv Akad Nauk USSR, Atmos Ocean Phys 8: 333-340

Zilitinkevich SS (1972) On the determination of the height of the Ekman boundary layer. Boundary-Layer Meteorol 3: 141-145

ATMOSPHERIC DISPERSION - OBSERVATIONS AND MODELS: COMMENTARY

Peter Manins*

Project Director, Latrobe Valley AirShed Study, c/- Environment Protection Authority, 546 Collins Street, Melbourne, Vic., 3000, Australia.

Summary. Practical problems of atmospheric dispersion cover a vast range of scales and are usually much more complex than can be modelled at present. The principles that have been learnt from observations and model building in more ideal studies, such as described in the lecture by Jeff Weil on 'Atmospheric Dispersion - Observations and Models', need to be kept firmly in mind. Then significant solutions to practical problems are possible.

INTRODUCTION

Weil (1988) has described the great advances in understanding of dispersion from single sources in the convective boundary layer and the nocturnal boundary layer over flat terrain. He pointed out the historical and ongoing importance of the simple Gaussian plume models of dispersion in the atmosphere; the great strides made in understanding convective dispersion by utilising laboratory water tank experiments; and the role highly detailed numerical solutions to the fundamental equations of fluid motion have in providing guidance and parameterisations for simple models which can be rapidly applied to practical problems. In this commentary I shall first discuss some questions about dispersion models which arose during discussion at the Symposium, and then turn to a practical problem with which I'm associated, the Latrobe Valley AirShed study.

FUTURE DEVELOPMENTS IN MODELLING

An important question which arose in the Symposium was 'How wise is it to make models simple?' Weil concluded that there will always be a need for simplified models no matter how good complex numerical codes become. Gaussian models do fairly well as compared with Large Eddy Simulations because, as the laboratory tank experiments of Deardorff and Willis show, the variance of concentrations in convective conditions is huge.

Another point, mentioned by Weil (1988), which is all too often overlooked is that the deviation between the prediction, which is an estimate of the ensemble mean concentration for given conditions, and an individual realisation - the observation - is still large, i.e., of the order of the prediction. He further noted that this has given a major impetus to the study of concentration fluctuations. This leads to the question of how good the agreement between experimental observations and models is ever likely to be. For instance, in terms of the CRSTER model how large is r^2 likely to get in the next few years as models improve? In discussion, Weil reported that in some Maryland studies

*on leave from CSIRO Division of Atmospheric Research, Aspendale, Vic., 3195, Australia.

r^2 was as high as approximately 0.6, which is probably as high as one could expect.

An important modelling problem is the extent to which the nature of the surrounding terrain influences regulations. Do models take account of trajectories over terrain, around obstacles, down valleys, and so on? Weil noted that there is a variety of models accredited by the U.S. EPA which account for terrain influences. They are, however, undergoing major changes following a series of field and laboratory studies. They do not yet address strongly convective conditions, though some do consider photochemically produced pollutants (in particular, ozone). (The convective wind tunnel undergoing development at Monash University is designed to study stack plumes in complex terrain and convective conditions.)

On a related question about the legal status of dispersion models in the U.S.A. - do they have status, and does the legal process bear any relation to reality - Weil observed that dispersion models are an important part of the legal process in air quality regulation in the U.S.A. There is an involved process of model accreditation by the U.S. Environment Protection Agency to ensure that models are reliable. A proponent may say that the circumstances demand an alternative model to those already accredited but the alternative must be put through the accreditation process and this may take a long time.

Finally, I note that the new and exciting work on atmospheric dispersion involving solution of the Langevin equation as an analog of Lagrangian dispersion was only mentioned in passing in the lecture by Weil (1988). This is a highly computationally intensive approach and will only find practical application through giving us a better understanding of turbulent dispersion, and new ideas for parameterisations. It is already having a strong impact, as was brought out in discussion where it was mentioned that the Lagrangian models now being developed use Eulerian statistics directly. There is no need for past assumptions about the transformations between Eulerian and Lagrangian statistics of the flows. We are very close to a complete description of, for example, convective flows in terms of a Lagrangian model.

THE LATROBE VALLEY AIRSHED STUDY

Perhaps the most comprehensive atmospheric dispersion study in Australia is the Latrobe Valley AirShed Study (Manins 1983, 1986). The Latrobe Valley is in southeastern Australia and contains the largest near-surface deposits of brown coal anywhere. The Valley is east-west in orientation; is approximately 150 x 40 km in area; has a population of about 110,000 people and is the centre of electricity generation for Victoria (5300 MW capacity).

The Study has spent seven years monitoring several air quality indicators at 15 places throughout the Valley and monitoring the meteorology of the region via towers, a daily radiosonde ascent, and Doppler acoustic sounders. Special campaigns have been undertaken with Dr John Carras and others to measure directly the stack plumes; particle types and compositions; the distribution of ozone and its precursor hydrocarbon species; and changes in sulphur dioxide and nitric oxide with travel distance. A full emissions inventory for pollutants over the 6,400 km^2 area has been carried out.

The major present objective of the Study is to answer the question:

> What will be the impact on the ambient Latrobe Valley air quality if changes are to be made to the emissions to air?

For the past two years, this has been addressed by model building with 12 scientist and engineers. The application of atmospheric dispersion models to the Latrobe Valley is a difficult task. The region is one of complex terrain, with mountains rising to over 500 m above the Valley floor, which slopes gently to the sea. Daytime dispersion is in mainly weakly stable conditions at the heights of the plumes, of which there are some twenty major ones. Rarely is the boundary layer strongly convective. There is a complex system of sea breezes in summer (Physick 1982). Nighttime is often characterised by calm surface conditions with extremely complex wind shears aloft due to interleaving katabatics and drainage winds (Manins 1983).

One of the difficulties being encounted in trying to model these features is the influence of the terrain on the mean winds, turbulence behaviour and convective layer height. These are particularly important influences for the production and dispersion of oxidants and fine particles - the pollutants of most concern in the Latrobe Valley.

My own Symposium poster paper[1] has been motivated by the Latrobe Valley work: it is based on a numerical and laboratory study (Kimura and Manins 1987).

When the air is thermally stably stratified, the wind over a valley may be very different from the situation in the neutral conditions discussed at length by Finnigan (1988) in his lecture 'Air Flow over Complex Terrain'. In moderate ambient stability and for valleys that are broad compared with the ridge heights, the flow may be excited into 'mountain lee waves'. Then the strongest winds may be found on the lee slopes of the hills (upstream slope to the valley) and light wind conditions, even recirculating winds, may be experienced on the forward slopes on the valley floor (the flow is 'blocked'). For a range of conditions there may be strong vertical velocities aloft in the large-amplitude 'lee waves', with high turbulence levels there.

At large thermal stabilities the blocked flow in the valley is fully stagnant and this may occur even in small-scale depressions, as well as in valleys the size of the Latrobe.

For the case of periodic valleys we found that while it exists, the blocked flow extends to a height of 0.7 times the ridge height. This is about the same as found for single valleys in the field (Manins and Sawford 1982) and in the laboratory (Baines and Hoinka 1985).

CONCLUSION

We have learnt through the Symposium of the impressive advances that have been made in generally idealised systems and this is certainly true of dispersion in the atmosphere. But, as was pointed out by Jeff Weil in particular, the real world can be much more complex. Application of these advances is not an easy process. However, the way to deal with such messy systems is to keep firmly in mind the principles that have been learnt from the more ideal studies. A direct solution from atmospheric researchers to some problems is often not possible. We should continue to do what Senator Puplick* so eloquently advocated in his address to the Symposium Dinner:

> Do research that is 'applicable', not necessarily 'applied'.
>
> For then the understanding is useable in many ways.

---

*Senator Christopher Puplick, Shadow Minister for the Environment, Arts and the Bicentennary, Australian Senate.

[1]P Manins (1987) Wind blocking in periodic valleys. Symposium poster paper

REFERENCES

Baines PG, Hoinka KP (1985) Stratified flow over two-dimensional topography in fluid of infinite depth: a laboratory simulation. J Atmos Sci 42: 1614-1630

Finnigan JJ (1988) Air flow over complex terrain. This volume, pp

Kimura F, Manins PC (1987) Blocking in peridic valleys. Boundary-Layer Meteorol, in press

Manins PC (1983) Nocturnal air circulation in the Latrobe Valley. Clean Air (Aust) 17: 29-32

Manins PC (1986) Air quality in the Latrobe Valley - State of knowledge June 1986. Latrobe Valley AirShed Study Steering Committee. Environment Protection Authority, 477 Collins St., Melbourne, Australia. ISBN 0 7306 0200 1

Manins PC, Sawford BL (1982) Mesoscale observations of upstream blocking. Q J R Meteorol Soc 108: 427-434

Physick WL (1982) Sea breezes in the Latrobe Valley. Aust Meteorol Mag 30: 255-263

Weil JC (1988) Atmospheric dispersion - observations and models. This volume, pp

Subject index

MIX
Papier aus verantwortungsvollen Quellen
Paper from responsible sources
FSC® C105338

If you have any concerns about our products,
you can contact us on
ProductSafety@springernature.com

In case Publisher is established outside the EU,
the EU authorized representative is:
Springer Nature Customer Service Center GmbH
Europaplatz 3, 69115 Heidelberg, Germany

Printed by Libri Plureos GmbH
in Hamburg, Germany